INTERNATIONAL

REVIEW OF CYTOLOGY

VOLUME 93

Genome Evolution in Prokaryotes and Eukaryotes

INTERNATIONAL

Review of Cytology

EDITED BY

G. H. BOURNE
St. George's University School of Medicine
St. George's, Grenada
West Indies

J. F. DANIELLI
(Deceased April 22, 1984)

ASSISTANT EDITOR

K. W. JEON
Department of Zoology
University of Tennessee
Knoxville, Tennessee

VOLUME 93

Genome Evolution in Prokaryotes and Eukaryotes

EDITED BY

D. C. REANNEY
Gene Link Australia Limited
South Melbourne, Victoria
Australia

PIERRE CHAMBON
Laboratoire de Génétique Moléculaire
des Eucaryotes du C.N.R.S.
Institut de Chimie Biologique
Faculté de Medecine
Strasbourg, France

ACADEMIC PRESS, INC. 1985
(Harcourt Brace Jovanovich, Publishers)
Orlando San Diego New York London
Toronto Montreal Sydney Tokyo

ACADEMIC PRESS, INC.
Orlando, Florida 32887

United Kingdom Edition published by
ACADEMIC PRESS INC. (LONDON) LTD.
24–28 Oval Road, London NW1 7DX

LIBRARY OF CONGRESS CATALOG CARD NUMBER: 52-5203
ISBN 0-12-364493-3

PRINTED IN THE UNITED STATES OF AMERICA

85 86 87 88 9 8 7 6 5 4 3 2 1

Contents

Section I. RNA and Early Genome Evolution

Self-Splicing RNA: Implications for Evolution

THOMAS R. CECH

Section II. DNA Reorganization and the Genetic Evolution of Subcellular Organelles

Mechanisms of DNA Reorganization in Bacteria

J. A. SHAPIRO

Genetic Organization of the Chloroplast

J. D. ROCHAIX

Transposable Elements in Eukaryotes

D. J. FINNEGAN

Splicing in Adenovirus and Other Animal Viruses

EDWARD B. ZIFF

Contributors

Numbers in parentheses indicate the pages on which the authors' contributions begin.

GIUSEPPE ATTARDI (93), *Division of Biology, California Institute of Technology, Pasadena, California 91125*

C. C. F. BLAKE (149), *Laboratory of Molecular Biophysics, University of Oxford, Oxford OX1 3PS, England*

THOMAS R. CECH (3), *Department of Chemistry, University of Colorado, Boulder, Colorado 80309*

D. J. FINNEGAN (281), *Department of Molecular Biology, University of Edinburgh, Edinburgh EH9 3JR, Scotland*

J. D. ROCHAIX (57), *Departments of Molecular Biology and Plant Biology, University of Geneva, 1211 Geneva 4, Switzerland*

JOHN H. ROGERS (187), *MRC Laboratory of Molecular Biology, Cambridge 2B2 2QH, England*

J. A. SHAPIRO (25), *Department of Microbiology, The University of Chicago, Chicago, Illinois 60637*

EDWARD B. ZIFF (327), *Kaplan Cancer Center and Department of Biochemistry, New York University Medical Center, New York, New York 10016*

Preface

Future historians of science may say that the past decade has produced some of the most startling discoveries in the history of modern biology—discoveries that challenged both conventional wisdom and the intuitive understanding of general genetics built up by the study of bacterial models such as *Escherichia coli.*

Perhaps the most unexpected of these discoveries was the finding that most genes in higher cells were "split" in the sense that coding modules (exons) were interrupted by ungrammatical sequences (introns) which, for the most part, appeared functionless. Equally unsettling was the accumulation of evidence for the view that the mobile genetic elements ("transposons") familiar to bacteriologists had counterparts in the chromosomes of plant and animal cells.

These new data have engendered a ferment among molecular biologists. The traditional view of phylogeny, which simplistically puts prokaryotes at the bottom and eukaryotes at the top, has been replaced by a growing conviction that the streamlined genomes of prokaryotes are a derived feature and the "split" structure is probably the original format for genes. The philosophical framework of the debate has been colored by the suggestion that transposons are a kind of molecular parasite: hence the epithet "selfish DNA" now commonly applied to any genetic unit with the capacity for autonomous multiplication within the genome.

The editors believe the time is now ripe for a "wide lens" view of the way this new information has restructured our thinking on evolution. There has been no lack to date of specialist reviews on isolated topics such as "transposition" or "mechanisms of RNA splicing" but we are unaware of any comprehensive interpretation of genome evolution in the light of the mass of new data now available on issues which go so directly to the crux of the problem. This volume then is an attempt to fill a major perceived need.

The volume has been divided into three sections. The book opens with T. Cech's provocative chapter on the implications of self-splicing RNA for the origin of genetic systems. This chapter occupies a section by itself at the beginning of this volume because an increasing number of biologists now appear to accept Crick's view that the self-replicating RNA "machines" were the original forms of life. This chapter also introduces the basic principles of "splicing," a central theme of the entire volume, in what is mechanistically their simplest format.

Section II turns to the modern repository of genetic information, DNA. The first chapter by J. Shapiro reviews mechanisms of DNA reorganization in bacteria in a way that makes the underlying unities and principles especially clear. The next two chapters deal with subcellular organelles—chloroplasts (J. Rochaix) and mitochondria (G. Attardi). While it is still reasonable to discuss the evolution of chloroplasts and mitochondria in terms of the endosymbiont theory (which holds that each organelle was originally a free-living "prokaryote") the present contributions illustrate the multiple way these small DNAs reflect the specialized requirements of their intracellular environments. Chloroplast and mitochondrial DNAs have been included in Section II because they combine features of the genetic organization of both prokaryotes and eukaryotes and hence form a bridge between the postulated "primitive" RNA-based splicing system and the more "typical" splicing processes which occur in the nuclei of eukaryotes. Genes in both subcellular organelles contain introns, i.e., both kinds of organelle can have an RNA splicing system.

Section III, dealing with the implications of RNA splicing for eukaryote evolution, opens with an interesting evaluation by Blake of the role of exons in protein evolution. This is followed by a cogent explanation of the mechanisms of RNA splicing that occur in the processing of transfer RNAs, mitochondrial and chloroplast RNAs, and nuclear RNAs by J. Rogers. The same author then goes on to review "retroposons," nuclear elements that appear to have been generated by reverse transcription of RNA sequences into DNA form. This section emphasizes the potential or actual mobility of many retroposons, thus setting the stage for the chapter by D. Finnegan which addresses itself specially to the issue of transposable genetic elements in the chromosomes of higher cells. The volume closes with a concise account by E. Ziff of the ways in which selection has exploited the versatility of the RNA splicing mechanism to accommodate the compacted expression of viral genes.

While this volume itself has set an ambitious goal the editors believe the standard of the contributions and the standing of the authors in their respective specialties go far toward achieving the desired purpose. The volume is not a collation of vaguely related data bundled between one set of covers; rather each author has been asked to organize his material around the evolutionary context set by the volume's theme. As an up-to-date treatment of genome evolution the volume should be invaluable not only to geneticists but to all biologists who wish to understand the molecular basis of evolution.

The volume does not, however, make any claim to produce "final" answers to the problems it addresses. Molecular biology is itself in a state of rapid evolution and new concepts must constantly adapt to the changing environment produced by novel data. If, over the next few years, this volume stimulates biologists to ask new questions and seek new solutions it will have fulfilled its function.

D. C. Reanney
P. Chambon

INTERNATIONAL

REVIEW OF CYTOLOGY

VOLUME 93

Genome Evolution in Prokaryotes and Eukaryotes

Section I

RNA and Early Genome Evolution

INTERNATIONAL REVIEW OF CYTOLOGY, VOL. 93

Self-Splicing RNA: Implications for Evolution

THOMAS R. CECH

Department of Chemistry, University of Colorado, Boulder, Colorado

I. Introduction: Ribosomal RNA Genes in *Tetrahymena*

Tetrahymena, a genus of ciliated protozoa, has some unique advantages for studies of rRNA genes and their expression. The rRNA genes (rDNA) reside in the nucleoli of the macronucleus as linear, extrachromosomal molecules (Gall, 1974; Enberg *et al.*, 1974; reviewed by Blackburn, 1982). Unlike chromosomal rDNA in many higher eukaryotes, where the rDNA is heterogeneous due to variations in the nontranscribed spacer, the *Tetrahymena* rDNA molecules are essentially homogeneous in size and nucleotide sequence. The rDNA is present in ≃10,000 copies per macronucleus. It comprises 2% of the total cellular DNA, so it is readily isolated either as purified DNA or as a minichromosome.

Each rDNA molecule contains about 20 kb (kilobase pairs). The molecules

ISBN 0-12-364493-3

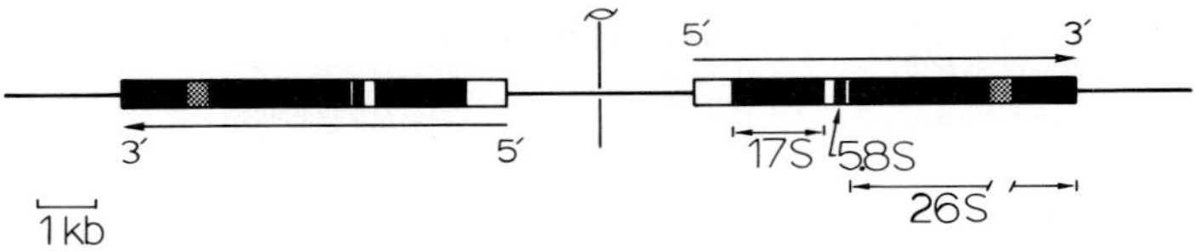

FIG. 1. Transcription map of the extrachromosomal rDNA of *Tetrahymena thermophila*. The wide bars represent the sequences coding for the primary transcript. Sequences present in mature rRNA are shown in black. Sequences removed from the precursor during processing are shown in white, except for the IVS which is stippled.

have a palindromic (inverted repeat) sequence symmetry (Karrer and Gall, 1976; Engberg *et al.*, 1976). As shown in Fig. 1, each half contains sequences coding for the 17 S, 5.8 S, and 26 S rRNAs.

Tetrahymena, like the other ciliated protozoa, contains two types of nuclei within the same cell. The transcriptionally active macronucleus is polyploid and divides amitotically during vegetative growth. The germinal micronucleus is diploid and divides mitotically. During conjugation, the old macronucleus is destroyed and a new one differentiates from a division product of the zygotic nucleus. The 10,000 extrachromosomal rDNA molecules in the macronucleus are formed by amplification of a nonpalindromic copy integrated in a micronuclear chromosome (Yao and Gall, 1977; Pan *et al.*, 1982).

Wild and Gall (1979) found that in some strains of *T. pigmentosa,* the 26 S rRNA coding region of the rDNA was split by a 0.4 kb intervening sequence (IVS, or intron). In any one strain, the rDNA was homogeneous: either all copies were interrupted, or all were continuous. Such homogeneity is a consequence of the amplification process. If the micronuclear copy that is amplified contains an IVS, all macronuclear copies will be IVS$^+$. *Tetrahymena thermophila, T. hyperangularis,* and *T. cosmopolitanis* all have a 0.4 kb IVS in their rDNA, while several other species that have been examined have uninterrupted rDNA (Din and Engberg, 1979).

The discovery of strains in which all copies of the rDNA were split by an IVS immediately led to the proposal that RNA splicing was taking place (Wild and Gall, 1979). This proposal was soon substantiated by the identification of an IVS-containing pre-rRNA *in vivo* (Cech and Rio, 1979; Din *et al.*, 1979) and the observation that excision of the IVS from pre-rRNA was a posttranscriptional event in isolated nuclei (Zaug and Cech, 1980; Carin *et al.*, 1980).

Splicing of pre-rRNA is not common. The only other nuclear pre-rRNA known to be transcribed and spliced is that of the slime mold *Physarum polycephalum* (Gubler *et al.*, 1980), which like *Tetrahymena* has extrachromosomal rDNA. While *Drosophila* and other dipteran flies have interrupted copies of rDNA, their genomes also contain uninterrupted copies; the interrupted units contribute little or nothing to rRNA production, and there is no real evidence that rRNA splicing takes place (reviewed by Beckingham, 1982). There are more examples of split rRNA genes in organellar genomes: chloroplast rDNA in *Chlamydomonas* (Rochaix and Malnoë, 1978) and mitochondrial rDNA in *Saccharomyces, Aspergillus, Neurospora,* and *Kluyveromyces* (Bos *et al.*, 1978; Netzker *et al.*, 1982; Burke and RajBhandary, 1982; Michel *et al.*, 1982). RNA splicing is clearly implicated in these cases, because the organelles contain only interrupted copies of rDNA. While more examples of split rRNA genes, nuclear and organellar, whill undoubtedly be found, it is probably significant that in all higher eukaryotes that have been studied (e.g., *Xenopus,* mouse, human) the nuclear and mitochondrial rRNA genes are not interrupted.

II. *Tetrahymena* Pre-rRNA Splicing Is Self-Catalyzed

By "self-catalyzed" I mean that the molecule lowers the activation energy for a chemical reaction in which that same molecule is a reactant. Lowering the activation energy results in a very great acceleration in the rate of the reaction. Those interested in the development of the idea of self-splicing RNA are directed to the papers by Cech *et al.* (1981) and Kruger *et al.* (1982). In this section, I will avoid such a historical account and discuss the most compelling evidence that we currently have for self-catalyzed splicing.

Unspliced pre-rRNA can be synthesized by incubation of isolated nuclei under conditions where the endogenous RNA polymerase I on the rDNA is active. Inhibition of splicing during transcription is accomplished by the inclusion of polyamines and Ca^{2+} in the transcription mixture and by keeping the concentration of monovalent cations low. Presumably these di- and polycations bind to the RNA and stabilize some splicing-incompetent conformation(s). The unspliced RNA is then purified by SDS-phenol extraction and sedimentation in a formamide-sucrose gradient (Cech *et al.*, 1981). When the resulting RNA is incubated in a solution containing only a monovalent cation (e.g., 200 m*M* NH_4^+ or Na^+), a divalent cation (5–10 m*M* Mg^{2+} or Mn^{2+}) and a guanosine compound ($\geq$ 1 μM guanosine or GTP), accurate RNA splicing occurs. In addition, the excised IVS RNA is converted to a covalent circular form. These reactions are described in detail in Section III.

The splicing and cyclization activities copurify with the RNA through gel electrophoresis in 8 *M* urea. In addition, the cyclization activity has been shown to copurify with the RNA through Cs_2SO_4 density gradient centrifugation. The activities are resistant to SDS-phenol extraction, boiling, treatment with 2% SDS, treatment with high concentrations of two nonspecific proteases, pronase and proteinase K, and combinations of the above treatments, such as boiling in SDS and then incubation with proteinase K and 37°C (Grabowski *et al.*, 1983). Such properties are clearly inconsistent with the splicing activity being a protein.

Final confirmation that the activity resides in the RNA was obtained using recombinant DNA. A portion of the *Tetrahymena* rRNA gene, containing 261 bp of the 5′ exon (sequences preceding the IVS), the entire 413 bp IVS, and 943 bp of the 3′ exon was cloned in an *E. coli* plasmid downstream from a promoter for *E. coli* RNA polymerase. RNA was transcribed from this template *in vitro* with purified *E. coli* RNA polymerase, thereby avoiding exposure of the RNA to any eukaryotic protein. When this RNA was deproteinized, it was found to undergo the same splicing reaction as characterized with the unspliced RNA from *Tetrahymena* nuclei (Kruger *et al.*, 1982).

III. Reactions Catalyzed by the RNA

A. Intramolecular Recombination

Any model for the mechanism of *Tetrahymena* pre-rRNA splicing must explain the following features of the reaction: (1) a guanosine nucleotide becomes attached to the 5′ end of the excised IVS RNA via a normal 3′,5′-phosphodiester bond; (2) cleavage occurs such as to produce 5′-phosphate and 3′-hydroxyl termini; (3) RNA ligation occurs without hydrolysis of ATP or any other energy cofactor; (4) covalent circularization of the excised IVS RNA requires only Mg^{2+} and a temperature ≥37°C. A mechanism that is consistent with these facts is shown in Fig. 2.

The mechanism involves phosphoryl transfer reactions:

$$\mathrm{R{-}OH + R'{-}O{-}\overset{\overset{\displaystyle O}{\|}}{\underset{\underset{\displaystyle O^{-}}{|}}{P}}{-}O{-}R'' \rightleftharpoons R{-}O{-}\overset{\overset{\displaystyle O}{\|}}{\underset{\underset{\displaystyle O^{-}}{|}}{P}}{-}O{-}R'' + R'{-}OH}$$

Because this is a transesterification reaction, it is expected to be highly reversible and should not require an energy cofactor. Splicing could take place by two such transesterification reactions. In the first, the nucleophile R—OH is guanosine. In the second, R—OH is the 5′-exon. IVS RNA circularization would require a third transfer, with the 3′-terminal guanosine residue serving as the R—OH.

An important prediction of the model was that RNA circularization must take place by an attack of the 3′-hydroxyl on some internal phosphodiester linkage, resulting in a circle that is smaller than the linear. The finding that the first 15 nt of the linear IVS RNA are released upon cyclization and that the circle has the 3′-terminal G joined to the A at position 16 (Zaug *et al.*, 1983) therefore provided strong support for the model.

Strand breakage, strand switching and reunion are the essence of nucleic acid

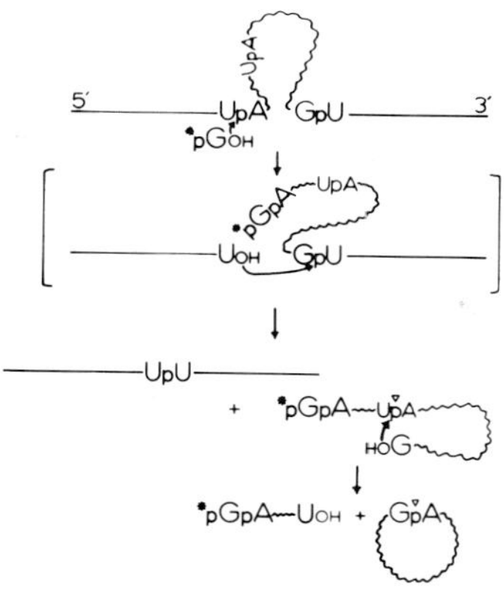

FIG. 2. Transesterification mechanism for *Tetrahymena* pre-rRNA splicing and IVS cyclization. Solid lines, exons; wavy line, IVS; asterisk, labeled guanosine cofactor, in this example 5′-GMP; ∇, phosphate that closes the circle. Square brackets denote a postulated intermediate that has not been isolated. (From Zaug *et al.*, 1983; reprinted by permission from *Nature* (London), **301,** 578–583, copyright (c) 1983 Macmillan Journals Limited.)

recombination. The reactions diagrammed in Fig. 2 involve exactly these steps and can therefore be considered to be intramolecular recombination reactions.

B. Intermolecular Recombination

The linear IVS RNA is capable of forming linear and circular dimers, trimers, and higher oligomers *in vitro* (Zaug and Cech, 1985). The reactions require prior denaturation of the RNA, a treatment which presumably facilitates the formation of intermolecular base-paired complexes. The ligation mechanism is an intermolecular analog of cyclization. That is, the 3′-terminal G of one molecule undergoes transesterification with the A at position 16 of another molecule, producing a molecule that is 15 nt smaller than a true dimer and releasing one 15-mer. Like the cyclization reaction, oligomerization requires only Mg^{2+}. Each oligomer still contains a 3′-terminal G—OH and a 15-mer at its 5′-end, and is therefore capable of circularization. None of these species has been detected in nuclear RNA from *Tetrahymena* grown at either 30 or 39°C (Brehm and Cech, 1983), so they are probably not important in normal RNA metabolism. It remains possible that such oligomerization could have occurred under prebiotic conditions.

C. Circle Autoreopening

When circular IVS RNA is incubated at 42°C at pH 7.5, it is converted to a linear form with a half-time of ~7 hours. At first we thought that the RNA was undergoing alkaline hydrolysis, but further investigation revealed that only one of 400 phosphodiester bonds in the RNA was being hydrolyzed, the bond that closed the circle (G_{414}–A_{16}). Furthermore, unlike alkaline hydrolysis, the RNA termini produced in circle reopening were 5′-phosphate and 3′-hydroxyl (Zaug *et al.*, 1984). The reaction is similar to a reversal of RNA cyclization, except the phosphoryl transfer occurs to water instead of to the 15-mer. The occurrence of this reaction indicates that the structure of the RNA is making one of the 400 phosphates in the molecule chemically special such as to make it preferentially reactive.

D. Is Self-Splicing Chemically Reasonable?

We must consider both thermodynamics, which determines whether the reaction is possible, and the activation energy barrier, which determines whether the reaction will occur at a reasonable rate.

As far as thermodynamics is concerned, the transesterification reactions are expected to be isoenergetic. Each transesterification produces no net change in the number of phosphodiester bonds. There is therefore no need for ATP or GTP as an energy source. The splicing reaction can be driven to completion by a high molar ratio of the guanosine cofactor (e.g., GTP) to pre-rRNA. *In vivo,* circularization or degradation of the IVS RNA would also help splicing go to completion by removing an end-product. The circularization reaction might be entropically driven, since it involves cleavage of one molecule into two.

RNA self-splicing is as thermodynamically reasonable as other transesterification reactions catalyzed by DNA topoisomerases, resolvases, and the λ *int* protein. These reactions are also ATP independent. (DNA gyrase is normally considered to be an ATP-dependent topoisomerase, but it requires ATP for the introduction of superhelical turns, not for its DNA nicking-closing activity). The enzymes involved in these DNA reactions of course do not alter the free energy of the reaction, but serve to lower the activation energy barrier.

In the case of self-splicing RNA, it is clearly the structure of the RNA molecule that lowers the activation energy for the transesterification reactions. The reactions do not occur in the presence of high concentrations of denaturing agents such as formamide or urea. IVS RNA circularization is completely inhibited by the presence of 30 μg/ml ethidium bromide, which intercalates in double helical regions (N. K. Tanner, unpublished data).

How does the structure of the RNA promote the reaction? First, it provides a binding site for the guanosine cofactor (Bass and Cech, 1984). The binding site has not been localized within the tertiary structure of the molecule, but presumably it places the guanosine with its 3′ hydroxyl group poised to attack the phosphate at the 5′ splice junction of the pre-rRNA. Many enzymes work by approximation, i.e., by binding substrates in proximity to each other and in a particular spacial orientation (Jencks, 1969). Thus, having a specific binding site for guanosine might explain much of the specificity and rate enhancement seen in ribozymic catalysis.

We now have evidence that the structure of the RNA is promoting the reaction in some additional manner. The circle autoreopening reaction occurs in the absence of free guanosine. The reaction is self-catalyzed: it has high specificity (1 of 400 phosphodiester bonds cleaved) and great rate enhancement relative to alkaline hydrolysis. Like the other self-catalyzed reactions, it requires the intact secondary–tertiary structure of the molecule. The chemical basis for the enhanced reactivity of this site is not yet known. One possibility is that a magnesium ion is specifically bound to the phosphate, coordinated by other groups in the RNA, and is increasing the electrophilicity of that particular phosphate. Site-specific binding of metal ions in folded RNA molecules is well known (Eichhorn, 1981; Brown *et al.*, 1983; Rubin and Sundaralingam, 1983).

IV. Generality of RNA-Mediated Splicing

A. Structural Homology with Fungal Mitochondrial Intervening Sequences

Many yeast mitochondrial mRNA introns contain several short sequence elements whose integrity is required for proper splicing. Mutations in these sequences are cis-dominant, which differentiates them from the trans-recessive mutations that interrupt the intronic open reading frames.

Two of the sequences that are the sites of cis-dominant mutations in intron 4 of the yeast cytochrome *b* (*cob*) gene, *box*9 and *box*2, are conserved in many other introns in the *cob* and cytochrome oxidase genes of yeast and other fungi. They are also found in the mitochondrial large rRNA gene introns in these organisms (Burke and RajBhandary, 1982; Michel *et al.*, 1982; Waring *et al.*, 1982). Remarkably, these same sequences occur in the introns of the nuclear rRNA genes of *Tetrahymena* and *Physarum*. The degree of homology is striking. The *Tetrahymena box*9 homolog matches the yeast *box*9 sequence in 10 of 12 positions, while the *box*2 homolog matches the corresponding yeast sequence in 11 of 12 positions. This is a higher degree of homology than shown by the yeast mitochondrial rRNA intron, which has 9/12 of *box*9 and 10/12 of *box*2.

Three additional mitochondrial sequence elements are defined in Table I. Elements A and B are conserved in primary sequence, while 9R′ is not conserved but is always complementary to the right (downstream) half of *box*9. The six sequence elements defined in Table I occur in the *Tetrahymena* IVS in the same 5′→3′ polarity as in the mitochondrial introns (Fig. 3). Sequences 9R′ and A are

TABLE I
SEQUENCE ELEMENTS CONSERVED IN MANY MITOCHONDRIAL INTRONS AND IN NUCLEAR rRNA INTRONS[a]

Element	Synonym	Consensus sequence[b]	Complementary to	Splicing-defective mutations
9R′	E	Not conserved	*box*9R	*cob* I4
A	P	AUGCUGGAAA	B	None found
B	Q	AAUCAGCAGG	A	None found
*box*9L	R	UCAGAGACUACA	*box*2	*cob* I4; I5; *oxi*3 I4
*box*9R	E′	Not conserved	9R′	*cob* I4
*box*2	S	AAGAUAUAGUCC	*box*9L	*cob* I4

[a]See Davies *et al.* (1982) and Cech *et al.* (1983) for details.
[b]The portion of each sequence that is proposed to be involved in base-pairing is underlined.

adjacent, as are the left and right elements of *box*9. Beyond that, the distance between elements is highly variable in both mitochondrial and nuclear systems.

The six sequence elements have been proposed to interact via pairing of complementary bases: 9R′·*box*9R, A·B, and *box*9L·*box*2. For the mitochondrial introns, the evidence for such a structure comes from phylogenetic sequence comparison and computer modeling (Michel *et al.*, 1982; Davies *et al.*, 1982). In addition, there is strong genetic evidence for the 9R′·*box*9R interaction in *cob* intron 4 (Weiss-Brummer *et al.*, 1983). For the *Tetrahymena* IVS, there is direct biochemical evidence for the structure shown in Fig. 4 (Cech *et al.*, 1983). The

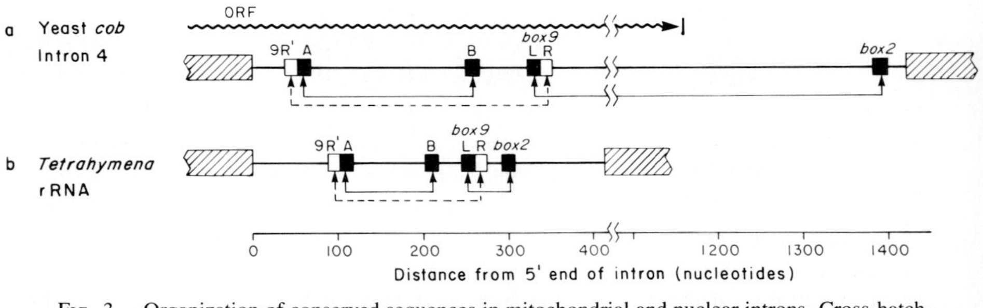

FIG. 3. Organization of conserved sequences in mitochondrial and nuclear introns. Cross-hatched areas, exons. Solid boxes, conserved intron sequences. Open boxes, intron sequences conserved only in their potential for base-pairing. Boxed sequences are enlarged slightly for clarity. Solid and dashed double-headed arrows show postulated secondary and tertiary base-pairing interactions. In (a), these interactions are based on phylogenetic sequence comparisons and genetic data, as described in the text. The pairings are therefore those expected to occur *in vivo,* where the RNA is presumably complexed with proteins. In (b), the interactions are based on direct biochemical analysis of the excised IVS RNA and on computer predictions (Cech *et al.*, 1983). The pairings are therefore those expected to occur in the deproteinized RNA. (From Cech *et al.*, 1983.)

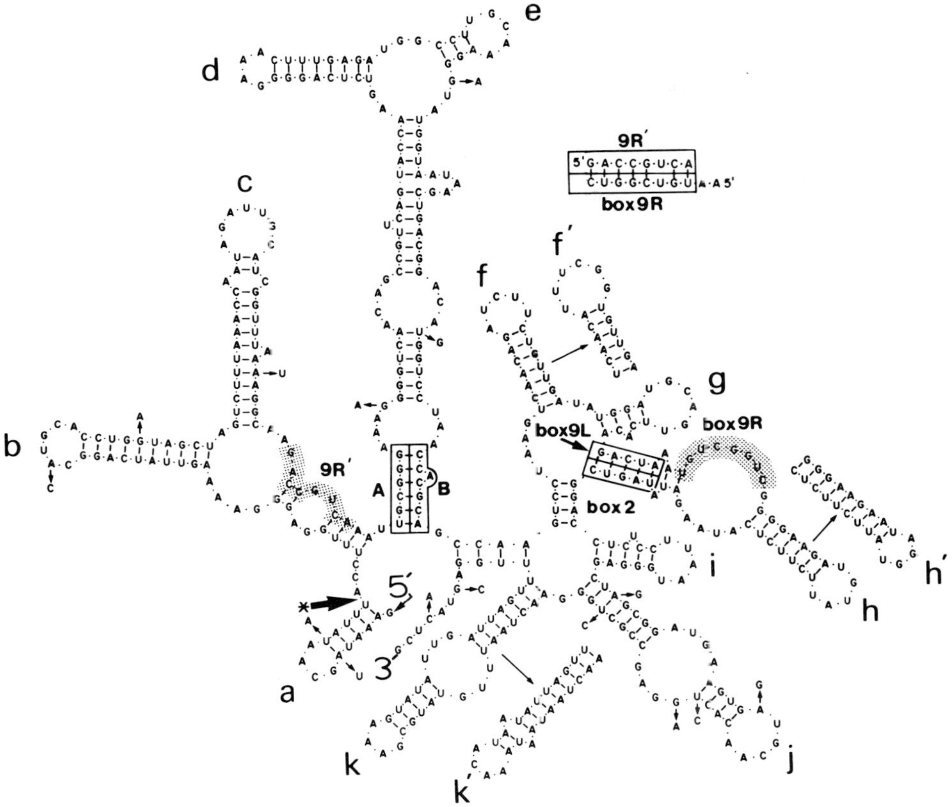

FIG. 4. Model of the secondary structure of the excised IVS RNA of *Tetrahymena*. IVS RNA labeled at its 5′ end was digested under nondenaturing conditions with RNase T1. Sites of cleavage were determined by sequencing gel electrophoresis. A computer program was used to predict the lowest-free-energy secondary structure that was consistent with the RNase T1 digestion data. The structure was subsequently found to be consistent with sites of cleavage by S1 nuclease and RNase T2 and by double-strand-specific cobra venom RNase (Cech *et al.*, 1983). Boxes denote base-paired portions of those sequences that are conserved between nuclear rRNA introns and mitochondrial introns. Shading denotes the *box*9R and 9R′ regions, which are conserved in position but not in nucleotide sequence. These may pair in a tertiary interaction, as detailed in the insert. Arrows designate single bases and hairpins that differ in the IVS of *T. pigmentosa*. (*) Site of addition of the 3′-end to form circular IVS with release of hairpin a. (From Cech *et al.*, 1983.)

A·B and *box*9L·*box*2 pairs form part of the secondary structure of the molecule. Nucleotides to the right of *box*9L are unpaired and are complementary to a sequence to the left of A, such that a *box*9R·9R′ tertiary interaction could be formed. Michel and Dujon (1983) and Waring *et al.* (1983) have proposed similar structures for the IVS region of the *Tetrahymena* pre-rRNA.

B. Some Mitochondrial Introns Are Self-Splicing

The *Tetrahymena* nuclear rRNA intron and many mitochondrial introns appear to be folded into very similar structures. In both systems the structure is essential for splicing. In spite of these similarities, it seemed unlikely that the mitochondrial introns could be self-splicing, because there was excellent genetic evidence for protein involvement in the process. Mitochondrial mRNA splicing is sensitive to trans-acting mutations that interrupt the intronic open reading frame. The occurrence of such mutants led to the proposal of intron-encoded "maturases" (Lazowska *et al.,* 1980). [There is also biochemical evidence for these intron-encoded polypeptides (Claisse *et al.,* 1980; Bechmann *et al.,* 1981; Jacq *et al.,* 1984).] In addition, there are nuclear mutants in both mitochondrial mRNA and rRNA splicing that make it seem likely that proteins other than the maturases are involved (Manella *et al.,* 1979; Dieckmann *et al.,* 1982).

The first evidence that mitochondrial RNA splicing occurred by the same mechanism as *Tetrahymena* rRNA splicing was provided by Garriga and Lambowitz (1983). They found that the intron excised from the *Neurospora* mitochondrial rRNA precursor, like that excised from the *Tetrahymena* rRNA precursor (Cech *et al.,* 1981), contains a 5′-terminal guanosine residue not encoded by the DNA. Tabak *et al.* (1984) made the same observation for the yeast mitochondrial rRNA intron, and in addition have evidence for a circular form of the intron that is missing the first few nucleotides of the linear form. These observations are a strong indication of a transesterification mechanism for splicing.

Recently, Garriga and Lambowitz (1984) have found that two *Neurospora* mitochondrial mRNA precursors undergo efficient self-splicing *in vitro* in reactions that involve covalent addition of guanosine to the excised introns. One of the introns is from the *cob* gene, and one is from an unidentified reading frame. It appears that a number of other *Neurospora* mitochondrial mRNA precursors may undergo self-splicing at a lower efficiency. Analogous studies have not yet been reported for the yeast mitochondrial pre-mRNAs.

How does one reconcile the observation of self-splicing of mitochondrial RNA *in vitro* with the known requirement for proteins *in vivo?* It seems likely that both the mitochondrial introns and the nuclear rRNA introns are basically self-splicing, but their structures are modulated and optimized for the reaction by the association of proteins *in vivo* (Cech *et al.,* 1983). The *Tetrahymena* rRNA splicing reaction proceeds at a reasonable rate in the absence of any proteins, but the rate could be enhanced perhaps 50-fold when proteins are bound. The proteins might be nothing more than the ribosomal proteins which are known to associate with the exons (mature rRNA sequences) during transcription. In the absence of proteins, many of the mitochondrial RNA introns may not be in the correct structure for self-splicing (Wollenzien *et al.,* 1983), such that the rate of self-catalyzed splicing is too low to be useful for RNA biosynthesis. The formation of the self-splicing structure is promoted by the association of one or more nuclear-encoded polypeptides and, in the case of some mRNA introns, the intro-encoded polypeptide. Note that because the intron-encoded "maturases" are for the most part intron-specific, it has been considered unlikely that they were splicing enzymes (Lazowska *et al.*, 1980).

In short, it appears that intron-encoded polypeptides help introns attain an active conformation for splicing. Such a circuit may provide a means for regulation of the splicing process and therefore of mitochondrial gene expression. Regulation need not be restricted to the gene which encodes the maturase. It has been proposed that the polypeptide encoded by intron 4 of the yeast *cob* gene is also required for processing the *oxi*3 pre-mRNA (DeLaSalle *et al.*, 1982).

V. How Were These Intervening Sequences Established in Evolution?

A. Progenitor rRNA Gene Had Intervening Sequence(s)

The original genes for large rRNA could have been formed by recombination within introns to bring together two or more exons, each of which encoded a functional domain of the rRNA. Such a model is analogous to that proposed by Gilbert (1978) for the piece-wise construction of mRNA genes. According to this view, ancestral rRNA genes were all split genes, and the introns have been largely removed through evolution. The introns in some species of *Tetrahymena* and in mitochondria would then be fossils which preserve the record of an outmoded form of rRNA gene organization.

If this were the case, we would expect all rRNA introns to interrupt the coding region at the same place. It is true that all rRNA introns occur in the 3′ half of the large rRNA gene. When we look at intron location at the nucleotide sequence level, however, we see that the 5 different introns that have been sequenced interrupt the large rRNA gene at 4 different sites. The *Tetrahymena* introns are located at base 1925, the *Physarum polycephalum* intron 2 at base 1949, the other *Physarum* intron and the various fungal mitochondrial rRNA introns all at base 2450, and the *Chlamydomonas* chloroplast rRNA intron at base 2593, the location being based on the sequence of *E. coli* large rRNA (Brosius *et al.*, 1980). [It is possible to make such a precise statement about the location of the introns because the sequence of the mature large rRNA is highly conserved between mitochondria, prokaryotes, and eukaryotes in this region (Gourse and Gerbi, 1980).] One can rescue the hypothesis of intron-containing progenitor rDNA by postulating that the progenitor had 4 or more introns, of which two remain in *Physarum* rDNA and only one remains in the other cases. While such a model cannot be discounted, it is more attractive to interpret the different location of introns as a reflection of different insertion events. The question of whether the ancestral rDNA did or did not have introns is considered in more detail by Gerbi *et al.* (1982).

B. Transposition by Reverse Splicing

One possibility for establishing introns by insertion will be considered. Introns could have had their origin as transposable elements that were able to enter both nuclear and mitochondrial compartments of a cell. Given that splicing activity is intrinsic to the IVS, more specific models for transposition can be formulated.

One model, where the insertion takes place at the RNA level, involves interaction of free linear IVS RNA (origin unknown) with mature rRNA or pre-rRNA transcribed from an intronless gene. An intron-containing rRNA would be produced, accompanied by release of the guanosine compound from the 5′ end of the IVS. This is precisely equivalent to a reverse splicing reaction, so it should be favored under conditions of low guanosine nucleotide concentration and high RNA concentration (consider Fig. 2 in reverse). Establishment of this change in the genome might occur by reverse transcription of the RNA followed by some sort of insertion or gene conversion event. This model therefore includes elements of the "processed gene" idea (Van Arsdell *et al.*, 1981; Hollis *et al.*, 1982).

In a second model, insertion takes place at the DNA level directly. The transposable element is a single-stranded DNA copy of the IVS. It inserts itself into an intronless gene at a time when the latter DNA is also single-stranded, e.g., at a replication fork. This produces a heteroduplex region in the DNA which might later be repaired to give an intron-containing double-stranded DNA. This model is attractive because of its simplicity. It must be noted, however, that currently there is no experimental evidence that a DNA copy of self-catalyzed RNA would have any similar activity. Given that the two nucleic acids are structurally nonequivalent (e.g., DNA forms B helices while RNA forms A helices) and chemically different (the 2′ hydroxyl group plays a role in many RNA reactions), the presence of an activity in the DNA cannot be assumed.

C. Transfer from Mitochondria to Nucleus

It is also possible that the introns were always present in various mitochondrial genomes, and are occasionally transferred to the nucleus. Transfer of genes or gene fragments from mitochondria to nucleus has recently been documented in *Podospora* (Wright and Cummings, 1983). There is also evidence that such transfer has occurred in yeast (Farrelly and Butow, 1983) and sea urchin (Jacobs *et al.*, 1983), so it may be a general phenomenon. It will be of interest to determine if any *Tetrahymena* mitochondrial genes contain introns and, if so, whether they resemble the intron of the macronuclear rDNA.

D. Gaining an Open Reading Frame

I have discussed several possible pathways for the evolution of the nuclear rRNA and mitochondrial introns. In this discussion, I have ignored one important difference between these two classes: the mitchondrial introns usually contain open reading frames (which in many cases are known to be translated), while the nuclear rRNA introns do not have open reading frames of any substantial length and do not appear to serve as mRNAs (see Brehm and Cech, 1983). It is therefore worth discussing when and how the mitochondrial open reading frames might have arisen.

In most of the mitochondrial introns, the conserved sequence elements are distributed along the entire length. Where open reading frames are present, the 9R′, A, B, and *box*9 sequences are usually within the reading frame, while the *box*2 sequence is always downstream from the termination codon near the 3′ splice site. (An example is that of yeast *cob* intron 4, Fig. 3.) Such interspersion points to an ancient relationship between the reading frame and the cis-acting splicing signals. The situation is different, however, for the yeast, *Neurospora* and *Aspergillus* mitochondrial rRNA introns. There, the conserved sequence elements flank the reading frame as if it had been acquired secondarily by some insertion event, in each case insertion between the *box*9 and *box*2 sequences. A third type of arrangement exists for intron 3 of the yeast *oxi*3 gene, where all the conserved sequence elements lie downstream from the termination codon of the open reading frame. Again, the sequence elements and the reading frame may have arisen independently.

Thus, it is possible to fit all the available data by postulating that the mitochondrial introns have gained open reading frames not present in the ancestral intron. In cases such as *cob* intron 4, the elimination of stop codons between the upstream exon and the open reading frame could later produce a chimeric reading frame that included some of the conserved splicing signals. By gaining such open reading frames, the mitochondrial introns could make proteins that allow better regulation of the splicing process.

VI. Why Does the *Tetrahymena* rRNA Intron Persist?

Having considered some possible pathways by which self-splicing introns could have arisen in genes, we must consider why they are maintained. Do they offer a selective advantage, or are they simply tolerated? I will discuss some ideas about these questions for the case of the nuclear rRNA introns. The considerations differ somewhat for the mitochondrial introns, many of which encode proteins.

A. Selfish RNA

There is no evidence that the rRNA introns confer any advantage to the species which have them. *T. pigmentosa* 8ALP, which has no IVS in its rDNA or encoded elsewhere in its genome (Wild, 1980), grows just as well as *T. pigmentosa* 6UM, which has interrupted rDNA (Wild and Gall, 1979).

In the absence of any evidence for a selective advantage, it is worth considering a minimal hypothesis. The IVS could be maintained because the following three conditions prevail: the IVS is innocuous if intact, because it splices itself from the pre-rRNA so efficiently; partial deletions of the IVS are lethal mutations, because they destroy the molecule's ability to splice itself; and precise deletion of the IVS (removal of the entire IVS and nothing more) is an unlikely occurrence. The first point seems likely. Splicing occurs so rapidly *in vivo* that it does not even add significantly to the time required for rRNA processing (Brehm and Cech, 1983). The second point has been clearly demonstrated *in vitro*. Small deletions (5–20 bp) of sequences from either of two internal sites result in complete abolition of splicing activity (J. Price, G. Kieft, E. Sievers, J. Kent, and T. Cech, unpublished data). The third point is speculative. One can in fact conceive of a pathway for precise deletion of an intron at the DNA level: the spliced RNA could serve as a template for DNA synthesis, and the resulting "processed gene" could be reintegrated into the chromosome. But this may be particularly unlikely in *Tetrahymena,* where the transcriptionally active macronucleus, which contains all the potential templates for reverse transcription, is segregated from the transcriptionally inactive micronucleus, which contains the genes which would have to be altered in order for there to be any heritable change in the rDNA.

Thus, it seems possible that the IVS might be maintained not because it has any positive role, but because it is innocuous only if kept intact. I call this the "selfish RNA" hypothesis. It differs somewhat (although not entirely) from the concept of selfish DNA (Doolittle and Sapienza, 1980; Orgel and Crick, 1980). Selfish DNA is maintained because of things it does at the DNA level, such as replicating preferentially and transposing copies of itself. It need not be transcribed into RNA to be maintained. The DNA that codes for selfish RNA, on the other hand, is maintained because of something that occurs at the RNA level, i.e., RNA splicing. Its maintenance is dependent upon its interrupting a transcription unit.

B. The Intron May Have a Function

The fact that some species of *Tetrahymena* have no intron in their rDNA is inconsistent with the intron having any obligatory function. It is certainly premature, however, to dismiss the possibility that the intron could serve a useful role, thereby giving a selective advantage to the intron-containing species. As mentioned above, the intron-containing and intronless strains of *T. pigmentosa* appear to grow equally well. But comparisons under laboratory conditions may have little relevance to the organism's ability to compete in its natural environment.

One can envision many possible roles for an intron, especially in rRNA maturation and rRNP assembly. For example, the intron could act as a spacer between two domains of the large rRNA, allowing them to combine with ribosomal proteins prior to their being brought together by the splicing event. A large variety of such scaffolding functions can be imagined. As another example, the intron-containing rRNA might have a structure that facilitated a particular nucleoside modification event. An example of this latter type of function has been described for the intron in a yeast tRNA precursor (Johnson and Abelson, 1983). If the rDNA intron serves some such function, what happens in the intronless species? The process is either accomplished by a different pathway, perhaps making use of a scaffolding protein instead of the intron, or it is simply accomplished less efficiently. In either case, there would be a phenotype that correlated with the presence of the intron. Thus, while these ideas are speculative, they are fortunately testable.

VII. Roles for RNA Catalysis

A. Early in Evolution

It is generally thought probable that RNA, not DNA, was the primordial genetic material. Information storage can be accomplished by either polynucleotide; there are many viruses with RNA genomes. Only RNA, however, has the ability to direct the synthesis of proteins. Single-stranded RNA folds into diverse structures, giving it a versatility of function. We can envision that such versatility would have been all the more important at life's early stages, when there were few proteins to facilitate biochemical processes. The more uniform double-helical structure and greater chemical stability of DNA make it better suited than RNA for long-term storage of genetic information. DNA may be a more recent advance, a way of protecting the useful nucleotide sequences that had been selected at the level of RNA. Consistent with such a view, biosynthesis of the ribonucleotide precursors of RNA begins with small molecules such as sugars, amino acids, and CO_2; biosynthesis of the deoxyribonucleotide precursors of DNA occurs by reduction of the corresponding ribonucleotides. These and other arguments in favor of RNA as the first genetic material have been put forth by Eigen *et al.* (1981) and others (see Ferris and Usher, 1983).

If the earliest genes were made of RNA, then RNA splicing can be considered to be the first form of genetic recombination. In a primordial system with a very limited number of different RNA molecules, such recombination might have been crucial for increasing the sequence diversity of RNA genes and producing new structural RNAs (Reanney, 1979). If some of these early genes consisted of self-splicing RNA, then the rearrangement process could have occurred before there were many proteins. Primitive self-splicing RNA might have been most useful if it rearranged itself inefficiently and imprecisely, producing many combinations of sequences. It time, the splicing system could become more refined, with an increasing tendency to catalyze those rearrangements that led to useful products. At the same time, the number of genes would be increasing, so that the ability to undergo multiple rearrangements would be less advantageous. In the end, rearrangement might become very precise, as in the *Tetrahymena* pre-rRNA.

Similar arguments can be made for the RNA transcribed from DNA genes. At a time when the number of genes was small, it would have been advantageous to be able to create a diversity of rearranged RNA products from a single transcript (Gilbert, 1978). Present day examples can be found in DNA viruses such as SV40 and adenovirus. These viruses use overlapping genes and alternative RNA splice sites to make efficient use of a genome that is limited in size.

Is it possible that RNA molecules once catalyzed reactions other than their own recombination? The catalytic potential of RNA is certainly less than that of proteins. RNA has only 4 types of "side chains" compared to the 20 found in polypeptides. RNA does not contain a group like histidine that has the useful property of being ionizable near neutral pH. On the other hand, RNA has some properties in common with enzymes. RNA bases contain hydrogen bond donors-acceptors that could potentially bind certain substrates. RNA molecules can fold into a limitless variety of structures, some of which might have hydrophobic pockets. RNA molecules, like metalloenzymes, can bind metal ions at specific sites (Eichhorn, 1981; Brown *et al.,* 1983; Rubin and Sundaralingam, 1983). Finally, 5 S rRNA contains some group that is titrated near neutral pH, resulting in a conformational switch in the molecule (Kao and Crothers, 1980). (This is in contrast to the general statement made above, that RNA bases normally do not undergo ionization near neutral pH.) Thus, it seems quite reasonable that RNA could catalyze a variety of biochemical reactions. Even catalysts with more modest turnover rate and substrate specificity than present-day enzymes would have been extremely valuable early in evolution.

There may even be vestiges of such ancient RNA catalysts, i.e., the nucleotide coenzymes (White, 1976). The most common coenzymes, including NAD^+, FAD, and coenzyme A, all contain adenosine monophosphate moieties. A possible pathway for the evolution of protein enzyme–coenzyme catalysts from early RNA catalysts is sketched in Fig. 5. While such a model is very exciting with respect to the idea of RNA catalysis, it should be noted that the adenylyl groups

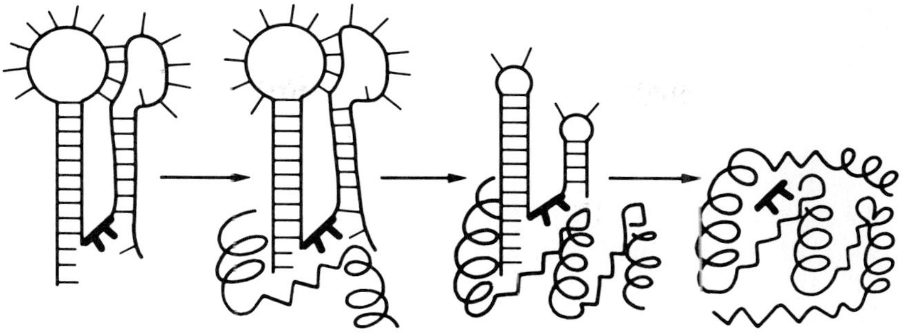

FIG. 5. A model for the evolution of coenzymes from polynucleotides, according to White (1982). RNA molecules with some catalytic activity, perhaps not very specific, may have existed before the advent of a reliable translation apparatus. Small polypeptides interacting with the RNA could enhance catalysis. As the translation apparatus becomes more advanced, larger polypeptides are synthesized and become responsible for more of the specificity of the RNP enzymes. Eventually, most of the polynucleotide is lost, leaving the former active site as a coenzyme. (From White, 1982, by permission of Academic Press, Inc.).

of the coenzymes do not appear to be involved directly in catalysis, but rather in binding the coenzyme to the enzyme. (In some coenzymes the part of the molecule that performs the chemistry, such as the flavin moiety of FAD or the nicotinamide moiety of NAD^+ or $NADP^+$, is a heterocyclic base that could be the evolutionary descendent of a nucleic acid base.)

B. TODAY

In present-day biological systems, the catalytic roles of RNA are not limited to self-catalyzed RNA splicing. RNase P, the enzyme that is responsible for maturation of the 5′ end of tRNA molecules in both prokaryotes and eukaryotes, has an RNA molecule as its catalytic subunit. There are many other candidates for catalytic RNA. The most fundamental of these is ribosomal RNA, which may catalyze a number of the reactions involved in protein synthesis.

RNase P is a ribonucleoprotein. Both a 120-kDa RNA molecule and a 18.5-kDa polypeptide are necessary for its activity *in vivo* (Kole *et al.*, 1980). Altman, Pace, and co-workers have recently found that the RNA moiety of RNase P catalyzes the accurate processing of transfer RNA precursors in the absence of the protein component *in vitro* (Guerrier-Takada *et al.*, 1983; Guerrier-Takada and Altman, 1984). The RNase P RNA acts as a true catalyst, being regenerated after the reaction such that each RNase P RNA molecule processes multiple pre-tRNA substrate molecules. The RNA catalyst from *E. coli* works well in solutions containing 60 m*M* Mg^{2+}, while that from *B. subtilis* has much higher salt and

Mg^{2+} concentration optima. Under such conditions, the RNA alone catalyzes the tRNA processing reaction at a rate similar to that of the holoenzyme under more physiological conditions. It seems likely that one function of the polypeptide subunit is to help the RNA fold into the optimum conformation for catalysis, in analogy to the role of intron-encoded polypeptides in mitochondrial mRNA splicing. The polypeptide subunit almost certainly has additional functions; for example, pre-4.5 S RNA, a minor substrate of RNase P, is not processed by the RNase P RNA alone, but is processed by the holoenzyme (Guerrier-Takada *et al.*, 1983).

The ribosome is the ultimate ribonucleoprotein, both in complexity and in importance. Until recently, it was commonly thought that the ribosomal RNAs might simply be providing a framework for the proper positioning of the ribosomal proteins. The finding that the rRNAs are so highly conserved not only in their structure but also in nucleotide sequence (particularly in nonhelical regions) has led to a reevaluation of this concept (Noller and Woese, 1981). The RNA components of the ribosomes are much more highly conserved than the protein components. One can therefore support the view that the proteins refine or tune the structure and activity of a translational apparatus that once was and may still be basically an RNA machine (Crick, 1968; Woese, 1972).

It has recently been found that single base changes in the rRNA can confer resistance to chloramphenicol, a drug which interferes with peptidyl transfer (Kearsey and Craig, 1981). Methylation of a single adenosine residue in 23 S rRNA prevents the action of thiostrepton, a peptide antibiotic that blocks the binding of EF-G and GTP to the ribosome (Thompson *et al.*, 1982). The finding that these drugs and others such as erythromycin (Sor and Fukuhara, 1982) act largely on the rRNA again leads one to think that the catalytic role of rRNA in protein synthesis may have been underestimated.

The peptidyltransferase activity of ribosomes has not been observed with isolated ribosomal proteins. It is certainly possible that the active site requires a special juxtaposition of proteins that is not readily formed except in the intact rRNP. It is also conceivable that the activity has not been observed because it is intrinsic to the RNA component of the ribosome. It would be ironic indeed if a process as ancient as RNA catalysis were still operating in such a central biosynthetic reaction as the synthesis of proteins on ribosomes.

Acknowledgments

I am grateful to the people in my laboratory, past and present, who had the ideas and did the experiments that led to the discovery and characterization of self-splicing RNA. I thank Phil Perlman for contributing ideas about open reading frames in introns. I am the recipient of Research Career Development Award No. CA00700 from the National Cancer Institute, DHHS.

REFERENCES

Bass, B. L., and Cech, T. R. (1984). *Nature (London)* **308,** 820–826.

Bechmann, H., Haid, A., Schweyen, R. J., Mathews, S., and Kaudewitz, F. (1981). *J. Biol. Chem.* **256,** 3525–3531.

Beckingham, K. (1982). *Cell Nucl.* **10,** 205–269.

Blackburn, E. H. (1982). *Cell Nucl.* **10,** 145–170.

Bos, J. L., Heyting, C., Borst, P., Arnberg, A. C., and Van Bruggen, E. F. J. (1978). *Nature (London)* **275,** 336–338.

Brehm, S. L., and Cech, T. R. (1983). *Biochemistry* **22,** 2390–2397.

Brosius, J., Dull, T. J., and Noller, H. F. (1980). *Proc. Natl. Acad. Sci. U.S.A.* **77,** 201–204.

Brown, R. S., Hingerty, B. E., Dewan, J. C., and Klug, A. (1983). *Nature (London)* **303,** 543–546.

Burke, J., and RajBhandary, U. L. (1982). *Cell* **31,** 509–520.

Carin, M., Jensen, B. F., Jentsch, K. D., Leer, J. C., Nielsen, O. F., and Westergaard, O. (1980). *Nucleic Acids Res.* **8,** 5551–5566.

Cech, T. R., and Rio, D. C. (1979). *Proc. Natl. Acad. Sci. U.S.A.* **76,** 5051–5055.

Cech, T. R., Zaug, A. J., and Grabowski, P. J. (1981). *Cell* **27,** 487–496.

Cech, T. R., Tanner, N. K., Tinoco, I., Weir, B. R., Zuker, M., and Perlman, P. S. (1983). *Proc. Natl. Acad. Sci. U.S.A.* **80,** 3903–3907.

Claisse, M., Slonimski, P. P., Johnson, J., and Mahler, H. R. (1980). *Mol. Gen. Genet.* **177,** 375–387.

Crick, F. H. C. (1968). *J. Mol. Biol.* **38,** 367–379.

Davies, R. W., Waring, R. B., Ray, J. A., Brown, T. A., and Scazzocchio, C. (1982). *Nature (London)* **300,** 719–724.

DeLaSalle, H., Jacq, C., and Slonimski, P. P. (1982). *Cell* **28,** 721–732.

Dieckmann, C. L., Pape, L. K., and Tzagoloff, A. (1982). *Proc. Natl. Acad. Sci. U.S.A.* **79,** 1805–1809.

Din, N., and Engberg, J. (1979). *J. Mol. Biol.* **134,** 555–574.

Din, N., Engberg, J., Kaffenberger, W., and Eckert, W. (1979). *Cell* **18,** 525–532.

Doolittle, W. F., and Sapienza, C. (1980). *Nature (London)* **284,** 601–603.

Eichhorn, G. L. (1981). *Adv. Inorg. Biochem.* **3,** 1–46.

Eigen, M., Gardiner, W., Schuster, P., and Winkler-Oswatitsch, R. (1981). *Sci. Am.* **244,** 78–94.

Engberg, J., Christiansen, G., and Leick, V. (1974). *Biochem. Biophys. Res. Commun.* **59,** 1356–1365.

Engberg, J., Andersson, P., Leick, V., and Collins, J. (1976). *J. Mol. Biol.* **104,** 455–470.

Farrelly, F., and Butow, R. A. (1983). *Nature (London)* **301,** 296–301.

Ferris, J. P., and Usher, D. A. (1983). *In* "Biochemistry" (G. Zubay, coord. ed.). pp. 1191–1241. Addison-Wesley, Reading, Massachusetts.

Gall, J. G. (1974). *Proc. Natl. Acad. Sci. U.S.A.* **71,** 3078–3081.

Gerbi, S. A., Gourse, R. L., and Clark, C. G. (1982). *Cell Nucl.* **10,** 351–386.

Gerriga, G., and Lambowitz, A. M. (1984). *Cell* **39,** 631–641.

Gilbert, W. (1978). *Nature (London)* **271,** 501.

Gourse, R. L., and Gerbi, S. A. (1980). *Nucleic Acids Res.* **8,** 3623–3637.

Grabowski, P. J., Brehm, S. L., Zaug, A. J., Kruger, K., and Cech, T. R. (1983). *In* "Gene Expression" UCLA Symp. Molec. Cell. Biol. (D. Hamer and M. Rosenberg, eds.), New Series Vol. VIII, pp. 327–342. Liss, New York.

Gubler, U., Wyler, T., Seebeck, T., and Braun, R. (1980). *Nucleic Acids Res.* **8,** 2647–2664.

Guerrier-Takada, C., and Altman, S. (1984). *Science* **223,** 285–286.

Guerrier-Takada, C., Gardiner, K., Marsh, T., Pace, N., and Altman, S. (1983). *Cell* **35,** 849–857.

Hollis, G. F., Heiter, P. A., McBride, O. W., Swan, D., and Leder, P. (1982). *Nature (London)* **296,** 321–325.

Jacobs, H. T., Posakony, J. W., Grula, J. W., Roberts, J. W., Xin, J. -H., Britten, R. J., and Davidson, E. (1983). *J. Mol. Biol.* **165,** 609–632.
Jacq, C., Banroques, J., Becam, A. M., Slonimski, P. P., Guiso, N., and Danchin, A. (1984). *EMBO J.* **3,** 1567–1572.
Jencks, W. P. (1969). "Catalysis in Chemistry and Enzymology." McGraw-Hill, New York.
Johnson, P. F., and Abelson, J. (1983). *Nature (London)* **302,** 681–687.
Kao, T. H., and Crothers, D. M. (1980). *Proc. Natl. Acad. Sci. U.S.A.* **77,** 3360–3364.
Karrer, K. M., and Gall, J. G. (1976). *J. Mol. Biol.* **104,** 421–453.
Kearsey, S. E., and Craig, I. W. (1981). *Nature (London)* **290,** 607–608.
Kole, R., Baer, M. F., Stark, B. C. and Altman, S. (1980). *Cell* **19,** 881–887.
Kruger, K., Grabowski, P. J., Zaug, A. J., Sands, J., Gottschling, D. E., and Cech, T. R. (1982). *Cell* **31,** 147–157.
Lazowska, J., Jacq, C., and Slonimski, P. P. (1980). *Cell* **22,** 333–348.
Mannella, C. A., Collins, R. A., Green, M. R., and Lambowitz, A. M. (1979). *Proc. Natl. Acad. Sci. U.S.A.* **76,** 2635–2639.
Michel, F., and Dujon, B. (1983) *EMBO J.* **2,** 33–38.
Michel, F., Jacquier, A., and Dujon, B. (1982). *Biochimie* **64,** 867–881.
Netzker, R., Köchel, H. G., Basak, N., and Küntzel, H. (1982). *Nucleic Acids Res.* **10,** 4783–4794.
Noller, H. F., and Woese, C. R. (1981). *Science* **212,** 403–410.
Orgel, L. E., and Crick, F. H. C. (1980). *Nature (London)* **284,** 604–607.
Pan, W. -C., Orias, E., Flacks, M., and Blackburn, E. H. (1982). *Cell* **28,** 595–604.
Reanney, D. (1979). *Nature (London)* **277,** 598–600.
Rochaix, J. D., and Malnoë, P. (1978). *Cell* **15,** 661–670.
Rubin, J. R., and Sundaralingam, M. (1983). *J. Biomolec. Struc. Dynam.* **1,** 639–646.
Sor, F., and Fukuhara, H. (1982). *Nucleic Acids Res.* **10,** 6571–6577.
Tabak, H. F., Van der Horst, G., Osinga, K. A., and Arnberg, A. C. (1984). In preparation.
Thompson, J., Schmidt, F., and Cundliffe, E. (1982). *J. Biol. Chem.* **257,** 7915–7917.
Van Arsdell, S. W., Denison, R. A., Bernstein, L. B., Weiner, A. M., Manser, T., and Gesteland, R. F. (1981). *Cell* **26,** 11–17.
Waring, R. B., Davies, R. W., Scazzocchio, C., and Brown, T. A. (1982). *Proc. Natl. Acad. Sci. U.S.A.* **79,** 6332–6336.
Waring, R. B., Scazzocchio, C., Brown, T. A., and Davies, R. W. (1983). *J. Mol. Biol.* **167,** 595–605.
Weiss-Brummer, B., Holl, J., Schweyen, R. J., Rodel, G., and Kaudewitz, F. (1983). *Cell* **33,** 195–202.
White, H. B., III (1976). *J. Mol. Evol.* **7,** 101–104.
White, H. B., III (1982). *In* "The Pyridine Nucleotide Coenzymes," pp. 1–17. Academic Press, New York.
Wild, M. A. (1980). PhD. dissertation, Yale University, New Haven, Connecticut.
Wild, M. A., and Gall, J. G. (1979). *Cell* **16,** 565–573.
Woese, C. R. (1972). *In* "Exobiology" (C. Ponnamperuma, ed.), pp. 301–341. North-Holland Publ., Amsterdam.
Wollenzien, P. L., Cantor, C. R., Grant, D. M., and Lambowitz, A. M. (1983). *Cell* **32,** 397–407.
Wright, R. M., and Cummings, D. J. (1983). *Nature (London)* **302,** 86–88.
Yao, M. -C., and Gall, J. G. (1977). *Cell* **12,** 121–132.
Zaug, A. J., and Cech, T. R. (1980). *Cell* **19,** 331–338.
Zaug, A. J., and Cech, T. R. (1985). In preparation.
Zaug, A. J., Grabowski, P. J., and Cech, T. R. (1983). *Nature (London)* **301,** 578–583.
Zaug, A. J., Kent, J. R., and Cech, T. R. (1984). *Science* **224,** 574–578.

Section II

DNA Reorganization and the Genetic Evolution of Subcellular Organelles

INTERNATIONAL REVIEW OF CYTOLOGY, VOL. 93

Mechanisms of DNA Reorganization in Bacteria

J. A. SHAPIRO

Department of Microbiology, The University of Chicago, Chicago, Illinois

I. Introduction

Classical cytogenetic studies have documented a wide variety of chromosomal rearrangements in all organisms that have been closely investigated. These rearrangements include programmed changes, such as chromatin diminution in *Ascaris* (Boveri, 1892), *Cyclops* (Beerman, 1977), and *Sciara* (Crouse, 1976) and macronuclear differentiation in ciliated protozoa (reviewed in Nanney, 1980), as well as less regular chromosome reorganizations such as fusions, duplications, deletions, inversions, terminal translocations, and transpositions. The advent of methods for directly characterizing DNA molecules extracted from viruses and from the cells of prokaryotes and eukaryotes has greatly expanded the catalogue of genomic reorganizations and documented events (such as chromosomal insertions of reverse-transcribed messenger RNA sequences; Hollis *et al.*, 1982) that were dismissed as fanciful only a few years ago. Some of the more recently recognized DNA reorganizations are also programmed—for example, changes at the *Saccharomyces* locus for mating-type (Nasmyth, 1982; Haber, 1983), expression of sequences encoding *Trypanosoma* variable surface glycoproteins (Borst, 1983), and joining of separate sequences encoding immunoglobulin domains (Tonegawa, 1983).

For many decades, most classes of genomic reorganization have been difficult to incorporate into a coherent theory of genome structure and function. They have traditionally been considered "accidents" that are part of the background of random variation which provides the raw material for selection. However, the documentation of programmed genome reorganizations makes such a viewpoint untenable. One major stumbling block to a more dynamic view of genome structure was the belief that the chemical stability of DNA forms the basis for the

ISBN 0-12-364493-3

reliable transmission of hereditary characteristics through cell lineages. However, as our knowledge of nucleic acid biochemistry and chromatin structure has increased in the past three decades, it has become apparent that DNA is subject to many kinds of biochemical activities that sequester, duplicate, and alter the sequence information contained in specific glycosidic and phosphodiester linkages. Thus, the problem of genomic stability and change can now be viewed as one of interaction between different macromolecular systems (involving proteins, nucleic acids, and membranes) subject to elaborate regulatory mechanisms rather than as a simple question of chemical events involving isolated covalent bonds.

II. Properties of Proteins Active on DNA

Before discussing specific kinds of DNA reorganization mechanisms active in bacteria, it will be helpful to review some of the known properties of proteins that act on DNA. One important point to note is that many of these proteins are not enzymes in the strict sense of the term. That is, many proteins active on DNA either do not catalyze specific chemical reactions or do not show turnover but rather function in a stoichiometric manner. The absence of turnover facilitates the control of DNA reorganizations because they can be restricted to a few reactions by limiting the amount of the proteins involved. Proteins which acted enzymatically might carry on repeated DNA reorganizations.

The following list of properties for DNA-active proteins is certainly incomplete. Moreover, not all of the properties have been well documented for proteins known to be involved in DNA reorganization. However, the basic objective here is to give some idea of the biochemical tools available to construct systems for DNA reorganization.

1. *Specific recognition of DNA structure.* Various proteins discriminate between structurally different DNA molecules with little or no nucleotide sequence specificity. There are proteins that bind to single-stranded but not double-stranded DNA (Chapter 9 of Kornberg, 1980, 1982). These include proteins involved in replication and recombination, such as phage T4 gene 32 protein (Alberts and Frey, 1970), the single-strand binding protein of *E. coli,* and the *recA* protein. By preferential binding to single-stranded regions, these so-called ''helix destabilizing proteins'' can facilitate the transition from duplex to unwound DNA structures. Nucleases are generally specific for single-stranded or duplex DNA and for exposed termini (exonucleases) or internal phosphodiester bonds (endonucleases) (Linn and Roberts, 1983). Some proteins recognize limited perturbations in the helix structure, such as nucleases and glycosidases specific for pyrimidine dimers (e.g., Braun and Grossman, 1974). Proteins also ''read'' specific modifications on DNA molecules, such as glycosylations and

methylations. One particularly interesting example is the methylase activity of type I restriction endonucleases, which shows a strong preference for molecules with a recognition sequence that already carries a methylation on one strand of the duplex (Yuan, 1981).

2. Proteins can recognize specific DNA sequences. There is a large literature on regulatory proteins, polymerases, and nucleases which act only on specific sequences. Restriction endonucleases are nowadays the best documented examples of sequence specificity, and a look at a catalogue describing these reagents will illustrate various degrees of specificity. Some enzymes only cleave a unique sequence of 4, 5, 6, or more nucleotides while other enzymes tolerate more than one nucleotide at particular positions in the recognition sequence. Type II restriction enzymes cleave at specific positions within the recognition sequence, but type I enzymes bind to the helix at a specific sequence and then migrate variable distances before cleavage (Yuan, 1981). Regulatory proteins such as repressors, the *E. coli* catabolite activator protein, and RNA polymerases recognize sequences that are all related to each other (and thus to a so-called "consensus sequence") but which are not necessarily identical (Rosenberg and Court, 1979). Other proteins, such as DNA gyrase, have preferred targets but will also act on many different sequences with no obvious homology to the favored one (Morrison and Cozzarelli, 1979). Some proteins bind to two different consensus sequences; the λ integrase is an example which will be described in more detail below.

3. Like other proteins, those active on DNA undergo allosteric transitions with various ligands, including DNA and other proteins as well as small molecules. These transitions play a key role in regulatory phenomena. Interactions between more than one molecule of the same protein permit cooperative effects in DNA binding such as those which form the basis for efficient repression of the λ prophage (Ptashne *et al.*, 1980; Johnson *et al.*, 1981). Protein–protein and DNA–protein affinities work together in the formation of three-dimensional structures, such as nucleosomes. The *recA* protein has a particularly large repertoire of allosteric interactions with DNA chains, nucleotides, and proteins that govern its catalytic activities both as a regulatory protease and as a recombination enzyme (reviewed in Little and Mount, 1982; Radding, 1982).

4. Proteins can cleave phosphodiester linkages (nucleases).

5. Proteins can reseal interrupted phosphodiester linkages (ligases).

6. Proteins can polymerize polynucleotides on a template strand (polymerases) as well as without a template (e.g., terminal transferases). Polymerization of RNA can proceed without a primer, but polymerization of DNA always begins at a 3′ hydroxyl terminus of either a DNA or RNA chain. Extension of a DNA chain within a duplex will displace the strand complementary to the template. Kornberg's book (1980) and the Supplement (1982) cover much of the work on polymerization activities.

7. Proteins can attach covalently and reversibly to the ends of DNA chains.

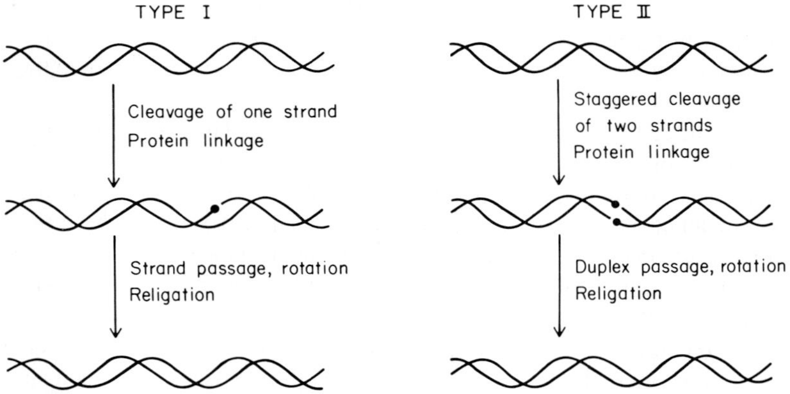

FIG. 1. Topoisomerase cleavage-religation reactions. The solid circles attached to interrupted DNA strands represent topoisomerase subunits covalently linked to the terminal phosphate group. The cleaved molecules can change topological conformation before reforming their original phosphodiester linkages. Rotation or passage of single- or double-stranded molecules across the breaks permits a wide variety of structural DNA changes without any alteration of nucleotide sequence information; these include loss or gain of supertwist density, knotting or untying of circular molecules, and formation or breakdown of catenanes (interlocked circular molecules). Clearly these activities play a key role in the movement of highly compacted DNA molecules in bacteria, where the length of a chromosomal molecule can be up to 1000 times the length of a cell (Cairns, 1963). For further details on the topological activities and molecular mechanisms of topoisomerases, see the reviews by Champoux (1978), Cozzarelli (1980), Gellert (1981), and Wang and Liu (1979).

Protein–DNA linkage provides a mechanism for preserving sites of cleavage so that they can be accurately resealed. The *cisA* protein of phage ϕX174 plays such a role in replication and circularization of viral single-stranded molecules (van Mansfeld *et al.*, 1980). Topoisomerases are proteins that specialize in transiently opening and closing DNA molecules to eliminate physical barriers (e.g., tangling) to DNA movement and to control superhelicity (Wang and Liu, 1979; Champoux, 1978; Cozzarelli, 1980; Gellert, 1981). Type I topoisomerases transiently open and then reseal single DNA strands while type II topoisomerases create transient double-stranded breaks (Fig. 1). The covalent attachment of protein can occur either at a 5′ phosphate (ϕX174 *cisA* protein and *E. coli* gyrase) or at a 3′ phosphate (eukaryotic type I topoisomerases and λ integrase; N. Craig and H. Nash, personal communication).

8. Proteins can catalyze the unwinding of DNA strands. Such enzymes have been called helicases (Geider and Hoffman-Berling, 1981), and are often different from the noncatalytic helix-destabilizing proteins.

9. Proteins can catalyze the annealing of complementary DNA strands (Weinstock *et al.*, 1979; reviewed in Radding, 1982). It is important to note that the *recA* protein can act as a helicase in the presence of certain cofactors (single-

stranded DNA, ATPγS) and as a reannealing enzyme in the presence of other cofactors (ATP, dATP).

10. Proteins can mediate the physical movement of intact duplex DNA molecules. One of the best-studied examples is the action of type I restriction endonucleases from *E. coli* (Yuan, 1981). These multicomponent enzymes bind to a specific sequence on the molecule, remain attached to that sequence, and translocate DNA segments on either side of the recognition site to create loops that grow larger or smaller depending on the direction of duplex movement. The relative translocation of regions on the same duplex occurs at the expense of ATP hydrolysis. Note that the net result of this kind of movement is a kind of one-dimensional "search" along the duplex that sequentially juxtaposes one site with many other sites on the same molecule.

III. Recombination Systems

Bacterial cells contain multiple systems for opening phosphodiester linkages in DNA molecules and resealing them in novel combinations. These systems differ in several respects, including their requirements for nucleotide sequence homology or specific sequences. It was fashionable to call many of these systems "extraordinary" (Franklin, 1967) or "illegitimate" (Weisberg and Adhya, 1977), but now it is clear that the distinction between "legitimate" and "illegitimate" recombination is purely arbitrary. Different recombination systems serve different functions, and it is important to remember that bacteria make use of all of them in different situations.

General Homologous Recombination. Because of its use in constructing linkage maps, this is the most widely known kind of recombination. General recombination can be observed between two duplexes which share overall or interspersed nucleotide sequence homology, and various recombinant chromosome structures can result from the exchanges (Fig. 2). Molecular models that explain how the different products of homologous recombination can arise from closely related biochemical events are frequently based on the original proposal of Holliday (1964). The central feature of this scheme is an intermediate "Holliday" structure containing at least one heteroduplex region where the complementary DNA strands come from different parental molecules. Two of the variations of this scheme are outlined in Fig. 3. The necessary biochemical activities include a duplex-specific single-strand cleavage activity to interrupt one or two parental strands, a helicase to separate parental strands, a reannealing function to form heteroduplex regions (in some versions), DNA polymerase to fill in gaps formed by strand separation, a cleavage activity to break or "resolve" the strands joining the four duplex "tails" of the Holliday structure (Mizuuchi *et al.*, 1982), and a ligation activity to seal up the recombinant products. Depending on how

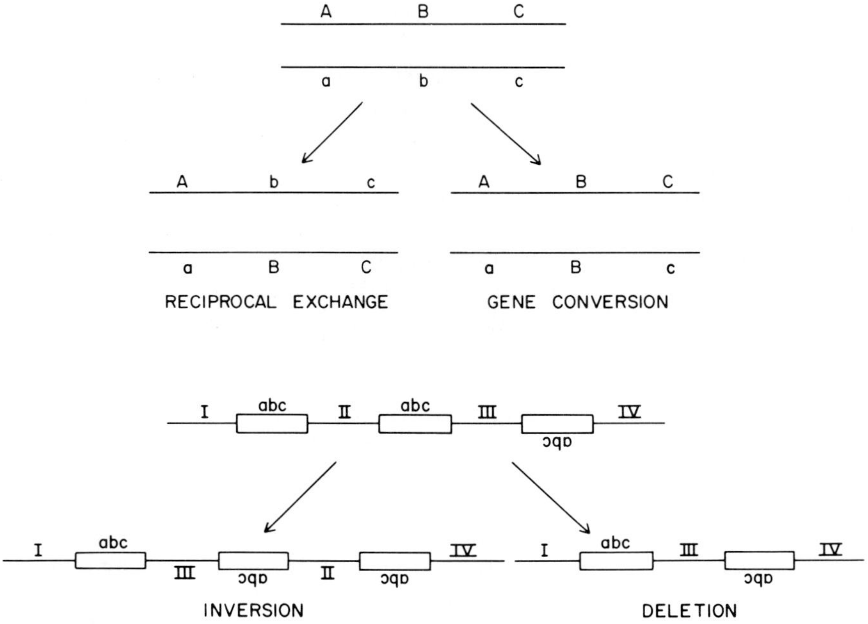

FIG. 2. Some of the results of general homologous recombination. In this figure, duplex DNA molecules are symbolized by single lines. The top panel shows two molecules which differ only at particular positions to give allelic differences (A/a, B/b, and C/c). It is important to note that these differences may be single nucleotide substitutions or more extensive variations (deletions, insertions, or nonhomologies of up to several thousand nucleotide pairs). Reciprocal exchange between these allelic differences results in new arrangements of markers with no loss or gain of information. Gene conversion can result in the loss of one allele and its replacement by its alternate. Reciprocal exchange of flanking markers is frequently observed with gene conversion of internal alleles (ABC + abc → ABc + aBC). The bottom panel shows some consequences of reciprocal exchanges between interspersed homologous sequences (open boxes) embedded in nonhomologous regions. Exchange between oppositely oriented homologies in one molecule results in inversion of the sequences flanked by the repeats. Exchange between parallel homologies in one molecule results in the deletion of the sequences flanked by the repeats and (not shown) their excision as a circle together with one copy of the repeat. Multiple events and exchanges between repeats on different DNA molecules can lead to more complex rearrangements, such as terminal and insertional translocations of sequences from one molecule to another.

the Holliday structure is resolved to two duplex molecules, the genetic products will or will not show recombination of outside markers, and the exact segregation of markers after replication will also depend on mismatch repair of heteroduplex regions. Recent studies on the gene conversion events that underly mating-type switching in yeast have indicated that double-strand breaks play an important role in general homologous recombination events (Nasmyth, 1982;

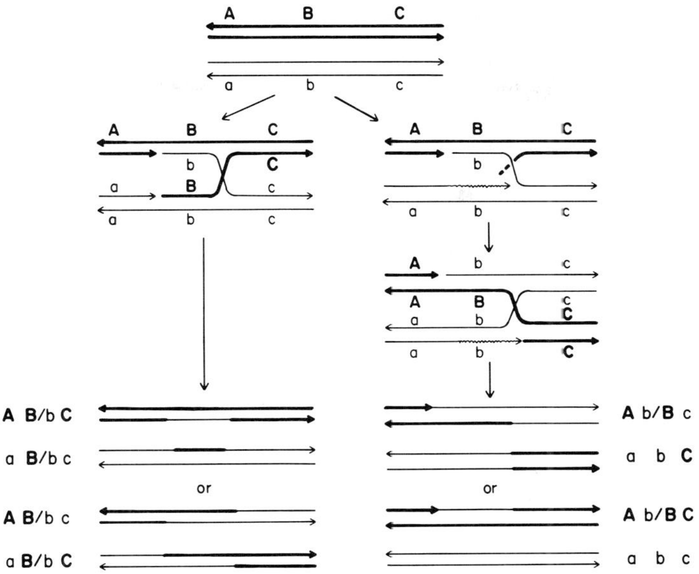

FIG. 3. Variations of the Holliday model for general homologous recombination (Holliday, 1964; Meselson and Radding, 1975). Arrowheads indicate the 5′ to 3′ polarity of each DNA strand. The left side shows Holliday's original proposal on the consequences of single-strand breaks in each parental molecule followed by reciprocal unwinding and reassociation of the interrupted strands. The right side shows the Meselson-Radding (1975) proposal where a single break in the first parental molecule is followed by polymerization (wavy line) and strand displacement on that molecule, strand invasion (D-looping), and digestion (---) of the displaced strand from the second parental molecule, and isomerization of the interlocked duplexes to generate a Holliday structure. The Holliday structure can be resolved in two different ways to generate heteroduplex molecules (potential gene conversion structures). Cleavage of previously uninterrupted chains produces reciprocal exchange of duplex regions outside the heteroduplex region, whereas cleavage of the exchanged strands regenerates the original configuration for outside markers.

Haber, 1983). A similar conclusion about the role of double-strand breaks comes from the observation that restriction enzyme cleavage of plasmid DNA enhances its recombinogenic activity in yeast transformation (see Szostak *et al.*, 1983, for a discussion).

So far, several biochemical activities of *E. coli* and its phages have been identified as the products of cistrons encoding general recombination functions (Table I). The best characterized of these is the *recA* product which can both act as a helix-destabilizing protein and also catalyze the reannealing of DNA strands to form heteroduplex regions between homologous segments of different molecules at the expense of ATP hydrolysis (see Radding, 1982, for an extensive recent review). Although most models postulate separate cleavage and ligation

TABLE I
SOME PATHWAYS FOR GENERAL HOMOLOGOUS RECOMBINATION IN *E. coli*

Pathway	Cistrons whose products are required for recombination	Comments
RecABC	*recA* *recB*, *recC* (coding sequences for exonuclease V subunits)	Responsible for ≥99% of chromosomal recombination in Hfr crosses (reviewed in Clark, 1980); stimulated by Chi sequences (Stahl, 1979)
RecAF	*recA* *recF* *recJ*	Activity enhanced at least 100-fold by loss of exonuclease I activity (*sbcB* mutation; reviewed in Clark, 1980); SOS inducible (Lovett and Clark, 1983); indifferent to Chi sequences
RecE	*recE* = *sbcA* (coding sequence for exonuclease VIII)	Activation of recombination system from the defective *rac* prophage (Kaiser and Murray, 1979) and may well involve activities other than ExoVIII (Gillen *et al.*, 1977); independent of *recA* product for plasmid and phage recombination (Gillen and Clark, 1974; Fishel *et al.*, 1981); indifferent to Chi sequences; may not be active on chromosomal sequences
Red	λβ and *exo*	Phage system (Smith, 1983); independent of *recA* product; indifferent to Chi sequences

events, combinations of a type I topoisomerase and *recA* protein can form covalently closed ''D loop'' structures between homologous circular duplex and single-stranded molecules (Cunningham *et al.*, 1981). Thus, concerted cleavage-ligation events may play an important role in homologous recombination. Several exonucleases have also been identified as recombination functions, but at least one of these (the exonuclease V product of the *recB* and *recC* cistrons) may function primarily as an ATP-dependent helicase (Taylor and Smith, 1980).

There are several alternative pathways for homologous recombination in *E. coli* cells already described (Table I). These do not act the same on all DNA molecules in the bacterial cell. For example, the *recE* pathway works well for plasmids and lambdoid phages but not for chromosomal DNA in conjugal crosses (Gillen and Clark, 1974; Fishel *et al.*, 1981). In addition, the alternative pathways for homologous exchange may yield different kinds of products so that there are mechanisms available to control what kind of recombination events occur. This point has not been well studied in bacteria, but in yeast there is extensive documentation of developmental stages and mutations which greatly alter the ratio of reciprocal exchange to gene conversion events (Roman, 1956;

Fabre and Roman, 1977; Game *et al.*, 1980; Magni and von Borstel, 1962; Malone and Esposito, 1980; reviewed in Kunz and Haynes, 1981).

General Nonhomologous Recombination. There are many recombination events which occur infrequently at a large number of genomic locations and which do not involve extensive homologous or repeated sequences. These include the formation of deletions, specialized transducing phages and some duplications, the precise excision of transposable element insertions, and the integration of circular DNA into other molecules (Fig. 4). In many cases, very short oligonucleotide repeats (5–20 base pairs) are preferred substrates for these recombinations (Farabaugh *et al.*, 1978; Efstratiadis *et al.*, 1980; Albertini *et al.*, 1982). The biochemical basis of these relatively nonspecific, rare events is not well understood, but two kinds of plausible mechanisms have been suggested. One is "replication slippage," where the polymerase "jumps" between short repeats (Efstratiadis *et al.*, 1980; Albertini *et al.*, 1982). The other is based on the experiments of Ikeda *et al.* (1981) which demonstrated that DNA gyrase plays a role in *in vitro* insertion of plasmids into the phage λ genome. This mechanism postulated that type II topoisomerases occasionally reseal transient double-strand breaks between two (or even three) sites rather than at a single site (Fig. 5). Since DNA gyrase makes a staggered double-strand break with a four base overhang (Morrison and Cozzarelli, 1979), short repeats may facilitate the resealing of breaks between two sites and so account for the observed preference in deletion endpoints.

Site-Specific Reciprocal Recombination. Several systems for reciprocal recombination between very highly defined DNA sequences mediated by specific proteins are known from genetic experiments (Fig. 6). These systems serve different functions, including regulation of transcription by promoter inversion (Simon *et al.*, 1980; Simon and Silverman, 1983), switching of polypeptide synthesis from a single promoter by coding sequence inversion (Toussaint and Resibois, 1983), prophage insertion and excision (Campbell, 1983), monomerization of plasmid dimers (Sternberg *et al.*, 1981; Austin *et al.*, 1981), and resolution of cointegrate intermediates in transposition (Reed, 1981; Heffron, 1983). Given this variety of function, we should expect the detailed protein–DNA interactions to show significant diversity, and this expectation is fulfilled for the two systems about which we have the most detailed information: λ integrative recombination and cointegrate resolution. Nonetheless, these two systems also show a fundamental similarity in the catalytic activity of the specific recombination proteins, suggesting that a single biochemical mechanism can be adapted to different purposes through regulatory networks controlling catalysis.

In λ insertion, two different but related segments of DNA called *attP* and *attB* recombine to form two new segments called *attL* and *attR* (respectively, the *p*hage, *b*acterial, *l*eft prophage and *r*ight prophage *att*achment sites). In excision, *attL* and *attR* recombine to form *attP* and *attB* (Fig. 7). All the recombination

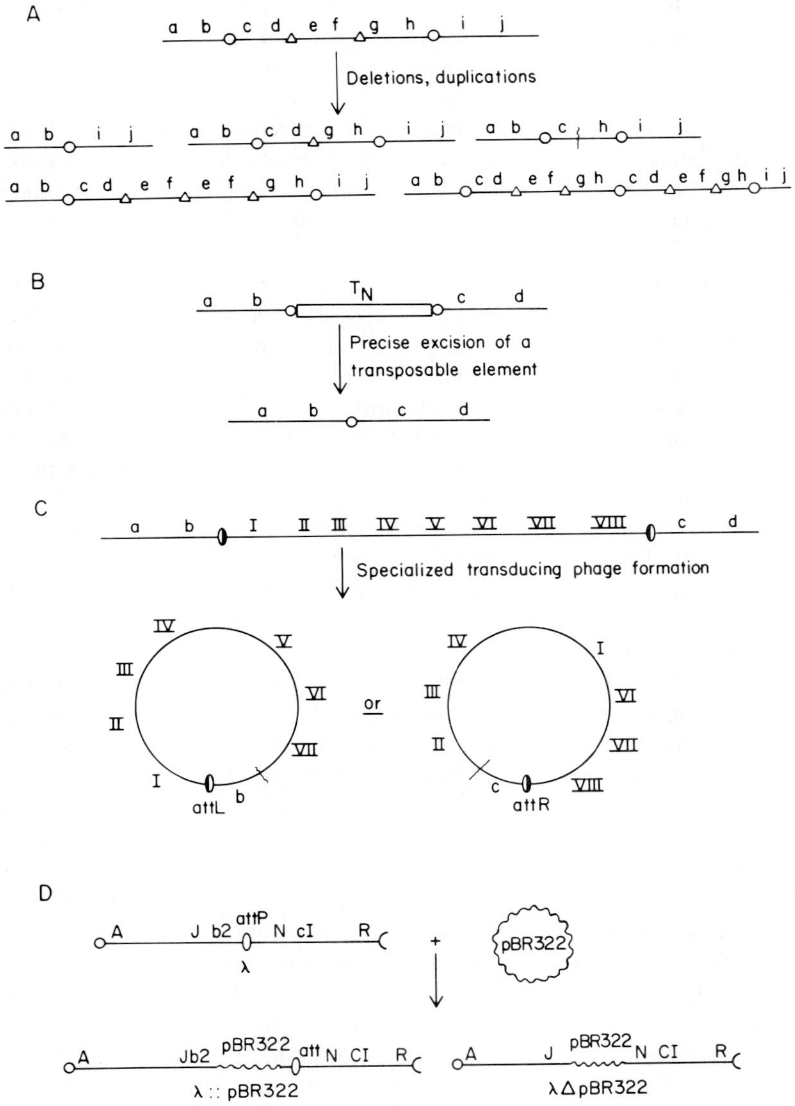
A
a b c d e f g h i j
Deletions, duplications
a b i j
a b c d g h i j
a b c h i j
a b c d e f e f g h i j
a b c d e f g h c d e f g h i j
B
a b T N c d
Precise excision of a transposable element
a b c d
C
a b I II III IV V VI VII VIII c d
Specialized transducing phage formation
IV V III VI II VII I attL b
or
IV I III VI II VII c attR VIII
D
A J b2 attP N cI R
λ
+
pBR322
A Jb2 pBR322 att N CI R
λ :: pBR322
A J pBR322 N CI R
λΔpBR322

normally occurs within a 15 base-pair "common core" region identical in all four *att* sequences, which differ outside the core by four "arms" symbolized P, P′, B, and B′ (see Nash, 1981, for a recent comprehensive review). Only the core, P and P′ sequences play a role in recombination. One of the indications for the functional differences between *attP, attB, attL,* and *attR* was the genetic observation that the insertion reaction requires only a single phage-encoded protein (integrase, the *int* cistron product) while the excision reaction requires (under most conditions) a second phage-encoded protein (the *xis* cistron product). These differential protein requirements provide directional control over prophage–chromosome recombination, and this directional control is exploited by an elaborate network controlling the differential expression of *int* and *xis* (see below). In addition to the *int* and *xis* proteins, the product of two bacterial chromosome sequences (*himA* and *himD*) are also required for λ integrative recombination (the combination being denoted "integration host factor" or IHF). Insertion and excision recombinations (*attP* × *attB* and *attL* × *attR*) can occur equally well between *att* sequences in all relative configurations (i.e., intermolecular recombination or intramolecular recombination between directly repeated or inverted *att* sites).

Both insertion and excision reactions can be studied in purifed *in vitro* systems. The requirements for insertion are a supercoiled *attP* substrate, the *int* and IHF proteins, and specific ionic conditions. No triphosphates or other external "high-energy" compound is needed, and the products of recombination between supercoiled substrates are still supercoiled. Thus, the energy to drive the reaction does not come from hydrolysis of high energy bonds or gross loss of superhelicity, and recombination does not involve polymerization. At least 10 inte-

---

FIG. 4. Some consequences of general nonhomologous recombination. Duplex molecules are shown as single lines. Open circles and triangles indicate repeated oligonucleotides (≤ 25 nucleotide pairs). A illustrates a molecule with two oligonucleotide repeats. Reciprocal exchange between these repeats within a single molecule can generate deletions, but exchange between these repeats in two molecules or two newly replicated daughter duplexes can create a duplication (see Weisberg and Adhya, 1977). Some exchanges do not involve oligonucleotide repetitions and so create novel sequences (wavy line) at the deletion end-points. B illustrates the frequently encountered situation where insertion of a transposable element has created a flanking oligonucleotide duplication. Reciprocal exchange between these repeats will excise the insert and regenerate the previous sequence. C illustrates the Campbell (1962) model for specialized transducing phage formation. In essence, this is the same kind of intramolecular exchange that leads to a deletion, but here we look at the excised DNA circle which we can isolate by virtue of its viral information that permits replication and encapsidation. The nature of the deletion joints (wavy line) in specialized transducing phages has not been adequately investigated. D illustrates some of the ways that λ phage can incorporate a small plasmid molecule. (The cases where this happens by the phage integrase-*att* system have not been included; see Pogue-Geile *et al.*, 1980). Sometimes the product has a simple plasmid insertion (λ::pBR322) but often plasmid incorporation is accompanied by loss of some λ sequences at the site of insertion (Ikeda *et al.*, 1981; Pogue-Geile *et al.*, 1980).

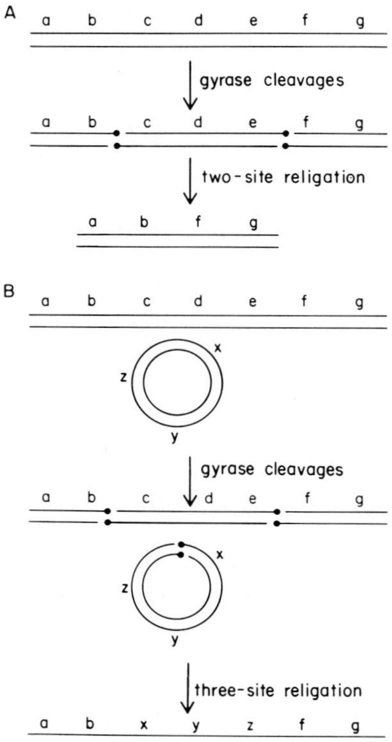

FIG. 5. Type II topoisomerase mechanism for general nonhomologous recombination. A illustrates deletion formation by religation of two double-strand cleavage sites on a single molecule. B illustrates how religation of three cleaved sites will incorporate a circular molecule into another molecule and also create a deletion at the site of incorporation. Note that this event requires the same number of religations as incorporation of a circular molecule with no accompanying deletion.

grase monomers per pair of attachment sites are required for a recombination, indicating that integrase acts stoichiometrically and binds cooperatively to the substrates. Direct evidence for multiple integrase binding sites at the common core, P and P′ regions has been obtained by "footprinting" studies (Fig. 7), and electron microscopy reveals the cooperative formation of condensed DNA–integrase structures ("intasomes") nucleated by binding to *attP* (Better *et al.*, 1982). It has been possible to identify two different "consensus" sequences that appear to be the determinants of integrase binding specificity (Hsu *et al.*, 1980). IHF also binds to *attP* sites interspersed between the integrase binding sites, but these host proteins are not required for formation of intasomes (Better *et al.*, 1982). The binding site in the P′ arm is probably the initial site of specific integrase–DNA interaction because (1) heparin removes integrase from all bind-

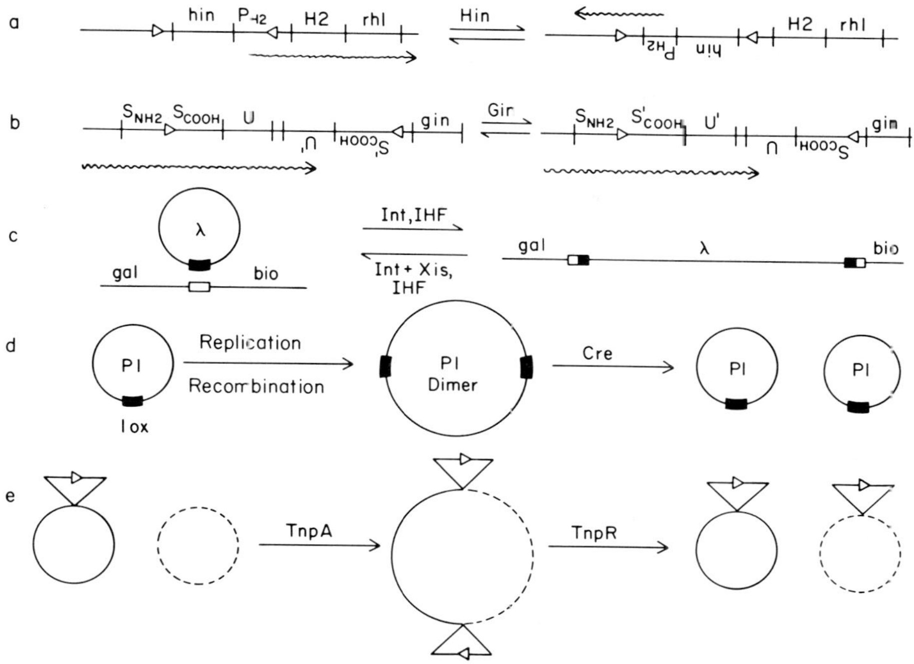

FIG. 6. Different genetic systems utilizing site-specific reciprocal recombination. Duplex DNA segments are represented by single lines and specific recombination sequences are represented by triangles (to indicate relative sequence orientation) or by squares. (a) The phase variation system of *Salmonella* (Simon and Silverman, 1983). The Hin function inverts the segment that includes the *hin* cistron and a promoter for transcription (in one orientation) of the h2 and *rh1* cistrons, so recombination controls H2 and H1 expression. (b) The host-range determinants of phage Mu (Toussaint and Resibois, 1983). The Gin function inverts the segment that includes two sets of alternative coding sequence: the U and U′ cistrons and the portions of the S and S′ cistrons encoding the carboxy-terminal segments of the corresponding proteins. The portion of the S and S′ cistrons encoding the amino-terminal segments lies outside the inverting region and is the same for both cistrons. Note that the specific recombination sequence also constitutes part of the S and S′ cistron. (c) The system for λ insertion and excision (Campbell, 1983). (d) Dimer monomerization in the plasmid prophage of phage P1 (Austin *et al.*, 1982). (e) Resolution of intermediates in transposition of Tn3-like transposons (Heffron, 1983). The first step is a distinct replicative recombination event catalyzed by the TnpA activity to generate a cointegrate structure where the duplicated transposons contain directly repeated recombination sequences that are substrates for the TnpR resolvase. Note that the Hin and Gin functions are interchangeable and that 12 nucleotides of the inverted 14 nucleotide recombination sequences of the *Salmonella* system are included in the 34 nucleotide inverted repeats of the Mu system. The *tnpR* coding sequence also shows significant amino acid sequence homology to *hin,* suggesting that the Tn3 resolvase system is related to the *Salmonella* and Mu systems.

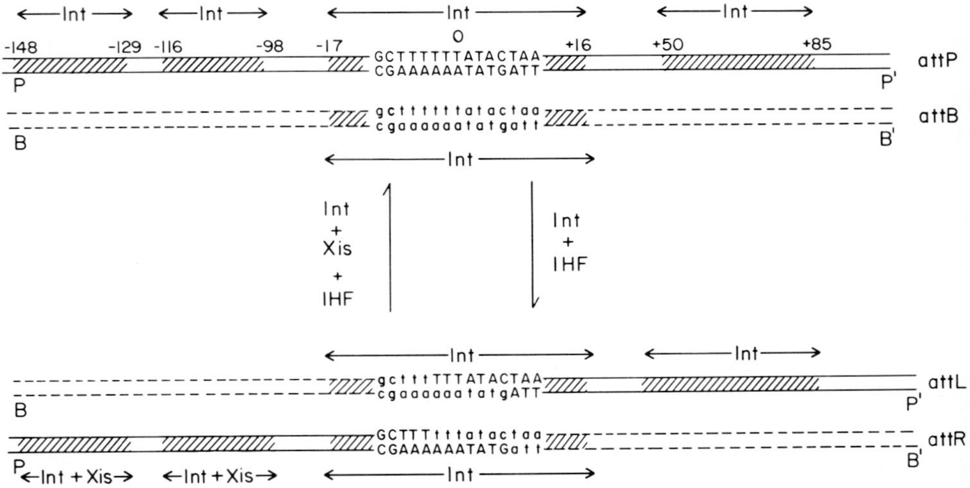

FIG. 7. Details of the λ integrative recombination system. The two attachment sites are schematized and are not to scale. The 15 nucleotide "common core" sequences are given to show the sites of cleavage and religation (Mizuuchi *et al.*, 1981; N. Craig and H. Nash, personal communication). The double-headed arrows and shaded regions indicate integrase binding regions in *attP* and *attB* as defined by footprinting experiments (Mizuuchi *et al.*, 1981). The binding sites for integrase and integrase plus *xis* protein in *attL* and *attR* are inferred from electron microscopy of DNA-protein complexes (Better *et al.*, 1983). The binding sites for integration host factor (IHF) are not shown.

ing sites except this one, and (2) the P′ site in *attL* permits the formation of an intasome, but the core and P sites in *attR* do not. The role of *xis* protein appears to be to facilitate cooperative binding because addition of *xis* proteins permits nucleation of intasomes at *attR* and facilitates the formation of joint *attL–attR* intasomes (Better *et al.*, 1983).

The positions of DNA strand exchange within the core sequence are known (Mizuuchi *et al.*, 1981; N. Craig and H. Nash, personal communication), but the exact biochemical mechanism has not been worked out. Although there is no direct evidence to identify either the *int, himA,* or *himD* product as the catalytically active protein, it is known that purified integrase has type I topoisomerase activity (Kikuchi and Nash, 1979). Thus, the *int* product is capable of cleaving and religating phosphodiester linkages. This topoisomerase activity is specific for those sequences which direct integrase binding to the core region of *att* (N. Craig and H. Nash, personal communication). If integrase is the catalytic protein, the roles of the *himA* and *himD* products need to be clarified. Whatever these roles are, the requirement for *himA* and *himD* expression makes it possible to tie λ insertion and excision into bacterial host regulatory circuits. For example, λ excision is normally a response to induction of the host cell SOS re-

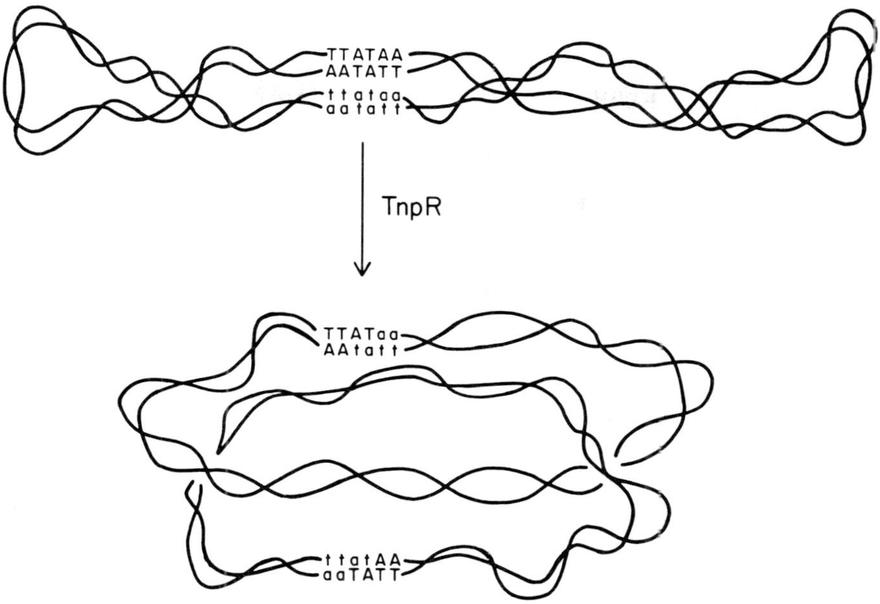

FIG. 8. Details of cointegrate resolution. The upper cartoon symbolizes a supertwisted molecule with directly repeated resolution sites. The palindromic hexanucleotide sequence where resolvase produces staggered double-strand cleavages with a two base overhang is shown. TnpR activity results in the formation of two catenated recombination products. These are supertwisted (not shown here for clarity) and generally have a single link between the two recombinant circles (Krasnow and Cozzarelli, 1983).

pair/mutagenesis system, and the *himA* cistron is known to be regulated by the SOS response (Miller *et al.*, 1981b).

The biochemistry of cointegrate resolution (Fig. 8) is simpler than λ integrative recombination. Only a single "resolvase" protein, the product of the *tnpR* cistron of transposable elements related to Tn3 and Tn1000 (formerly called γδ), is required *in vitro* (Reed, 1981). The reaction is similar to λ in the requirement for supercoiled substrate, the supercoiled nature of the recombinant products, the absence of a requirement for high-energy cofactors, and the high recombinase to substrate ratio (Reed, 1981; Krasnow and Cozzarelli, 1983). Moreover, resolvase (like integrase) has type I topoisomerase activity, suggesting that successive single-strand cleavages followed by religations are the important biochemical steps in cointegrate resolution (Krasnow and Cozzarelli, 1983). However, in the absence of divalent cations, it is also possible to isolate a DNA species with a staggered double-strand break at the site of recombination with resolvase monomers covalently bound to the exposed terminal 5′ phosphate groups (Reed and Grindley, 1981), but it is not yet clear whether this structure is a true reaction intermediate or a byproduct of two partial reactions.

One major difference between cointegrate resolution and λ integrative recombination lies in the arrangement of the specific substrate sequences. For the λ case, all arrangements are functional, but this is not so for resolution. *In vivo,* resolution is a virtually irreversible reaction (Muster *et al.,* 1983a). Since the recombining sequences are duplicates and hence identical, the products are also identical. Thus, the only way to ensure irreversibility is to build in a requirement for a particular topological arrangement of substrates, and this has been observed both *in vivo* and *in vitro:* only direct repeats of the resolution sites on the same molecule recombine at all but the highest levels of resolvase (Reed, 1981). This arrangement is also required for the topoisomerase activity of resolvase, thus indicating that effective binding of resolvase only occurs on molecules with direct substrate repeats (Krasnow and Cozzarelli, 1983). The resolution site itself, like the λ attachment sites, appears to have a complex structure. Genetic experiments indicate that sequences about 80 base pairs from the crossover point are required for efficient recombination (Kostriken *et al.,* 1981), and footprinting studies reveal at least three resolvase binding sites (Grindley *et al.,* 1982). However, resolvase is also a transcriptional regulatory protein, and it is not yet known whether all three binding sites play a role in site-specific recombination (Heffron, 1983).

There are basically two classes of hypothesis to account for the requirement for specific substrate arrangements in resolvase-mediated recombination: (1) the first resolvase complex to bind to one resolution site "tracks" along the DNA until it encounters a second properly oriented site, at which point an effective DNA-topoisomerase/recombinase structure is formed, or (2) effective resolvase binding occurs only when two resolution sites are properly juxtaposed within a single supertwisted molecule (the supertwisting making possible more intimate contact between the two sites than would occur in a bimolecular pairing). Whatever, the final explanation turns out to be, directional specificity in both the λ and transposon recombination systems illustrates an important general principle in protein–DNA interactions: namely, sequence-specific binding often does not occur independently of protein–protein interactions and these interactions, in turn, depend on the relative positions of the recognized sequences in DNA molecules.

A third site-specific reciprocal recombination system—the *lox*-Cre system of phage P1 (Sternberg *et al.,* 1981; Fig. 6)—has recently become amenable to *in vitro* analysis (Abremski *et al.,* 1983). There is an interesting and probably biologically relevant difference between the P1 recombination on the one hand and the λ and Tn3 systems on the other hand. The products of intermolecular recombination mediated by integrase or resolvase are catenated (cf. Fig. 8) whereas the products of analogous events mediated by the P1 Cre activity are often separate DNA circles. Since the *lox*-Cre system apparently serves to facilitate segregation of monomeric P1 plasmids (Austin *et al.*, 1981), it is appropriate that this system generate physically separate recombination products.

Replicative Recombination. Many of the recombination events typical of transposable elements result in two sequence duplications, one of the "donor" element itself and the other of a short oligonucleotide at the "target" site (Grindley and Sherratt, 1979; Shapiro, 1979). These recombination events can be summarized into two possible general forms as illustrated in Fig. 9. While transposition simply results in the movement of a transposable element to a new site, genome rearrangements encompass a variety of events that create new relative sequence organizations. It is important to note that this kind of recombination involves three "sites" (viz. each end of the transposable element and the target oligonucleotide) rather than two as in most of the other recombination events considered above. The specificity for transposable element termini is usually absolute, whereas there can be all degrees of specificity for the target sequence. While there is yet no detailed biochemical data about these "replicative" recombinations, the structures of the products imply a number of cleavage, ligation, and polymerization steps which have been incorporated into molecular models published in the last 4 years (Grindley and Sherratt, 1979; Shapiro, 1979; Galas and Chandler, 1981), and two of these are schematized in Fig. 10. The major difference between these models concerns the timing and orientation of the cleavage-ligation events that join strands of the donor element to the staggered breakpoints of the target oligonucleotide.

In one model, there are symmetrical cleavages prior to the replication on each complementary donor strand at the transposable element termini. This symmetry in cleavage was postulated (in part) because of the generally observed inverted repeat symmetry bounding transposable elements. Symmetrical cleavages followed by ligation to the cleaved target sequence provide a physical maker in the recombining DNA for the precise termination of the polymerization step. The symmetrical cleavage-ligation model only generates recombinant ("DNA rearrangement") structures and thus requires a second reciprocal exchange step to generate "full transposition" products (the Tn3 transposition pathway of Fig. 9). In the case of Tn3, Tn1000, and selected elements, this reciprocal exchange is mediated by an independent resolvase-catalyzed site-specific resolution reaction. For elements which have no independent resolvase-mediated recombination system, the symmetrical model must postulate a reciprocal exchange linked mechanistically to the replicative recombination step in order to explain full transposition in cells lacking general homologous recombination functions. One possibility is that the exchange involves the segregation of newly replicated transposable elements (Shapiro, 1980); another is that a single, infrequently synthesized protein catalyzes both the replicative recombination and resolution reactions (R. Jaskunas, personal communication). There have been several reports in the liter-

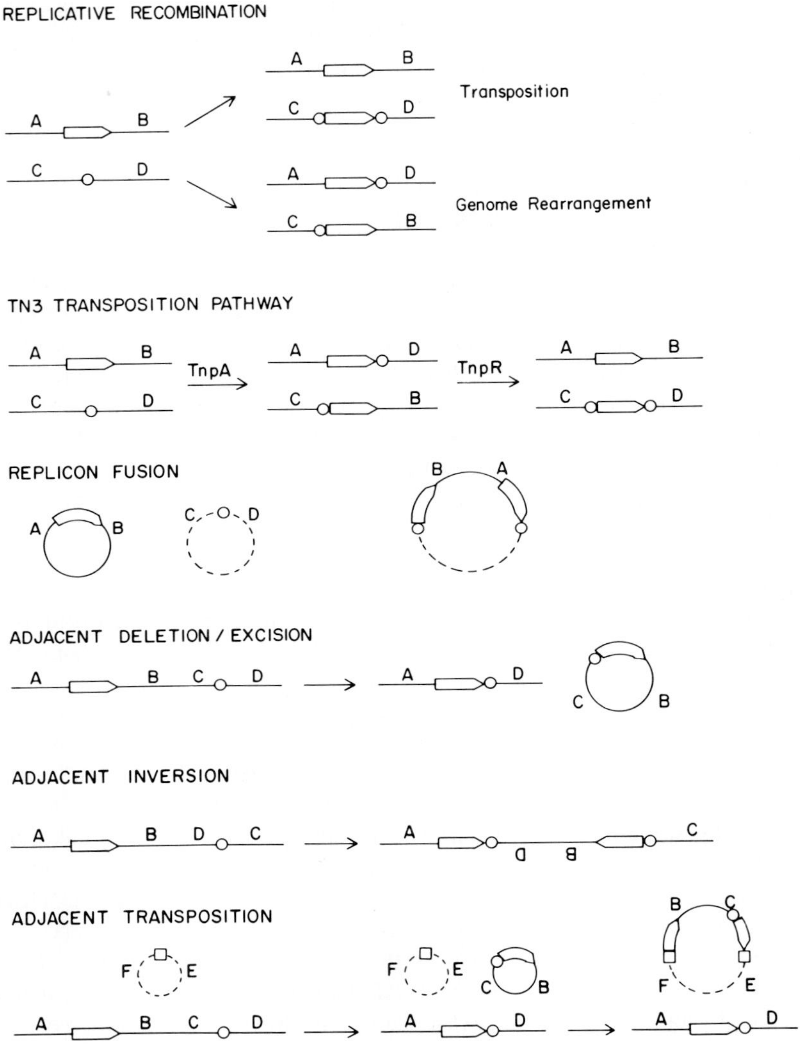

FIG. 9. Replicative recombination events. Duplex molecules are represented by single lines. The pointed rectangles indicate copies of a transposable element and the small open circles indicate a short oligonucleotide sequence at the target site that is duplicated in the recombination event. The bottom four panels show how different structures result from genome rearrangements depending on how regions A, B, C, and D are connected to each other.

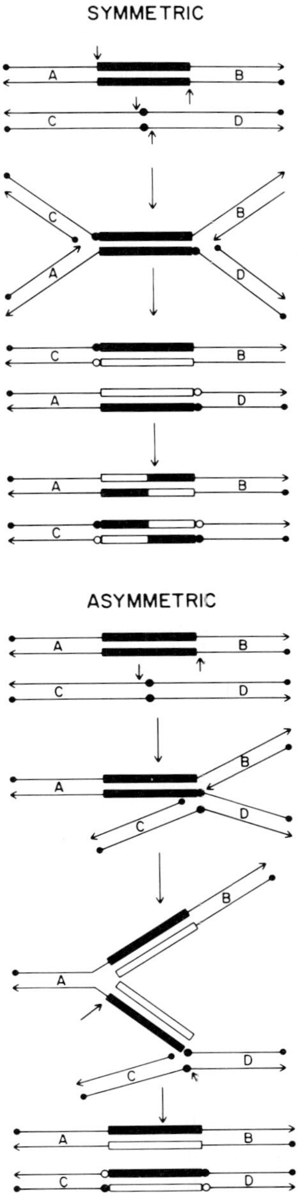

FIG. 10. Symmetrical and asymmetrical models for replicative recombination. The balls indicate 5′ phosphate groups, and the arrowheads indicate 3′ hydroxyl groups at the ends of DNA chains. The boxes represent a transposable element, and the large circles represent the target oligonucleotide sequence. Closed boxes and circles are parental, and open boxes and circles are newly synthesized in the recombination events. The short arrows indicate the positions of cleavage and ligation events at the extremities of the transposable element and target oligonucleotide. More detailed discussions of these two models are Shapiro (1979, 1980) and Galas and Chandler (1981).

ature of apparent reciprocal exchanges in Rec^- cells between directly repeated transposable elements that do not direct the synthesis of a Tn3-like resolvase (see Iida *et al.*, 1983; Muster *et al.*, 1983a, and references therein), but the relationship of these observations with the normal pathway to transposition for these elements remains to be worked out.

The asymmetrical model was formulated to obviate the need for a resolution step in full transposition. It postulates that only a single cleavage at one extremity of the donor element is involved in the initial steps of replicative recombination, that replication terminates precisely at the end of the donor element, and that an asymmetrical cleavage (i.e., on the same strand as the first cleavage) occurs at the other extremity of the donor element before ligation to the target site. While this model eliminates the need for a resolution activity in full transposition, it requires two additional functions: one to terminate the replication fork (because there is no physical marker in the donor DNA to accomplish this), and another to choose which donor strand to cleave at the end of replication.

It is known from genetic studies of a variety of transposable elements that specific element-encoded proteins are necessary for replicative recombination to occur (Toussaint and Resibois, 1983; Iida *et al.*, 1983; Heffron, 1983). These proteins are frequently called "transposases," but that term is likely to be a misnomer in most cases. In the case of the best studied of these proteins, the *tnpA* cistron product of Tn3, it is required for replicative recombinations (such as cointegrate formation) but will not promote "transposition" without the resolvase product of the *tnpR* cistron. The *tnpA* product has been purified and shows extremely high affinity for single-stranded DNA; it binds more tightly in a noncooperative fashion than does the *E. coli* single-stranded binding protein cooperatively (Fennewald *et al.*, 1981). So far no sequence specificity has been demonstrated for the *tnpA* protein, and its precise biochemical role in replicative recombination remains to be worked out.

Restriction Enzymes. The use of restriction endonucleases, ligases, and other enzymes extracted from bacteria for *in vitro* DNA reorganization is very widely known and needs only the briefest discussion here. The range of cleavage specificities of restriction enzymes endows them with great versatility as reagents for creating new DNA structures, as evidenced by the products of human genetic engineering. It is important to keep in mind that we are aware only of the restriction enzymes with the most limited recognition sequences because of the way these enzymes are identified (Linn and Roberts, 1983); low cleavage specificity might not produce as digestion products the defined fragments needed to class an endonuclease as a restriction enzyme. Thus, it is possible that many of the DNA reorganizations which occur in bacteria are biochemically analogous to *in vitro* genetic engineering. Clearly, the role of restriction enzyme systems for creating new DNA structures *in vivo* requires more serious attention than it has yet received.

IV. Pathways of DNA Reorganization

From what we have seen of the known biochemical systems for recombining homologous and nonhomologous DNA segments, bacterial cells have a wide repertoire of mechanisms for creating new genomic organizations. Some of these known systems are relatively simple (e.g., site-specific reciprocal recombination) whereas others must involve multiple sequential biochemical activities (e.g., general homologous recombination and replicative recombination). In this section, it will be useful to trace some of the better studied examples of genomic reorganization that involve combinations of these recombination systems. There are two reasons for this: (1) to illustrate the kinds of reorganizations we already know to be possible, and (2) to show how natural it is in bacteria for multiple biochemical systems to collaborate in building new DNA structures.

Insertion, Excision, and Reinsertion Cycles. Before the advent of restriction endonuclease technology, bacterial genetic engineering was based on the use of "episomes," genetic elements capable of inserting into and detaching from the bacterial chromosome (Jacob and Wollman, 1961; Campbell, 1962). These included conjugative sex factors like F and temperate bacteriophages like λ. Normal λ insertion and excision mediated by the integrase site-specific reciprocal recombination system is described above. F insertion to form Hfr strains is a more complicated process. The F plasmid has a region containing several transposable elements: two copies of IS3, one IS2, and one Tn1000. Either by homologous exchange with such elements already resident in the chromosome or by a replicative recombination event (replicon fusion or cointegrate formation), F can insert into multiple chromosomal locations where it will be flanked by direct repeats of the transposable element (Fig. 11). Excision to regenerate the original F plasmid can then occur by reciprocal recombination between the flanking repeats.

One of the main utilities of episomes resides in their ability to undergo excision events that lead to incorporation of adjacent chromosomal segments into the autonomously replicating form. This was first described for λ in the formation of λ*gal* specialized transducing phage (Campbell, 1962) and for F in the formation of F′ variants, such as F *lac* (Jacob and Wollman, 1961). This is a kind of *in vivo* cloning. For λ, the formation of specialized transducing phage is generally thought to occur by general nonhomologous recombination between sites inside one end of the prophage and outside the other end, thus creating a substitution of viral sequences by host sequences (Fig. 4; Campbell, 1962, 1983). For F, a similar process is possible, but there are two salient differences between F and λ with regard to these excisions. Because the products of λ excision are encapsidated into virions, there are size limitations on the amount of host material that can be incorporated into a specialized transducing phage; such limitations do not exist for F. Moreover, F contains several transposable elements that provide

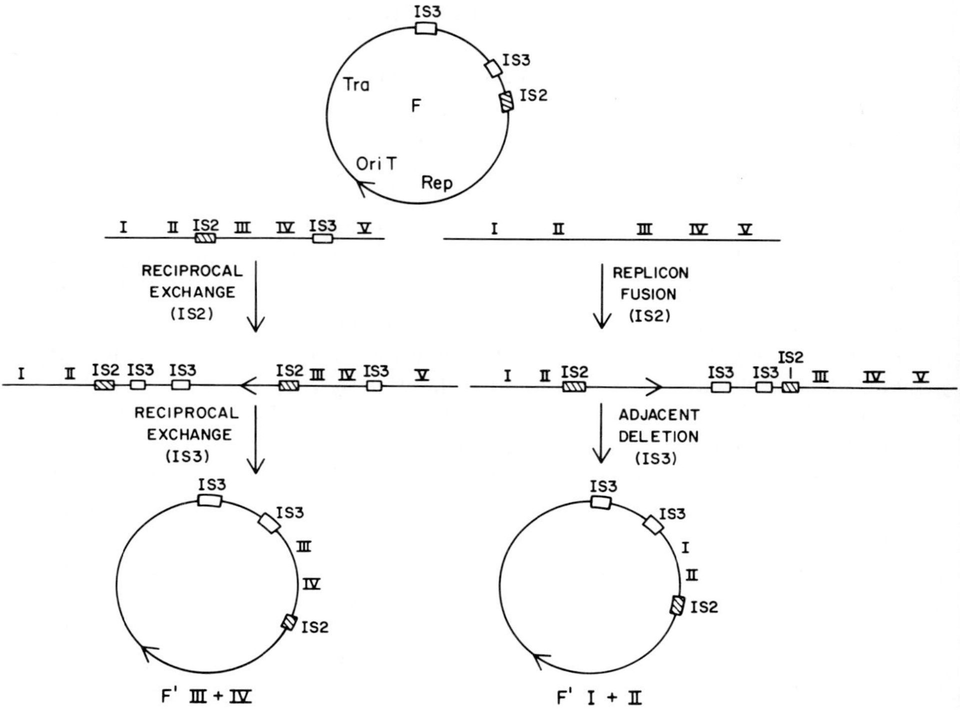

FIG. 11. Insertion and excision of F to form F′ plasmids. The figure illustrates two of the several possible pathways of F-chromosome recombination to generate F′ molecules of the same general structure. The pathway on the left depends on reciprocal exchange between interspersed IS2 and IS3 homologies in F and the host chromosome. This appears to be the most common pathway for insertion because *recA* mutations reduce Hfr formation. The pathway on the right depends on homology-independent replicative recombination events. Note that in both cases, the incorporated chromosomal segment substitutes the same F segment flanked by an IS2 and an IS3.

alternative excision routes, either by homologous exchange or replicative recombination (Fig. 11). In both cases, the incorporated host sequences will be flanked by transposable elements and so subject to further movement in the genome independent of F sequences.

Once a specialized transducing phage or F′ plasmid has been formed, it can reinsert by two different mechanisms: homologous exchange between the incorporated fragment and the corresponding chromosomal region or by the system(s) available for normal λ and F insertion. In the first case, a tandem duplication is created around the reinserted episome, and Hfrs of this type are thought to account for the high level of specific chromosome transfer mediated by F′ cells. Generally, these tandem duplications are unstable in cells capable of general homologous recombination. In the second case (reinsertion by the episome-specific system), the result is frequently a transposition of the incorporated host fragment to a new genomic location. Such transpositions can be seen, for example, with F′ factors where the chromosomal homology has been deleted (Cuzin and Jacob, 1964; Beckwith *et al.*, 1966) and with specialized transducing phage where λ was originally inserted at a secondary attachment rather than at *attB* (Shimada *et al.*, 1972).

Replicative Recombination Cycles. The transposition of certain elements, like Tn3, involves more than one recombination mechanism (replicative followed by site-specific reciprocal) which results in the movement of a specific, highly evolved DNA structure to a new location without disturbing other linkages in the genome (Fig. 9). Replicative recombination is capable, however, of generating several other kinds of genomic reorganizations, and successive cycles of replicative recombination can create transposons which have incorporated previously nonmobile sequences (Fig. 9). One of the most elegant examples of these transpositions is the phenomenon of mini-Muduction (Fig. 12), which combines the replicative recombination cycle and viral encapsidation mechanisms of phage Mu to create transposons for any genetic marker that can be selected in *E. coli* (Toussaint and Resibois, 1983). In the presence of a helper phage genome to supply morphogenetic functions, a deleted Mu prophage (mini-Mu) amplifies itself by repeated cycles of replicative recombination. The result is a dispersion of mini-Mu genomes throughout the genome. At some frequency, two directly repeated mini-Mus flank a short host sequence. Since Mu encapsidation starts approximately 50–100 base pairs upstream of the *c* terminus of any Mu prophage and packages about 38 kilobase pairs of DNA without sequence specificity, the entire mini-Mu–host sequence–Mini-Mu structure can end up in a virion and then be injected into another bacterial cell, where it can behave as a single transposable element, inserting into the genome independently of any homology.

A Specific Constructional Genealogy. In the course of constructing strains for specific tasks, *in vivo* genetic engineers often follow elaborate genealogies involving multiple recombination systems. These genealogies are instructive because they illustrate how novel genomic structures can be formed in a stepwise fashion and provide typical solutions to problems which must arise repeatedly in evolution. Figure 13 illustrates an example which utilizes only *in vivo* mechanisms: the formation of λ*plac*5 (Shapiro *et al.*, 1969; Ippen *et al.*, 1970).

V. Regulation of DNA Reorganization

The increase in our awareness of multiple systems for recombination has led one molecular biologist to state that "DNA is no longer a molecule you can count on." Since DNA is subject to alteration by so many biochemical activities, the maintenance of genome organization cannot reside solely in the stability of phosphodiester and glycosidic linkges but rather is a function of the regulatory

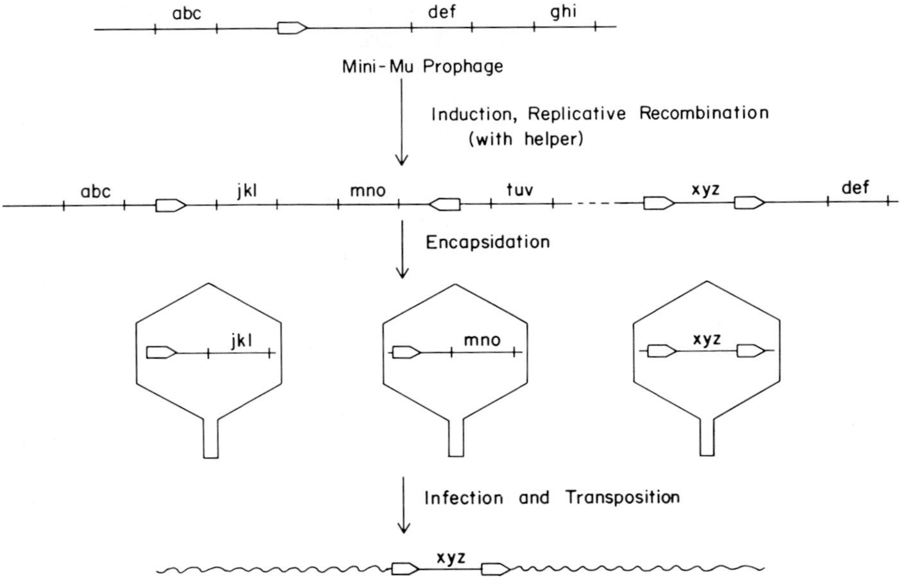

FIG. 12. Mini-Muduction. When a strain carrying a mini-Mu prophage (short arrow) is induced in the presence of a helper plaque-forming Mu genome, replication and packaging occur. The mini-Mu replication/recombination cycle spreads mini-Mu prophages through the genomic DNA, where some sequences are flanked by direct mini-Mu repeats. The mini-Mu extremity serves as an encapsidation sequence and occasionally mini-Mu-flanked sequences are packaged into phage particles. After infection, these sequences can transpose into the recipient cell genome by homology-independent Mu-specific recombination events.

FIG. 13. Construction geneaology of λ *plac*5 (Shapiro *et al.*, 1969; Ippen *et al.*, 1970). Nine steps were required to produce a λ phage carrying a defined portion of the *lac* operon. The isolation of F′ *lac* and F *ts lac* are described in Jacob and Wollman (1961) and Cuzin and Jacob (1964), respectively. The selection for a Lac^+/42°C phenotype at step (5) demands both insertion of F *ts lac* into the chromosome and inactivation of *galK* because phosphorylated galactose is toxic to *E. coli* $galE^-$ mutants. The λ cI857 prophage introduced at step (6) has a thermolabile repressor and so kills lysogenic cells at 42°C unless all lethal functions have been deleted. Thus, selection for $lacI^-$ (constitutive) clones at 42°C selects for deletions that remove part of *lacI* (but not *lacPOZY*) and the adjacent λ prophage. The nature of all the recombination events involved is not known, but the likeliest sequence is as follows: (1) and (2) reciprocal exchange between interspersed IS homologies, as in Fig. 11; (5) replicon fusion mediated by an IS element on F *ts lac* into the *gal* operon; (6) λ integrative recombination; (7) general nonhomologous recombination; (8) λ integrative recombination; (9) general nonhomologous recombination.

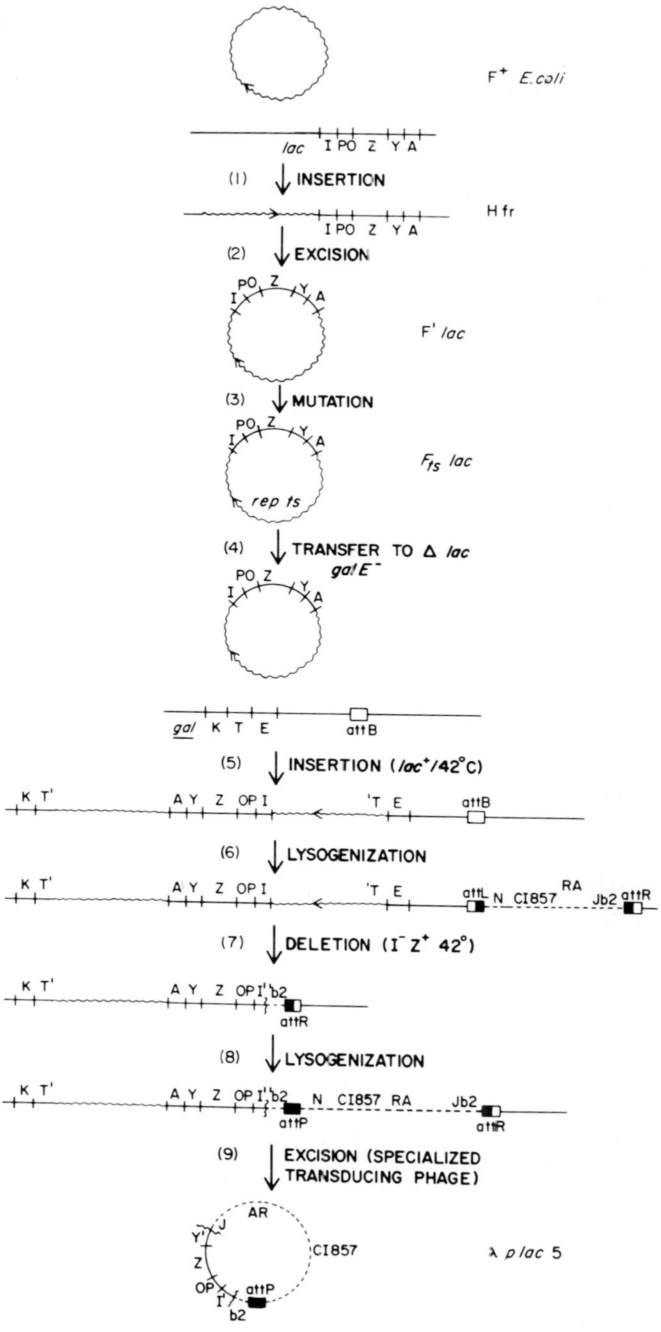
F⁺ E. coli
lac I PO Z Y A
(1) INSERTION
Hfr
I PO Z Y A
(2) EXCISION
F' lac
(3) MUTATION
Fts lac
rep ts
(4) TRANSFER TO Δ lac galE⁻
gal K T E attB
(5) INSERTION (lac⁺/42°C)
K T' A Y Z OP I 'T E attB
(6) LYSOGENIZATION
attL N CI857 RA Jb2 attR
(7) DELETION (I⁻ Z⁺ 42°)
K T' A Y Z OP I' 'b2 attR
(8) LYSOGENIZATION
N CI857 RA Jb2 attP attR
(9) EXCISION (SPECIALIZED TRANSDUCING PHAGE)
AR J Y' Z OP I' b2 attP CI857
λ plac 5

networks that control and coordinate these biochemical activities. Our knowledge of how these networks operate is still in the earliest stages, but since the regulation of genome reorganization is clearly a central problem in understanding evolution, I will briefly mention some of the kinds of control phenomena of which we are aware.

1. Control by the existing genome distribution of recombination substrates. Two classical observations illustrate how critical is the role present genomic structure plays in determining which new genomic structure are likely to emerge: (1) the limitation of λ specialized transduction to the *gal-attB-bio* region unless λ has inserted at a secondary attachment site in a cell lacking *attB* and (2) the increased probability of *gal* deletion by several orders of magnitude when the operon contains an IS1 insertion (Reif and Saedler, 1975). Anyone who has faced the problem of strain construction can think of many similar cases, where it is necessary to create one structure with two particular sequences in close proximity before another structure can arise. I call this feature of genomic variation ''genealogical information'' to indicate that existing genome organization influences the possible pathways to new organizations.

2. Control by the stoichiometric nature of many DNA reorganization activities. If proteins like λ integrase and the Mu A cistron product (Pato and Reich, 1982) can only act once to promote a particular kind of recombination event, then that class of events is subject to tighter control by the mechanisms that govern synthesis of those reorganizational activities.

3. Control of transcription of sequences encoding reorganizational activities. This control can be very elaborate, as in the case of phage λ, where there are several kinds of interactive regulatory mechanisms (Miller *et al.*, 1981a: Campbell, 1983). Transcription of *int* occurs after infection from the P_I promoter that is subject to positive activation by the viral *cII* product and host *himA* protein (also required for integrase activity). In turn, *cII* expression is controlled by the λ *cI* repressor, the host cyclic AMP system, and the *cIII* cistron which protects the *cII* product from degradation by the host *hfl* degradation activity. By means of these controls, integrase transcription is sensitive to the state of host physiology and to the viral lysis/lysogeny decision in the newly infected cell. Transcription of *xis* (which might ultimately lead to excision of a newly inserted prophage) is prevented under the same conditions by positioning of the P_I promoter so that it overlaps the first triplet in *xis*. Effective transcription of *int* from the P_L promoter during viral lytic growth is prevented by the presence of an RNase III-sensitive *sib* site downstream of *int* and *attP;* this site leads to functional inactivation of the transcript for *int* translation and so blocks prophage insertion during the lytic cycle. However *int* is effectively transcribed from the P_L promoter after prophage derepression because the insertion event at *attP* separates the *sib* sequence from the P_L operon and so relieves the block to *int* translation. It is interesting to note here the self-regulatory consequences of this recombination event.

The SOS system is an elaborate control network governing *E. coli*'s response to radiation and other treatments that interfere with normal chromosome replication (Witkin, 1976; Little and Mount, 1982). This system controls transcription of sequences for DNA repair functions, cell division inhibitors, and a variety of DNA reorganization activities. These activities include the *umuC* mutator function, the λ excision system, the *himA* activity, and the *recF* pathway for general homologous recombination (Lovett and Clark, 1983).

4. Control of translation of transcripts encoding reorganizational activities. The translational control function of the λ *sib* sequence has just been described. Tn10 has a different mechanism of translational control. One IS10 element in Tn10 encodes a polypeptide essential for transposition and replicative recombination. The initial portion of the sequence encoding this polypeptide can be transcribed off the noncoding strand from an internal IS10 promoter to produce an RNA complementary to the mRNA which prevents its translation (Kleckner, 1983).

5. Direct inhibition of recombination activity by protein. This kind of regulation is now well-documented for Tn5 (Reznikoff, 1982). The active IS50 element of Tn5 encodes two distinct polypeptides from a single open reading frame utilizing two different initiation codons. The longer of these polypeptides is required for Tn5 transposition, and the shorter polypeptide inhibits transposition so that its overproduction effectively shuts down Tn5 replicative recombination.

6. The primary structure (nucleotide sequence) of the substrates for recombination activities. Where, and with what efficiency, various DNA reorganizations occur depends in large measure on specific sequence recognition. Thus, the sequence of an *att* site, the termini of a transposable element, or a target site for replicative recombination can determine how frequently that particular location takes part in a recombination event when the requisite proteins are available. Sometimes these specificities involve more than one site. The termini of phage Mu are an example. Mu termini do have inverted symmetry, like other transposable elements, but the symmetrical sequences are interrupted by nonhomologous regions to introduce an asymmetry into the Mu system (Kahmann and Kamp, 1979). As a consequence, the replicative recombination requires one *c* terminus and one β terminus; two *c* or two β termini in inverted orientation are not substrates for Mu replicative recombination functions (Toussaint and Resibois, 1983). Other elements, like Tn7 (Lichtenstein and Brenner, 1982) and the joining regions in the immunoglobulin V and J regions (Tonegawa, 1983), show similar interrupted inverse homologies at recombination sites.

7. Substrate secondary and higher order structures. The topology or biochemical activity of substrate sequences can influence their participation in recombination events. For example, superhelicity is needed for λ integrative recombination and Tn5 transposition (Nash, 1981; Isberg and Syvanen, 1982), and target molecules must be replicating in order to undergo Tn1-mediated replicative recombination (Muster *et al.*, 1983b).

8. Interactions between mobile genetic elements. There is a set of poorly understood regulatory phenomena where the presence of one mobile genetic element influences the behavior of a second element. These phenomena are diverse and so will have distinct biochemical mechanisms, but they are worth mentioning together because they illustrate regulatory systems dealing with a higher level of genetic organization than the simple "consensus sequence" systems of current regulatory models. I will choose three of the better-documented examples.

First are interactions between the two directly repeated IS1 elements that flank the chloramphenicol resistance determinant of Tn9. Normally one of these IS1s participates much more frequently in replicative recombination events than the other one, but when internal sequence alterations lower the activity of the first IS1, then the second IS1 increases its participation in recombination events both on an absolute as well as a relative scale (Machida *et al.*, 1982). Thus, it appears that one IS1 can sense the activity of another IS1 and adjust its behavior accordingly.

The second example concerns the phenomenon known as "transposition immunity" (reviewed in Heffron, 1983). This refers to the observation that certain transposons (such as Tn3 and Tn501) can insert at a large number of target sites within one DNA molecule, but once one copy of such an element is inserted at a particular site, it makes all other potential insertion sites unavailable for a second related element, even though those sites may be tens of thousands of base-pairs distant from the first insertion. The phenomenon is not a general block to insertion in a particular cell because other molecules remain perfectly good targets for transposition, and intramolecular replicative recombinations still take place. In this case, one element can act at a distance along one DNA molecule to prevent its recombination with an outside sequence.

The third example concerns the target specificity of replicative recombination. Tn1 and Tn501 do not show transposition immunity to each other. When Tn501 inserts into a naturally occurring plasmid that harbors Tn1, the Tn1 forms a preferred region for Tn501 insertion. If the Tn1 is deleted from that plasmid, then new preferred Tn501 insertion regions are found. When Tn1 is reinserted into the deleted plasmid at a new location, it no longer serves as a preferred Tn501 insertion site. These data indicate that the mechanism for choosing the Tn501 target is capable of discriminating features of DNA organization that are more elaborate than simple comparisons between localized DNA sequences (Grinsted *et al.*, 1978).

VI. Summary: Biochemical Systems for DNA Reorganization

Nucleotide sequences in DNA molecules constitute one of the main systems for information storage in living cells. As organisms encounter varying external environments and create new internal environments for their cells, they often need to change the stored information. If bacteria, for example, had no way of receiving, maintaining, and expressing DNA sequences encoding enzymes that inactivate antibiotics, they would not be able to survive the challenges of human or microbial antagonists. Given the requirement for genomic flexibility in the real world, it ought not to surprise us that many biochemical systems exist precisely for the purpose of creating new DNA structures. To return to our example of how bacteria acquire antibiotic resistance, we can only marvel at the panoply of biochemical components involved in such a supposedly "simple" genetic change. These can include, for example, all the activities involved in the creation of a transposable resistance determinant and its transposition to a transmissible plasmid, the activities required for plasmid replication and segregation, and the activities required for conjugal plasmid transfer—in all, dozens (maybe even hundreds) of specific proteins and cell structural components.

The aspects of DNA reorganization which we have considered above in some detail form only a small but essential part of the bacterial apparatus governing genome behavior. As our knowledge of DNA biochemistry advances, many questions about genetic change disappear because we see the specific biochemical mechanisms at work and so have a way to understand how various novel structures could have arisen. Indeed, the problem of hereditary variation now seems to have been turned upside down. Rather than ask how genetic changes occur, we must now ask what keeps them from occurring too often for organisms to be capable of reproducing themselves. In other words, we must now seek to understand the regulatory mechanisms which protect the information stored in DNA molecules so that it can be utilized with such great efficiency and reliability when it is needed.

Acknowledgments

I thank all my colleagues who have communicated unpublished results and given patient lessons on the DNA reorganization systems they study. I am grateful to Chan Stroman for manuscript preparation. My research is supported by grants from the National Institute of General Medical Sciences (GM 24960), the National Cancer Institute (CA 19265), and the National Science Foundation (PCM 8200971).

References

Abremski, K., Hoess, R., and Sternberg, N. (1983). *Cell* **32,** 1301–1311.
Albertini, A. M., Hofer, M., Calos, M. P., and Miller, J. H. (1982). *Cell* **29,** 319–328.
Alberts, B. M., and Frey, L. (1970). *Nature (London)* **227,** 1313–1318.
Austin, S., Ziese, M. and Sternberg, N. (1981). *Cell* **25,** 729–736.

Beckwith, J. R., Singer, E., and Epstein, W. (1966). *Cold Spring Harbor Symp. Quant. Biol.* **31,** 393–401.
Beerman, S. (1977). *Chromosoma* **60,** 297–344.
Better, M., Lu, C., Williams, R. C., and Echols, H. (1982). *Proc. Natl. Acad. Sci. U.S.A.* **79,** 5837–5841.
Better, M., Wickner, S., Auerbach, J., and Echols, H. (1983).
Borst, P. (1983). *In* ''Mobile Genetic Elements'' (J. A. Shapiro, ed.), pp. 621–659. Academic Press, New York.
Boveri, Th. (1892). *S. B. Ges. Morphol. Physiol. Munich* **8,** 114–125.
Braun, A., and Grossman, L. (1974). *Proc. Natl. Acad. Sci. U.S.A.* **71,** 1838–1842.
Cairns, J. (1963). *Cold Spring Harbor Symp. Quant. Biol.* **28,** 43–46.
Calos, M. P., and Miller, J. H. (1980). *Cell* **20,** 579–595.
Campbell, A. (1962). *Adv. Genet.* **11,** 101–145.
Campbell, A. (1983). *In* ''Mobile Genetic Elements'' (J. A. Shapiro, ed.), pp. 65–103. Academic Press, New York.
Casadaban, M. J., and Cohen, S. N. (1979). *Proc. Natl. Acad. Sci. U.S.A.* **76,** 4530–4533.
Champoux, J. J. (1978). *Annu. Rev. Biochem.* **47,** 449–479.
Clark, A. J. (1980). pp. 891–899.
Cozzarelli, N. R. (1980). *Cell* **22,** 327–328.
Craig, N. L. and Roberts, J. W. (1980). *Nature (London)* **283** 26–29.
Craig, N. L., and Roberts, J. W. (1981). *J. Biol. Chem.* **256,** 8039–8044.
Crouse, H. (1966). *Chromosoma* **18,** 230–254.
Cunningham, R. P., Wu, A. M., Shibata, T., Das Gupta, C., and Radding, C. M. (1981). *Cell* **24,** 213–223.
Cuzin, F., and Jacob, F. (1964). *C. R. Acad. Sci. Paris* **258,** 1350–1352.
Efstratiadis, A., Posakony, J. W., Maniatis, T., Lawn, R. M., O'Connel, C., Spritz, R. A., De Riel, J. K., Forget, B. G., Weissman, S. M., Slightom, J. L., Blechl, A. E., Smithies, O., Baralle, F. E., Shoulders, C. C., and Proudfoot, N. J. (1980). *Cell,* **21,** 653–668.
Fabre, F., and Roman, H. (1977). *Proc. Natl. Acad. Sci. U.S.A.* **74,** 1667–1671.
Farabaugh, P. J., Schmeissner, U., Hofer, M., and Miller, J. H. (1978). *J. Mol. Biol.* **126,** 847–857.
Fennewald, M., Genard, S., Chou, J., Casadaban, M., and Cozzarelli, N. (1981). *J. Biol. Chem.* **256,** 4687–4690.
Fishel, R. A., James, A. A., and Kolodner, R. (1981). *Nature (London)* **294,** 184–186.
Franklin, N. C. (1967). *Genetics* **55,** 699–707.
Galas, D., and Chandler, M. (1981). *Proc. Natl. Acad. Sci. U.S.A.* **78,** 4858–4862.
Game, M. C., Zamb, T. J., Braun, R. J., Resnick, M., and Roth, R. M. (1980). *Genetics* **94,** 51–68.
Geider, K., and Hoffmann-Berling, H. (1981). *Annu. Rev. Biochem.* **50,** 233–260.
Gellert, M. (1981). *Annu. Rev. Biochem.* **50,** 879–910.
Gillen, J. R., and Clark, A. J. (1974). *In* ''Mechanisms in Recombination'' (R. F. Grell, ed.), pp. 123–136. Plenum, New York.
Gillen, J. R., Karu, A. E., Nagaishi, H., and Clark, A. J. (1977). *J. Mol. Biol.* **113,** 27–41.
Gillen, J. R., Willis, D. K., and Clark, A. J. (1981). *J. Bacteriol.* **145,** 521–532.
Grindley, N. and Sherratt, D. (1978). *Cold Spring Harbor Symp. Quant. Biol.* **43,** 1257–1260.
Grindley, N., Laush, M., Wells, R., Wityk, R., Salvo, J., and Reed, R. (1982). *Cell* **30,** 19–27.
Grinsted, J., Bennett, P. M., Higginson, S., and Richmond, M. H. (1978). *Mol. Gen. Genet.* **166,** 313–320.
Haber, J. E. (1983). *In* ''Mobile Genetic Elements'' (J. A. Shapiro, ed.), pp. 559–619. Academic Press, New York.

Heffron, F. (1983). *In* "Mobile Genetic Elements" (J. A. Shapiro, ed.), pp. 223–260. Academic Press, New York.
Holliday, J. E. (1964). *Genet. Res.* **5,** 282–304.
Hollis, G. F., Hieter, P. A., McBride, O. W., Swan, D., and Leder, P. (1982). *Nature (London)* **296,** 321–325.
Hsu, P. -L., Ross, W., and Landy, A. (1980). *Nature (London)* **285,** 85–91.
Iida, S., Meyer, J., and Arber, W. (1983). *In* "Mobile Genetic Elements" (J. A. Shapiro, ed.), pp 159–221. Academic Press, New York.
Ikeda, H., Moriya, K., and Matsumoto, T. (1981). *Cold Spring Harbor Symp. Quant. Biol.* **45,** 399–408.
Ippen, K., Shapiro, J. A., Beckwith, J. R. (1971). *J. Bacteriol.* **108,** 5–9.
Isberg, R. R., and Syvanen, M. (1982). *Cell* **30,** 9–18.
Jacob, F., and Wollman, E. (1961). "Sexuality and the Genetics of Bacteria." Academic Press, New York.
Johnson, A. D., Poteete, A. R., Lauer, G., Sauer, R. T., Ackers, G. K., and Ptashne, M. (1981). *Nature (London)* **294,** 217–223.
Kahman, R., and Kamp, D. (1979). *Nature (London)* **280,** 247–250.
Kaiser, K., and Murray, N. E. (1979). *Mol. Gen. Genet.* **175,** 159–174.
Kikuchi, Y., and Nash, H. A. (1979). *Proc. Natl. Acad. Sci. U.S.A.* **76,** 3760–3764.
Kleckner, N. (1981). *Annu. Rev. Genet.* **15,** 341–404.
Kleckner, N. (1983). *In* "Mobile Genetic Elements" (J. A. Shapiro, ed.), pp. 261–298. Academic Press, New York.
Kornberg, A. (1980). "DNA Replication." Freeman, San Francisco California; and Supplement (1982).
Kostriken, R., Morita, C., and Heffron, F. (1981). *Proc. Natl. Acad. Sci. U.S.A.* **78,** 4041–4045.
Krasnow, M. A., and Cozzarelli, N. R. (1983). *Cell,* **32,** 1312–1324.
Kunz, B. A., and Haynes, R. H. (1981). Phenomenology and genetic control of mitotic recombination in yeast. *Annu. Rev. Genet.* **15,** 57–89.
Lichtenstein, C., and Brenner, S. (1982). *Nature (London)* **297,** 601–603.
Linn, S. M., and Roberts, R. J., eds. (1983). *Cold Spring Harbor Lab Monogr.* **14.**
Little, J. W., and Mount, D. W. (1982). *Cell* **29,** 11–22.
Lovett, S. T., and Clark, A. J. (1983). *J. Bacteriol.* **153,** 1471–1478.
McEntee, K., Weinstock, G. M., and Lehman, I. R. (1979). *Proc. Natl. Acad. Sci. U.S.A.* **76,** 2615–2619.
Machida, Y., Machida, C., Ohtsubo, H., and Ohtsubo, E. (1982). *Proc. Natl. Acad. Sci. U.S.A.* **79,** 277–281.
Magni, G. E., and Von Borstel, R. C. (1962). *Genetics* **47,** 1097–1108.
Malone, R. E., and Esposito, R. E. (1980). *Proc. Natl. Acad. Sci. U.S.A.* **77,** 503–507.
Meselson, M. S., and Radding, C. M. (1975). *Proc. Natl. Acad. Sci. U.S.A.* **72,** 358–361.
Miller, H. I., Abraham, J., Benedik, M., Campbell, A., Court, D., Echols, H., Fischer, R., Galindo, J. M., Guarneros, G., Hernandez, T., Mascarenhas, D., Montanez, C., Schindler, D., Schmeissner, U., and Sosa, L. (1981a). *Cold Spring Harbor Symp. Quant. Biol.* **45,** 439–445.
Miller, H. I., Krk, M., and Echols, H. (1981b). *Proc. Natl. Acad. Sci. U.S.A.* **78,** 6754–6758.
Mizuuchi, K., Weisberg, R., Enquist, L., Mizuuchi, M., Buraczynska, M., Foeller, C., Hsu, P. L., Ross, W., and Landy, A. (1981). *Cold Spring Harbor Symp. Quant. Biol.* **45,** 429–437.
Mizuuchi, K., Kemper, B., Hays, J., and Weisberg, R. A. (1982). *Cell* **29,** 357–365.
Morrison, A., and Cozzarelli, N. R. (1979). *Cell* **17,** 175–184.
Muster, C. J., MacHattie, L. A., and Shapiro, J. A. (1983a). *J. Bacteriol.* **153,** 976–990.
Muster, C. J., Shapiro, J. A., and MacHattie, L. A. (1983b). *Proc. Natl. Acad. Sci. U.S.A.* **80,** 2314–2317.

Nanny, D. L. (1980). "Experimental Ciliatology." Wiley, New York.
Nash, H. A. (1981). *Annu. Rev. Genet.* **15,** 143–167.
Nasmyth, K. (1982). *Annu. Rev. Genet.* **16,** 439–500.
Pato, M. L. (1982). *Cell* **29,** 219–225.
Pogue-Geile, K. L., Das Sarma, S., King, S. R., and Jaskunas, S. R. (1980). *J. Bacteriol.* **142,** 992–1003.
Prakash, S., Prakash, L., Burke, W., and Montelone, B. A. (1980). *Genetics* **94,** 31–50.
Ptashne, M., Jeffrey, A., Johnson, A. D., Maurer, R., Meyer, B. J., Pabo, C. O., Roberts, T. M., and Sauer, R. T. (1980). *Cell* **19,** 1–11.
Radding, C. M. (1978). *Cold Spring Harbor Symp. Quant. Biol.* **43,** 1315–1316.
Radding, C. M. (1981). *Cell,* **25,** 3–4.
Radding, C. M. (1982). *Annu. Rev. Genet.* **16,** 405–437.
Reed, R. R. (1981). *Cell* **25,** 713–719.
Reed, R. R., and Grindley, N. D. F. (1981). *Cell* **25,** 721–728.
Reif, H. -J., and Saedler, H. (1975). *Mol. Gen. Genet.* **137,** 17–28.
Reznikoff, W. S. (1982). *Cell* **31,** 307–308.
Roman, H. (1956). *Cold Spring Harbor Symp. Quant. Biol.* **21,** 175–183.
Rosenberg, M., and Court, D. (1979). *Annu. Rev. Genet.* **13,** 319–353.
Shapiro, J. A. (1979). *Proc. Natl. Acad. Sci. U.S.A.* **76,** 1933–1937.
Shapiro, J. A. (1980). *In* "Plasmids and Transposons: Environmental Effects and Maintenance Mechanisms" (C. Stuttard and K. R. Rozee, eds.), pp. 1–19. Academic Press, New York.
Shapiro, J., MacHattie, L., Eron, L., Ihler, G., Ippen, K., Beckwith, J., Arditti, R., Reznikoff, W., and MacGillivray, R. (1969). *Nature (London)* **224,** 768–774.
Shimada, K., Weisberg, R. A., and Gottesman, M. E. (1972). *J. Mol. Biol.* **63,** 483–503.
Simon, M., and Silverman, M. (1983). *In* "Mobile Genetic Elements" (J. A. Shapiro, ed), pp. 537–557. Academic Press, New York.
Simon, M., Zieg, J., Silverman, M., Mandel, G., and Doolittle, R. (1980). *Science* **209,** 1370–1374.
Smith, G. R. (1983). *In* "The Bacteriophage Lambda" (R. Hendrix, J. Roberts, F. Stahl, and R. Weisberg, eds.), Cold Spring Harbor Laboratory, Cold Spring Harbor, New York, in press.
Stahl, F. W. (1979). *Annu. Rev. Genet* **13,** 7–24.
Sternberg, N., Hamilton, D., Austin, S., Yarmolinsky, M., and Hoess, R. (1981). *Cold Spring Harbor Symp. Quant. Biol.* **45,** 297–309.
Szostak, J. W., Orr-Weaver, T. L., Rothstein, R. J., and Stahl, F. W. (1983). *Cell* **33.**
Taylor, A., and Smith, G. R. (1980). *Cell* **22,** 447–457.
Tonegawa, S. (1983). *Nature (London)* **302,** 575–581.
Toussaint, A., and Resibois, A. (1983). *In* "Mobile Genetic Elements" (J. A. Shapiro, ed.), pp. 105–158. Academic Press, New York.
van Mansfeld, A. D., Langeveld, S. A., Baas, P. D., Jansz, H. S., van der Marel, G. A., Veeneman, G. H., and van Boom, J. H. (1980). *Nature (London)* **288,** 561–566.
Wang, J. C., and Liu, L. F. (1979). *Mol. Genet.* **3,** 65–88.
Weinstock, G. M., McEntee, K., and Lehman, I. R. (1979). *Proc. Natl. Acad. Sci. U.S.A.* **76,** 126–130.
Weisberg, R. A., and Adhya, S. (1977). *Annu. Rev. Genet.* **11,** 451–473.
Witkin, E. M. (1976). *Bacteriol. Rev.* **40,** 869–907.
Yuan, R. (1981). *Annu. Rev. Biochem.* **50,** 285–315.

INTERNATIONAL REVIEW OF CYTOLOGY, VOL. 93

Genetic Organization of the Chloroplast

J. D. ROCHAIX

Departments of Molecular Biology and Plant Biology, University of Geneva, Geneva Switzerland

I. Introduction

Chloroplasts are highly complex organelles capable of performing a variety of important tasks. These include primarily the conversion of light energy into chemical energy which is used to reduce carbon dioxide into carbohydrates. The reducing power generated during photosynthesis is also utilized for the conversion of nitrate and sulfate into organic substances and for the biosynthesis of several amino acids and fatty acids.

It is well documented that chloroplasts possess their own genetic system and protein-synthesizing apparatus which cooperate closely with the nucleocytoplasm in the biosynthesis of organelles. Although our knowledge on the informational content of chloroplast DNA is still limited, it appears that a large number of chloroplast components are encoded by the chloroplast genome. During the past years it has become increasingly apparent that numerous features of chloroplast genetic systems resemble those of bacterial systems. However, certain structural features of chloroplast genes are more like those of eukaryotic nuclear genes. One aim of this article is to cover those lines of research in chloroplast molecular genetics which bear on the dual prokaryotic and eukaryotic nature of chloroplasts. Rather than being exhaustive this review will focus on specific recent developments in the field which are related to this subject. Chlo-

ISBN 0-12-364493-3

roplast genetics is likely to be important for understanding chloroplast gene regulation. Therefore a section of this review describes recent attempts at correlating the genetic and physical chloroplast DNA maps. Multiple aspects of chloroplast genetic systems have been discussed in several reviews which have appeared recently (Bedbrook and Kolodner, 1979; Bohnert *et al.*, 1982; Edelman, 1981; Gillham *et al.*, 1984; Gray and Doolittle, 1982; Herrmann and Possingham, 1980).

II. Conformation and Organization of Chloroplast DNA

Most chloroplast DNAs consist of cirles in the size range between 120 and 190 kb (Bedbrook and Kolodner, 1979). One exception is *Acetabularia* where linear chloroplast DNA molecules larger than 300 kb have been detected (Green *et al.*, 1977). The use of fluorochromes, displaying high specificity for DNA and high fluorescent yield upon binding, has revealed that the chloroplast DNA is not uniformly distributed within the organelle, but that it forms several discrete nucleoids in higher plants and several algae (Chiang *et al.*, 1981; Coleman, 1978, 1979; James and Jope, 1978). Nucleoids appear to consist of chloroplast DNA associated with a protein body. Up to 36 polypeptides with molecular weights ranging from 12,000 to 18,000 have been found in the spinach chloroplast DNA–protein complex (Briat *et al.*, 1982a). Electron microscope examination of this complex after glutaraldehyde fixation shows a highly condensed structure with supertwisted DNA and bound proteins that produced beaded substructures (Briat *et al.*, 1982a). A 17-kDa polypeptide immunologically related to the *E. coli* HU protein which is known to condense DNA (Rouvière-Yaniv and Gros, 1975) has recently been found in spinach (Briat and Rouvière-Yaniv, unpublished results).

Mild lysis of chloroplasts with Triton X-100 has allowed for the isolation of transcriptionally active DNA–protein complexes from *Euglena* (Hallick *et al.*, 1976; Rushlow *et al.*, 1980; Schiemann *et al.*, 1977) and spinach (Briat *et al.*, 1979; Briat and Mache, 1980). Transcription is not inhibited by rifampicin, α-amanitin, nor even heparin. Although the chloroplast ribosomal DNA is preferentially transcribed, many regions of the spinach chloroplast DNA are also expressed in this system (Blanc *et al.*, 1981). Modifications of ionic strength or addition of heparin drastically affect the transcription pattern and the size of the RNA chains (Blanc *et al.*, 1981).

A striking feature of the majority of the chloroplast genomes examined is the presence of an inverted repeat which contains the chloroplast rRNA genes. This arrangement has been found in 150 species of angiosperms distributed over 40 families, in one species of gymnosperms, in four fern species (Palmer and Thompson, 1982; Palmer and Stein, 1982; Whitfeld and Bottomley, 1983), in one green alga (Rochaix, 1978), and also in the cyanelle of the photoautotrophic flagellate *Cyanophora paradoxa* (Bohnert and Löffelhardt, 1982). Cyanelles are photosynthetic organelles which resemble both prokaryotic organisms (their peptidoglycan cell wall is similar to that of cyanobacteria) and chloroplasts (their DNA resembles chloroplast DNA in size and organization). It has been proposed that cyanelles may be a "missing link" between cyanobacteria and plastids (Gray and Doolittle, 1982).

A few legume species contain a single chloroplast ribosomal unit possibly because of the loss of one segment of the inverted repeat (Chu and Tewari, 1982; Koller and Delius, 1980; Palmer and Thompson, 1981). The alga *Euglena* displays still a different organization in which its chloroplast ribosomal units are tandemly arranged in three to five copies (Gray and Hallick, 1978; Jenni and Stutz, 1978; Koller and Delius, 1982b; Rawson *et al.*, 1978).

Extensive sequence comparisons between several angiosperm chloroplast DNAs have revealed that chloroplast DNA rearrangements occur relatively often during phylogeny in those genomes which do not have an inverted repeat, whereas the organization of the genomes with both segments of the inverted repeat is remarkably conserved (Palmer and Thompson, 1982). Among the chloroplast genomes of five disparate angiosperm families (corn, petunia, cucumber, spinach, and mung bean) that contain inverted repeats and have evolved separately for over 60 million years, only two major rearrangements are apparent (Palmer and Thompson, 1982). While the chloroplast genomes of spinach, petunia, and cucumber appear to be colinear, they differ from that of mung bean by a large inversion of 50 kb and from that of corn by a second 20 kb inversion. In contrast, numerous rearrangements are seen when the chloroplast genomes of either pea or broad bean, both of which lack the inverted repeat, are compared to mung bean (Palmer and Thompson, 1982).

Although the alfalfa chloroplast DNA lacks the inverted repeat, its genome organization does not appear to be rearranged relative to the conserved mung bean genome except for the deletion of one ribosomal unit (Palmer *et al.*, 1984). This finding implies that the loss of the inverted repeat occurred before the chloroplast genome rearrangements in pea and broad bean. It is especially interesting that the broad bean chloroplast genome can be derived from an alfalfa-like ancestral genome by only two major sequence inversions (Palmer *et al.*, 1984). More generally, all characterized major chloroplast DNA rearrangements appear to be inversions.

The existence of a correction mechanism between the two segments of the inverted repeat appears likely since they are undistinguishable by restriction enzyme analysis and, more importantly, because identical spontaneous and induced deletions have been found in both segments in spinach (Herrmann *et al.*, 1980) and in *Chlamydomonas reinhardii* (Grant *et al.*, 1980; Myers *et al.*, 1982). Intramolecular recombination between the two parts of the chloroplast

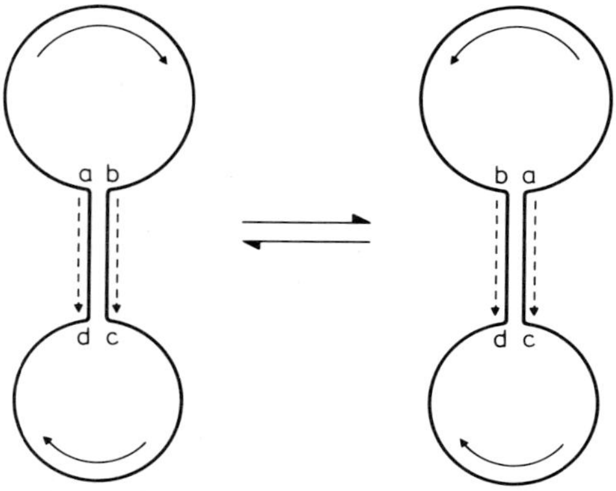

FIG. 1. Isomerization of chloroplast DNA by intramolecular recombination within the inverted repeat. The two segments of the inverted repeat are drawn parallel to each other (broken arrows). The upper and lower loops represent the two single copy regions. Intramolecular recombination inverts one single copy region relative to the other.

inverted repeat has recently been demonstrated in common bean (Palmer, 1983) and in the cyanelle DNA of *Cyanophora paradoxa* (Bohnert and Löffelhardt, 1982). It remains to be seen whether the inversion of one single copy region relative to the other by this intramolecular recombination (Fig. 1) has any functional significance.

Palmer *et al.* (1984) have developed a model of chloroplast genome evolution based on the postulated existence of a site-specific chloroplast recombination system. It is assumed that the molecular mechanisms for chloroplast genome isomerization by intramolecular recombination may be similar to those operating in well characterized systems such as inversion heterogeneity in yeast 2 μ DNA (Broach *et al.,* 1982) and herpes simplex virus (Roizman, 1979), and phase variation in *Salmonella* (Scott and Simon, 1982). A short specific sequence, perhaps less than 20 bp, located within the chloroplast inverted repeat would act as the preferential site for intramolecular recombination. The latter would be mediated by a specialized system which acts independently from any generalized cellular recombination system. Additional sites, with sequences slightly different than the strong sites in the inverted repeat, are assumed to be dispersed throughout the chloroplast genome and to serve as substrate for the site specific recombination system at considerably reduced frequencies. Genomes containing inverted repeats are stable because all the site-specific recombination activity acts preferentially on the strong sites in the inverted repeat. After loss of the latter, the secondary sites are used for recombination which results in chloroplast genome inversions and deletions. The stability of the alfalfa chloroplast genome can be explained by the loss of the recombinational activity which accompanied or followed shortly the loss of the inverted repeat. Palmer *et al.* (1984) point out that this model can be tested by examining the regions at the ends of inversions and deletions for sequence homologies. An interesting task for the future will be to characterize this chloroplast recombinational activity.

III. Chloroplast Genes

A. Ribosomal RNA Genes

The arrangement, size, and sequence of the rRNA genes of chloroplasts and bacteria are very similar (Table I). In all cases studied the gene order is 5′-16 S rDNA-spacer-23 S rDNA-5 S rDNA (Fig. 2). All chloroplast ribosomal spacers which have been examined contain the genes coding for $tRNA^{Ile}$ and $tRNA^{Ala}$ (Fig. 2) as found in several *E. coli* rRNA operons (Lund *et al.*, 1976; for review cf Weil *et al.*, 1981). Surprisingly, these chloroplast tRNA genes contain large introns ranging between 707 and 949 bp in maize (Koch *et al.*, 1981) and tobacco (Takaiwa and Sugiura, 1982a). The introns account for most of the spacer region. These tRNA genes are uninterrupted in the ribosomal spacers of *Euglena* (Graf *et al.*, 1980; Orozco *et al.*, 1980) and *Chlamydomonas reinhardii* (Schneider and Rochaix, unpublished results).

Another distinctive feature of higher plant ribosomal units is the presence of a small 4.5 S rRNA gene between the 23 S and 5 S rRNA genes (Whitfeld *et al.*, 1978a; Bowman and Dyer, 1979) which is homologous to the sequence of the 3′ terminal region of bacterial 23 S rRNA (MacKay, 1981; Machatt *et al.*, 1981; Edwards *et al.*, 1981). This 4.5 S rRNA gene is missing in *Euglena* and *C. reinhardii* whose 23 S rRNA 3′ end resembles that of bacteria (Fig. 2).

The chloroplast 23 S rRNA gene of *Chlamydomonas reinhardii* displays two

TABLE I
Chloroplast rRNAs[a]

rRNA	Organism	Size (bp)	H (%)	Reference
23 S	*Zea mays*	2950	71	Edwards and Kössel (1981)
	Tobacco	2804	67	Takaiwa and Sugiura (1982b)
16 S	*Zea mays*	1491	74	Schwarz and Kössel (1981)
	Tobacco	1486	74	Tohdoh and Sugiura (1982)
	Euglena gracilis	1491	72	Graf *et al.* (1982)
	Chlamydomonas reinhardii	1475	76	Dron *et al.* (1982)
5 S	Tobacco	121	55	Takaiwa and Sugiura (1980)

[a]Only one 5 S rRNA sequence is indicated although chloroplast 5 S rRNA sequences from several plants have been determined which bear a strong resemblance to one another (Dyer and Bowman, 1978). H refers to the sequence homology with the corresponding *E. coli* rRNAs.

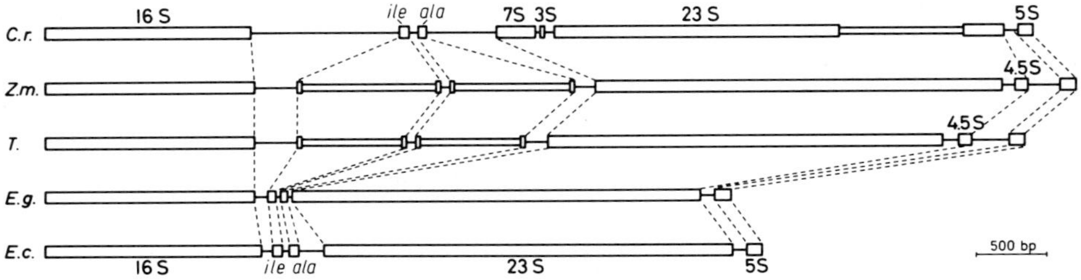

FIG. 2. Comparison of the organization of the ribosomal RNA genes from chloroplasts and *E. coli*. Coding regions and introns are shown by large and thin white bars, respectively. Dotted lines indicate the homologous regions between the different ribosomal units. C.r., *Chlamydomonas reinhardii* (Allet and Rochaix, 1979; Dron *et al.*, 1982b; Rochaix and Darlix, 1982; Rochaix and Malnoe, 1978). Z.m., *Zea mays* (Edwards and Kössel, 1981; Edwards *et al.*, 1981; Koch *et al.*, 1981; Schwarz and Kössel, 1980). T., tobacco (Takaiwa and Sugiura, 1980, 1982a,b; Tohdoh and Sugiura, 1982). E.g. *Euglena gracilis* (Graf *et al.*, 1980, 1982; Orozco *et al.*, 1980). E.c., *E. coli* (Brosius *et al.*, 1981). The genes for $tRNA^{Ile}$ and $tRNA^{Ala}$ are also indicated.

unusual features. It contains an 888 bp intron near its 3′ end and it is preceded at the 5′ end by two small genes coding for 3 S and 7 S rRNA, both of which are associated with the large chloroplast ribosomal subunit (Rochaix and Malnoe, 1978). Sequencing of the 7 S and 3 S rRNAs as well as their genes and neighboring regions has shown that the 7 S and 3 S rRNA genes are 282 and 47 bp long, respectively, and that their sequences are homologous to the 5′ termini of prokaryotic and other chloroplast 23 S rRNAs (Rochaix and Darlix, 1982). Partial nucleotide sequence complementarity exists between the 5′ ends of the 7 S and 3 S rRNAs and the 23 S rRNA sequences which flank the ribosomal intron (Rochaix and Darlix, 1982; Allet and Rochaix, 1979). This observation is compatible with a role for 3 S rRNA (and perhaps 7 S rRNA) in the splicing reaction, although a direct proof will have to rely on an analysis of mutants that are defective in the processing of the 23 S rRNA precursor.

The chloroplast ribosomal units from maize (Edwards and Kössel, 1981; Edwards *et al.*, 1981; Koch *et al.*, 1981; Schwarz and Kössel, 1980; Schwarz *et al.*, 1981b) and tobacco (Takaiwa and Sugiura, 1982a,b; Tohdoh and Sugiura, 1982) have been entirely sequenced. Large portions of the ribosomal units of *Euglena* (Graf *et al.*, 1980; Orozco *et al.*, 1980, 1982) and *Chlamydomonas reinhardii* (Allet and Rochaix, 1979; Dron *et al.*, 1982b; Rochaix and Darlix, 1982) have also been sequenced. These studies have revealed that chloroplast and bacterial rRNAs (Brosius *et al.*, 1981) are highly homologous. For example, maize 16 S and 23 S rRNA are 74 and 71% homologous to their *E. coli* counterparts (cf. Table I; Edwards and Kössel, 1981; Schwarz and Kössel, 1980). In contrast, the ribosomal spacer regions of higher plants and *E. coli* appear to be unrelated although these regions have been considerably conserved in tobacco, maize, and spinach chloroplasts (Briat *et al.*, 1982b; Schwarz *et al.*, 1981b; Tohdoh and Sugiura, 1982). A tRNA^{Val} gene is located about 300 bp upstream of the 5′ end of the 16 S rRNA gene in these higher plants. In addition, three discrete sites of this leader region of maize that are protected from DNAse I digestion after *E. coli* RNA polymerase binding (Schwarz *et al.*, 1981b) are well conserved in spinach (Briat *et al.*, 1982b). As in *E. coli,* the 16 S rRNA genes of maize (Schwarz *et al.*, 1981b), tobacco (Tohdoh and Sugiura, 1982), and *Chlamydomonas reinhardii* (Dron *et al.*, 1982b) are flanked by complementary sequences which may form cleavage sites for processing enzymes. In *Euglena* however, the leader sequence of the 16 S rRNA gene is nearly 70% homologous to the DNA sequence of the 16 S to 23 S rDNA spacer. The homology includes the 3′ end of the 16 S rRNA gene, the adjacent spacer, and the tRNA^{Ile} gene which appears as a tRNA^{Ile} pseudogene with 64% sequence homology in the leader region (Orozco *et al.*, 1980). In addition, this pseudogene is flanked by sequences resembling partial tRNA sequences. These data suggest that the two 16 S rRNA gene-flanking regions may have arisen through some sort of duplication before they evolved independently (Orozco *et al.*, 1980). An extra 16 S rRNA gene is present about 3 kb upstream of the first ribosomal unit of *Euglena* (Jenni and Stutz, 1979). Its sequence is 98% homologous to the functional 16 S rRNA genes. The major difference is a deletion of 9 bp which does not appear to affect the secondary structure of the 16 S rRNA (Roux *et al.*, 1983). Sequences about 200 bases upstream (including the pseudo tRNA structures) and 15 bases downstream of the extra 16 S rRNA and of the functional 16 S rRNA genes are highly homologous. Since no stable transcript of the extra 16 S rRNA gene is detectable, it can be concluded that the gene is either not transcribed or that its product is rapidly degraded (Roux *et al.*, 1983).

B. Transfer RNA Genes

Fractionation of total chloroplast tRNA from higher plants or algae by two-dimensional polyacrylamide gel electrophoresis yields between 30 and 40 spots, most of which have been identified as tRNAs by aminoacylation (Burkard *et al.*, 1980; Orozco and Hallick, 1982a). It is therefore probable that chloroplasts contain a full set of tRNAs. A large number of tRNA genes—in some cases up to 21—have been mapped on the chloroplast genomes of spinach (Driesel *et al.*, 1979), *Euglena* (Hallick *et al.*, 1983), maize, pea, and on the cyanelle DNA of *Cyanophora parodoxa* (for review cf. Weil *et al.*, 1982). At least six chloroplast isoacceptor tRNAs have been shown to be encoded by different genes (Burkard *et al.*, 1979; Weil *et al.*, 1982). In higher plants most of the tRNA genes are located in the large single copy region of the chloroplast genome, several (up to 9) are in the inverted repeat, and only a few have been found in the small single copy region (Weil *et al.*, 1982). *Euglena* chloroplast DNA contains at least 8–9 tRNA gene loci (El-Gewely *et al.*, 1981; Orozco and Hallick, 1982a). Six of these loci containing 20 different tRNA genes have been sequenced (Graf *et al.*, 1980; Hollingsworth and Hallick, 1982; Karabin and Hallick, 1983; Orozco and Hallick, 1982b; Orozco *et al.*, 1980; Karabin and Hallick, unpublished observations; Nickoloff and Hallick, unpublished observations). At each tRNA coding locus in *Euglena* the genes are tightly clustered and separated by short AT-rich spacers that lack obvious secondary structure. An arrangement of this sort is compatible with the idea that some genes may be transcribed into polycistronic precursors. It is noteworthy that a cluster of 3 tRNA genes (Val, Asn, Arg) has been shown to be transcribed as a polycistronic precursor by HeLa cell RNA polymerase III (Gruissem *et al.*, 1982). Several tRNA genes of higher plants have also been sequenced (Deno *et al.*, 1982; Schwarz *et al.*, 1981a; Steinmetz *et al.*, 1982). In contrast to *E. coli,* all chloroplast tRNA genes examined do not code for the 3′ CCA terminus. It has to be added posttranscriptionally as for many eukaryotic cytoplasmic and mitochondrial tRNAs.

The picture which emerges from these studies is that chloroplast tRNA sequences are more homologous to the corresponding prokaryotic than to the eukaryotic tRNA sequences (Orozco and Hallick, 1982a; Karabin and Hallick, 1983). Comparison of 17 chloroplast tRNA sequences from *Euglena* and *E. coli* reveals an average sequence homology of 70% with a range between 62 and 78%. The homology between 8 chloroplast tRNAs of *Euglena* and eukaryotic cytoplasmic tRNAs is 58% with a range between 47 and 65%. The homology with mitochondrial tRNAs of fungi is even lower, 49%, with a range between 42 and 66%. The sequence homology of *Euglena* tRNAs is highest with 10 homologous chloroplast tRNAs of higher plants, 76%, with a range between 49 and 93% (Orozco and Hallick, 1982a). For some chloroplast tRNAs, however, the distinction between prokaryotic and eukaryotic character is difficult to draw. Spinach chloroplast $tRNA_3^{Thr}$ (Kashdan *et al.*, 1980) has only slightly more sequence homology to *E. coli* $tRNA_4^{Thr}$ (69%) than to yeast cytoplasmic $tRNA_{1A}^{Thr}$ (65%). The latter shares a stretch of 19 bases, which includes the TΨC arm, with spinach $tRNA_4^{Thr}$. Similarly, the sequence homology of *Euglena* $tRNA^{Asn}$ with $tRNA^{Asn}$ from *E. coli* and from a eukaryote are almost the same (64 and 60%, respectively; Orozco and Hallick, 1982a).

A remarkable feature is the homology between different *Euglena* chloroplast tRNA genes. The most striking example is $tRNA^{Val}$ which is 78% homologous to $tRNA^{Ala}$, 74% homologous to $tRNA^{Gly}$, and 68% homologous to $tRNA^{Phe}$ (Orozco and Hallick, 1982a) *Euglena* $tRNA^{Val}$ is also highly homologous to its *E. coli* counterpart (73%). Interestingly, Holmquist *et al.* (1973) have discovered that eight out of nine pairs of *E. coli* tRNAs with more than 66% homology involve a $tRNA^{Val}$. It has been suggested that *E. coli* and *Euglena* chloroplast valine tRNAs resemble an ancestral tRNA sequence from which other tRNAs were derived by gene duplication and subsequent gene divergence (Holmquist *et al.*, 1973; Orozco and Hallick, 1982b). Recently synthetic oligonucleotides complementary to the highly conserved chloroplast tRNA regions have been constructed and used successfully as probes to detect tRNA genes on chloroplast restriction fragments (Nickoloff and Hallick, 1982).

In conclusion, chloroplast tRNA genes do display about 70% sequence homology with prokaryotic tRNA genes. However, chloroplast tRNA genes lack the terminal 3′ CCA sequence and in higher plants several of them contain long introns (458 to 949 bp) that are considerably larger than those of eukaryotic nuclear tRNA genes (cf. Section III,D).

C. Protein Genes

After subtraction of the chloroplast rRNA and tRNA genes from the whole chloroplast genome, one is still left with a large amount of potential genetic information which could encode more than 100 polypeptides with an average molecular weight of 30,000. When proteins are labeled in isolated chloroplasts in the presence of light and fractionated by two-dimensional gel electrophoresis, about 80 spots can be detected by autoradiography (Ellis *et al.*, 1977). Considerable efforts have been and are still being spent in several laboratories to identify and to map chloroplast protein genes. A powerful approach has been to use cloned chloroplast restriction fragments either as templates in an *in vitro* coupled or linked transcription–translation system (Bottomley, 1982; Coen, 1982) or as probes for selecting chloroplast mRNA by hybridization (Westhoff *et al.*, 1981). The *in vitro* synthesized products are then identified by immunoprecipitation with specific antibodies. Other approaches have relied on the sequence relatedness between homologous genes in chloroplasts and in *E. coli* (Watson and Surzycki, 1982, 1983). Table II lists the chloroplast polypeptides whose genes have been mapped on the chloroplast DNAs of several organisms. Several of these genes have also been sequenced.

Sequencing of several chloroplast protein genes has revealed that in some cases the sequence homology with corresponding prokaryotic genes is considerable, as with rRNA and tRNA genes. In higher plants the gene sequences of the α and β subunits of the ATP synthase are 55 and 58% homologous to their *E. coli* counterparts, respectively (Deno *et al.*, 1983; Zurawski *et al.*, 1982a). The corresponding amino acid sequence homologies are 54 and 67%. Less amino acid sequence homology is observed between the chloroplast and bacterial ϵ and CFo-III subunits of ATP synthase I (dicyclohexylcarbodiimide binding protein), 26 and 25%, respectively (Zurawski *et al.*, 1982a; Howe *et al.*, 1982a).

In contrast to mitochondria, chloroplasts follow the universal code (Dron *et al.*, 1982a; Howe *et al.*, 1982a; McIntosh *et al.*, 1980; Zurawski *et al.*, 1981, 1982a). Codons terminating with A and U are highly favored over those ending in G and C. In one instance however an open reading frame (ORF) of 91 codons

TABLE II
GENES OF CHLOROPLAST POLYPEPTIDES: MAPPING AND SEQUENCING

Polypeptide	Gene		References
Large subunit of ribulose bisphosphate carboxylase	rbcL	M[a]	Bowman *et al.*, 1981 (wheat); Coen *et al.*, 1977 (maize); Link, 1981 (mustard); Malnoë *et al.*, 1979 (*Chlamydomonas reinhardii*); Palmer, 1982 (*Atriplex triangularis, Cucumis sativa*); Palmer and Stein, 1982 (fern); Seyer *et al.*, 1981 (tobacco); Stiegler *et al.*, 1982 (*Euglena gracilis*); van Ee *et al.*, 1982 (Spirodela); Whitfeld and Bottomley, 1980 (spinach)
		S	Dron *et al.*, 1982a (*Chlamydomonas reinhardii*); McIntosh *et al.*, 1980 (maize); Shinozaki and Sugiura, 1982a (tobacco)
P700 apoprotein of photosystem I		M	Westhoff *et al.*, 1983b (spinach)
Photosystem II polypeptides, 32-kDa polypeptide	psbA	M	Bedbrook *et al.*, 1978 (maize); Driesel *et al.*, 1980 (spinach); Keller *et al.*, 1982 (*Euglena gracilis*); Link, 1981 (mustard); Malnoë *et al.*, 1979 (*Chlamydomonas reinhardii*); Palmer, 1982 (*Atriplex triangularis, Cucumis sativas*); Palmer and Stein, 1982 (fern); Van Ee *et al.*, 1982 (*Spirodela*)
D2-polypeptide[b]		S	Zurawsky *et al.*, 1982b (spinach); Erickson, Rahire, and Rochaix, unpublished results (*Chlamydomonas reinhardii*)
		S	Dron, Rahire, and Rochaix, unpublished results (*Chlamydomonas reinhardii*)
PSII-1, PSII-2 polypeptides cytochrome b_6/f complex		M	Westhoff *et al.*, 1983c (spinach)
Cytochrome *f*		M	Willey *et al.*, 1983 (pea); Alt *et al.*, 1983b (spinach)
Cytochrome b_6		M	Alt *et al.*, 1983b (spinach)
Subunit IV		M	Alt *et al.*, 1983b (spinach)
ATP synthase complex α subunit	atpA	M	De Heij *et al.*, 1983 (Spirodela); Palmer and Stein, 1982 (fern); Westhoff *et al.*, 1981 (spinach)
		S	Deno *et al.*, 1983 (tobacco)
β and ε subunits	atpB, atpE	M	De Heij *et al.*, 1983 (*Spirodela*); Howe *et al.*, 1982b (wheat); Palmer and Stein, 1982 (fern); Westhoff *et al.*, 1981 (spinach)
		S	Krebbers *et al.*, 1982 (maize); Zurawski *et al.*, 1982a (spinach)
CFo-I subunit	atpF	S	Westhoff *et al.*, 1983a (spinach)

TABLE II (*Continued*)

Polypeptide	Gene		References
CFo-III subunit (dicyclohexylcarbodiimide binding protein)	atpH	S	Alt *et al.*, 1983a (spinach); Howe *et al.*, 1982a (wheat)
Components of the protein synthesizing system[c]			
Elongation factor T	tufA	M	Watson and Surzycki, 1982 (*Chlamydomonas reinhardii*)

[a]M, Mapping on the chloroplast genome; S, sequence of the gene.

[b]The D2 polypeptide of *Chlamydomonas reinhardii* is translated on chloroplast ribosomes (Chua and Gillham, 1977).

[c]At least 6 proteins of the small ribosomal and 5 proteins of the large ribosomal subunit are synthesized within pea chloroplasts (Eneas-Filho *et al.*, 1981). Their genes are therefore likely to be located on the chloroplast genome.

has been found in *Euglena* which contains a UGA codon (Orozco and Hallick, 1982b). While this codon is read as tryptophan in mitochondria (Barrell *et al.*, 1979; Macino *et al.*, 1979) there is no indication that the same holds for chloroplasts. All tryptophan codons of functional chloroplast genes examined are UGG. This ORF in *Euglena* could therefore be a pseudogene. There is a putative ribosome binding site 15 bp upstream of the AUG initiation codon. The gene is furthermore preceded by sequences which resemble prokaryotic promoters and it is followed by a sequence with a structure similar to prokaryotic terminators.

While no introns have yet been found in the chloroplast protein genes of higher plants, the genes of the large subunit of ribulosebisphosphate carboxylase of *Euglena* and of the thylakoid polypeptide gene (psbA in Fig. 3) of *Euglena* and *Chlamydomonas reinhardii* contain several introns (Stiegler *et al.*, 1982; Hallick, unpublished observations; Erickson, Rahire, and Rochaix, unpublished observations, cf. Section III,D).

The arrangement of genes on the chloroplast genomes of higher plants has been remarkably conserved even among distantly related plants such as the monocotyledonous *Spirodela* and the dicotyledonous *Petunia* (de Heij *et al.*, 1983; Westhoff *et al.*, 1981). Figures 3 and 4 compare the gene arrangements on the chloroplast DNA maps of spinach, which is typical for higher plants, and of a green unicellular alga, *Chlamydomonas reinhardii*.

Several polypeptides encoded by chloroplast DNA are part of chloroplast multisubunit complexes (Table II). From the limited data available, it appears that some of these genes are scattered while others are clustered on the chloroplast genome. Genes of polypeptides of photosystem II are located at sites which are distant from each other on the chloroplast genomes of spinach (Fig. 3)

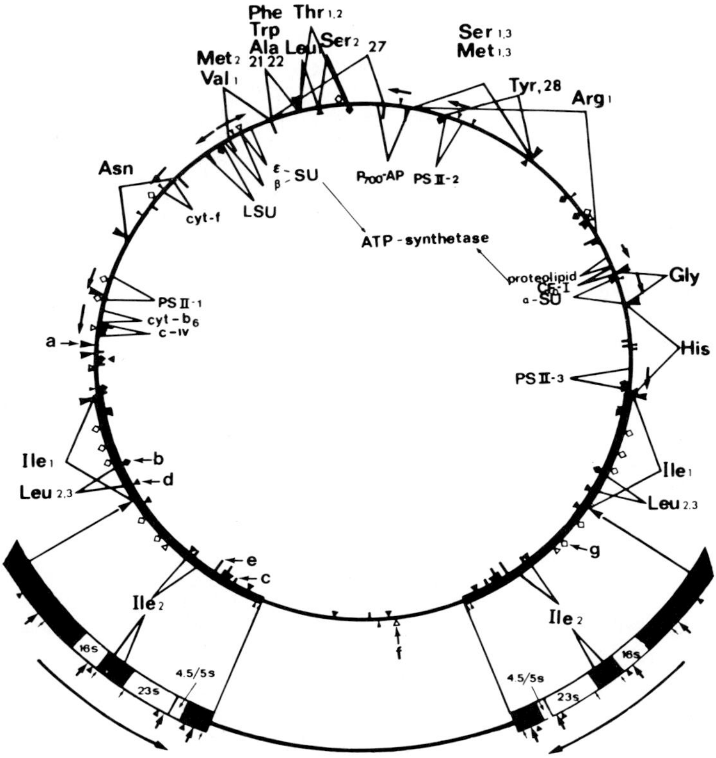

FIG. 3. Chloroplast DNA map of spinach (Courtesy of Dr. R. G. Herrmann). Restriction cleavage sites are specified by the symbols a, *Sal;* b, *Pst*I; c, *Kpn*I; d, *Xma*I; e, Xho: f, *Bgl*I; g, *Pvu*I. The inverted repeat is indicated by thicker lines and the two rDNA operons (Whitfeld *et al.*, 1978b) are shown on the extended part on the lower side of the figure. The genes which have been mapped are those of the α, β, ϵ subunits (Westhoff *et al.*, 1981), the CFo-I subunit (Westhoff *et al.*, 1984), the CFo-II subunit (Alt *et al.*, 1983a) of the ATP synthase, those of the large subunit of ribulose bisphosphate carboxylase, LSU (Whitfeld and Bottomley, 1980), of the 32-kDa polypeptide, PSII-3 (Driesel *et al.*, 1980), of two other photosystem II polypeptides, PSII-1 and PSII-2 (Westhoff *et al.*, 1983b), of the P700 apoprotein (Westhoff *et al.*, 1983a), of cytochrome *f,* cytochrome b_6; and of subunit IV of the cytochrome b_6/f complex (Alt *et al.*, 1983b). The mapped tRNA genes are also indicated (Driesel *et al.*, 1979). Transcription polarities are marked by arrows.

and of *Chlamydomonas reinhardii* (Fig. 4). In spinach the genes of cytochrome b6 and of subunit IV of the cytochrome b^6/f complex lie close to each other (Fig. 3, Alt *et al.*, 1983b). Whereas in *E. coli* the genes of all eight subunits of the ATP synthase are organized into one single operon (Saraste *et al.*, 1981), the corresponding chloroplast genes are arranged in two clusters CFo-III, CFo-I, α and β, ϵ, respectively (Fig. 3; Westhoff *et al.*, 1981, 1984; Alt *et al.*, 1983a).

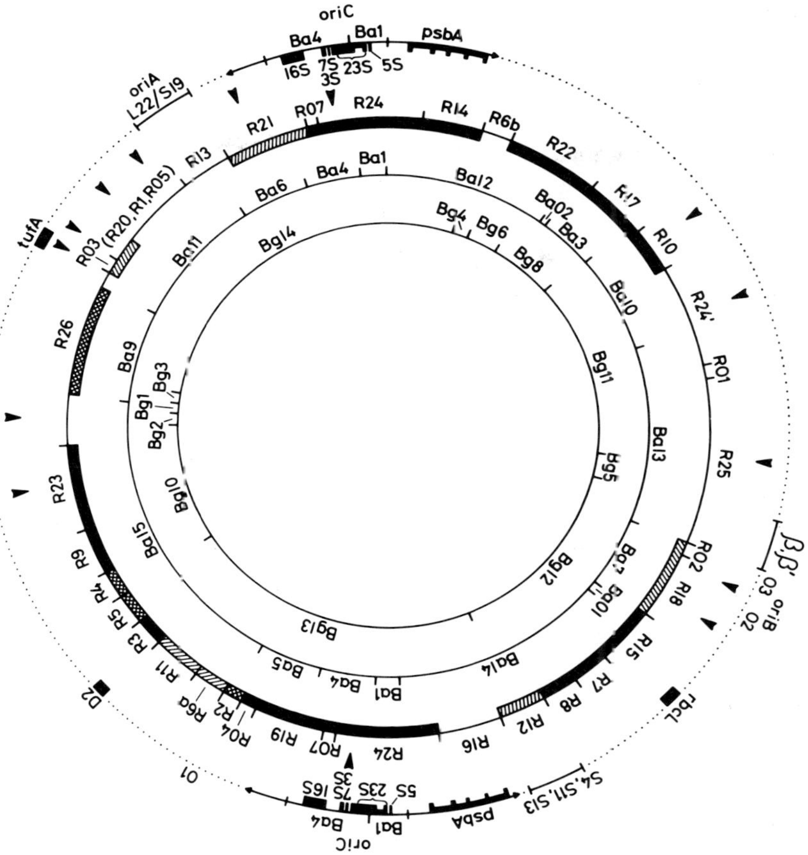

FIG. 4. Chloroplast DNA map of *Chlamydomonas reinhardii*. The three inner circles from the outside to the inside represent the *Eco*RI, *Bam*HI, and *Bgl*II restriction maps (Rochaix, 1978, and unpublished results). Dark wedges indicate the positions of the 4S RNA genes (Malnoe and Rochaix, 1978). The two segments of the inverted repeat are drawn on the outside of the map. They contain the rRNA genes (Rochaix and Malnoe, 1978) and the gene of the 32-kDa membrane polypeptide, psbA (Malnoe *et al.*, 1979; Erickson, Rahire, and Rochaix, unpublished results). The introns in the 23 S rRNA gene and in psbA are drawn in thinner lines relative to the coding sequences. D2 is the gene for another photosystem II polypeptide (Rochaix *et al.*, 1983b). The genes for the large subunit of ribulose bisphosphate carboxylase, rbcL (Malnoe *et al.*, 1979) and for the elongation factor EF-Tu, tufA (Watson and Surzycki, 1982) are also indicated. The other gene locations should be considered as tentative since they are based only on heterologous hybridizations with specific probes for the *E. coli* genes of the ribosomal proteins L22 and/or S19, for S4 and/or S11 and/or S13 and for the genes of the β and β′ subunits of *E. coli* RNA polymerase (Watson and Surzycki, 1983). The chloroplast DNA regions whose transcripts are present in large (solid area), medium (striped area), and low amounts (hatched area) are shown. The three identified chloroplast ARS sequences are indicated by 01, 02, and 03 (Rochaix *et al.*, 1983a). The three chloroplast DNA sequences promoting autonomous replication in *Chlamydomonas* are marked by ori A, ori B, and ori C (Rochaix *et al.*, 1984). Waddell *et al.* (1984) have mapped a chloroplast replication origin on the same *Eco*RI fragment as oriA.

The γ, δ, and CFo-II subunits of the chloroplast ATP synthase are coded for by nuclear genes (Nelson, 1981). In spinach and maize the β and ϵ genes have a 4 bp overlap between the stop codon of the β subunit gene and the start of the ϵ subunit gene and appear to be transcribed together into a dicistronic mRNA since probes specific for either of these two genes hybridize to the same RNA (Zurawski *et al.*, 1982a). Overlapping termination and initiation codons have also been observed twice in the *E. coli* trp operon (Oppenheim and Yanofsky, 1980; Platt and Yanofsky, 1975), in ϕX174 for the D and J protein genes (Barrell *et al.*, 1976) and in ϕ29 where the gene for the 13.9 kDa early protein overlaps a second open reading frame (Murray and Rabinowitz, 1982). In all these cases the genes overlap by only one base, i.e., the last base of the stop codon is the first nucleotide of the initiation codon. Oppenheim and Yanofsky (1980) have proposed that this gene arrangement allows translational coupling. A ribosome terminating translation of the first gene transcript is able to immediately start translation of the second mRNA, thereby allowing equimolar synthesis of the two gene products. This type of coordinate synthesis appears to be useful for polypeptides that are part of a multicomponent enzyme complex. It is not clear why this tight coupling is restricted to the synthesis of the β and ϵ subunits.

Another unusual gene arrangement occurs in the chloroplast DNA of maize where the 5′ ends of a $tRNA^{His}$ gene and of a gene corresponding to a 1.6 kb RNA, that are oppositely oriented, overlap by a few nucleotides (Schwarz *et al.*, 1981a). Recently, two overlapping divergent ORFs encoding two polypeptides of 132 and 118 amino acids have been found about 1 kb upstream of the 16 S rRNA gene of *C. reinhardii* (Schneider and Rochaix, unpublished observations). The identity of these polypeptides has not yet been determined.

The 5′ untranslated regions of prokaryotic mRNAs contain a Shine Dalgarno sequence complementary to the 3′ end of ribosomal 16 S rRNA which is located 3 to 7 nucleotides upstream of the initiator codon (Shine and Dalgarno, 1974). Short sequences complementary to the 3′ end of chloroplast 16 S rRNA which may act as ribosome binding sites have been found in chloroplasts several nucleotides upstream of the genes of the large subunit of ribulose bisphosphate carboxylase (Dron *et al.*, 1982a; McIntosh *et al.*, 1980; Zurawsky *et al.*, 1981) of the β subunit (Zurawsky *et al.*, 1982a) and of subunit III of the ATP synthase (Howe *et al.*, 1982a). However, these chloroplast "Shine-Dalgarno sequences" are not identical with each other as shown for example in maize where these sequences differ because they pair with different regions of the 3′ end of 16 S rRNA (Krebbers *et al.*, 1982).

There are sequence homologies between prokaryotic transcriptional regulatory elements such as the Pribnow box and the −35 region (Rosenberg and Court, 1979) and the 5′ upstream regions of several chloroplast genes such as the gene for the large subunit of ribulosebisphosphate carboxylase (Dron *et al.*, 1982a;

Gatenby *et al.*, 1981; McIntosh *et al.*, 1980; Shinozaki and Sugiura, 1982a; Zurawsky *et al.*, 1981), the 32-kDa protein gene (Zurawsky *et al.*, 1982b), the β subunit ATP synthase gene (Krebbers *et al.*, 1982; Shinozaki *et al.*, 1982; Zurawsky *et al.*, 1982a), the rRNA genes (Briat *et al.*, 1982b; Schwarz *et al.*, 1981; Tohdoh *et al.*, 1981), and several tRNA genes (Kato *et al.*, 1981; Steinmetz *et al.*, 1982).

The termination sites of transcription of bacteria contain a stretch of U residues which is preceded by a hairpin structure (Rosenberg and Court, 1979). Similar structures have also been found at the end of chloroplast transcripts (Dron *et al.*, 1982a; Zurawsky *et al.*, 1981).

D. Chloroplast Introns

Introns in the large ribosomal RNA genes of chloroplasts have only been found in *Chlamydomonas reinhardii*. The 23 S rRNA gene of this alga contains an 888 bp intron near its 3′ end (Rochaix and Malnoe, 1978). Sequencing of the intron has revealed an open reading frame (ORF) which could encode a polypeptide of 163 amino acids (Rochaix, unpublished observations). No sequence homology is apparent between this ORF and those found in mitochondrial introns of yeast (Dujon, 1980) and *Aspergillus nidulans* (Netzker *et al.*, 1982). Michel and Dujon (1982) have performed an extensive analysis of intron sequences of fungal mitochondrial genes. The remarkable results of this study is that all these sequences fall into two distinct groups, neither of which follows the GU...AG rule of eukaryotic pre-messenger RNAs (Breathnach *et al.*, 1978). Members of the same group share short, but distinctive sequence elements and can be folded into conserved secondary structures. Recently, these two groups of introns have been shown to include the ribosomal introns of two protists (group I) and two chloroplast tRNA introns of maize (group II; cf. below). Group I introns start after a U and end with a G. The major feature is the presence of two short conserved complementary sequences near the 5′ and 3′ ends of the introns. These two sequences appear to play an important role in the splicing of the fourth intron of the yeast mitochondrial cytochrome *b* gene where they correspond to the box 9 and box 2 mutations (De la Salle *et al.*, 1982; Michel and Dujon, 1982). Elements corresponding to the box 9 and box 2 consensus sequences are present in the 23 S rRNA intron of *C. reinhardii* (Rochaix and Michel, unpublished results). The intron, which follows the U and G rule, can be folded into a secondary structure which agrees with the general model proposed for group I introns (Michel and Dujon, 1982b).

Several chloroplast tRNA genes of higher plants contain large introns (cf. Table III). In contrast, none of 20 sequenced tRNA genes of *Euglena* (Hallick, 1982; R. B. Hallick, unpublished observations) and of two sequenced tRNA

TABLE III
CHLOROPLAST INTRONS[a]

Gene	Organism	Number of introns	Size (bp)	ORF	References
rRNA genes					
23 S rRNA	*Chlamydomonas reinhardii*	1	888	163	Rochaix and Malnoë (1978); Allet and Rochaix (1979); Rochaix (unpublished results)
tRNA genes					
$tRNA^{Ile}$	*Zea mays*	1	949	45	Koch *et al.* (1981)
$tRNA^{Ala}$	*Zea mays*	1	806	123	Koch *et al.* (1981)
$tRNA^{Leu}$	*Zea mays*	1	458	−60	Steinmetz *et al.* (1982)
$tRNA^{Ile}$	Tobacco	1	707	71	Takaiwa and Sugiura (1982a)
$tRNA^{Ala}$	Tobacco	1	710	31	Takaiwa and Sugiura (1982a)
$tRNA^{Val}$	Tobacco	1	571	−64 (O)	Deno *et al.* (1982)
Protein genes					
rbcL	*Euglena gracilis*	>8			Stiegler *et al.* (1982); Hallick (unpublished results)
psbA	*Euglena gracilis*	>2			Hallick (unpublished results)
psbA	*Chlamydomonas reinhardii*	4	1250 1350 1100 1700		Erickson, Rahire, and Rochaix (unpublished results)

[a]rbcL and psbA are the genes coding for the large subunit of ribulose bisphosphate carboxylase and for the 32-kDa chloroplast membrane polypeptide, respectively. The number of amino acids encoded by each open reading frame (ORF) contained within the introns is given. The minus sign indicates that the ORF is oriented in opposite direction relative to the gene; (O) indicates that the ORF overlaps with an exon.

genes of *Chlamydomonas reinhardii* (Schneider and Rochaix, unpublished observations) are split.

The $tRNA^{Ile}$ and $tRNA^{Ala}$ genes in the ribosomal spacer of maize and tobacco both contain large introns (Koch *et al.*, 1981; Takaiwa and Sugiura, 1982a). These genes appear to be transcribed and processed into functional mature tRNAs since $tRNA_2^{Ile}$ from maize hybridizes uniquely to the spacer region (Guillemaut and Weil, 1982). The recently established sequence of this tRNA has made it possible to determine that the splicing site is located two bases away from the anticodon (Guillemaut and Weil, 1982). The $tRNA^{Leu}$ gene of maize also contains an intron of 458 bp which starts after the first base of the anticodon (Steinmetz *et al.*, 1982). In tobacco, the chloroplast $tRNA^{Val}$ coding sequence is interrupted by a 571 bp intron (Deno *et al.*, 1982). Thus, no general rule appears to be valid for the splicing site in chloroplast tRNA genes, as opposed to nuclear tRNA genes, where the introns are located in an invariant position one base after the anticodon (Abelson, 1979). An intriguing question is whether the specificity of the splicing reaction is governed by other still unknown parameters in the chloroplast tRNA precursor or whether several chloroplast tRNA splicing enzymes exist.

A significant sequence homology exists between the introns of the $tRNA^{Ile}$ and $tRNA^{Ala}$ genes of the ribosomal spacer in maize and in tobacco, suggesting that these introns may have originated from a common ancestor by duplication (Koch *et al.*, 1981). These intron sequences conform to the general model of group II introns (Michel and Dujon, 1983). A small sequence homology is apparent between these introns and the intron of the $tRNA^{Val}$ gene of tobacco (Deno *et al.*, 1982). Regions of homology include the 5′ and 3′ ends and two short internal intron regions. Most of these regions belong to the highly conserved sequence stretches in the secondary structure model of group II introns (Michel and Dujon, 1983). In contrast, the sequence of the intron of the $tRNA^{Leu}$ gene from maize is not related in any obvious way to the other chloroplast tRNA introns (Steinmetz *et al.*, 1982). Several of these chloroplast tRNA introns contain ORFs (cf. Table III). Whether some of these are translated into functional polypeptides is still unknown.

Little is known about the intron sequences of chloroplast protein genes since they have been discovered only recently in two algae. The gene of the large subunit of ribulose bisphosphate carboxylase of *Euglena* contains at least 8 introns (Stiegler *et al.*, 1982; Hallick, unpublished observations). The genes of the 32-kDa polypeptide of *Euglena* and *Chlamydomonas reinhardii* are also split (Hallick, unpublished observations; Erickson, Rahire, and Rochaix, unpublished observations). In *Chlamydomonas reinhardii* this gene is located in the inverted repeat. It is 6.7 kb long although the coding sequence, which is 90% homologous to the corresponding sequence in spinach, contains only 1056 bp. The four introns, from the 5′ to the 3′ end, are 1.25, 1.35, 1.1, and 1.7 kb long. Although the sequence data of these introns are still fragmentary, the box 2 consensus sequence is present near the 3′ end of the third and fourth introns suggesting that these introns belong to group I.

The presence of short 4 to 5 bp repeats near the ends of the chloroplast ribosomal intron (Allet and Rochaix, 1979) and some of the tRNA introns (Koch *et al.,* 1981) may be fortuitous. Alternatively the possibility exists that these introns are remnants of transposable elements. It is well documented that the latter are capable of moving from one site to another on the chromosomes and episomes of prokaryotic cells and that each integration event is accompanied by the duplication of a short stretch from the target sequences (Calos and Miller, 1980). A more likely hypothesis is that the chloroplast introns reflect some ancient gene structures. It is surprising that chloroplast tRNA introns have been found only in higher plants, not in algae. The opposite holds for introns in rRNA and protein coding genes which have only been detected in algae, but not in higher plants. Admittedly, the data are still too limited to allow general conclusions.

IV. Chloroplast Transcription

Chloroplast genomes are differentially transcribed in different cell types and under different physiological conditions. In maize the transcript of the gene of the large subunit of ribulose bisphosphate carboxylase is detectable in bundle sheath cells, but it is almost or completely absent in mesophyll cells (Link *et al.,* 1978). In contrast, the transcript of the neighboring β subunit gene is present in equal amounts in both types of cells (Jolly *et al.,* 1981). The chloroplast RNA polymerase activity rises sharply during light induced plastid maturation of dark grown seedlings (Apel and Bogorad, 1976). Major changes occur in the plastid mRNA pools, especially for the 32-kDa membrane polypeptide during light induction (Bedbrook *et al.,* 1978; Edelman and Reisfeld, 1978). In mustard the increase in the steady state level of this transcript appears to be controlled by phytochrome (Link, 1982). Transcription studies of the chloroplast genome of *Chlamydomonas reinhardii* during synchronous cell growth and at various stages of the greening process in a yellow mutant have also revealed differential gene expression (Matsuda and Surzycki, 1980).

How is chloroplast gene expression regulated? What is the structure of chloroplast RNA polymerase? Are there stage specific regulatory factors? Only partial answers have been obtained. A chloroplast DNA-dependent RNA polymerase has been purified from maize plastids (Smith and Bogorad, 1974) and also from whole cell homogenates (Kidd and Bogorad, 1980). The maize enzyme, which is resistant to α-amanitin and to rifamycin SV, consists of at least

14 polypeptides (180, 140, 121, 110, 100, 95, 85, 75, 70, 55, 42, 40, 38, and 27 kDa). A polypeptide of 27.5 kDa from maize plastids, designated S factor, preferentially enhances transcription *in vitro* by chloroplast RNA polymerase, of chloroplast DNA sequences in supercoiled form (Jolly and Bogard, 1979). While the S factor has no effect on transcription by *E. coli* RNA polymerase, a factor with σ like activity has been isolated from *Chlamydomonas reinhardii* cells (Surzycki and Shellenbarger, 1976). This factor of 51 kDa is capable of replacing the *E. coli* σ factor, associating with *E. coli* core RNA polymerase to form rifampicin-resistant preinitiation complexes. It is noteworthy that, in contrast to higher plant chloroplast RNA polymerases, the *C. reinhardii* enzyme is sensitive to rifampicin (Guertin and Bellemare, 1979; Surzycki, 1969; Surzycki and Rochaix, 1971). Recently the spinach chloroplast RNA polymerase has also been partially purified and shown to consist of 7 major polypeptides of 150, 110, 102, 80, 75, 38, and 25 kDa (Lerbs *et al.*, 1983). The 80 and 75 kDa subunits bind strongly to DNA (Lerbs *et al.*, 1983). The enzyme is resistant to rifamycin and α-amanitin and it is strongly inhibited by heparin. It will be interesting to examine by immunological means how the subunits of the maize and spinach chloroplast RNA polymerases are related to each other and also how they compare with the subunits of prokaryotic and nuclear eukaryotic RNA polymerases.

The maize RNA polymerase has been used successfully to transcribe *in vitro* the genes of the large subunit of ribulose bisphosphate carboxylase (Jolly *et al.*, 1981) and of a $tRNA^{His}$ gene (Schwarz *et al.*, 1981a). In both cases the *in vitro* initiation site for transcription was determined by Sl nuclease mapping and shown to be indistinguishable from the 5′ end of the *in vivo* transcript. However it has not yet been demonstrated whether the 5′ ends of these transcripts result from the authentic initiation of transcription or whether they have been produced by specific processing events. One way to solve this problem is to label the 5′ ends of chloroplast transcripts with vaccinia guanylyltransferase since this enzyme is capable of adding GMP to di- or triphosphate terminated polyribonucleotides (Monroy *et al.*, 1978). Although the exact role of the S factor in the initiation of transcription is not yet clear, it appears to promote differential expression of chloroplast genes *in vitro* since in its presence the ratio of transcripts from the genes of the large subunit of ribulose bisphosphate carboxylase and the β subunit is three, while in its absence this ratio is close to one (Jolly *et al.*, 1981).

Binding studies with *E. coli* RNA polymerase and spinach chloroplast DNA have revealed several strong chloroplast DNA binding sites, some of which map in the same regions as known protein coding sequences (Zech *et al.*, 1981). Fragments containing the chloroplast rRNA genes bind poorly to the bacterial enzyme although they carry the strongest chloroplast promoters (Zech *et al.*, 1981). As long as the relationship between bacterial and chloroplast RNA polymerases is not elucidated, results obtained in a heterologous system should be interpreted with caution.

V. Correlation between Genetic and Physical Chloroplast DNA Maps

It was already recognized at the beginning of this century that plastids of higher plants are inherited in a maternal–uniparental fashion (Baur, 1909; Correns, 1909). In the few species which exhibit biparental transmission of cell organelles (Tilney-Bassett, 1978), there is no evidence for recombination between parternal and maternal chloroplast DNA molecules (Metzlaff *et al.*, 1981). The only organism in which chloroplast gene recombination has been observed is *Chlamydomonas reinhardii,* a green alga containing a single chloroplast (Sager, 1977; Gillham, 1978). Since the isolation of the first uniparental mutant of *Chlamydomonas* by Sager (1954), a large number of these mutants have been obtained and characterized (Gillham, 1978). A distinctive feature of uniparental genes is that they are transmitted mainly by the mt^{+} parent to the offspring. The *Chlamydomonas* chloroplast genetic analysis would not have been possible without the occasional occurrence of biparental zygotes which inherit uniparental genes from both parents. After the two parental chloroplasts fuse during the early stages of meiosis, the two chloroplast genomes mix with each other and can undergo recombination. Most uniparental mutations involve antibiotic resistance or acetate autotrophy (Sager, 1977; Gillham, 1978). All uniparental mutants examined appear to fall into a single linkage group (Sager, 1977, Gillham, 1978). Unitl recently the uniparental linkage group and the physical chloroplast DNA of *Chlamydomonas reinhardii* were two separate entities. The identity between uniparental and chloroplast genes has only been proven in the last years. Using defined chloroplast DNA deletions as markers, Grant *et al.* (1980) could demonstrate unambiguously the uniparental inheritance of chloroplast DNA in *Chlamydomonas reinhardii.* Similar results were obtained from interspecies crosses between *Chlamydomonas moewusii* and *Chlamydomonas eugametos* (Lemieux *et al.*, 1981; Mets, 1980). These results did not establish however that the chloroplast DNA contains the uniparental markers. Myers *et al.* (1982) were able to map the uniparental locus *ac-u-c,* using deletion mutants, on the chloroplast DNA of *Chlamydomonas reinhardii.* This locus appears to code for the β subunit of ATP synthase (Woessner *et al.*, 1984). Mapping of this locus relative to the uniparental linkage group will establish a correlation site between the genetic and physical chloroplast DNA maps. Spreitzer and Mets (1980) isolated a uniparental mutant that lacks ribulose bisphosphate carboxylase activity and has an altered large subunit isoelectric point. A comparative sequence analysis of the wild-type and mutant large subunit genes has revealed a single base pair change (Dron *et al.*, 1983). The corresponding Gly to Asp amino acid substitution is near the first active site of the enzyme. Since the location of the large subunit gene is known on the physical chloroplast DNA map (Malnoe *et al.*, 1979) and since the locus of this mutation has been shown to be linked to other genetic markers in the established uniparental linkage group (Mets and Geist, 1983), this result provides the first direct link between the chloroplast DNA and the uniparental linkage group. Since the active site peptides of the large subunit have been well characterized (Lorimer, 1981; Schloss *et al.*, 1978; Stringer and Hartman, 1978) and since they are spread over a large region of the large subunit polypeptide, it may be possible to perform a fine structure recombination analysis of chloroplast DNA in this region.

The gene coding for the 32-kDa chloroplast membrane polypeptide in *Chlamydomonas reinhardii* is another promising marker for correlating chloroplast genetics and biochemistry. The sequence of this protein has been remarkably conserved between higher plants and algae (Zurawsky *et al.*, 1982b; Erickson *et al.*, 1984a). This protein appears to be the target of triazine-type herbicides (Pfister *et al.*, 1981; Steinback *et al.*, 1981). Further support comes from a comparison of the gene sequence of this protein from atrazine-resistant and sensitive biotypes of the weed *Amaranthus hybridus*. A single base pair change has been found which gives rise to a single amino acid substitution (Hirshberg and McIntosh, 1983). Since these two biotypes have arisen independently, a polymorphic change cannot be ruled out completely. Diuron-resistant uniparental mutants have been isolated in *Chlamydomonas reinhardii* (Galloway and Mets, 1982; Erickson *et al.*, 1984b). Recently the genes of the 32-kDa membrane polypeptide from wild type and a diuron resistant mutant of *Chlamydomonas reinhardii* have been sequenced (Erickson *et al.*, 1984b). The two genes differ by a base pair change which results in a serine to alanine substitution. Interestingly, the same Ser residue is changed into Gly in the atrazine resistant weed biotype (Hirschberg and McIntosh, 1983).

Another approach for correlating uniparental genetic loci with chloroplast DNA regions has relied on interspecific hybrids of *Chlamydomonas eugametos* and *Chlamydomonas moewusii* (Lemieux *et al.*, 1983). Analysis of recombinant chloroplast restriction patterns from the hybrid progeny indicates linkage between a streptomycin-sensitivity locus and a chloroplast restriction fragment which hybridizes to the 16 S rRNA gene region of *Chlamydomonas reinhardii* (Lemieux *et al.*, 1983). Among the uniparental mutations conferring resistance to streptomycin, kanamycin, erythromycin, or chloramphenicol in *Chlamydomonas reihardii* (Gillham, 1978; Bennoun *et al.*, 1981), several may well be located within the chloroplast rRNA genes. Since chloroplast and mitochondrial rRNA sequences are highly related to each other (Dujon, 1980; cf. Section III, A), and since several antibiotic-resistant mutations have been located in mitochondrial rRNAs (Dujon, 1980), uniparental mutations of this sort may be identified relatively easily by DNA sequencing and they may provide valuable additional correlation sites. Once a correlation between genetic and physical maps is established it would be possible to determine the approximate physical sites of known uniparental mutations. These sites may be characterized further by transcript mapping and by DNA sequencing. In addition, this correlation may help to assess the physical parameters in chloroplast gene recombination. It can be anticipated, however, that the correlation between the genetic and physical map may not be extended over the entire chloroplast genome because of intramolecular recombination within the inverted repeat (Palmer, 1983, Mets and Geist, 1983).

VI. Chloroplast DNA Replication

Density transfer experiments have shown that the replication of chloroplast DNA in *Chlamydomonas* and *Euglena* is semiconservative (Chiang and Sueoka, 1967; Manning and Richards, 1972a). In corn and pea, replication of chloroplast DNA initiates with the formation of two displacement loops, 7.1 kb apart, which expand until the two displacing strands elongate past each other (Kolodner and Tewari, 1975a). Replication then proceeds bidirectionally by a Cairns replication mechanism until the other side of the DNA molecule is reached (Kolodner and Tewari, 1975b). At or near the site where Cairns replication terminates, a rolling-circle type of replication is initiated (Gilbert and Dressler, 1968), as demonstrated by the presence of circular chloroplast DNA molecules with tails longer than the unit size circles. Replicating chloroplast molecules have also been described in *Euglena* (Manning and Richards, 1972b).

By taking advantage of the known restriction map of *Euglena gracilis* chloroplast DNA (Hallick, 1984), Koller and Delius (1982a) and Ravel-Chapuis *et al.* (1982) have independently mapped the chloroplast replication origin at a site 5 to 6.8 kb upstream of the 5′ end of the extra 16 S rRNA gene (cf. Section III,A). This site is at or very near one end of the region of variable size in the *Euglena gracilis* chloroplast DNA (Jenni *et al.*, 1981), which is AT rich and flanked by two short inverted repeats (Schlunegger *et al.*, 1983). It appears that DNA replication proceeds unidirectionally away from the ribosomal region for about 5 kb before the replication becomes bidirectional (Koller and Delius, 1982).

Similar experiments have been performed recently with *Chlamydomonas reinhardii* by Waddel *et al.* (1984). They were able to locate two chloroplast replication origins about 10 and 16.5 kb, respectively, upstream of the 5′ end of one of the two 16 S rRNA genes. Interestingly, among those chloroplast DNA sequences which are capable of promoting autonomous replication in *Chlamydomonas reinhardii* (Fig. 3), one maps in the same region (Rochaix *et al.*, 1984).

The close proximity of the replication origins to the rRNA genes is not only valid for the chloroplasts of *Euglena* and *Chlamydomonas reinhardii,* but also for several mitochondrial DNAs (Anderson *et al.*, 1981; Bibb *et al.*, 1981; Kroon *et al.*, 1977; Ramirez and Dawid, 1978). It could explain the selective retention of rRNA genes during streptomycin-induced mutagenesis in *Euglena* (Heizmann *et al.*, 1982).

A small mitochondrial restriction fragment of *Xenopus laevis* containing the replication origin has been shown to promote autonomous replication in yeast (Zakian, 1981). Screening of the chloroplast DNA of *Chlamydomonas reinhardii* has revealed at least 8 distinct regions in this genome which promote autonomous replication in yeast (ARS sequences) (Rochaix *et al.*, 1983; Loppes and Denis, 1983; Vallet *et al.*, 1984, cf. Fig. 4). One of these regions which spans 400 bp has been sequenced. It is very AT rich (77%) and contains a short 21 bp GC-rich island. Striking features of this sequence are multiple direct and inverted repeats, 8 to 10 bases long. The relationship of these chloroplast ARS sequences to authentic chloroplast replication origins remains to be explored. It is noteworthy that to date eukaryotic DNA, but not prokaryotic DNA, has been shown to contain ARS sequences (Stinchcomb *et al.*, 1980).

Chloroplast DNA synthesis has been shown to occur in isolated chloroplasts and in crude chloroplast extracts of maize (Zimmermann and Weissbach, 1982) and of *Chlamydomonas reinhardii* (Keller and Ho, 1981). In the latter case, *in organello* chloroplast DNA synthesis is semiconservative, dependent on exogenously added ribonucleotides, partially inhibited by rifamycin and sensitive to nalidixic acid, suggesting that chloroplasts contain topoisomerase activity. Indeed, topoisomerase I activity has been partially purified from spinach chloroplasts and shown to have prokaryotic properties (Siedlecki *et al.*, 1983). The enzyme has an apparent molecular weight of 115,000, requires Mg^{2+}, catalyzes a unit change in the linkage number of supercoiled DNA and cannot relax positive supercoiled DNA.

A DNA polymerase with a molecular weight of 105,000 has been extracted from spinach chloroplasts (Sala *et al.*, 1980). The enzyme resembles the animal mitochondrial DNA polymerase by its resistance to aphidicolin and its preference for the synthetic primer template $(dT)_{12-18}$ poly(rA). In contrast, a DNA polymerase extracted from *Chlamydomonas reinhardii* chloroplasts is unable to use the same primer template. The apparent molecular weight of this DNA polymerase activity depends on the ionic strength. At 0.8 *M* NaCl the activity runs at 180 kDa whereas at 0.05 *M* NaCl, activities are detectable at 180, 80, and 40 kDa. SDS–polyacrylamide gel electrophoresis reveals a major polypeptide of 38 kDa and two minor polypeptides of 75 and 15 kDa (Keller and Ho, 1981). Since the specific activity of the enzyme is low, it may represent a subunit of a larger enzyme complex. Interestingly, a 40-kDa DNA polymerase has been recovered from a whole cell extract of *Chlamydomonas reinhardii* that has the same catalytic properties and thermal stability as the chloroplast enzyme (Keller and Ho,

1981). An intriguing possibility is that the DNA polymerase complexes from the nucleus and chloroplast share common subunits. This idea is compatible with the finding that a chloroplast DNA fragment containing a chloroplast replication origin (Waddell *et al.,* 1984) is capable of promoting autonomous replication in *Chlamydomonas reinhardii* (Rochaix *et al.,* 1984). It is likely that in this case replication occurs in the nucleocytoplasmic compartment rather than in the chloroplast. Similar observations have been made in yeast where mitochondrial replication origins are active in the nucleo-cytoplasm (Blanc and Dujon, 1981).

In view of the remarkable sequence relatedness of the components of the protein synthesizing machineries of chloroplasts and prokaryotes, a challenging task for the future is to examine whether this homology also holds for the DNA synthesis enzymes. It appears that a high-molecular-weight polypeptide of 125 kDa responsible for DNA chain elongation is conserved in both prokaryotic and eukaryotic organisms (Hübscher *et al.,* 1981). The polypeptide is sensitive to proteolysis, which generates a number of catalytically active fragments that have similar sizes among different organisms (Hübscher *et al.,* 1981).

VII. Migratory Chloroplast DNA

In 1905 Mereschkowsky proposed that chloroplasts may have arisen from free living organisms, related to present day blue green algae, which invaded primitive eukaryotic cells. This endosymbiont theory has since been discussed by others (Margulis, 1967; Gray and Doolittle, 1982) and has become rather popular in recent years because of the remarkable sequence homology observed between several chloroplast and bacterial genes (cf. Section III). The endosymbiont hypothesis proposes that the prokaryotic intruders gradually lost their genetic autonomy and became increasingly dependent on the nuclear activity of their host.

The alternative hypothesis assumes that in the uncompartmentalized primitive ancestor cell, the genes involved in specialized cellular functions such as energy production, for example, segregated into one or more clusters surrounded by membranes (Sonneborn, 1967; Bogorad, 1975). After this compartmentalization occurred, these genetic systems were able to evolve independently although gene transfers between the organellar and the nuclear genomes must have occurred occasionally.

A final decision between these two hypotheses is difficult and may even be beyond the domain of science. In both cases, however, the transfer of genes between cellular compartments is postulated. These gene migrations from organelles to nucleus or vice versa might reach an intermediate stage when, after transfer, the old and the new gene copy are still active. Later, one copy may be inactivated through mutations and in some cases it may even be completely removed. Evidence for gene movements between mitochondrial, chloroplast, and nucleocytoplasmic compartments has appeared recently. Fragments of the yeast mitochondrial gene of the ribosomal var 1 protein and of cytochrome *b* apoprotein together with a putative mitochondrial replication origin have been found in the yeast nuclear genome (Farrelly and Butow, 1983). Since these fragments are not contiguous on the mitochondrial genome, rearrangements must have occurred. From the high sequence homology, 80%, between the nuclear DNA and the homologous mitochondrial DNA sequences, Farrelly and Butow (1983) estimate that these nuclear pseudo genes diverged from the mitochondrial genes about 25 million years ago. Mitochondrial DNA sequences have not only been found in the nucleus of yeast and another fungus *Podospora anserina* (Wright and Cummings, 1983), but also in the chromosomal DNA of sea urchins (Jacobs *et al.*, 1983; Fox, 1983) and locusts (Gellissen *et al.*, 1983). The results in the later two cases are that mitochondrial DNA sequences are not only present, but also repeated in the nuclear genome. In *Neurospora* the active gene of the ATPase subunit 9 is located in the nucleus (Sebald and Wachter, 1978) and an apparent inactive copy of the same gene has been found in the *Neurospora* mitochondrial genome (Van den Boogart *et al.*, 1982).

Similarly, several chloroplast DNA sequences hybridize with nuclear DNA. An 660 bp probe internal to the intron of the chloroplast 23 S rRNA gene of *Chlamydomonas reinhardii* hybridizes to at least five distinct nuclear non-ribosomal DNA fragments (Rochaix, 1981). Watson and Surzycki (1983) have shown that both the chloroplast and nuclear genomes of *Chlamydomonas reinhardii* share homology with *E. coli* genes for transcriptional and translational components. Probes specific for the genes of the β and β′ subunits of RNA polymerase of *E. coli* hybridize to both chloroplast and nuclear DNA (Watson and Surzycki, 1983; Hong and Surzycki, unpublished results). The stronger hybridization signal to the nuclear fragment obtained with the β probe suggests that the active gene is in the nucleus and that the weaker chloroplast signal may be due to a pseudogene. Indirect evidence based on the use of antibiotics (Ellis and Hartley, 1971) and on heat inactivation of chloroplast ribosomes (Bunger and Feierabend, 1980) indicates that the chloroplast RNA polymerase is encoded by the nuclear genome. The interesting possibility that both the nuclear and chloroplast genes are active—perhaps under different physiological conditions—cannot be ruled out, however. A definite answer to these questions will have to await the sequence analysis of the corresponding nuclear and chloroplast genes. A probe specific for the *spc* operon of *E. coli* which includes the genes of the ribosomal proteins L15, L30, and S5 hybridizes mainly to a nuclear DNA fragment and produces a faint signal with a chloroplast DNA region of *Chlamydomonas reinhardii* (Watson and Surzycki, 1983). Interestingly, the *E. coli* probe for the gene of the elongation factor EF-G hybridizes uniquely to nuclear DNA suggesting that this protein is encoded by a nuclear gene in *Chlamydomonas reinhardii* (Watson and Surzycki, 1983) as in *Euglena* (Breitenberger *et al.*, 1979). In contrast, in spinach, this gene is located in the chloroplast genome (Ciferri *et al.*, 1979). It appears therefore that the distribution of genes between chloroplast and nuclear genomes is not universal. Similarly, a probe specific for the gene of the small subunit of ribulose bisphosphate carboxylase hybridizes to a region near the gene of the large subunit in the cyanelle DNA of *Cyanophora paradoxa* (Heinhorst and Shively, 1983; Bohnert *et al.*, 1983). This surprising result needs to be confirmed by an appropriate sequence analysis since in all plants studied, the gene of the small subunit is encoded by a nuclear multigene family (Berry-Lowe *et al.*, 1982).

Stern and Lonsdale (1982) have recently shown that a 12 kb DNA sequence of maize mitochondrial DNA is related, with more than 90% base sequence homology, to a portion of the inverted repeat of the maize chloroplast genome. This sequence includes the 16 S rRNA gene and the genes of $tRNA^{Ile}$ and $tRNA^{Val}$. It therefore appears that the chloroplast DNA segment was incorporated into the mitochondrial genome after a duplication and transposition event. Extensive sequence rearrangements in the region of the mitochondrial genome that has homology with chloroplast DNA have occurred in cytoplasmic male sterile lines of maize (Stern and Lonsdale, 1982). Even more surprising is the finding of a DNA sequence in the mitochondrial genome of maize homologous to the chloroplast gene coding for the large subunit of ribulose bisphosphate carboxylase (Lonsdale *et al.*, 1983). *In vitro* transcription–translation of this cloned mitochondrial DNA region in an *E. coli* cell-free system produces a 21-kDa polypeptide that can be immunoprecipitated with antibodies to wheat ribulose bisphosphate carboxylase (Lonsdale *et al.*, 1983). The remarkable sequence conservation of these integrated chloroplast DNA fragments implies that the events occurred relatively recently on an evolutionary time scale or that chloroplast and mitochondrial genomes may interact through chloroplast mitochondrial fusions whose occurrence still remains to be demonstrated unambiguously. The presence of chloroplast DNA sequences in mitochondrial genomes is not unique to maize and has now been demonstrated in 11 species of angiosperms (Stern *et al.*, 1983). Interactions between the two organelles have already been proposed at the protein level because in several cases chloroplast mutations appear to affect not only the chloroplast, but also the mitochondrial compartment (Bennoun, 1981; Conde *et al.*, 1975; Surzycki and Gillham, 1971).

VIII. Conclusions

A remarkable feature which has emerged from the study of the organization and structure of chloroplast genes is the extensive sequence homology between chloroplast and bacterial coding sequences. This homology is highest among the rRNA genes and it is still considerable for several tRNA and protein coding genes (cf. Section III). Other prokaryotic properties include the existence of putative ribosome binding sites preceding the AUG initiation codons and of sequences located upstream of the transcription initiation sites which bear a strong resemblance to the bacterial Pribnow box and to the −35 region (Rosenberg and Court, 1979). Another property shared by chloroplasts and prokaryotic organisms is the sequence overlap between some genes (cf. Section III). It is difficult to envisage these overlaps as a solution for space economy since chloroplast genomes contain noncoding regions. It is more likely that this particular gene arrangement plays some role in the regulation of chloroplast gene expression although no experimental proof is available yet.

Homologies between the *E. coli* and chloroplast RNA polymerases are also apparent. In *Chlamydomonas reinhardii* the enzyme is sensitive to rifampicin and a σ like polypeptide has been isolated (cf. Section IV). The hybridization of probes specific for the genes of the β and β′ subunits of *E. coli* RNA polymerase with nuclear and chloroplast DNA fragments of *Chlamydomonas reinhardii* also suggests structural homology (Watson and Surzycki, 1983) although the coding function of these DNA fragments remains to be demonstrated. An open field for the future is to explore how closely the chloroplast and bacterial enzymes are structurally related and whether the enzymes differ in terms and sequence requirements for specific initiation of transcription. The chloroplast DNA is associated with proteins to form a complex whose structure may be related to the nucleoid of bacteria. At least one polypeptide from this complex is immunologically related to the *E. coli* HU protein (Briat and Rouvière-Yaniv, unpublished observations).

Several properties of chloroplast genomes are however not shared by bacterial genomes. The most striking example is the presence of split chloroplast genes. To date, introns have been found in the rRNA genes of *Chlamydomonas reinhardii* and in some protein genes of *Euglena* and *Chlamydomonas reinhardii,* but not in similar genes of higher plant chloroplasts (cf. Section III). Conversely, large introns have been detected in several chloroplast tRNA genes from higher plants, but not in those of algae where more than 20 tRNA genes have been examined (cf. Section III). It is still too early to assess the validity of this rule because of the limited number of genes considered. Chloroplast introns do not follow the GU...AG rule of higher eukaryotic introns, but they appear to conform—at least to some extent—to the intron secondary structure models of mitochondrial introns of lower eukaryotes (Michel and Dujon, 1983). It is intriguing that the sites of the chloroplast tRNA introns in the anticodon loop are variable amongst different chloroplast tRNA genes in contrast to the constant splice site in nuclear tRNA (cf. Section III).

The ability of chloroplast DNA sequences to promote autonomous replication in yeast is also a trait shared by eukaryotic nuclear DNA, but not by prokaryotic DNA (Stinchcomb *et al.*, 1980). Finally, the presence of short repetitive elements interspersed throughout several, but not all, chloroplast genomes, (Bedbrook and Kolodner, 1978; Rochaix, 1972, 1981) is also reminiscent of eukaryotic features.

Whatever the origin of chloroplasts may be, these properties clearly show that chloroplast genomes share both prokaryotic and eukaryotic features. The view that they are simply the remnants of entrapped prokaryotic genomes appears to be an oversimplification. Because the selective pressures experienced by cellular organelles must have been considerably different from those operating on free living organisms, the evolution of organellar genomes is likely to have followed a unique course.

The recent discovery of chloroplast DNA sequences in the nuclear and mitochondrial genomes raises challenging questions. At what rate do these transfers between different cellular compartments occur? Do they represent a continuation of an evolutionary trend of gene compartmentalization in eukaryotic cells or do they have a functional role during plant development? In what direction did the transfer occur originally? For example, were all the mitochondrial DNA sequences homologous to chloroplast DNA transferred from the chloroplast to the mitochondria or did some of them originate in the mitochondria before being moved to the chloroplast? What are the mechanisms which promote transfer of DNA between organelles and nucleocytoplasm? The presence of chloroplast DNA in mitochondrial genomes also raises new questions on the genetic interactions between chloroplasts and mitochondria. Are there genes which are normally present and active in both organelles? Results obtained in the next few years should bring some interesting and exciting answers to these questions, but will undoubtedly, in turn, raise new questions.

Acknowledgments

I am grateful to R. B. Hallick, H. J. Bohnert, R. G. Herrmann, C. Lemieux, D. M. Lonsdale, and J. D. Palmer for communicating unpublished results. I thank J. Erickson, M. Goldschmidt-Clermont, R. J. Spreitzer, J. Surzycki, and S. J. Surzycki for many helpful suggestions. This work was supported by Grant 3.258.082 from the Swiss National Science Foundation.

References

Abelson, J. (1979). *Annu. Rev. Biochem.* **48,** 1035–1069.

Allet, B., and Rochaix, J. D. (1979). *Cell* **18,** 55–60.

Alt, J., Winter, P., Sebald, W., Moser, J. G., Schedl, R., Westhoff, P., and Herrmann, R. G. (1983a). *Curr. Genet.* **7,** 129–138.

Alt, J. Westhoff, P., Sears, B. B., Nelson, N., Hurt, E., Hauska, G., and Herrmann, R. G. (1983b). *EMBO J.,* **2,** 979–986.

Anderson, S., Bankier, A. T., Barrell, B. G., de Bruijn, M. H. L., Coulson, A. R., Drouin, J., Eperon, I. C., Nierlich, D. P., Roe, B. A., Sanger, F., Schreier, P. H., Smith, A. J. H., Staden, R., and Young, I. G. (1981). *Nature (London)* **290,** 457–465.

Apel, K., and Bogorad, L. (1976). *Eur. J. Biochem.* **67,** 615–620.

Barrell, B. G., Air, G. M., and Hutchinson, C. A. (1976). *Nature (London)* **264,** 34–41.

Barrell, B. G., Bankier, A. T., and Drouin, J. (1979). *Nature (London)* **282,** 189–194.

Baur, E. (1909). *Z. Vererbungs.* **1,** 330–351.

Bedbrook, J. R., and Bogorad, L. (1976). *Proc. Natl. Acad. Sci. U.S.A.* **73,** 4309–4313.

Bedbrook, J. R., and Kolodner, R. (1979). *Annu. Rev. Plant Physiol.* **30,** 593–620.

Bedbrook, J., Link, G., Coen, D. M., Bogorad, L., and Rich, A. (1978). *Proc. Natl. Acad. Sci. U.S.A.* **75,** 3060–3064.

Bennoun, P. (1981). *FEBS Lett.* **136,** 1–2.

Bennoun, P., Delepelaire, P., and Delosme, M. (1981). *Curr. Genet.* **3,** 251–253.

Berry-Lowe, S. L., McKnight, T. D., Shah, D. M., and Meagher, R. (1982). *J. Mol. Appl. Genet.* **1,** 483–498.

Bibb, M. J., Van Etten, R. A., Wright, C. T., Walberg, M. W., and Clayton, D. A. (1981). *Cell* **26,** 167–180.

Blanc, H., and Dujon, B. (1981). *In* "Mitochondrial Genes" (P. P. Slonimski *et al.,* eds.). Cold Spring Harbor Laboratory, Cold Spring Harbor, New York, in press.

Blanc, M., Briat, J. -F., and Laulhère, J. P. (1981). *Biochim. Biophys. Acta* **655,** 374–382.

Bogorad, L. (1975). *Science* **188,** 891–898.

Bohnert, H. J., and Löffelhardt, W. (1982). *FEBS Lett.* **150,** 403–406.

Bohnert, H. J., Driesel, A. J., Crouse, E. J., Gordon, K., Herrmann R. G., Steinmetz, A., Mubumbila, M., Keller, M., Burkard, G., and Weil, J. H. (1979). *FEBS Lett.* **103,** 52–56.

Bohnert, H. J., Crouse, E. J., and Schmitt, J. M. (1982). *In* "Encycl. Plant Physiology: Nucleic Acids and Proteins in Plants" (D. Boulter and B. Parthier eds.), pp. 475–530. Vol. 14B. Springer-Verlag, Berlin and New York.

Bohnert, H. J., Michalowski, C., Koller, B., Delius, H., Mucke, H., and Löffelhardt, W. (1983). *In* "Endocytobiology II" (H. E. A. Schenk and W. Schwemmler, eds.). De Gruyter, New York, in press.

Bottomley, W. (1982). *In* "Methods in Chloroplast Molecular Biology" (M. Edelman, R. B. Hallick, and Chua, N. H., eds.), pp. 581–604. Elsevier, Amsterdam.

Bowman, C. M., and Dyer, T. (1979). *Biochem. J.* **183,** 605–613.

Bowman, C. M., Koller, B., Delius, H., and Dyer, T. A. (1981). *Mol. Gen. Genet.* **183,** 93–101.

Breathnach, R., Benoist, C., O'Hare, K., Gannon, F., and Chambon, P. (1978). *Proc. Natl. Acad. Sci. U.S.A.* **75,** 4853–4857.

Breitenberger, C. A., Graves, M. C., and Spremulli, L. L. (1979). *Arch. Biochem. Biophys.* **194,** 265–270.

Briat, J. -F., and Mache, R. (1980). *Eur. J. Biochem.* **111,** 503–509.

Briat, J. -F., Laulhère, J.-P., and Mache, R. (1979). *Eur. J. Biochem.* **98,** 285–292.

Briat, J. -F., Gigot, C., Laulhère, J. -P., and Mache, R. (1982a). *Plant Physiol.* **69,** 1205–1211.

Briat, J. -F., Dron, M., Loiseaux, S., and Mache, R. (1982b). *Nucleic Acids Res.* **10,** 6865–6877.

Broach, J. R., Guarascio, V. R., and Jayaram, M. (1982). *Cell* **29,** 227–234.

Brosius, J., Dull, T. J., Sleeter, D. D., and Noller, H. F. (1981). *J. Mol. Biol.* **148,** 107–127.

Bunger, W., and Feierabend, K. (1980). *Planta* **149,** 163–169.

Burkard, G., Canaday, J., Crouse, E., Guillemaut, P., Imbault, P., Keith, G., Keller, M., Mubumbila, M., Osario, L., Sarantoglou, V., Steinmetz, A., and Weil, J. H. (1980). *In* "Genome Organization and Expression in Plants" (J. Leaver, ed.), pp. 313–320. Plenum, New York.

Calos, M. P., and Miller, J. H. (1980). *Cell* **20,** 579–595.

Chiang, K. S., and Sueoka, N. (1967). *Proc. Natl. Acad. Sci. U.S.A.* **57,** 1506–1513.

Chiang, K. S., Friedman, E., Malavasic, M. J., Lin Feng, M., Eves, E., Feng, T. -Y., and Swinton, D. C. (1981). *Ann. N.Y. Acad. Sci.* **361,** 219–247.
Chu, N. M., and Tewari, K. K. (1982). *Mol. Gen. Genet.* **186,** 23–32.
Chu, N. M., Oishi, K. K., and Tewari, K. K. (1981). *Plasmid* **6,** 279–292.
Chua, N. H., and Gillham, N. W. (1977). *J. Cell Biol.* **74,** 441–452.
Ciferri, O., di Pasquale, G., and Tiboni, O. (1979). *Eur. J. Biochem.* **102,** 331–335.
Coen, D. M. (1982). *In* "Methods in Chloroplast Molecular Biology" (M. Edelman, R. B. Hallick, and N. H. Chua, eds.), pp. 581–604. Elsevier, Amsterdam.
Boen, D. M., Bedbrook, J. R., Bogorad, L., and Rich, A. (1977). *Proc. Natl. Acad. Sci. U.S.A.* **74,** 5487–5491.
Coleman, A. W. (1978). *Exp. Cell Res.* **114,** 95–100.
Coleman, A. W. (1979). *J. Cell Biol.* **82,** 299–305.
Conde, M. F., Boynton, J. E., Gillham, N. W., Harris, E. H., Tingle, C. L., and Wang, W. L. (1975). *Mol. Gen. Genet.* **140,** 183–220.
Correns, C. (1909). *Z. Vererbungs.* **1,** 291–329.
De Heij, J. T., and Groot, G. S. P. (1981). *FEBS Lett.* **134,** 6–10.
De Heij, J. T., Lustig, H., Moeskops, D. J., Bovenberg, W. A., Bisanz, C., and Groot, G. S. P. (1983). *Curr. Genet.* **7,** 1–6.
De La Salle, H., Jacq, C., and Slonimski, P. P. (1982). *Cell* **28,** 721–731.
Deno, H., Kato, A., Shinozaki, K., and Sugiura, M. (1982). *Nucleic Acids Res.* **10,** 7511–7520.
Deno, H., Shinozaki, K., and Sugiura, M. (1983). *Nucleic Acids Res.* **11,** 2185–2191.
Driesel, A. J., Crouse, E. J., Gordon, K., Bohnert, H. J., Herrmann, R. G., Steinmetz, A., Mubumbila, M., Keller, M., Burkard, G., and Weil, J. H. (1979). *Gene* **6,** 285–306.
Driesel, A. J., Speirs, J., and Bohnert, H. J. (1980). *Biochem. Biophys. Acta* **610,** 297–310.
Dron, M., Rahire, M., and Rochaix, J. -D. (1982a). *J. Mol. Biol.* **162,** 775–793.
Dron, M., Rahire, M., and Rochaix, J. -D. (1982b). *Nucleic Acids Res.* **10,** 7609–7620.
Dron, M., Rahire, M., Rochaix, J. D., and Mets, L. (1983). *Plasmid* **9,** 321–324.
Dujon, B. (1980). *Cell* **20,** 185–197.
Dyer, T. A., and Bowman, M. (1979). *Biochem. J.* **183,** 595–604.
Edelman, M. (1981). *In* "The Biochemistry of Plants, Proteins and Nucleic Acids" (A. Marcus ed.), Vo. 6, pp. 249–301. Academic Press, New York.
Edelman, M., and Reisfeld, A. (1978). *In* "Chloroplast Development" (G. Akoyunoglou *et al.*, eds.), pp. 641–652. Elsevier, Amsterdam.
Edwards, K., and Kössel, H. (1981). *Nucleic Acids Res.* **9,** 2853–2869.
Edwards, K., Bedbrook, J., Dyer, T., and Kössel, H. (1981). *Biochem. Int.* **2,** 533–538.
El-Gewely, M. R., Lomax, M. L., Lau, E. T., Helling, R. B., Farmerie, W., and Barnett, W. E. (1981). *Mol. Gen. Genet.* **181,** 296–305.
Ellis, R. J., and Hartley, M. R. (1971). *Nature* (*London*) New Biol. **233,** 193–196.
Ellis, R. J., Highfield, P. E., and Silverthorne, J. (1977). *Proc. Int. Cong. Photosynth., 4th.* p. 497.
Eneas-Filho, J., Hartley, M. R., and Mache, R. (1981). *Mol. Gen. Genet.* **184,** 484–488.
Erickson, J. M., Schneider, M., Vallet, J. M., Dron, M., Bennoun, P., and Rochaix, J. D. (1984a). *In* "Advances in Photosynthesis Research" (C. Sybesma, ed.), Vol. IV, pp. 491–500. M. Nijhoff and W. Junk, The Hague.
Erickson, J. M., Rahire, M., Bennoun, P., Delepelaire, P., Diner, B., and Rochaix, J. D. (1984b). *Proc. Natl. Acad. Sci. U.S.A.* **81,** 3617–3621.
Farrelly, F., and Butow, R. A. (1983). *Nature* (*London*) **301,** 296–301.
Fox, T. (1983). *Nature* (*London*) **301,** 371–372.
Galloway, R. E., and Mets, L. (1982). *Plant Physiol.* **70,** 1673–1677.
Gatenby, A. A., Castleton, J. A., and Saul, N. W. (1981). *Nature* (*London*) **291,** 117–121.
Gellissen, G., Bradfield, J. Y., While, B. N., and Wyatt, G. R. (1983). *Nature* (*London*) **301,** 631–634.

Gilbert, W., and Dressler, D. (1968). *Cold Spring Harbor Symp. Quant Biol.* **33,** 473–484.
Gillham, N. W. (1978). "Organelle Heredity." Raven, New York.
Gillham, N. W., Boynton, J., and Harris, E. H. (1984). *In* "DNA and Evolution: Natural Selection and Genome Size" (T. Cavalier-Smith, ed.). Wiley, New York, in press.
Graf, L., Kössel, H., and Stutz, E. (1980). *Nature (London)* **286,** 908–910.
Graf, L., Roux, E., and Stutz, E. (1982). *Nucleic Acids Res.* **10,** 6369–6381.
Grant, D. M., Gillham, N. W., and Boynton, J. E. (1980). *Proc. Natl. Acad. Sci. U.S.A.* **77,** 6067–6071.
Gray, M. W., and Doolittle, W. F. (1982). *Microbiol. Rev.* **46,** 1–42.
Gray, P. W., and Hallick, R. B. (1978). *Biochemistry* **17,** 284–289.
Green, B. R., Muir, B. L., and Padmanabhan, U. (1977). *In* "Progress in Acetabularia Research" (C. F. L. Woodcock, ed.), pp. 107–122. Academic Press, New York.
Gruissem, W., Prescott, D. M., Greenberg, B. M., and Hallick, R. B. (1982). *Cell* **30,** 81–92.
Guertin, M., and Bellemare, G. C. (1979). *Eur. J. Biochem.* **96,** 125–129.
Guillemaut, P., and Weil, J. H. (1982). *Nucleic Acids Res.* **10,** 1653–1659.
Hallick, R. B. (1985). *In* "The Biology of Euglena IV" (D. E. Buctow, ed.). Academic Press, New York.
Hallick, R. B., Lipper, C., Richards, O. C., and Rutter, W. J. (1976). *Biochemistry* **15,** 3039–3040.
Hallick, R. B., Greenberg, B. M., Gruissem, W., Hollingsworth, M. J., Karabin, G. D., Narita, J. O., Nickoloff, J. A., Passavant, C. W., and Stiegler, G. L. (1983). *In* "Structure and Function of Plant Genomes" (O. Ciferri, ed.), pp. 155–166. Plenum, New York.
Heinhorst, S., and Shively, J. (1983) *Nature (London)* **304,** 373–374.
Heizmann, P., Hussein, Y., Nicolas, P., and Nigon, V. (1982). *Curr. Genet.* **5,** 9–15.
Herrmann, R. G., and Possingham, J. V. (1980). *In* "Chloroplasts" (J. Reinert, ed.), pp. 45–96. Springer-Verlag, Berlin and New York.
Herrmann, R. G., Seyer, P., Schedl, R., Gordon, K., Bisanz, C., Winter, P., Hildebrandt, J. W., Wlaschek, M., Alt, J., and Sears, B. B. (1980). *Colloq. Ges. Biol. Chem.* **31,** 1–21.
Hirschberg, J., and McIntosh, L. (1983). *Science* **222,** 1346–1349.
Hollingsworth, M. J., and Hallick, R. B. (1982). *J. Biol. Chem.* **257,** 12795–12799.
Holmquist, R., Jukes, T. H., and Pangburn, S. (1973). *J. Mol. Biol.* **78,** 91–116.
Howe, C. J., Auffret, A. D., Doherty, A., Bowman, C. M., Dyer, T. A., and Gray, J. C. (1982a) *Proc. Natl. Acad. Sci. U.S.A.* **79,** 6903–6907.
Howe, C. J., Bowman, C. M., Dyer, T. A., and Gray, J. C. (1982b). *Mol. Gen. Genet.* **186,** 525–530.
Hübscher, U., Spanos, A., Albert, W., Grummt, F., and Banks, G. R. (1981). *Proc. Natl. Acad. Sci. U.S.A.* **78,** 6771–6775.
Jacobs, H. T., Posakony, J. W., Grula, J. W., Roberts, J. W., Xin, J. -H., Britten, R. J., and Davidson, E. H. (1983). *J. Mol. Biol.* **165,** 603–632.
James, T. W., and Jope, C. (1978). *J. Cell Biol.* **79,** 623–630.
Jenni, B., and Stutz, E. (1978). *Eur. J. Biochem.* **88,** 127–134.
Jenni, B., and Stutz, E. (1979). *FEBS Lett.* **102,** 95–99.
Jenni, B., Fasnacht, M., and Stutz, E. (1981). *FEBS Lett.* **125,** 175–179.
Jolly, S. O., and Bogorad, L. (1979). *Proc. Natl. Acad. Sci. U.S.A.* **77,** 822–826.
Jolly, S. O., McIntosh, L., Link, G., and Bogorad, L. (1981). *Proc. Natl. Acad. Sci. U.S.A.* **78,** 6821–6825.
Karabin, J., and Hallick, R. B. (1983). *J. Biol. Chem.* **258,** 5512–5518.
Kashdan, M. A., Pirtle, R. M., Pirtle, I. L., Calagan, J. L., Vreman, H. J., and Dudock, B. S. (1980). *J. Biol. Chem.* **255,** 8831–8835.
Kato, A., Shimada, H., Kusuda, M., and Sugiura, M. (1981). *Nucleic Acids Res.* **9,** 5601–5607.
Keller, M., Burkhard, G., Bohnert, H. J., Mubumbila, M., Gordon, K., Steinmetz, A., Heiser, D., Crouse, E. J., and Weil, J. H. (1980). *Biochem. Biophys. Res. Commun.* **95,** 47–54.

Keller, M., Rutti, B., and Stutz, E. (1982). *FEBS Lett.* **149,** 133–137.
Keller, S. J., and Ho, C. (1981). *Int. Rev. Cytol.* **69,** 157–190.
Kidd, G., and Bogorad, L. (1980). *Biochim. Biophys. Acta* **609,** 14–30.
Koch, W., Edwards, K., and Kössel, H. (1981). *Cell* **25,** 203–213.
Koller, B., and Delius, H. (1980). *Mol. Gen. Genet.* **178,** 261–269.
Koller, B., and Delius, H. (1982a). *EMBO J.* **1,** 995–998.
Koller, B. and Delius, H. (1982b). *Mol. Gen. Genet.* **188,** 305–308.
Kolodner, R., and Tewari, K. K. (1975a). *J. Biol. Chem.* **250,** 8840–8847.
Kolodner, R., and Tewari, K. K. (1975b). *Nature* (*London*) **256,** 708–711.
Kolodner, R., and Tewari, K. K. (1979). *Proc. Natl. Acad. Sci. U.S.A.* **76,** 41–45.
Krebbers, E. T., Larrinua, I. M., McIntosh, L., and Bogorad, L. (1982). *Nucleic Acids Res.* **10,** 4985–5002.
Kroon, A. M., Pepe, G., Bakker, H., Hultrop, M., Bollen, J. E., van Bruggen, E. F. J., Cantore, P., Terpstra, P., and Saccone, C. (1977). *Biochim. Biophys. Acta* **739,** 114–121.
Lemieux, C., Turmel, M., and Lee, R. W. (1981). *Curr. Genet.* **2,** 139–147.
Lemieux, C., Turmel, M., Seligy, V. L., and Lee, R. W. (1984). *Proc. Natl. Acad. Sci. U.S.A.* **81,** 1164–1168.
Lerbs, S., Briat, J. -F., and Mache, R. (1983). *Plant Molec. Biol.* **2,** 67–74.
Link, G. (1981). *Nucleic Acids Res.* **9,** 3681–3694.
Link, G. (1982). *Planta* **154,** 81–86.
Link, G., Coen, D. M., and Bogorad, L. (1978). *Cell* **15,** 725–731.
Lonsdale, D. M., Hodge, T. P., Howe, C. J., and Stern, D. B. (1983). *Cell,* **34,** 1007–1014.
Loppes, R., and Denis, C. (1983). *Current Genet.* **7,** 473–480.
Lorimer, G. H. (1981). *Biochemistry* **20,** 1236–1240.
Lund, E., Dahlberg, J. E., Lindahl, L., Jaskunas, S. R., Dennis, P. P., and Nomura, M. (1976). *Cell* **7,** 165–177.
Machatt, M. A., Ebel, J. P., and Branlant, C. (1981). *Nucleic Acids Res.* **9,** 1533–1549.
Macino, G., Coruzzi, G., Nobrega, F. G., Li, M., and Tzagaloff, A. (1979). *Proc. Natl. Acad. Sci. U.S.A.* **76,** 3784–3785.
MacKay, R. M. (1981). *FEBS Lett.* **123,** 17–18.
McIntosh, L., Poulsen, C., and Bogorad, L. (1980). *Nature* (*London*) **288,** 556–560.
Malnoe, P. M., and Rochaix, J. -D. (1978). *Mol. Gen. Genet.* **166,** 269–275.
Malnoe, P. M., Rochaix, J. -D., Chua, N. H., and Spahr, P. -F. (1979). *J. Mol. Biol.* **133,** 417–434.
Manning, J. E., and Richards, O. C. (1972a). *Biochim. Biophys. Acta* **259,** 285–296.
Manning, J. E., and Richards, O. C. (1972b). *Biochemistry* **11,** 2036–2043.
Margulis, L. (1967). *J. Theor. Biol.* **14,** 22–25.
Matsuda, Y., and Surzycki, S. J. (1980). *Mol. Gen. Genet.* **180,** 463.
Mereschkowsky, C. (1905). *Biol Zentralbl.* **25,** 593–604.
Mets, L. J. (1980). *Curr. Genet.* **2,** 193–209.
Mets, L. J., and Geist, L. J. (1983). *Genetics* **105,** 559–579.
Metzlaff, M., Börner, T., and Hageman, R. (1981). *Theor. Appl. Genet.* **60,** 37–41.
Michel, F., and Dujon, B. (1982). *Biochimie* **64,** 867–881.
Michel, F., and Dujon, B. (1983). *EMBO J.* **2,** 33–38.
Monroy, G., Spencer, E., and Hurwitz, J. (1978). *J. Biol. Chem.* **253,** 4490–4498.
Murray, C. L., and Rabinowitz, J. C. (1982). *J. Biol. Chem.* **257,** 1053–1062.
Myers, M., Grant, D. M., Rabert, D. K., Harris, E. H., Boynton, J. E., and Gillham, N. W. (1982). *Plasmid* **7,** 133–151.
Nelson, N. (1981). *Curr. Top. Bioenerg.* **11,** 1–33.
Netzker, R., Köchel, H. G., Basak, N., and Küntzel, H. (1982). *Nucleic Acids Res.* **10,** 4783–4794.

Nickoloff, J. A. and Hallick, R. B. (1982). *Nucleic Acids. Res.* **10,** 8191–8210.
Oppenheim, D. S., and Yanofsky, C. (1980). *Genetics* **95,** 785–795.
Orozco, E., and Hallick, R. . (1982a). *J. Biol. Chem.* **257,** 3258–3264.
Orozco, E., and Hallick, R. B. (1982b). *J. Biol. Chem.* **257,** 3265–3275.
Orozco, E. M., Rushlow, K. E., Dodd, J. R., and Hallick, R. B. (1980). *J. Biol. Chem.* **255,** 10997–11003.
Palmer, J. D. (1982). *Nucleic Acids Res.* **10,** 1593–1605.
Palmer, J. D. (1983). *Nature (London)* **301,** 92–93.
Palmer, J. D., and Stein, D. B. (1982). *Curr. Genet.* **5,** 165–170.
Palmer, J. D., and Thompson, W. F. (1981). *Proc. Natl. Acad. Sci. U.S.A.* **78,** 5533–5537.
Palmer, J. D., and Thompson, W. F. (1982). *Cell* **29,** 537–550.
Palmer, J. D., Aldrich, K. J., and Thompson, W. F. (1984). Submitted.
Pfister, K., Steinback, K. E., Gardner, G., and Arntzen, C. J. (1981). *Proc. Natl. Acad. Sci. U.S.A.* **78,** 981–985.
Platt, T., and Yanofsky, C. (1975). *Proc. Natl. Acad. Sci. U.S.A.* **72,** 2399–2403.
Ramierz, J. L., and Dawid, I. B. (1978). *J. Mol. Biol.* **119,** 133–146.
Ravel-Chapuis, P., Heizmann, P., and Nigon, V. (1982). *Nature (London)* **300,** 78–81.
Rawson, J. R., Kushner, S. R., Vapnek, D., Alton, N. K., and Boerma, C. L. (1978). *Gene* **3,** 191–209.
Rochaix, J. D. (1972). *Nature (London)* New Biol. **238,** 76–78.
Rochaix, J. D. (1978). *J. Mol. Biol.* **126,** 597–617.
Rochaix, J. D. (1981). *Experientia* **37,** 323–332.
Rochaix, J. D., and Darlix, J. L. (1982). *J. Mol. Biol.* **159,** 383–395.
Rochaix, J. D., and Malnoe, P. M. (1978). *Cell* **15,** 661–670.
Rochaix, J. D., Dron, M., Rahire, M., Boissel, J. M., and van Dillewjin, J. (1983a). *In* "Structure and Function of Plant Genomes" (O. Ciferri, ed.), pp. 205–212. Plenum, New York.
Rochaix, J. D., Dron, M., Schneider, M., Vallet, J. M., and Erickson, J. (1983b). *Miami Winter Symp., Adv. Gene Technol. Mol. Genet. Plants Anim., 15th,* 81–100.
Rochaix, J. D., Rahire, M., and van Dillewijn, J. (1984). *Cell* **36,** 925–931.
Roizman, J. R., Guarascio, V. R., and Jayaram, M. (1982). *Cell* **29,** 227–234.
Rosenberg, M., and Court D. (1979). *Annu. Rev. Genet.* **13,** 319–353.
Rouvière-Yaniv, J., and Gros, F. (1975). *Proc. Natl. Acad. Sci. U.S.A.* **72,** 3428–3432.
Rouz, E., Graf, L., and Stutz, E. (1983). *Nucleic Acids Res.* **11,** 1957–1968.
Rushlow, K. E., Orozco, E. M., Lipper, C., and Hallick, R. B. (1980). *J. Biol. Chem.* **255,** 3786–3792.
Sager, R. (1954). *Proc. Natl. Acad. Sci. U.S.A.* **40,** 356–363.
Sager, R. (1977). *Adv. Genet.* **19,** 287–340.
Sala, F., Amileni, A. R., Parisi, B., and Spadari, S. (1980). *Eur. J. Biochem.* **112,** 211–217.
Saraste, M., Gay, N. J., Eberle, A., Runswick, M. J., and Walker, J. E. (1981). *Nucleic Acids Res.* **9,** 5287–5296.
Schiemann, J., Wollgiehn, R., and Parthier, B. (1977). *Biochem. Physiol. Pflanzen* **171,** 474–478.
Schloss, J. V., Stringer, C. D., and Hartman, F. C. (1978). *J. Biol. Chem.* **253,** 5707–5711.
Schlunegger, J., Fasnacht, M., Stutz, E., Koller, B., and Delius, H. (1983). *Biochim. Biophys. Acta* **478,** 128–145.
Schwarz, Z., and Kössel, H. (1980). *Nature (London)* **283,** 739–742.
Schwarz, Z., Jolly, S. O., Steinmetz, A., and Bogorad, L. (1981a). *Proc. Natl. Acad. Sci. U.S.A.* **78,** 3423–3427.
Schwarz, Z., Kössel, H., Schwarz, E., and Bogorad, L. (1981b). *Proc. Natl. Acad. Sci. U.S.A.* **78,** 4748–4752.
Scott, T. N., and Simon, M. I. (1982). *Mol. Gen. Genet.* **188,** 313–321.

Sebald, W., and Wachter, E. (1978). *Mosbacher Colloq. Energy Conversion Biol. Membr. 29th* p. 228.
Seyer, P., Kowallik, K. V., and Herrmann, R. G. (1981). *Curr. Genet.* **3,** 189–204.
Shine, J., and Dalgarno, L. (1974). *Proc. Natl. Acad. Sci. U.S.A.* **71,** 1342–1346.
Shinozaki, K., and Sugiura, M. (1982a). *Gene* **20,** 91–102.
Shinozaki, K., and Sugiura, M. (1982b). *Nucleic Acids Res.* **10,** 4923–4934.
Shinozaki, K., Deno, H., Kato, A., and Sugiura, M. (1983). *Gene* **24,** 147–155.
Siedlecki, J., Zimmermann, W., and Weissbach, A. (1983). *Nucleic Acids Res.* **11,** 1523–1536.
Silverthorne, J., and Ellis, R. J. (1980). *Biochem. Biophys. Acta* **607,** 319–330.
Smith, H. J., and Bogorad, L. (1974). *Proc. Natl. Acad. Sci. U.S.A.* **71,** 4839–4842.
Sonneborn, T. M. (1967). *In* "Heritage of Mendel" (R. A. Brink, ed.), p. 375.
Spreitzer, R., and Mets, L. J. (1980). *Nature (London)* **285,** 114–115.
Sprinzl, M. and Gauss, D. H. (1982a). *Nucleic Acids Res.* **10,** r1–r55.
Sprinzl, M., and Gauss, D. H. (1982b). *Nucleic Acids Res.* **10,** r57–r81.
Steinback, K. E., McIntosh, L., Bogorad, L., and Arntzen, C. J. (1981). *Proc. Natl. Acad. Sci. U.S.A.* **78,** 7463–7467.
Steinmetz, A., Gubbins, E. J., and Bogorad, L. (1982). *Nucleic Acids Res.* **10,** 3027–3037.
Stern, D. B., and Lonsdale, D. M. (1982). *Nature (London)* **299,** 698–702.
Stern, D. B., Palmer, J. D., Thompson, W. F., and Lonsdale, D. M. (1983). *UCLA Symp. Mol. Cell. Biol. New Ser.* **12,** 445–446.
Stiegler, G. L., Matthews, H. M., Bingham, S. E., and Hallick, R. B. (1982). *Nucleic Acids Res.* **10,** 3427–3444.
Stinchcomb, D. T., Thomas, M., Kelly, J., Selker, E., and Davis, R. W. (1980). *Proc. Natl. Acad. Sci. U.S.A.* **77,** 4559–4563.
Stringer, C. D., and Hartman, F. C. (1978). *Biochem. Biophys. Res. Commun.* **80,** 1043–1048.
Surzycki, S. J. (1969). *Proc. Natl. Acad. Sci. U.S.A.* **63,** 1327–1334.
Surzycki, S. J., and Gillham, N. W. (1971). *Proc. Natl. Acad. Sci. U.S.A.* **68,** 1301–1306.
Surzycki, S. J., and Rochaix, J. D. (1971). *J. Mol. Biol.* **62,** 89–109.
Surzycki, S. J., and Shellenbarger, D. L. (1976). *Proc. Natl. Acad. Sci. U.S.A.* **73,** 3961–3965.
Takaiwa, F., and Sugiura, M. (1980). *Mol. Gen. Genet.* **180,** 1–4.
Takaiwa, F., and Sugiura, M. (1982a). *Nucleic Acids Res.* **10,** 2665–2676.
Takaiwa, F., and Sugiura, M. (1982b). *Eur. J. Biochem.* **124,** 13–19.
Tilney-Bassett, R. A. E. (1978). *In* "The Plastids" (J. T. Kirk and R. A. E. Tilney-Bassett eds.), p. 251. Elsevier, Amsterdam.
Tohdoh, N., and Sugiura, M. (1982). *Gene* **17,** 213–218.
Tohdoh, N., Shinozaki, K., and Sugiura, M. (1981). *Nucleic Acids Res.* **9,** 5399–5406.
Vallet, J. M., Rahire, M., and Rochaix, J. D. (1984). *EMBO J.* **3,** 415–421.
Van den Boogaart, P., Samallo, J., and Agsteribbe, E. (1982). *Nature (London)* **298,** 187–189.
Van Ee, J. H., Vos, Y. J., Bohnert, H. J., and Planta, R. J. (1982). *Plant Mol. Biol.* **2,** 117–131.
Waddell, J., Wang, W. -M., and Wu, M. (1984). *Nucleic Acids Res.* **12,** 3843–3856.
Watson, J. C., and Surzycki, S. J. (1982). *Proc. Natl. Acad. Sci. U.S.A.* **79,** 2264–2267.
Watson, J. C., and Surzycki, S. J. (1983). *Curr. Genet.* **7,** 201–210.
Weil, J. H., Guillemaut, P., Burkhard, G., Canaday, J., Mubumbila, M., Osorio, M. L., Keller, M., Gloecker, R., Steinmetz, A., Keith, G., Heiser, D., and Crouse, E. J. (1981). *In* "Photosynthesis" (G. Akoyunoglov, ed.), Vol. 5, pp. 777–786. Balbien, International Science, Philadelphia, Pennsylvania.
Weil, J. H., Mubumbila, M., Kuntz, M., Keller, M., Steinmetz, A., Crouse, E. J., Burkard, G., Guillemaut, P., Selden, R., McIntosh, L., Bogorad, L., Löffelhardt, W., Mucke, H., and Bohnert, H. J. (1982). *In* "Cell Function and Differentiation" (G. Akoyunoglou, ed.), pp. 321–331. Liss, New York.

Westhoff, J. C., Nelson, N., Bünemann, H., and Herrmann, R. G. C. (1981). *Curr. Genet.* **4,** 109–120.

Westhoff, P., Alt, J., Nelson, N., Bottomley, W., Bünemann, H., and Herrmann, R. G. (1983a). *Plant Mol. Biol.* **2,** 95–107.

Westhoff, P., Alt, J., and Herrmann, R. G. (1983b). *EMBO J.* **2,** 2229–2237.

Westhoff, P., Alt, J., Nelson, N., and Herrman, R. G. (1984). *Mol. Gen Genet.,* in press.

Whitfeld, P. R., and Bottomley, W. (1980). *Biochem. Int.* **1,** 172–178.

Whitfeld, P. R., and Bottomley, W. (1983). *Annu. Rev. Plant Physiol.* **34,** 279–310.

Whitfeld, P. R., Leaver, C. J., Bottomley, W., and Atchison, B. A. (1978a). *Biochem. J.* **175,** 1103–1112.

Whitfeld, P. R., Herrmann, R. G., and Bottomley, W. (1978b). *Nucleic Acids Res.* **5,** 1741–1751.

Willey, D. L., Huttly, A. K., Philips, A. L., and Gray, J. C. (1983). *Mol. Gen. Genet.* **189,** 85–89.

Woessner, J. P., Masson, A., Harris, E. H., Bennoun, P., Gillham, N. W., and Boynton, J. E. (1984). *Plant Molec. Biol.,* in press.

Wolstenholme, D. R., Goddard, J. M., and Fauron, C. M. R. (1979). *In* ''Extrachromosomal DNA'' (D. J. Cummings, P. Borst, I. B. Dawid, S. M. Weissman, and C. F. Fox, eds.), pp. 409–425. Academic Press, New York.

Wright, R. M., and Cummings, D. J. (1983). *Nature (London)* **302,** 86–88.

Zakian, V. (1981). *Proc. Natl. Acad. Sci. U.S.A.* **78,** 3129–3132.

Zech, M., Hartley, M. R., and Bohnert, H. J. (1981). *Curr. Genet.* **4,** 37–46.

Zimmermann, W., and Weissbach, A. (1982). *Biochemistry* **21,** 3334–3343.

Zurawski, G., Perrot, B., Bottomley, W., and Whitfeld, P. R. (1981). *Nucleic Acids Res.* **9,** 3251–3270.

Zurawski, G., Bottomley, W., and Whitfeld, P. R. (1982a). *Proc. Natl. Acad. Sci. U.S.A.* **79,** 6260–6264.

Zurawski, G., Bohnert, H. J., Whitfeld, P. R., and Bottomley, W. (1982b). *Proc. Natl. Acad. Sci. U.S.A.* **79,** 7699–7703.

Animal Mitochondrial DNA: An Extreme Example of Genetic Economy

GIUSEPPE ATTARDI

Division of Biology, California Institute of Technology, Pasadena, California

I. Introduction

The segregation in the special environment of mitochondria of a portion of the eukaryotic genome under the control of the nuclear genome represents a unique situation in nature. It is not surprising, therefore, that since its discovery more than 30 years ago (Ephrussi *et al.*, 1949; Mitchell and Mitchell, 1952), this genetic system has attracted the attention of many investigators. Thus, questions concerning the essential role of mitochondrial DNA (mtDNA) in the assembly of a functional inner mitochondrial membrane, the coordination of the nuclear and the mitochondrial genomes in the biogenesis of mitochondria, and the origin and evolution of mtDNA have been intensively pursued over the past three decades in a variety of systems, from yeast to man. It is, however, only in the last few years that the use of the new powerful DNA, RNA, and protein technologies has permitted a detailed analysis of the structure and expression of the mitochondrial

ISBN 0-12-364493-3

genome, providing some insights into the fundamental principles which govern its operation and evolution.

Although the mtDNA of only relatively few organisms has been analyzed in detail so far, a general picture has already emerged from these investigations. Thus, it is clear that the essential genetic functions of mtDNA have been in great part preserved in a vast spectrum of organisms at different evolutionary levels in the eukaryotic world. Table I lists the genes which have been identified so far in *Saccaromyces cerevisiae* and mammalian mtDNA. The sets of known genes in the two types of organisms are identical, with the exception that the yeast genome has a few more isoaccepting tRNA genes and the genes for subunit 9 of the ATPase complex and for a ribosome associated protein, which are absent in mammalian mtDNA. A substantial fraction of the informational content of the mitochondrial genome from both types of organisms is represented by unidentified reading frames (URFs). Most of the *S. cerevisiae* URFs are contained in introns of split genes and appear to code for proteins required for mRNA splicing, the so-called maturases. The URFs of the mammalian mtDNA code for proteins of as yet unknown function.

In contrast to the general conservation of genetic content, the structure and gene organization of the mitochondrial genome and the mechanisms of its expression have evolved in strikingly diverse ways in different organisms. Within this variety of forms, the mitochondrial genomes of all animal cells appear to

TABLE I
GENETIC CONTENT OF *S. cerevisiae* AND MAMMALIAN MITOCHONDRIAL DNA

	S. cerevisiae	Mammals
Ribosomal RNA genes	21 S rRNA 15 S rRNA	16 S rRNA 12 S rRNA
tRNA genes	~25 tRNA species	22 tRNA species
Identified protein coding genes		
Cytochrome *c* oxidase complex	Subunit I Subunit II Subunit III	Subunit I Subunit II Subunit III
Cytochrome *b*–c_1 complex	Apocytochrome *b*	Apocytochrome *b*
ATPase complex	Subunit 6 Subunit 9 aapl product?	Subunit 6 — URFA6L product?
Small ribosomal subunit	Ribosome-associated protein (var-1)	?
Unidentified reading frames (URFs)	≥10 (including at least two mRNA maturases)	8

have preserved the same basic pattern. In this review, I will present an up-to-date account of what is known about the structure, function, and evolution of animal mtDNA.

II. Unique Features and Diversity of the Mitochondrial Genetic Code

A dramatic example of the evolutionary individuality of the mitochondrial genetic system is provided by the unusual features of the genetic code and of the codon recognition pattern utilized in mitochondria from different organisms (Barrell *et al.*, 1980; Heckman *et al.*, 1980; Bonitz *et al.*, 1980; Clary and Wolstenholme, 1983a,b,c; de Bruijn, 1983). The most conspicuous among such features, one which has been found in all mitochondrial genomes analyzed in detail, i.e., those from animal cells, yeast, and filamentous fungi, is the use of UGA as a tryptophan codon rather than as a stop codon. Furthermore, there are genetic code differences between the individual mitochondrial genetic systems, even within the same philogenetic group (Fig. 1). Thus, in mitochondria from mammalian cells (Barrell *et al.*, 1980), *Xenopus laevis* (B. Roe, personal communication), *Drosophila* (Clary and Wolstenholme, 1983a, de Bruijn, 1983), and *S. cerevisiae* (Hudspeth *et al.*, 1982), but not in mitochondria from *Neurospora crassa* and *Aspergillus nidulans* (Browning and RajBhandary, 1982; Grisi *et al.*, 1982; Netzker *et al.*, 1982), AUA codes for methionine, rather than for isoleucine. In *S. cerevisiae* (Bonitz *et al.*, 1980), but not in filamentous fungi (Heckman *et al.*, 1980; Grisi *et al.*, 1982; Netzker *et al.*, 1982), the CUN triplets code for threonine instead of leucine. In mammalian, but not in fungal mitochondria, besides AUG, also AUA, AUU, and AUC can function as initiator codons; in this role, AUU and AUC may indeed be read by *N*-formyl-methionyl-tRNA and thus code for *N*-formyl-methionine, like AUG and AUA. There is in fact evidence that at least some of the mitochondrially synthesized polypeptides in mammalian cells (Steffens and Buse, 1979; Chomyn *et al.*, 1981), as in *S. cerevisiae* (Bianchetti *et al.*, 1971), are initiated with *N*-formyl-methionine, suggesting the possibility that initiation of protein synthesis in mitochondria requires *N*-formyl-methionyl-tRNA. In *Drosophila* also, AUA or AUAA and AUU can function as initiator codons in place of AUG (Clary *et al.*, 1982; Clary and Wolstenholme, 1983a,b; de Bruijn, 1983). A unique feature of the mammalian and *X. laevis* mitochondrial genetic code is the use of AGA and/or AGG as stop codons rather than as arginine codons (Anderson *et al.*, 1981, 1982; Bibb *et al.*, 1981; B. Roe, personal communication). By contrast, in *Drosophila* mitochondria, AGA appears to code for serine (Clary and Wolstenholme, 1983b,c; de Bruijn, 1983). The above discussed diversity of the mitochondrial genetic code in different organisms clearly indicates that the evolution of this genetic system has been subject to rather loose constraints.

Genetic code	Met$_m$ Anti-codon	Met$_m$ Codon	Met$_f$ Anti-codon	Met$_f$ Codon	Ile Anti-codon	Ile Codon	Leu Anti-codon	Leu Codon	Thr Anti-codon	Thr Codon	Arg Anti-codon	Arg Codon	Trp Anti-codon	Trp Codon	Ser Anti-codon	Ser Codon	Lys Anti-codon	Lys Codon	Termination codon
Universal	CAU	AUG	CAU	AUG	GAU	AUU AUC	UAA	UUA UUG	GGU	ACU ACC	GCG	CGU CGC	CCA	UGG	GGA	UCU UCC	UUU	AAA AAG	UAA UAG UGA
					*CAU	AUA	GAG	CUU CUC	UGU	ACA ACG	UCG	CGA CGG			UGA	UCA UCG			
							UAG	CUA CUG			UCU	AGA AGG			GCU	AGU AGC			
Mammals	CAU	AUA AUG	CAU	(AUU) (AUC) AUA AUG	GAU	AUU AUC	*UAA	UUA UUG	UGU	ACU ACC ACA ACG	UCG	CGU CGC CGA CGG	*UCA	UGA UGG	UGA	UCU UCC UCA UCG	*UUU	AAA AAG	UAA UAG AGA AGG
							UAG	CUU CUC CUA CUG							GCU	AGU AGC			
Drosophila	CAU	AUA AUG	CAU	(AUU) AUA(A) AUG	GAU	AUU AUC	UAA	UUA UUG	UGU	ACU ACC ACA ACG	UCG	CGU CGC CGA CGG	UCA	UGA UGG	UGA	UCU UCC UCA UCG	CUU	AAA AAG	UAA UAG
							UAG	CUU CUC CUA CUG							GCU	AGU AGC ? AGA			
Yeast	CAU	AUA AUG	CAU	AUG	GAU	AUU AUC	UAA	UUA UUG	UGU	ACU ACC ACA ACG	UCG	CGU CGC CGA CGG	*UCA	UGA UGG	UGA	UCU UCC UCA UCG	UUU	AAA AAG	UAA UAG
									UAG	CUU CUC CUA CUG	UCU	AGA AGG			GCU	AGU AGC			
Neurospora and Aspergillus	CAU	AUG	CAU	AUG	GAU	AUU AUC	*UAA	UUA UUG	UGU	ACU ACC ACA ACG	UCG	CGU CGC CGA CGG	*UCA	UGA UGG	UGA	UCU UCC UCA UCG	UUU	AAA AAG	UAA UAG
															GCU	AGU AGC			
					CAU	AUA	UAG	CUU CUC CUA CUG											

III. An Economical Decoding Mechanism

Another striking feature of the mitochondrial genetic system is its use of a new decoding mechanism, which allows reading of all codons by a smaller number of tRNAs than the minimum of 32 tRNAs required to read the genetic code according to the wobble mechanism (Crick, 1966). This new mechanism involves the existence of a "two out of three" base interaction between codon and anticodon (Lagerkvist, 1978) in the four-codon family boxes. In mammalian cells (Barrell *et al.*, 1980), *X. laevis* (B. Roe, personal communication), *S. cerevisiae* (Bonitz *et al.*, 1980), and *N. crassa* (Heckman *et al.*, 1980), in the eight family boxes with four codons for one amino acid, only one specific mitochondrial tRNA has been found, instead of two, as in the universal decoding mechanism (Fig. 1). In each case, this single tRNA has a U in the first position of the anticodon; it is assumed this U can pair with all four bases in the third position of the codons in each family box. This mechanism could not obviously operate in the nonfamily boxes, because it would lead to misreading. The observation that, in *N. crassa*, mitochondrial tRNA species for the family boxes have an unmodified U in the first position of the anticodon, while tRNAs specific for the two codons ending in purines in the non-family boxes, like $tRNA^{Leu}_{UUR}$ and $tRNA^{Trp}_{UGR}$, have an unknown modified U in the same position, has led to the proposal that this modification may be the mechanism preventing misreading of the two codons ending in pyrimidines in the nonfamily boxes (Heckman *et al.*, 1980). Support for this mechanism has come from the observation of a modified U (probably, 5-methoxy-carbonylmethyluridine) in the wobble position of the mitochondrial $tRNA^{Trp}_{UGR}$ from *S. cerevisiae* (Sibler *et al.*, 1980), $tRNA^{Lys}_{AAR}$ and $tRNA^{Trp}_{UGR}$ from rat liver (Randerath *et al.*, 1981), $tRNA^{Lys}_{AAR}$ from hamster cells (HsuChen *et al.*, 1984), and $tRNA^{Lys}_{AAR}$, $tRNA^{Trp}_{UGR}$, $tRNA^{Leu}_{UUR}$, and $tRNA^{Glu}_{GAR}$ from bovine liver (B. Roe, personal communication).

Twenty-four tRNAs would be required to translate the mitochondrial genetic code according to the decoding mechanism described above. The number of tRNA genes found in mammalian and fungal mtDNA is close to this figure. In the mammalian mitochondrial genetic code, a further simplification has occurred because of the total absence of AG^{A}_{G} arginine codons in the protein coding genes and the corresponding absence of a tRNA gene with an anticodon for these codons in the mtDNA sequence (Anderson *et al.*, 1981). Furthermore, only one $tRNA^{Met}$ gene has been found in mammalian and, so far, also in *Drosophila yacuba* mtDNA (D. Wolstenholme, personal communication). It is believed that a single gene can

FIG. 1. Distinctive features of the genetic code and pattern of codon recognition used in mitochondria from mammals, *Drosophila*, yeast, and filamentous fungi. The asterisked C and U represent modified bases. See text for details.

specify by secondary modification(s) of the primary transcript the methionine-tRNA ($tRNA^{Met}$) and the *N*-formyl-methionine-tRNA ($tRNA^{F\text{-}Met}$): both species of methionine-tRNA have been detected in mammalian mitochondria (Lynch and Attardi, 1976; Aujame and Freeman, 1979). Therefore, in mammalian, and probably, in general, animal cell mitochondria, a set of 22 tRNA genes can specify the tRNAs needed to support the organelle specific protein synthesis.

There is a case among mammalian mitochondrial tRNAs where a C in the first position of the anticodon, rather than a U, enables it to pair with A and G in third codon position. This is the case of the $tRNA^{Met}$ and the $tRNA^{F\text{-}Met}$, which, as mentioned above, are presumably products of the same tRNA gene. This gene has CAT as anticodon, and it must be assumed that an unmodified C or a modified C makes these tRNAs capable of reading both AUG and AUA; CAU may be able to read also AUU and AUC, if the $tRNA^{F\text{-}Met}$ is the only initiator tRNA used in mammalian mitochondria. The same base-pairing rule may possibly account for the capacity of the single $tRNA^{Lys}$ so far detected in *Drosophila* (Clary and Wolstenholme, 1983a; de Bruijn, 1983) and mosquito mitochondria (HsuChen *et al.*, 1983), which has the anticodon CUU, with the C apparently unmodified, to recognize both AAG and AAA.

Another unusual type of base pairing may be involved in the recognition of the AGA codon in insect mitochondria. With almost 80% of the *D. yacuba* mtDNA already sequenced (D. Wolstenholme, personal communication), no evidence yet has been found for a tRNA with the anticodon UCU corresponding to the putative serine codon AGA. It has been suggested that the anticodon GCU of the $tRNA^{Ser}_{AGY}$ may recognize AGA besides AGY (Clary and Wolstenholme, 1983c); a similar suggestion has been made for the $tRNA^{Ser}_{AGY}$ from mosquito mitochondria (Dubin *et al.*, 1983). The atypical structural features of this tRNA (see below) could presumably allow an unusual "G-A" wobble.

IV. Size and Physical Structure of Animal mtDNA

The mtDNA of all animal cells so far analyzed is in the form of circular molecules of relatively uniform length (Borst, 1972). A range of sizes between 15.7 and 19.5 kb has been reported for different species (Brown, 1983), with the variation within the same taxonomical group (for example, *Drosophila*) being as large as that between different taxonomical groups. In contrast to the relative constancy in size and structure, animal mtDNAs vary greatly in GC content, between a minimum of 21% in *Drosophila* species and a maximum of 46% in chicken (Brown, 1983). Furthermore, a strand bias in GT content, of various degree in different species, has been observed. This strand bias allows the separation of the two strands in an alkaline CsCl, with the density difference between the two strands being the largest (31–43 mg/ml) in warm-blooded vertebrates and the least in invertebrates (5–10 mg/ml) (Brown, 1983). An interesting feature of animal mtDNAs is the presence of a significant number of ribonucleotides incorporated throughout the genome (0.1–0.2% in vertebrate mtDNA). Although preferred sites of ribonucleotide substitution occur at the replication origins, presumably a remnant of RNA primer events, one finds alkali labile sites at other positions in the genome: the latter may be due to ribonucleotides misincorporated during replication (Brennicke and Clayton, 1981).

V. The Mammalian Mitochondrial Genome

A. General Gene Organization

The diversity in structure and gene organization of the mitochondrial genome in various organisms is best illustrated by a comparison of *S. cerevisiae* and mammalian mtDNA. *S. cerevisiae* mtDNA has a very loose gene organization, with its genes being separated by AT-rich spacers (Prunell and Bernardi, 1974, 1977); furthermore, several of these genes are discontinuous, with introns interrupting the coding sequences (Tabak *et al.*, 1983). The tRNA genes are in their majority clustered within 12% of the genome (Tabak *et al.*, 1983). This pattern contrasts strikingly with the extremely compact gene organization of the mammalian mitochondrial genome (Anderson *et al.*, 1981, 1982; Bibb *et al.*, 1981; Saccone *et al.*, 1981; Grosskopf and Feldmann, 1981a; Kobayashi *et al.*, 1981; Pepe *et al.*, 1983), as exemplified by the human mtDNA (Fig. 2). With the exception of a segment around the origin of replication, which corresponds to 5 to 7% of the genome and includes the D-loop, the entire length of mammalian mtDNA is saturated by genes situated on the two strands. In particular, the heavy (H)-strand, which is the main coding strand, contains the genes for the two rRNA species, 14 tRNA species and 12 significant (≥ 120 nt) reading frames, while the light (L)-strand contains the genes for 8 tRNA species and one significant reading frame.

In the H-strand, the rRNA and tRNA genes and the significant reading frames are in most cases butt-jointed to each other or separated by a few nucleotides, and there is a nearly complete absence of noncoding stretches. Thus, the initiator codon of each reading frame, which can be either AUG or AUA or AUU or AUC, follows immediately, or with an interval of a few nucleotides, the gene adjacent to it on its 5′ side; furthermore, most reading frames lack a termination codon and exhibit either a T or a TA following the last sense codon and immediately preceding the adjacent gene on the 3′ side. As will be discussed below (Section V,D,4), in these cases, completion of the termination (ochre) codon occurs at the time of RNA processing by polyadenylation of the mRNAs (Anderson *et al.*, 1981; Ojala *et al.*, 1981a). Two of the H-strand reading frames

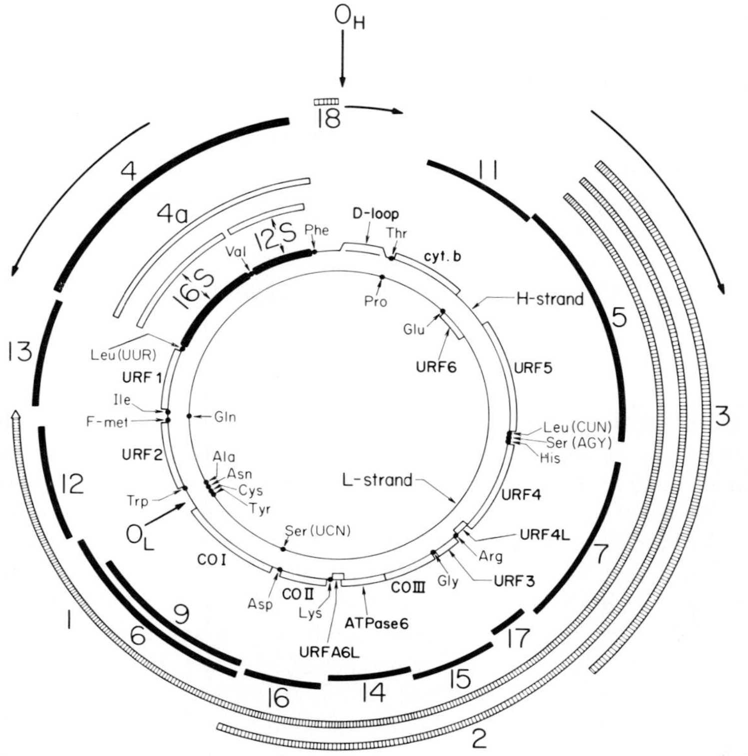

FIG. 2. Genetic and transcription maps of the HeLa cell mitochondrial genome. The two inner circles show the positions of the two rRNA genes, as derived from mapping and RNA sequencing experiments (Ojala *et al.*, 1980; Crews and Attardi, 1980), and those of the reading frames and tRNA genes, as derived from the mtDNA sequence (Anderson *et al.*, 1981). Mapping positions of the oligo(dT)-cellulose bound and nonbound H-strand transcripts are indicated, respectively, by black and white bars, those of the oligo(dT)-cellulose bound L-strand transcripts by hatched bars. Left and right arrows indicate the direction of H- and L-strand transcription, respectively. The vertical arrow, marked O_H, and the rightward arrow at the top indicate the location of the origin and the direction of H-strand synthesis; the arrow marked O_L indicates the origin of L-strand synthesis. COI, COII, and COIII: subunits I, II, and III of cytochrome *c* oxidase; URF, unidentified reading frame. Modified from Chomyn *et al.* (1983).

overlap their respective adjacent reading frame; in particular, URFA6L overlaps out of frame (by 46 nt in human mtDNA) the ATPase 6 gene, and URF4L overlaps out of frame (by 7 nt in human mtDNA) URF4. Overlapping by one nucleotide of contiguous tRNA genes has been found in mouse mtDNA for the H-strand $tRNA^{Ser}_{AGY}$ and $tRNA^{Leu}_{CUN}$ genes (Bibb *et al.*, 1981), and in human mtDNA for the L-strand $tRNA^{Tyr}$ and $tRNA^{Cys}$ genes (Anderson *et al.*, 1981). Between the H-strand genes and the L-strand genes there is no or very limited overlapping (by a few nucleotides); an exception is represented by the overlapping of URF5 and URF6, by 14 nt in mouse mtDNA (Bibb *et al.*, 1981) and by 17 nt in bovine mtDNA (Anderson *et al.*, 1982).

In contrast to the mitochondrial genes of yeast and filamentous fungi, the mammalian mitochondrial genes are continuous. Another distinctive feature of the mammalian mitochondrial gene organization is the scattered distribution of the tRNA genes, which again contrasts with the clustered arrangement of the majority of the tRNA genes in *S. cerevisiae* mtDNA. In the H-strand of the mammalian mtDNA, the tRNA genes separate with nearly absolute regularity the rRNA and protein coding genes.

B. Mitochondrial Genes and Their Products

1. *Transfer RNA Genes*

Twenty-two tRNA genes have been identified in human mtDNA (Anderson *et al.*, 1981), as well as in the other mammalian mtDNAs so far sequenced (Bibb *et al.*, 1981; Anderson *et al.*, 1982; Saccone *et al.*, 1981; Grosskopf and Feldmann, 1981b; Kobayashi *et al.*, 1981; Cantatore *et al.*, 1982; Pepe *et al.*, 1983). Their specificities agree with those of the charged tRNAs previously found to hybridize to human mtDNA (Lynch and Attardi, 1976) and rat mtDNA (Aujame and Freeman, 1979). The mammalian mitochondrial tRNAs are generally smaller than their cytoplasmic counterparts, ranging in size from 59 to 75 nt in man. In all cases, the 3′ terminal -CCA is not encoded in the DNA and must be added posttranscriptionally.

The structure of mammalian mitochondrial tRNAs reveals only a partial agreement with the pattern which is considered to be constant in all nonmitochondrial tRNAs (Sprinzl *et al.*, 1980), both as concerns the secondary structure and the invariant nucleotides. With one exception ($tRNA^{Ser}_{AGY}$), all mammalian mitochondrial tRNAs can be folded into a basic cloverleaf structure. Conserved are the 7 bp acceptor stem, the 5 bp anticodon stem, and the 7 nt anticodon loop; however, there are several instances of mismatches in the above mentioned stems, as well as in the "DHU" and "TΨC" stems. Some of the invariant tRNA nucleotides are preserved in mammalian mitochondrial tRNAs, like the presence of a U in the position preceding the anticodon (except in the $tRNA^{F\text{-}met}$, where there is a C), of a pyrimidine in the first position of the anticodon loop and of a purine in the position immediately 3′ to the anticodon. More numerous, however, are the deviations from the invariant structure. Thus, most tRNAs lack the constant G_{15}, the sequence G_{18}–G_{19} found so far in all tRNAs active in protein synthesis, and the universal G-T-Ψ-C-Pu-A sequence in the "TΨC" loop. The stem of the latter loop varies between 3 and the standard 5 bp; the loop itself deviates in general from the constant length of 7 nt, varying between 3 and 9 nt. Also, the "DHU" loop varies considerably in size, between 3 and 10 nt, while the "DHU" stem varies between 3 and 4 bp. The most extreme example of structure variability in the "DHU" loop region of the molecule is exhibited by the tRNA$^{Ser}_{AGY}$, which lacks completely the "DHU" loop and stem, a feature which is conserved in all mammalian mitochondrial tRNA$^{Ser}_{AGY}$ so far sequenced (de Bruijn *et al.*, 1980; Bibb *et al.*, 1981; Grosskopf and Feldman, 1981b; Anderson *et al.*, 1982). Bovine tRNA$^{Ser}_{AGY}$ has been shown to be chargeable with serine (Arcari and Brownlee, 1980; de Bruijn *et al.*, 1980). The invariant interaction between G_{15} and the constant pyrimidine at position 48 is presumably not possible in mammalian mitochondrial tRNAs. These tRNAs, therefore, appear to be stabilized by fewer tertiary interactions than the standard tRNAs, suggesting a greater evolutionary flexibility; the tRNA$^{Ser}_{AGY}$ represents an extreme case of this flexibility.

In human and hamster mitochondrial tRNAs, the level of methylation (all in bases) is about 30% of that of cytoplasmic tRNAs and with significant differences in methylation pattern (Attardi and Attardi, 1971; Davenport *et al.*, 1976).

Figure 3 shows three examples of human tRNAs, the tRNA^{Phe} (Crews and Attardi, 1980), which is representative of the majority of mitochondrial tRNA species, the tRNA$^{Leu}_{UUR}$ (Eperon *et al.*, 1980), which shows the least degree of divergence from the invariant pattern, and the tRNA$^{Ser}_{AGY}$ (de Bruijn *et al.*, 1980), which exhibits the most striking deviation from the standard pattern.

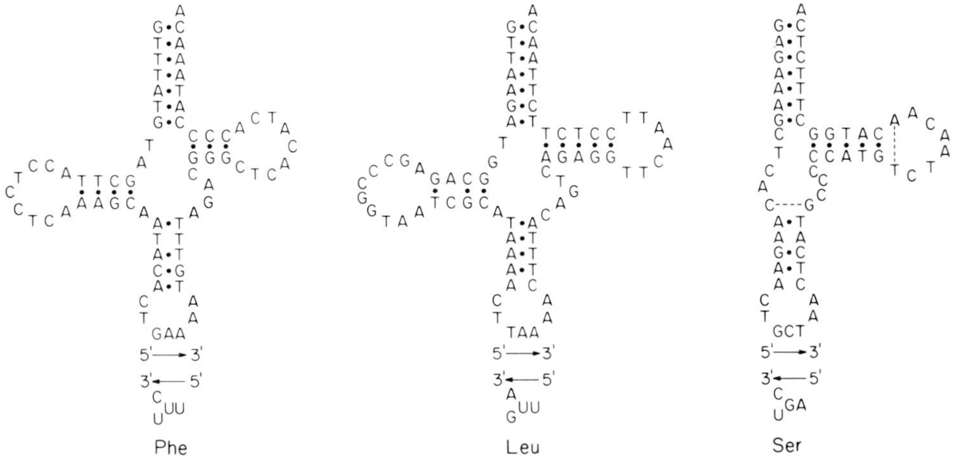

FIG. 3. The DNA sequences of the human mitochondrial tRNA^{Phe}, tRNA$^{Leu}_{UUR}$, and tRNA$^{Ser}_{AGY}$ genes are shown, arranged in the cloverleaf configuration.

2. *Ribosomal RNA Genes*

The two rRNA species encoded in mammalian mtDNA, 12 S rRNA and 16 S rRNA, are the smallest known high molecular weight rRNA species (954 and 1559 nt, respectively, in man), if one excludes the rRNAs of *Trypanosomatids* mitochondria (Simpson and Simpson, 1978; Hoeijmakers *et al.*, 1981). The 3′ ends of the mitochondrial rRNAs of hamster (Dubin *et al.*, 1981; Kotin and Dubin, 1984), human (Dubin *et al.*, 1982) and mouse (Van Etten *et al.*, 1983) origin have been analyzed in detail, and have revealed unusual features. Thus, both 12 S and 16 S rRNAs are oligoadenylated, exhibiting between one and 5 to 10 A residues at their 3′ ends. Furthermore, the 3′ ends of the 16 S rRNA from the three species mentioned above are characteristically ragged: although the majority of these ends correspond to one or two nucleotides immediately upstream of the $tRNA^{Leu}_{UUR}$ gene, a minor portion maps within the $tRNA^{Leu}_{UUR}$ gene itself (Dubin *et al.*, 1982; Van Etten *et al.*, 1983; Kotin and Dubin, 1984), up to 7 nt downstream of the 5′ end in the case of the mouse 16 S rRNA gene. The significance of these observations with regard to the mode of formation of the mammalian mitochondrial rRNAs will be discussed below (Section V,D,5).

Human and hamster mitochondrial rRNAs are methylated (Attardi and Attardi, 1971; Dubin and Taylor, 1978; Dubin *et al.*, 1978). In hamster cells, the level of this methylation [only in ribose residues in the large rRNA (on the average, about 3 methyl groups per molecule), in bases in the small rRNA (about 7 methyl groups per molecule)] is considerably lower than that of cytoplasmic rRNAs (77 and 45 methyl groups per molecule in the large and small rRNA species, respectively) (Dubin and Taylor, 1978; Dubin *et al.*, 1978). No pseudouridine has been detected in the hamster mitochondrial rRNA species, whereas 41 and 29 pseudouridines per molecule have been found, respectively, in the large and small cytoplasmic rRNA species from the same organism (Dubin and Taylor, 1978).

No 5 S rRNA equivalent has been found in the human, and in general animal, mitochondrial ribosomes. However, a 23-nt sequence is present at the 3′ end of the human 16 S rRNA molecule, that exhibits a 68% sequence homology and a similarity in structure to a portion of the *Bacillus subtilis* 5 S rRNA; these features and the location of the corresponding coding sequence relative to the mitochondrial rRNA genes, which is analogous to the mapping position of the 5 S rRNA gene in bacteria, have led to the suggestion that this segment represents a truncated 5 S rRNA gene which has become a part of the 16 S rRNA gene (Nierlich, 1982).

3. *Protein Coding Genes*

Thirteen significant reading frames, coding for proteins which cover a size range from about 70 to 610 amino acids, have been identified in all mammalian mtDNA so far sequenced; 12 of them are located in the H-strand and one in the

L-strand. Although other shorter reading frames have been found in one or the other mtDNA, they are not conserved in all mammalian mtDNAs, a finding which argues against their significance. Five of the 13 reading frames have been identified as to the specificity of the polypeptides they code for by a correlation of the DNA or RNA sequence with protein sequence data or with known yeast gene sequences: these are the genes for cytochrome *c* oxidase subunits I (COI), II (COII), and III (COIII), ATPase subunit 6 and cytochrome *b*. Table II lists the human mtDNA reading frames, their specificity, where this is known, the expected molecular weight of the encoded proteins, and the corresponding putative mRNAs. The unique features of these mRNAs will be described below (Section V,D,2.). About 60% of the protein coding sequences of mammalian mtDNA pertain to reading frames whose products have not been identified with any protein of known function, the so-called unidentified reading frames (URFs). With the exception of the smallest URF, URFA6L, the mitochondrial URFs have no obvious amino acid sequence homology to any of the yeast mtDNA reading frames. On the contrary, there is a partial homology between some of the human mitochondrial URFs (URF1, URF3, URF4, URF5, and URFA6L) and some of the URFs of mtDNA from filamentous fungi (Grisi *et al.*, 1982; Netzker *et al.*, 1982; Brown *et al.*, 1983; C. Scazzocchio, personal communication). It seems likely that the mammalian mtDNA URFs code for proteins specified by nuclear genes in yeast. It is interesting that, in mammalian mitochondria, the use of codons other than AUG (AUA, AUU, and AUC) as initiator codons has so far been found exclusively in the URFs.

TABLE II
CORRESPONDENCE BETWEEN READING FRAMES OF HUMAN MITOCHONDRIAL DNA AND PUTATIVE mRNAs

Coding strand	Reading frame	Molecular weight of expected protein product	Putative mRNA
Heavy strand	URF 1	35,600	RNA 13
	URF 2	38,900	RNA 12
	CO I	57,000	RNA 9
	CO II	25,500	RNA 16
	URF A6L	7,900	RNA 14
	ATPase 6	24,800	RNA 14
	CO III	30,000	RNA 15
	URF 3	13,200	RNA 17
	URF 4L	10,700	RNA 7
	URF 4	51,400	RNA 7
	URF 5	66,600	RNA 5
	Cytochrome *b*	42,700	RNA 11
Light strand	URF 6	18,600	RNA 1,2,3?

The observation that the URFs are conserved in different mammalian mtDNAs, and that they correspond to putative mRNAs with the same structural and metabolic properties as the mRNAs of well identified genes gave support to the idea that these URFs are expressed in mammalian mitochondria (Anderson *et al.*, 1981, 1982; Bibb *et al.*, 1981; Montoya *et al.*, 1981; Ojala *et al.*, 1981a, Gelfand and Attardi, 1981). Recently, this idea has been directly verified by the identification of several of the protein products of the human URFs: this was achieved by the use of antibodies directed against peptides predicted from the DNA sequence (Mariottini *et al.*, 1983; Chomyn *et al.*, 1983), and, in one case, by a comparison of the partial protease cleavage pattern with that expected from the DNA sequence (Oliver *et al.*, 1983).

Figure 4 (left lane) shows the electrophoretic pattern in an SDS–urea–poly-

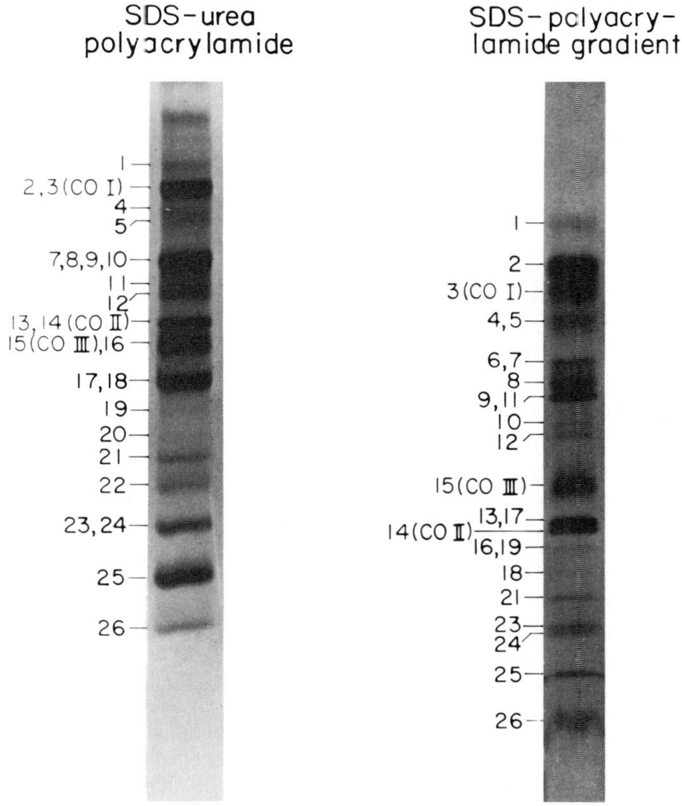

FIG. 4. Fluorograms, after electrophoresis through an SDS–8 *M* urea–polyacrylamide slab gel or an SDS–15 to 25% polyacrylamide gradient slab gel, of the proteins from the mitochondrial fraction of HeLa cells labeled for 2 hours with [^{35}S]methionine in the presence of 100 μg/ml emetine. Modified from Ching and Attardi (1982).

acrylamide gel of the HeLa cell mitochondrial translation products, which have been labeled *in vivo* with [^{35}S]methionine in the presence of emetine (to inhibit cytoplasmic protein synthesis). One can see bands of various intensities, some of which correspond to more than one component. In an SDS–polyacrylamide gradient system, shown also in Fig. 4 (right lane), the various mitochondrial translation products have different absolute and relative mobilities. By a bidimensional fractionation in the two gel systems, it has thus been possible to resolve the individual components of the composite bands. A total of up to 26 components have been reproducibly identified among the HeLa cell mitochondrial translation products (Ching and Attardi, 1982). These products cover the range of apparent molecular weights between about 3,500 and 51,000. Considering the bias in the size estimates based on electrophoretic mobilities in SDS–polyacrylamide gels of the mitochondrially synthesized hydrophobic proteins, the above range of molecular weights is in agreement with the size range expected for the proteins encoded in the mtDNA reading frames. The fact that the number of protein components is larger than that predicted from the number of the reading frames strongly suggests that some of the mitochondrially synthesized polypeptides detected in HeLa cells may be related to the primary translation products by secondary modification, or proteolytic cleavage, or premature termination. Figure 4 also shows that the three largest subunits of the cytochrome *c* oxidase, which are mitochondrially synthesized, correspond to components 3 (COI), 14 (COII), and 15 (COIII), as determined by electrophoretic analysis of the subunits of the highly purified human enzyme complex (Hare *et al.*, 1980; Ching and Attardi, 1982).

Figure 5 illustrates an example of the use of the antipeptide antibody approach for the identification of the proteins encoded in the human mtDNA reading frames. Antibodies raised against the COOH-terminal nonapeptide of the putative protein product of URFA6L, and, less efficiently, also antibodies against the NH_2-terminal octapeptide of the same protein, immunoprecipitate specifically component #25 of the HeLa cell mitochondrial translation products (Mariottini *et al.*, 1983). The size of protein 25, as estimated from its electrophoretic mobility, is compatible with that expected for the putative product of URFA6L. Furthermore, the identification of protein 25 as the URFA6L gene product has been confirmed by a fingerprinting analysis, after trypsin digestion, of protein 25. The same approach has been applied successfully to the identification, among the HeLa cell mitochondrial translation products, of the proteins encoded in other URFs as well as of the ATPase 6 subunit (Chomyn *et al.*, 1983). Table III lists the present status of the assignment of the various mitochondrial translation products from HeLa cells to individual mtDNA reading frames.

Nothing is known about the physiological role of the products of the mammalian mtDNA URFs, except that, with the exception of the URFA6L product, they are hydrophobic proteins (Kyte and Doolittle, 1982), associated presumably

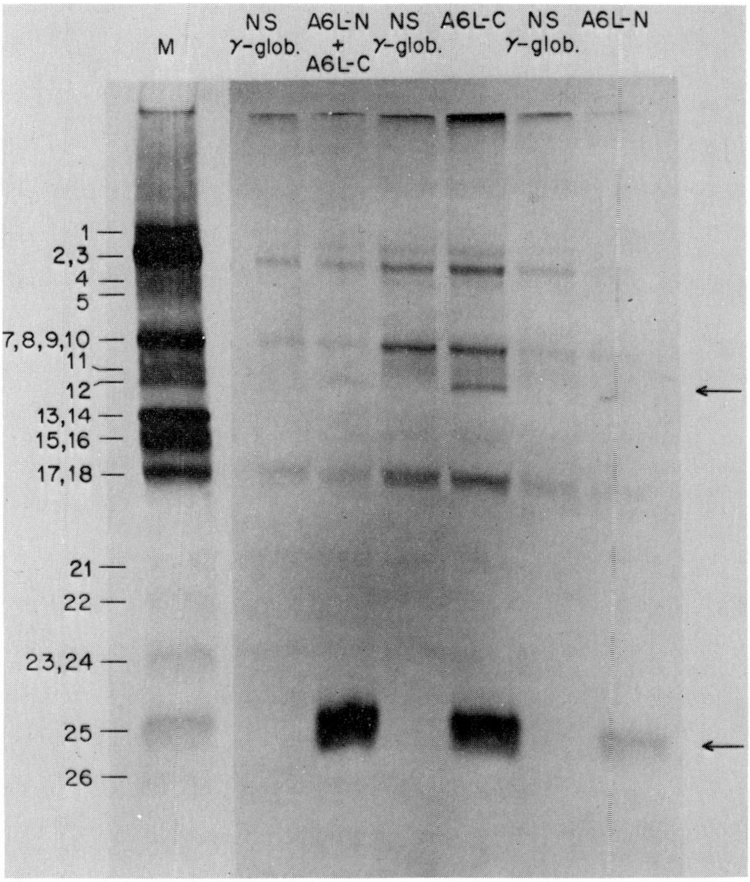

FIG. 5. Fluorogram, after electrophoresis through an SDS–8 *M* urea–polyacrylamide gel, of the HeLa cell mitochondrial translation products (labeled *in vivo* with [^{35}S]methionine in the presence of emetine) immunoprecipitated by antibodies against the NH_2-terminal octapeptide (A6L-N) or the COOH-terminal nonapeptide (A6L-C) of the putative polypeptide encoded in URFA6L. The arrows indicate the positions of the bands of the specifically immunoprecipitated products. NS, Normal serum; M, pattern of mitochondrial translation products. Taken from Mariottini *et al.* (1983).

with the inner mitochondrial membrane. There are, however, some hints about the possible functional identity of the product of the smallest URF, URFA6L. There is in fact evidence suggesting that this gene is homologous to the recently identified *S. cerevisiae* mtDNA gene, aapl (Macreadie *et al.*, 1983). The aapl gene codes for a protein 48 amino acids long, and it has been suggested that this protein, which has significant homology (~ 50%) to a protein of the same size encoded in URFx of *A. nidulans* mtDNA (Grisi *et al.*, 1982), is associated with

TABLE III
CORRESPONDENCE BETWEEN READING FRAMES OF HUMAN MITOCHONDRIAL DNA AND MITOCHONDRIAL TRANSLATION PRODUCTS

Reading frame	Mitochondrial translation product[a]
URF 1	#12
CO I	#3
CO II	#14 (#20)
URF A6L	#25
ATPase 6	#17 (#21)
CO III	#15
URF 3	#24

[a]Numbered according to Ching and Attardi (1982). The numbers in parentheses indicate components which also reacted with peptide-specific antibodies.

the yeast oligomycin sensitive ATPase as an integral part of the complex (subunit 8?) (Macreadie *et al.*, 1983). In *S. cerevisiae,* mutations in the aapl gene affect the assembly of a functional ATPase complex (Macreadie *et al.*, 1983). The mammalian URFA6L product is longer than the aapl and URFx gene products (66 to 68 amino acids in the four species so far analyzed). Although the overall homology between the mammalian and the fungal proteins is rather limited (20–25%) (Fig. 6), there are striking elements of structural similarities between them. In particular, the first four amino acids are identical in all the four mammalian and the two fungal proteins, and there is another short stretch which exhibits a reasonable degree of homology between the two sets of proteins at a position corresponding to the middle of the aapl and URFx products. Furthermore, all these proteins exhibit a concentration of basic amino acids (mostly lysine) in the COOH-end proximal half of the molecule (Macreadie *et al.*, 1983). Although the mammalian URFA6L products, in contrast to the fungal proteins, have a mean hydropathy value falling within the range characteristic of soluble proteins (Kyte and Doolittle, 1982), a closer look shows that the NH_2-end proximal half of these polypeptides has a degree of hydrophobicity similar to that of the aapl and URFx products. In *Drosophila* species and in *X. laevis,* the URFA6L products have an intermediate size between those of the fungal and of the mammalian proteins (53 and 55 amino acids, respectively) (see Sections VI and VII). Although their homology to the fungal proteins is not significantly greater than observed for the mammalian products, they share with the latter the above mentioned structural similarities to the fungal proteins. It should be noted that URFA6L is the most rapidly diverging mitochondrial gene in animal cells (see Section VIII,B,3).

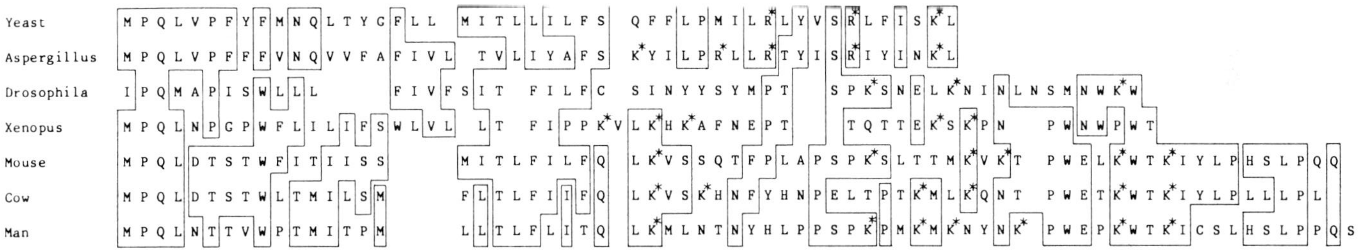

FIG. 6. Amino acid sequence comparison of the products of the yeast aapl gene and *Aspergillus nidulans* URFx with the products of URFA6L of *Drosophila yacuba, Xenopus laevis,* mouse, cow, and human mtDNA. The sequences have been visually aligned, with introduction of gaps to maximize the homology. The lysine (K) and arginine (R) residues are marked with an asterisk. The sequences have been taken from Macreadie *et al.* (1983; yeast), Grisi *et al.* (1982; *A. nidulans*), Clary and Wolstenholme (1983a; *D. yacuba*), Bibb *et al.* (1981; mouse), Anderson *et al.* (1982; cow), and Anderson *et al.* (1981; man). The *X. laevis* URFA6L sequence was kindly provided by Dr. B. Roe.

The mammalian (and in general animal cell) URFA6Ls share with aapl and URFx the close proximity in mapping position to the ATPase 6 gene, which they all precede in the direction of transcription. This arrangement may reflect a need for a coordinate expression of the two genes. In animal cells, the relationship between the two genes has evolved into one in which the two reading frames overlap out of frame. The significance of the latter unusual organization may again be related to the mode of expression of these genes. As shown in Fig. 2, in human cells, a single mRNA, RNA 14, serves the two overlapping reading frames; the same situation holds in mouse cells (Bibb *et al.*, 1981) and, presumably, in all animal cells. RNA 14 has the unique structural features at the 5′ end and at the 3′ end and the characteristic metabolic properties which have been described for human mitochondrial mRNAs (Montoya *et al.*, 1981; Ojala *et al.*, 1981a; Gelfand and Attardi, 1981). Moreover, the failure to detect any other RNA mapping in the same region (Ojala *et al.*, 1980) strongly suggests that RNA 14 is utilized not only for the translation of URFA6L, which is not surprising in view of its typical 5′ end structure, but also for the translation of the ATPase 6 gene.

In prokaryotic and eukaryotic systems, several examples of overlapping reading frames are known, and various mechanisms have been proposed to account for their expression (Sanger *et al.*, 1977; Godson *et al.*, 1978; Kozak, 1981; Kastelein *et al.*, 1982). In the case of the overlapping URFA6L and ATPase 6 gene, the lack of a ribosome attachment site, analogous to that present in 7 S RNA (Ojala *et al.*, 1981b), in the segment of RNA 14 upstream of the ATPase 6 reading frame speaks against an independent entry site for ribosomes destined to translate the downstream reading frame. A possible mechanism operating in the translation of the ATPase 6 gene is a frameshift of ribosomes translating the upstream reading frame, which would result in premature termination at out-of-phase stop codons, followed by reinitiation at the ATPase 6 start codon. Such a mechanism has been proposed to be responsible for the translation of the lysis cistron of MS2 RNA, which overlaps out of phase on its 5′ end the coat protein gene (Kastelein *et al.*, 1982). Ribosome slippage on the mRNA has been previously suggested as a possible mechanism accounting for the relatively frequent (~ 5%) phenotypic reversion of a frameshift mutation in yeast mtDNA (Fox and Weiss-Brummer, 1980).

C. Origins of Replication

The replication of the mammalian, and in general animal mtDNA is unidirectional and highly asymmetric, with the synthesis of the daughter L-strand starting only after about two-thirds of the daughter H-strand have been synthesized (Robberson *et al.*, 1972). The position of the origin of H-strand synthesis (O_H in Fig. 2) is marked by a displacement loop (D-loop), formed by the synthesis and stable association with the template of a rapidly turning over H-strand initiation sequence (Kasamatsu *et al.*, 1971; Arnberg *et al.*, 1971). The D-loop is several hundred to more than 1000 nt long, with large size differences among species [~ 680 nt in man (Crews *et al.*, 1979)]. The D-loop occupies a major portion of the mtDNA segment between the $tRNA^{Phe}$ and $tRNA^{Pro}$ genes (Fig. 2). This region does not contain any long reading frame and is the most rapidly diverging portion of vertebrate mtDNA (see below, Section VIII,B,4). However, in human cells, a relatively stable polyadenylated RNA (7 S RNA) is encoded in the L-strand in the region between $tRNA^{Phe}$ and O_H (Ojala *et al.*, 1981b), and, in *X. laevis,* an H-strand specified 4 S RNA maps in the same region (Rastl and Dawid, 1979). This region also contains the initiation sites for H-strand and L-strand transcription (see Section V,D,5), and presumably, the *cis*-acting control sequences for DNA replication and transcription. The origin of L-strand synthesis (O_L) is located at approximately map position 67/100 [relative to O_H taken as 100(0)/100] in the clockwise direction, between the L-strand $tRNA^{Cys}$ and $tRNA^{Asn}$ genes (Fig. 2). The H-strand in this region can be folded into a characteristic hairpin with an 11–12 bp stem and a 12–13 nt loop. This hairpin is conserved among the vertebrate species so far analyzed (Anderson *et al.*, 1981, 1982; Bibb *et al.*, 1981; Pepe *et al.*, 1983; B. Roe, personal communication), in spite of primary sequence divergence, an observation which points to a probable physiological role of this structure. It is believed that L-strand synthesis starts when the daugher H-strand, growing clockwise from O_H, reaches O_L and exposes the H-strand template [for details concerning mtDNA replication in animal cells, see a recent review on this topic (Clayton, 1982)].

D. RNA Synthesis and Processing in Mammalian Mitochondria

It is not surprising that the evolutionary emergence of the animal mtDNA has been accompanied by the development of unusual mechanisms of gene expression to match the unique features of gene organization of this genome. The mitochondrial DNA transcription and RNA processing events have been most extensively investigated in HeLa cells, and this system will be used as a model in the following description, with references to other systems where appropriate.

1. *The Symmetric Transcription of HeLa Cell Mitochondrial DNA*

The most striking feature of mtDNA transcription in HeLa cells is its symmetry (Aloni and Attardi, 1971a). Both strands are transcribed over their entire length (Aloni and Attardi, 1971b; Murphy *et al.*, 1975). Pulse-labeling experiments have indicated that the L-strand is transcribed at an overall rate which is two or three times higher than that of the H-strand transcription (Cantatore and Attardi, 1980). However, the majority of the L-strand transcripts have a half-life which is much shorter than that of the H-strand transcripts and do not accumulate in mitochondria to any significant extent (Aloni and Attardi, 1971a). The completely symmetrical transcription of HeLa cell mtDNA has to be contrasted with the highly asymmetrical distribution of informational content between the two strands. For its possible significance, see below (Section V,D,7).

2. *Discrete Mitochondrial DNA Transcripts*

Chromatography of mitochondrial RNA through oligo(dT)-cellulose separates an oligo(dT)-cellulose bound and a nonbound fraction (Amalric *et al.*, 1978). The oligo(dT)-cellulose bound components carry at their 3′ end a poly(A) tail of ~ 55 residues, which is added posttranscriptionally (Hirsch and Penman, 1973; Ojala and Attardi, 1974). Among the oligo(dT)-cellulose nonbound components, the majority of the molecules of the two rRNA species have a short stretch of between one and 5 to 10 A residues at their 3′ ends, while the tRNAs lack completely 3′ terminal A residues (Dubin *et al.*, 1982). A high resolution of mitochondrial RNA species has been achieved by electrophoresis through agarose slab gels in the presence of the strongly denaturing agent methylmercuric hydroxide (CH_3HgOH) (Amalric *et al.*, 1978). Figure 7 shows the autoradiogram, after electrophoresis in this gel system, of samples of the oligo(dT)-cellulose bound (lane 1) and unbound (lane 2) mitochondrial RNA fractions from cells labeled for 4 hours with [^{32}P]orthophosphate in the presence of 0.1 μg/ml actinomycin D (to inhibit nuclear RNA synthesis). The bound fraction shows a typical pattern with multiple discrete components, which have been previously identified and characterized (Amalric *et al.*, 1978; Ojala *et al.*, 1980; Gelfand and Attardi, 1981). The 18 discrete poly(A)-containing RNAs cover a range of molecular sizes between ~ 215 and ~ 10,400 nt, and are numbered in order of decreasing sizes. Among them, the three largest species (RNAs 1, 2, and 3) and the smallest one (RNA 18) are L-strand coded, while the others are encoded in the H-strand (Amalric *et al.*, 1978). The nonbound fraction exhibits, besides the prominent 16 S and 12 S rRNA bands, a series of minor bands which, in their majority, align well with the bands of the poly(A)-containing RNAs, and which very probably represent RNA species differing from the latter for the absence or shorter length of the poly(A) tail.

With the exception of RNAs 4, 6, and 10, the polyadenylated transcripts of the H-strand, for their perfect correspondence to significant reading frames (Fig. 2 and Table II), their presence in polysomes and their relative metabolic stability (Table IV), are presumably the mRNAs of mitochondrially synthesized polypeptides. Human mitochondrial mRNAs have distinctive structural properties. In contrast to other eukaryotic mRNAs, they lack a "cap" structure at the 5′ end (Grohmann *et al.*, 1978). Furthermore, they do not exhibit another typical attribute of eukaryotic mRNAs, namely the presence of a 5′-noncoding stretch, but start at or very near to the initiator codon (AUG or AUA or AUU) (Montoya *et al.*, 1981; Chomyn *et al.*, 1981). This finding poses interesting questions con-

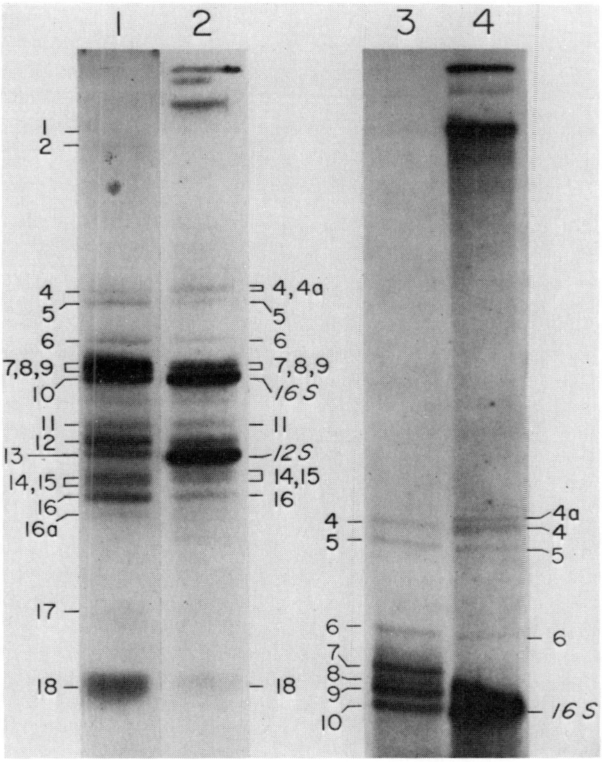

FIG. 7. Electrophoretic analysis of oligo(dT)-cellulose bound (lanes 1 and 3) and nonbound (lanes 2 and 4) mitochondrial RNA from HeLa cells labeled for 4 hours with [^{32}P]orthophosphate in the presence of 0.1 μg/ml actinomycin D. The samples were run through an agarose-CH_3HgOH slab gel for 6 hours (lanes 1 and 2) or 9 hours (lanes 3 and 4). Modified from Montoya *et al.* (1983).

cerning the mechanism whereby mitochondrial ribosomes attach themselves to these mRNAs. Also at their 3′ end, human mitochondrial mRNAs have unique structural features. Namely, most mRNAs lack a 3′-noncoding stretch (the only exceptions being RNAs 5, 9, and 16); furthermore, these mRNAs have an incomplete stop codon, an ochre codon being generated by the poly(A) addition step (see Section V,D,4) (Ojala *et al.*, 1981a).

Polyadenylated RNA 6, on account of its mapping position (Fig. 2) and its relatively short half-life (Table IV), is probably a precursor of RNA 9 (COI mRNA). RNA 10 represents a small fraction, polyadenylated, of the 16 S rRNA (Amalric *et al.*, 1978; Montoya *et al.*, 1983). The nature of RNA 4 will be discussed below (Section V,D,5).

Among the L-strand coded polyadenylated RNAs, species 1, 2, and 3, which

TABLE IV
STEADY-STATE AMOUNT, RATE OF SYNTHESIS, AND HALF-LIFE OF INDIVIDUAL MITOCHONDRIAL RNA SPECIES[a]

RNA species	Steady-state amount (number of molecules/cell)	Rate of synthesis (number of molecules/min/cell)	Half-life (in minutes)	
			Kinetics of incorporation of label	Decay of label after cordycepin block
Oligo(dT)-unbound				
16 S rRNA	n.d.	n.d.	215	282
12 S rRNA	34,000	265	208	n.d.
Oligo(dT)-bound				
2	n.d.	n.d.	n.d.	7
4	44	0.8	n.d.	39
5	165	3.4	n.d.	87
6	125	4.4	n.d.	16
7	960	15.0	47	112
9 (COI mRNA)	950	10.5	67	116
10 (16 S rRNA)	560	2.1	185	n.d.
11 (Cyt. *b* mRNA)	570	7.0	53	56
12	720	6.1	73	51
13	650	18.0	n.d.	n.d.
14 (ATPase 6 mRNA)	770	6.8	59	141
15 (COIII mRNA)	980	9.4	71	
16 (COII mRNA)	1190	10.0	77	191
17	225	4.5	n.d.	n.d.
18 (7 S RNA)	1900	7.3	67	n.d.

[a]Modified from Attardi *et al.* (1982).

are very short-lived, with a half-life estimated to be less than 10 minutes (Gelfand and Attardi, 1981; Table IV), may be intermediates in the formation of L-strand coded tRNAs. The observation that their common 5′-end proximal 500 nt stretch corresponds precisely to an unidentified reading frame of the L-strand (URF6) (Fig. 2) strongly suggests that these RNA species, or some as yet unidentified derivative of them, function as mRNA for this reading frame.

RNA 18 (also called 7 S RNA on the basis of its sedimentation constant) is a small polyadenylated RNA which maps in the region immediately upstream of the origin of H-strand synthesis (Ojala *et al.*, 1981b). Besides the rRNAs and tRNAs, it is, on a molar basis, the most abundant mtDNA transcript; it is the only polyadenylated RNA encoded in the L-strand which accumulates, having a stability comparable to that of well characterized mitochondrial mRNAs (Table IV). The 7 S RNA contains, near its 3′ terminus, a small reading frame for a polypeptide of 23 or 24 amino acids; however, the lack of conservation of this sequence in bovine (Anderson *et al.*, 1982), rat (Sekiya *et al.*, 1980), and mouse mtDNA (Bibb *et al.*, 1981) raises questions about its significance. The most intriguing feature of this RNA is the presence of an 11-nt stretch that shows perfect base sequence complementarity to a stretch near the 3' terminus of the 12 S rRNA (Ojala *et al.*, 1981b), in a way that is strikingly reminiscent in its details of the interaction between the Shine-Dalgarno sequence and bacterial mRNAs (Steitz and Jakes, 1975). The existence of this 11-nt stretch suggests that 7 S RNA has *in vivo* an interaction with 12 S rRNA either free or incorporated into the small ribosomal subunit. The location of the 7 S RNA near the origin of replication may be significant for its function, possibly in connection with the initiation of mtDNA replication or with the complete transcription of the L-strand.

3. *Organization of Mitochondrial DNA Transcripts*

The extreme compactness of gene organization of the human mitochondrial genome is reflected in the tight organization of its transcripts (Ojala *et al.*, 1980) (Fig. 2). All the identified H- and L-strand transcripts are colinear with the DNA, pointing to the lack of intervening sequences in the genes. On the H-strand, the sequences coding for the discrete transcripts (including the expected tRNA gene transcripts) occupy the entire length of the strand, with the exception of a portion, corresponding to about 7% of the genome, around the origin of replication, from coordinate 98/100 to coordinate 5/100. Excluding RNA 4 (see Section V,D,5), which corresponds in mapping position to the two rRNA species (12 S and 16 S), the rRNA precursor (RNA u4a, see Section V,D,5) and RNA6, which overlaps the COI mRNA, there is no apparent overlapping in the H-strand of the sequences coding for the various discrete transcripts. An identical compact organization of mtDNA transcripts has been observed in mouse cells (Van Etten *et al.*, 1982).

In human mtDNA, a perfect correspondence between individual reading frames of the H-strand and the H-strand coded putative mRNAs has been demonstrated by high resolution S1 mapping experiments (Ojala *et al.*, 1980) and RNA sequencing analysis (Montoya *et al.*, 1981; Ojala *et al.*, 1981a). In nearly all cases, each mRNA contains only one reading frame; the two exceptions are RNA 14, which contains a 5′-end proximal reading frame of 207 nt (URFA6L), overlapping out of phase by 46 nt the reading frame for the ATPase subunit 6 polypeptide, and RNA 7, which contains a 5′-end proximal unidentified reading frame of 297 nt (URF4L), overlapping out of phase by 7 nt the unidentified reading frame URF4. The existence of two pairs of reading frames overlapping out of phase, each pair being represented by a single mRNA, is preserved in the bovine (Anderson *et al.*, 1982), mouse (Bibb *et al.*, 1981), and rat (Grosskopf and Feldmann, 1981a; Pepe *et al.*, 1983) mitochondrial genomes. This intriguing situation raises the question as to the mechanism whereby the two proteins encoded in each pair of overlapping reading frames are synthesized from the same mRNA (see Section V,B,3)

4. *The "tRNA Punctuation" Model of Mitochondrial RNA Processing*

The above mentioned S1 mapping and sequence analysis of the human polyadenylated transcripts of the H-strand has revealed that these transcripts, together with the tRNA sequences, form a continuum at the nucleotide level, starting at a position ~ 400 bp downstream of the origin of replication in the direction of H-strand transcription and terminating at the distal end of the D-loop (Ojala *et al.*, 1980, 1981a; Montoya *et al.*, 1981). These observations strongly suggest that the transcription of the H-strand which results in the synthesis of the polyadenylated RNA species occurs in the form of a single large molecule. From this polycistronic transcript the mature transcripts would derive by precise endonucleolytic cleavages occurring in most cases before and after a tRNA sequence. A considerable amount of evidence indicates that this processing occurs on nascent molecules still attached to the template (Fig. 8a) (Cantatore and Attardi, 1980). It seems likely that the tRNA sequences which are interspersed with the rRNA and structural genes represent the signals for the processing enzymes (Ojala *et al.*, 1981a). Presumably, they acquire the cloverleaf structure while still

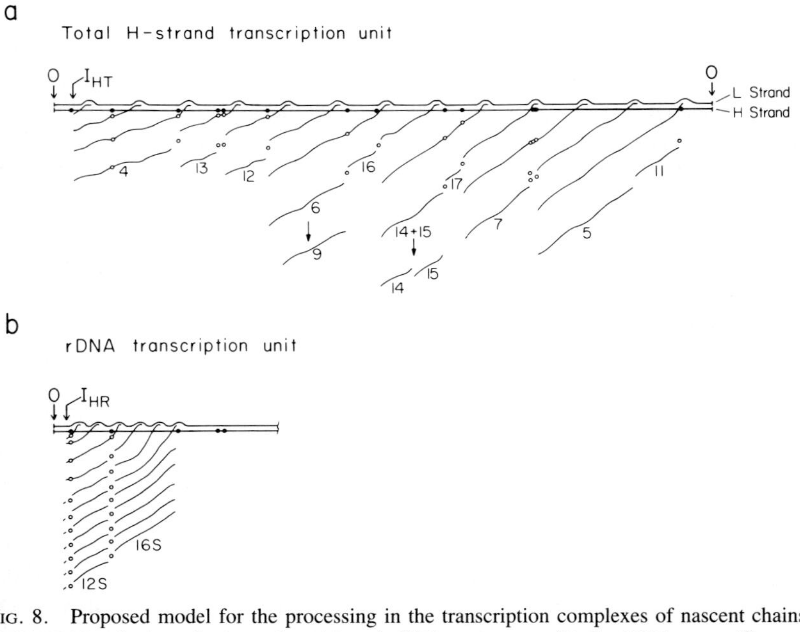

FIG. 8. Proposed model for the processing in the transcription complexes of nascent chains of total H-strand polycistronic transcripts (a) and rRNA precursors (b). ●, tRNA gene; ○, mature tRNA. Modified from Ojala *et al.* (1981a).

in the polycistronic transcript, and it is such a structure that is recognized by the processing enzymes. At least four enzymatic activities must be involved in this processing. First, there must be an enzymatic activity, analogous to that of the bacterial RNase P (Robertson *et al.*, 1972), which cuts the nascent RNA precisely on the 5′-end side of the tRNA. Then, there must be an activity which cuts precisely on the 3′-end side of the tRNA. Furthermore, since every RNA species different from tRNA deriving from the processing of the polycistronic transcript is polyadenylated, a process of adenylation may be intimately associated with the endonucleolytic step releasing the 3′ end of these RNA species. Similarly, -CCA addition must occur at the 3′ end of the tRNAs, subsequently to or concomitantly with the endonucleolytic step which releases this end. It seems possible that the four above mentioned enzymatic activities are assocated to form a single complex.

There are a few processing sites in the H-strand transcripts where no tRNA sequences have been found, i.e., at the site where RNA 9 is cleaved out of RNA 6, and at the borders between RNA 14 and RNA 15 and between RNA 5 and RNA 11 (Fig. 2). It is conceivable that the processing enzyme(s) recognizes a secondary structure which shares some critical features with the cloverleaf structure of the tRNA [as RNase P in bacteria, which can recognize the precursor of the 4.5 S RNA and a bacteriophage ϕ80 induced M_3 RNA (Bothwell *et al.*, 1976)]. A stem-loop structure which occurs at the border between the ATPase 6 and COIII reading frames in mouse mtDNA has been suggested as one which could replace a tRNA as a processing signal (Bibb *et al.*, 1981). Furthermore, it should be mentioned that a sequence complementary to an L-strand coded tRNA, which should therefore be susceptible to folding into a cloverleaf-like structure, occurs in the RNA 6 segment just preceding the cleavage site for RNA 9 and in the RNA 5 segment adjoining the 5′ end of RNA 11 (Fig. 2).

The common 5′ end and the 3′ termini of the three large L-strand transcripts 1, 2, and 3, as mapped by S1 protection experiments and on the basis of electrophoretic size estimates, fall very near to L-strand tRNA genes ($tRNA^{Glu}$, $tRNA^{Ser}$, and $tRNA^{Gln}$) or to the anti-$tRNA^{Arg}$ sequence (Fig. 2). Thus, it seems likely that tRNA sequences play a role as processing signals also for the L-strand transcripts.

The rare occurrence on the same strand of contiguous tRNA genes overlapping by one nucleotide, which was mentioned above (i.e., the $tRNA^{Ser}_{AGY}$ and $tRNA^{Leu}_{CUN}$ genes on the H-strand of mouse mtDNA, and the $tRNA^{Tyr}$ and $tRNA^{Cys}$ genes on the L-strand of human mtDNA), would prevent a single H-strand or L-strand polycistronic transcript from being processed to produce both tRNAs. Therefore, one must assume either the occurrence of a unique cleavage in each transcript with the formation of two tRNAs, one complete and the other lacking a nucleotide, but both functional, or the posttranscriptional repair of the incomplete tRNA, or the existence of two alternative modes of processing of the polycistronic transcripts: the lower efficiency of utilization of the primary transcripts in the last mentioned case could conceivably be permitted by the excess of transcription activity over the cellular needs (see Section V,D,6).

RNA processing and polyadenylation play a crucial role in the formation of the termination codons of the reading frames of mammalian mtDNA (Anderson *et al.*, 1981; Ojala *et al.*, 1981a). As mentioned above, all human mitochondrial mRNAs, with the exception of RNA 5 (URF5 mRNA), RNA 9 (COI mRNA), and RNA 16 (COII mRNA), lack a stop codon at the end of the coding sequence and terminate with a U or UA, which follows the last sense codon. An ochre

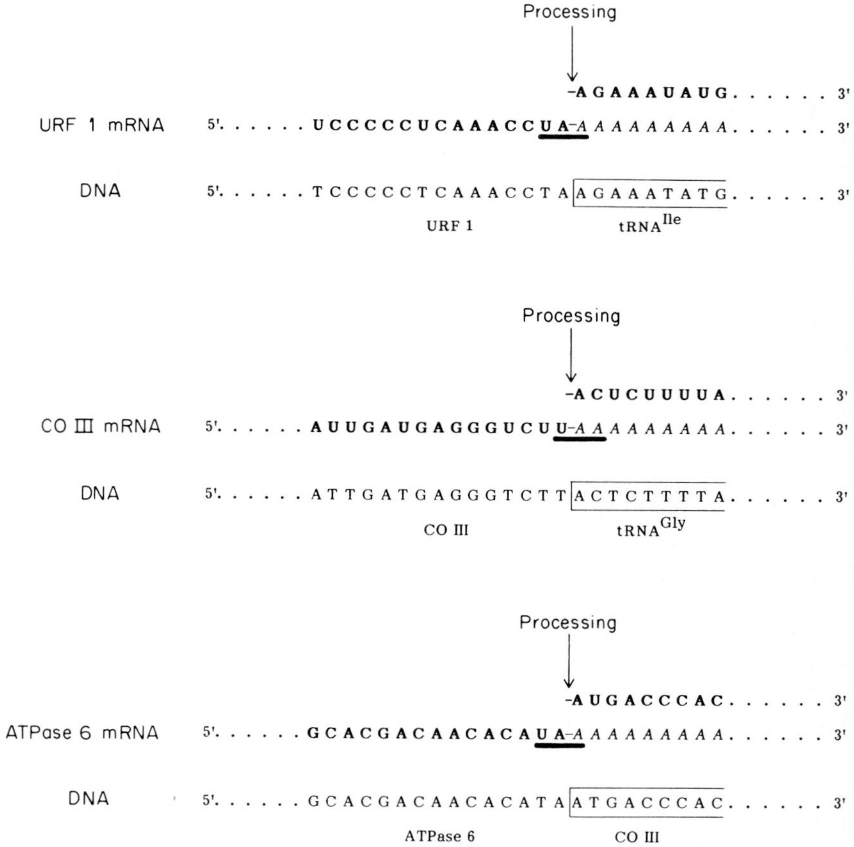

FIG. 9. A 3′-end proximal segment of three mtDNA reading frames (URF1, COIII, and ATPase 6) and a 5′-end proximal segment of the adjacent tRNA gene or reading frame are shown, aligned with the corresponding segments of the polycistronic transcript: these are shown cleaved and polyadenylated at the 3′ end of the mRNAs. The ochre codons created by the polyadenylation are underlined. See text for details.

codon is created at the time of addition of a poly(A) tail to the 3′ end of the mRNA (Fig. 9). The three exceptions to this rule, RNAs 5, 9, and 16, do possess a complete termination codon (UAA, AGA, and UAG, respectively) at some distance from the 3′ terminus.

5. *Alternative Pathways of Transcription of the rDNA Region*

What is the relationship of rRNA formation to the polycistronic H-strand transcript discussed above is a question which has received recently considerable attention. The model of H-strand transcription in the form of a single giant molecule places the rRNA genes, tRNA genes, and structural genes in a single transcription unit. This is a situation without precedent in nature, and raises the question of how a differential expression of these genes is achieved in such a system. In particular, one must explain how the higher rate of synthesis of the rRNAs relative to that of the individual mRNAs (15- to 60-fold), that is observed in HeLa cell mitochondria (Table IV), is produced. Two types of observations have been crucial in elucidating the process involved in the regulation of rRNA and mRNA synthesis. The first type of observations has been the identification of the initiation sites for mtDNA transcription in HeLa cells (Montoya *et al.*, 1982). For this analysis, the 5′ ends of mitochondrial RNA molecules terminating with tri- or di-phosphates, therefore presumably deriving from transcription initiation, were labeled with [α-^{32}P]GTP and the enzyme guanylyltransferase (capping enzyme) and mapped on the mitochondrial genome; the 5′ ends of nascent mitochondrial RNA molecules isolated from transcription complexes were similarly mapped. The two types of mapping experiments gave consistent results, identifying two initiation sites for H-strand transcription. As shown in Fig. 10, one of these sites corresponds to a position very near to the 5′ end of the 12 S rRNA gene (I_{HT}) and the other, to a position 20 to 40 bp upstream of the tRNA^{Phe} gene (I_{HR}).

Evidence that the two H-strand transcription initiation sites described above are implicated in two functionally different transcription events has been provided by the identification and characterization of the primary transcripts of the rDNA region (Montoya *et al.*, 1983). Among the polyadenylated mitochondrial RNA species encoded in HeLa cell mtDNA, there is a species (RNA 4) (Fig. 2), whose mapping position corresponds to that of the two rRNA genes (Ojala *et al.*, 1980). Two other rDNA transcripts have recently been identified in the oligo-(dT)-cellulose nonbound RNA fraction from HeLa cells labeled with [^{32}P]orthophosphate in the presence of 0.1 μg/ml actinomycin D (Montoya *et al.*, 1983). As one can see in Fig. 7 (lane 2), the electrophoretic pattern of this fraction reveals the presence of a fairly pronounced band with the same apparent mobility as that of the oligo(dT)-cellulose bound RNA 4 (lane 1) and of a fainter, somewhat more slowly moving band (RNA 4a), which is not well resolved from the oligo(dT)-cellulose nonbound RNA 4. The latter two species (which will be

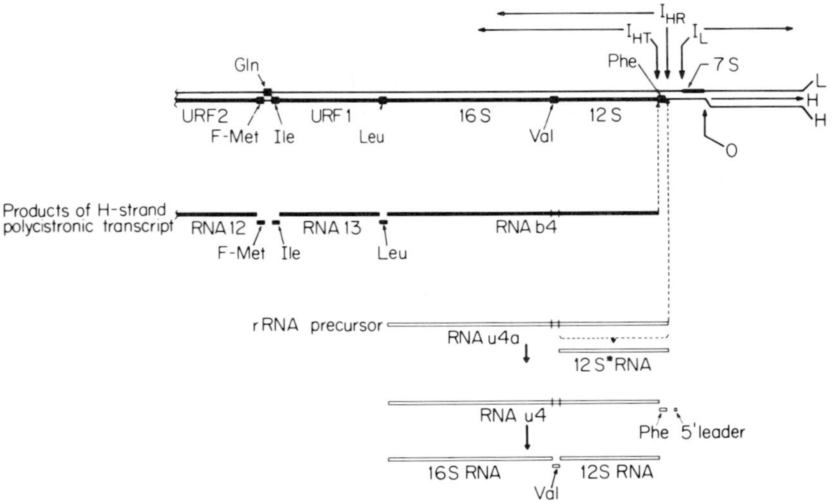

FIG. 10. Portion of the HeLa cell mtDNA genetic and transcription maps illustrating the rRNA gene region and the adjacent regions. In the upper portion of the diagram, the leftward and rightward arrows indicate the direction of H- and L-strand transcription, respectively, and the downward arrows, the initiation sites for H-strand (I_{HT} and I_{HR}) and L-strand transcription (I_L); the upward arrow indicates the location of the origin of H-strand synthesis (O). The lower portion of the diagram shows the discrete transcripts which have been identified in the rDNA region and adjacent region of the H-strand. Modified from Gaines and Attardi (1984).

referred to as RNAs u4a and u4, to distinguish them from the oligo(dT)-cellulose bound RNA 4, designated hereafter as RNA b4) are better resolved after a longer electrophoretic run (Fig. 7, lane 4). S1 protection mapping experiments have shown that the 5′ ends of RNA b4 and RNA u4 map close to or at the 5′ terminus of the 12 S rRNA gene, while the 5′ end of RNA u4a maps about 95 bp upstream of that position (Fig. 10): these two sites correspond precisely to the two H-strand transcription initiation sites described above. At the 3′ end, the RNA species b4, u4, and u4a map in correspondence to the 3′ end of the 16 S rRNA gene, as the mature 16 S rRNA.

In order to establish which of the rDNA transcripts RNAs b4, u4, and u4a, is on the pathway of rRNA synthesis, the kinetics of labeling of these species after short pulses of [5-^{3}H]uridine has been investigated (Montoya *et al.*, 1983). These experiments have shown clearly that RNAs u4a and u4 are the most rapidly labeled RNA species after the 12 S rRNA; RNA b4 becomes labeled much more slowly, at a rate comparable to that of the other polyadenylated RNA species encoded in the H-strand regions lying downstream of the rRNA region. These findings are consistent with the idea that RNAs u4a and u4 are on the pathway of rRNA synthesis, RNA u4a being the putative primary transcript of this pathway, and RNA u4, an intermediate from which the tRNA^{Phe} sequence and the 25 nt 5′ leader have been removed (Fig. 10). However, the observation that the 12 S rRNA is the most rapidly labeled RNA species, even after a 1 minute [5-^{3}H]uridine pulse, strongly suggests that the bulk of 12 S rRNA, and presumably also 16 S rRNA, is processed from nascent chains (Fig. 8b). RNAs u4a and u4 would thus represent only a minor part of the primary rRNA transcripts, which may accumulate to a small extent under conditions of slowed processing, like in the presence of actinomycin D. RNA b4, on the other hand, on account of its kinetic properties, appears to be a part of the polycistronic transcript of the entire H-strand, which is destined to yield all the polyadenylated RNAs and most of the tRNAs encoded in the H-strand (Fig. 10).

The above described mapping and kinetic experiments have thus established a clear correlation between the two mtDNA H-strand transcription initiation sites and two overlapping transcription events (Montoya *et al.*, 1983). Of these, one starts at the upstream initiation site (I_{HR}) and terminates at the 3′ end of the 16 S rRNA gene, being responsible for the synthesis of the rRNAs (Fig. 11). Evidence for termination of the rRNA transcripts at or near the 3′ end of the 16 S rRNA gene is mainly based on the sharp decrease in the rate of transcription of

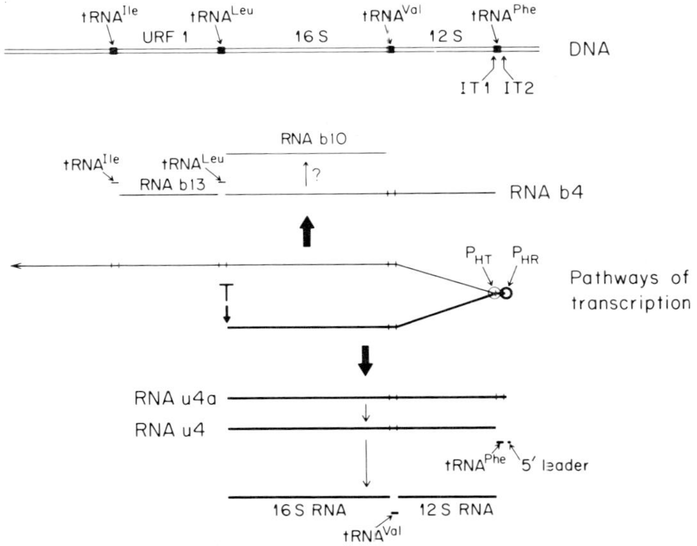

FIG. 11. Diagram representing the proposed model of alternative pathways of transcription of the rDNA region of human mtDNA. In the upper portion of the diagram, the segment of H-strand corresponding to the rRNA gene region and the adjacent regions is shown; the arrows marked I_{HT} and I_{HR} indicate the initiation sites for H-strand transcription. P_{HT}, Putative promoter for the transcription of the total H-strand; P_{HR}, putative promoter for rRNA synthesis. From Montoya *et al.* (1983). IT1 and IT2 correspond to I_{HT} and I_{HR} in Figs. 8 and 10.

the H-strand past the 16 S rRNA gene (Montoya *et al.*, 1983). It is interesting that the 3′ terminal region of the 16 S rRNA gene of several mammalian mtDNAs, including human, can be folded into a structure resembling, in a rudimentary form, the hairpin-oligo(U) signal postulated for bacterial termination–attenuation (Dubin *et al.*, 1982). Additional support for a termination of transcription near the 3′ end of the 16 S rRNA gene comes from the discovery, mentioned above, that the 3′ ends of the mammalian 16 S rRNA are ragged, and that a substantial fraction of them "invade" the adjacent $tRNA^{Leu}$ gene (see Section V,B,2). These observations point to a certain imprecision of the process leading to their formation, in contrast to the absolute precision of the processing event leading to the release of the 3′ ends of the mRNAs. The other transcription event starts at the downstream initiation site (I_{HT}) near the 5′ end of the 12 S rRNA gene, proceeds beyond the 3′ end of the 16 S rRNA gene and results in the synthesis of a polycistronic molecule corresponding to almost the total H-strand (Fig. 11). RNA b4, which results from processing of this transcript, may have no physiological role other than that of joining the promoter near the origin of replication to the main portion of the transcription unit lying downstream with respect to the rRNA genes.

An implication of the model of two alternative pathways of transcription of the rDNA region of mammalian mtDNA is that the $tRNA^{Phe}$ and $tRNA^{Val}$ genes are transcribed at a much higher rate than the majority of the H-strand tRNA genes, leading to a larger accumulation or turnover of the corresponding tRNA species. A quantitative analysis of the individual mitochondrial tRNAs in HeLa cells has revealed that the steady-state amounts of $tRNA^{Phe}$ and $tRNA^{Val}$ are indeed two to four times higher than those of several other H-strand coded tRNAs (including the $tRNA^{Leu}_{UUR}$ encoded in the mtDNA segment lying immediately downstream of the 16 S rRNA gene), without bearing any relationship to the expected codon usage for mitochondrial translation (King and Attardi, unpublished observations).

Further support for the existence of two independent, but overlapping, pathways of mtDNA transcription leading to rRNA and mRNA synthesis has come from the results of an investigation of mtDNA transcription and RNA processing in isolated organelles (Gaines and Attardi, 1984). Isolated HeLa cell mitochondria, when incubated in the presence of an [α-^{32}P]NTP and an appropriate buffer, support transcription of their DNA in a way closely resembling the *in vivo* process. As shown in Fig. 12, the pattern of *in vitro* synthesized mitochondrial RNAs is substantially identical to that of *in vivo* synthesized products. However, the processing of the rRNA species is considerably slowed down in isolated organelles, as shown by the accumulation of RNAs u4a and u4 and by the appearance of a processing intermediate, termed here 12 S*, which comigrates in the gel with mRNA 12. S1 mapping experiments have shown that this processing intermediate is equivalent to 12 S rRNA still carrying the $tRNA^{Phe}$ and the 25 nt

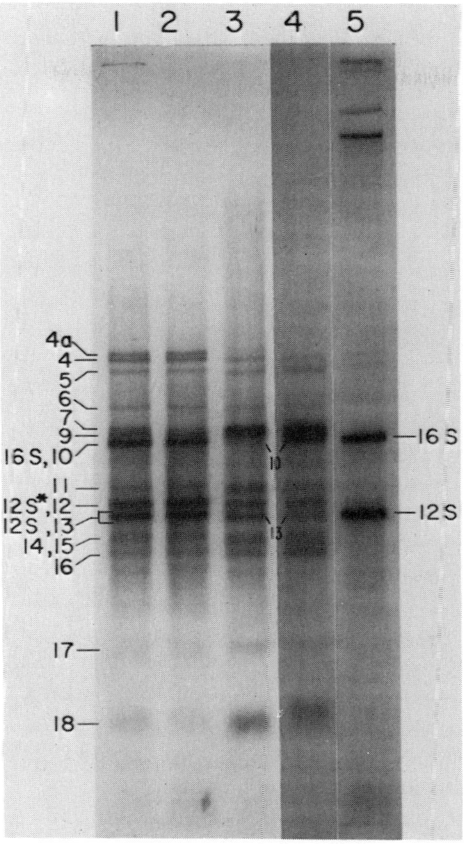

FIG. 12. Comparison of mtDNA transcription in isolated mitochondria and *in vivo*. Mitochondrial RNA from organelles incubated *in vitro* in an appropriate buffer in the presence of [α-^{32}P]UTP (lanes 1–3) and *in vivo* labeled RNA (lanes 4, 5) were fractionated into oligo(dT)-cellulose bound and nonbound RNA, and the various fractions were run through an agarose-CH_3HgOH slab gel. Lanes 1–3: total (lane 1), oligo(dT)-cellulose nonbound (lane 2), and bound (lane 3) RNA labeled *in vitro*. Lanes 4, 5: oligo(dT)-cellulose bound (lane 4) and nonbound (lane 5) RNA labeled *in vivo*. Taken from Gaines and Attardi (1984).

leader at its 5′ end (Fig. 10). The slowing down of the rRNA processing is especially marked at temperatures below 37°C or in the presence of ethidium bromide or proflavine. It is interesting that the latter two intercalating drugs have a preferential inhibitory effect on rRNA labeling over mRNA labeling, a result which supports the existence of a separate control of rRNA and mRNA synthesis, implicit in the model of overlapping independent transcription units (Gaines and Attardi, 1984).

6. *Control of rRNA and mRNA Levels in Human Mitochondria*

The existence in the H-strand of human mtDNA of two transcription units with distinct promoters, one used for the synthesis of rRNAs, and the other, for the synthesis of the mRNAs and most of the tRNAs encoded in the H-strand is presumably justified by the need for the cell to control independently the rate of synthesis of the rRNAs and of the other two mitochondrial RNA classes. A difference in the efficiency of the two promoters is probably the main factor determining the different rates of rRNA and mRNA synthesis in human mitochondria. However, an important aspect of this control is that each of the two kinds of transcripts of the rDNA region must have its fate determined by having started at a specific site. The different behavior of the rRNA precursor and of the polycistronic transcript of the entire H-strand at the optional termination site at the 3′ end of the 16 S rRNA gene is undoubtedly the most intriguing feature of this system. The two transcripts differ only by the presence of an extra 95 nt stretch at the 5′ end of the rRNA precursor. Conceivably, the two transcripts are synthesized through the activity of two different RNA polymerases, only one of which recognizes the signal at the 3′ end of the 16 S rRNA gene. Alternatively, the same polymerase is involved in the synthesis of the two transcripts, and their different behavior is due to a signal associated with the extra 95 nt stretch. In any case, since most of the rRNA precursors are processed while being synthesized (Montoya *et al.*, 1983), one must assume that the 95 nt stretch at the 5′ end of the rRNA precursor does not usually see the 3′ end where termination occurs. Therefore, if the polymerase itself is not involved in the choice between termination and nontermination, there must be a long-range effect of the 5′ end of the rRNA precursor. This long-range effect could involve some form of polar transmission of a signal for termination, as by sequential formation of secondary structures and/or sequential binding of ribosomal proteins. The understanding of the factors which control initiation and potential termination of synthesis of the two types of rDNA transcripts in human mitochondria will depend on the development of suitable "open" *in vitro* mtDNA transcription systems.

Apart from the regulation at the level of initiation of transcription, there is good evidence for a control of expression of the mitochondrial rRNA genes and structural genes at the level of stability of the transcripts. An analysis of the metabolic stability of the mitochondrial RNAs in HeLa cells has revealed, surprisingly, that not only the mRNA species are metabolically unstable (half-life of 30–60 minutes), but also the rRNA species turn over (half-life of 3–4 hours) (Table IV). Since all mRNAs have similar half-lives, and since the rRNAs are also unstable, it seems unlikely that this instability reflects the requirements for rapid changes in individual gene expression, as in bacteria. A more plausible interpretation is that the turnover of mitochondrial mRNAs and rRNAs is the consequence of the high rate of mtDNA transcription, which exceeds the normal cellular needs under the conditions of *in vitro* growth, thereby providing ample capacity for adaptation to increased demands. According to this interpretation, in HeLa cells and other cells cultured *in vitro,* the stability of the rRNA and mRNA species is linked to their utilization. Under conditions of increased respiratory demands, the utilization of the mtDNA transcripts and, correspondingly, their stability would increase. This view would place at least one type of control of expression of the mitochondrial genes on an adjustment of the rate of decay of the mitochondrial RNA species in relationship to the degree of their utilization. In this regulation, it seems possible that proteins synthesized in the cytoplasm play the main, if not exclusive, role. In agreement with this idea are the results of experiments which have indicated a remarkable increase in stability of mitochondrial mRNA and rRNA species in cultured mammalian cells under conditions causing a block of synthesis of the organelle RNA (Lansman and Clayton, 1975; England *et al.,* 1978).

7. *L-Strand Transcription*

The mapping study of the transcription initiation sites which was mentioned above (Section V,D,5) has also revealed the occurrence of an initiation site for L-strand transcription very near to the 5′ end of 7 S RNA (Montoya *et al.,* 1982) (I_L, Fig. 10). A scanning of the mtDNA sequence in the region between the 5′ ends of the 12 S rRNA and the 7 S RNA coding sequences has shown the presence, within the tRNA^{Phe} gene, of a 15 nt stretch at a position 21 nt upstream of the 12 S rRNA gene, which exhibits a 12/15 homology to another 15 nt stretch at a position 28 nt upstream of the 7 S RNA coding sequence:

L-strand 5′....AATGT TTAGACGGG....21 nt.... 12 S rRNA
H-strand 5′ ...AATGTGTTAGTTGGG....28 nt.... 7 S rRNA

These two sequences do not occur anywhere else in the H-strand and L-strand of the whole mtDNA. Though statistically not significant ($p \sim 0.2$), the homology of the two nucleotide stretches may be meaningful because of their similar position relative to the two transcriptional initiation sites lying downstream. It is plausible, in fact, that the two sequences are recognition signals for initiation of transcription at these sites.

It is not known what relationship exists between the L-strand transcription initiation site and the synthesis of 7 S RNA and/or the complete transcription of the L-strand or the synthesis of the RNA primer for H-strand synthesis (Clayton, 1982). It seems likely that the large L-strand-coded polyadenylated transcripts 1, 2, and 3 derive from processing of molecules initiated near the origin of H-strand synthesis. If the initiation site near the 5′ end of 7 S RNA is involved in the formation of these large transcripts, and if 7 S RNA derives from processing of these transcripts, such processing must be a relatively infrequent event; in fact, the rate of formation of 7 S RNA is relatively low (Table IV), as compared to the overall rate of transcription of the L-strand, which is two to three times higher than that of the H-strand (Cantatore and Attardi, 1980).

The known genes located in the L-strand, together with the coding sequence for 7 S RNA, represent only about 7% of the total length of the strand. Since these genes are scattered throughout the L-strand, the complete transcription of this strand (Murphy *et al.,* 1975), presumably under the control of a single promoter (Montoya *et al.*, 1982), could in principle be viewed as an economical mechanism for the expression of these genes, utilizing the same processing devices operating in the expression of the H-strand genes. However, this interpretation would not explain the elevated rate of transcription of the L-strand; furthermore, it would not account for the transcription of the portion of the L-strand between the $tRNA^{Gln}$ and the initiation site for L-strand transcription, which is devoid of any known genes. It is thus conceivable that the complete L-strand transcription plays some additional role, possibly related to the expression of the H-strand genes. Knowledge concerning the timing of L-strand transcription relative to that of the H-strand in the same mtDNA molecule and concerning its rate in different regions of the genome is needed before a plausible interpretation of its significance can be advanced. The analysis of the transcription processes in invertebrate mtDNAs, which have a different gene order and/or polarity as compared to mammalian mtDNAs, should also provide valuable information on this question.

VI. The Mitochondrial Genome of Other Vertebrates

Recently, the complete sequence of *Xenopus laevis* mtDNA has been determined (B. Roe, personal communication). This study has revealed that the detailed gene organization of this mtDNA is identical to that of the mammalian mtDNAs, thus confirming the earlier electron microscopical observations on the tRNA and rRNA gene arrangement in this genome (Ohi *et al.,* 1978; Ramirez and Dawid, 1978). Among the minor differences, it is worth mentioning that, in *X. laevis* mtDNA, all reading frames use AUG as initiator codon, and URFA6L overlaps the ATPase gene to a significantly lower extent than in the mammalian mtDNA (10 nt). The transcription map of *X. laevis* mtDNA has also been found to mirror closely in number and position of transcripts that of mammalian mtDNAs (Rastl and Dawid, 1979), strongly suggesting that the mechanisms of RNA synthesis and processing in *Xenopus* mitochondria are similar in their general features to those observed in human mitochondria.

Although no data are available, as yet, concerning mitochondrial gene organization and expression in other vertebrates, it seems very likely that the general patterns found in mammals are preserved in all classes of vertebrates.

VII. The Mitochondrial Genome of Invertebrates

The mtDNA of invertebrates has revealed striking differences in gene organization from the basic pattern observed in mammalian mtDNA. The most extensively investigated among the invertebrate mtDNAs is that of the genus *Drosophila*. The mitochondrial genome of species of this genus is similar in size and structure to that of other animal cells, except for the presence of an AT-rich region. This region varies in size in the mtDNAs of different species from 1.0 kb in *D. ananassae* to 5.1 kb in *D. melanogaster*, accounting for the difference in size of the mtDNA from these species (from 15.9 to 19.5 kb, respectively) (Wolstenholme *et al.*, 1979). In all *Drosophila* species analyzed (6), the AT-rich region contains the origin of replication of the first replicated strand (Goddard and Wolstenholme, 1978, 1980). In *D. melanogaster*, an early electron microscopical analysis of RNA–DNA hybrids had shown that this region maps close to the position of the two rRNA genes (Klukas and Dawid, 1976). mtDNA replication in *Drosophila* species is unidirectional and highly asymmetric, as in vertebrates, with synthesis of one strand being up to 99% complete before synthesis of the complementary strand is initiated (Goddard and Wolstenholme, 1978, 1980).

Recently, the determination of the nearly complete sequence of *D. yacuba* mtDNA (Clary *et al.*, 1982, 1983; Clary and Wolstenholme, 1983a–c) and of an approximately 5 kb fragment of *D. melanogaster* mtDNA (de Bruijn, 1983) has revealed, besides the variations in genetic code which were mentioned earlier (Section II), a striking difference in gene arrangement with respect to mammalian mtDNA. As shown in Fig. 13, the compact gene organization and the interspersion of tRNA genes with the rRNA and structural genes which are characteristic of mammalian mtDNA are also found in *Drosophila* mtDNA. However, the distribution of genes between the two strands and the detailed gene order are markedly different in this mtDNA. Thus, the genes are less unequally distributed between the two strands. In the strand which is transcribed in the direction opposite to that of replication (counterclockwise in Fig. 13), the genes are arranged in a tightly packed fashion in two groups, a major group occupying almost the entire first half of the strand downstream of the AT-rich region, and a minor group located between map positions 25/100 and 35/100. The major group contains URF2, the COI and COII genes, URFA6L, the ATPase 6 and COIII genes, and URF3: these genes are arranged in the same order as in the mammalian genome, but the tRNA genes which separate them only in part correspond to those found at the same positions in mammals. The minor group of counterclockwise transcribed genes comprises, in order, the genes for tRNA^{Thr}, cytochrome *b*, and $\text{tRNA}^{\text{Ser}}_{\text{UCN}}$. In the strand transcribed in the same direction as that of replication (clockwise in Fig. 13), most of the genes are clustered in two groups, one, including in order the 12 S and 16 S rRNA genes, located immedi-

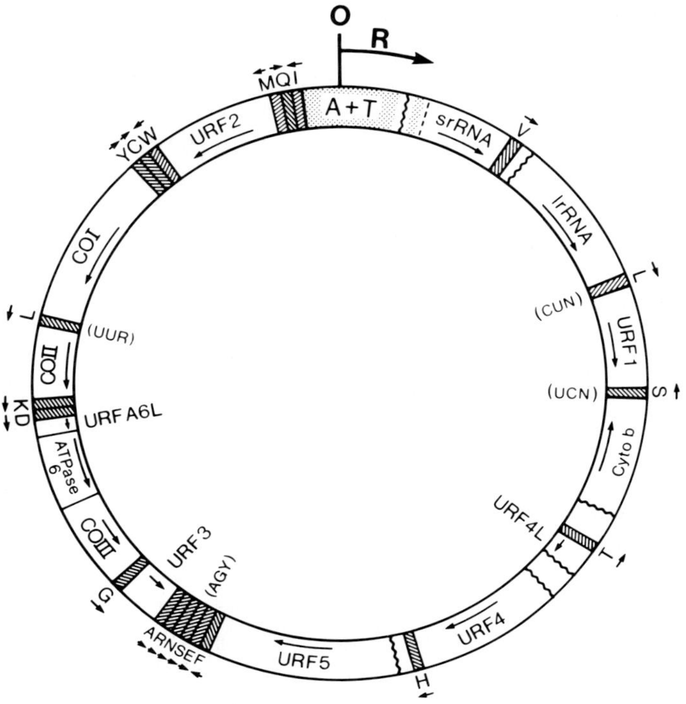

FIG. 13. Genetic map of the *Drosophila yacuba* mitochondrial genome. In the map, the large and small rRNA genes are designated as lrRNA and srRNA, respectively, the tRNA genes by their amino acid specificities, the reading frames as in Fig. 1, and the AT-rich region as A+T. Wavy lines indicate imprecise gene limits. The arrow in or close to each gene indicates the direction of transcription. At the top, the position of the origin and the direction of replication are indicated by O and the rightward arrow (R), respectively. (Diagram kindly supplied by Dr. D. Wolstenholme.)

ately downstream of the AT-rich region, and the other located between map positions 35/100 and 57/100. In the first group, the two rRNA genes are separated by the tRNA^{Val} gene, as in the mammalian genome; however, in contrast to the latter, the small rRNA gene is not flanked on its 5′ side by the tRNA^{Phe} gene or any other tRNA gene, and the large rRNA gene is separated from URF1 by the tRNA$^{Leu}_{CUN}$ gene instead of the tRNA$^{Leu}_{UUR}$ gene. The location of the rRNA genes on the side of the AT-rich region which is replicated first differs from the arrangement found in vertebrate mtDNA, and accounts for the observation that, in *Drosophila* mtDNA, replication proceeds toward the rRNA genes rather than away from them (Goddard and Wolstenholme, 1980).

A comparison of the mammalian and *Drosophila* mtDNA gene maps suggests that multiple events of translocation and inversion must have occurred to account for the present-day differences in gene order and polarity between the mitochondrial genomes from the two classes of organisms. In particular, the different location and polarity of the rRNA genes, URF1 and adjacent tRNA genes relative to URF2, the COI and COII genes, URFA6L, and the ATPase 6 and COIII genes (Fig. 2) must have resulted from a major translocation and inversion event. However, it is not possible to say whether the *Drosophila* or the mammalian mtDNA gene organization is closer to the primitive pattern. At the DNA sequence level, also in *Drosophila* mtDNA, most genes are butt-jointed to each other or separated by a few nucleotides. As in the mammalian mitochondrial genome, URFA6L overlaps the ATPase 6 gene, although by only 7 nt. The reading frames lack a significant 5′ noncoding region; several of them also lack a 3′ untranslated region, and exhibit a truncated ochre codon (T or TA) after the last sense codon; the latter observation indicates that the polyadenylation model for the formation of the ochre codon operating in mammalian mitochondria (Anderson *et al.*, 1981; Ojala *et al.*, 1981a) is also functioning in insect mitochondria.

Eleven oligo(dT)-cellulose bound RNA species and one nonbound have been identified in mitochondrial RNA Preparations from *D. melanogaster* (Merten and Pardue, 1981). An identical set of species with the same binding properties to oligo(dT)-cellulose has been observed in *D. virilis* (Merten and Purdue, 1981) and in *Aedes albopictus* (Eaton and Randlett, 1978). In the three organisms, the oligo(dT)-cellulose nonbound mitochondrial RNA species is the small rRNA component, while one of the oligo(dT)-cellulose bound RNA species is the large rRNA component. The almost quantitative binding to oligo(dT)-cellulose of the large mitochondrial RNA species from *Drosophila* and mosquito does not seem to be due to its high AT content (> 82% in *D. melanogaster*); in fact, a poly(A)-tail of ~36 nt has been detected at the 3′ end of the large rRNA species from *A. albopictus* (Dubin *et al.*, 1982). By contrast, the bulk of the small rRNA species from the same source contains one or two 3′-terminal As (Dubin and HsuChen, 1983). A comparison with the *D. yacuba* mtDNA sequence suggests that some at least of the 3′ terminal A residues in the small rRNA species are added post-transcriptionally, like in the 12 S rRNA from mammalian sources.

Electron microscopy of RNA–DNA hybrids and RNA transfer hybridization experiments have shown that the two rRNA genes map next to each other in the mtDNA of *D. melanogaster*, with a gap of about 160 bp between them, and that the small rRNA gene may overlap the AT-rich region by as much as 200 bp (Klukas and Dawid, 1976). Furthermore, RNA–DNA hybridization experiments, after digestion of appropriate restriction fragments with 3′-end- or 5′-end-specific exonucleases, have indicated that transcription, in this organism too, proceeds from the small to the large RNA (Klukas and Dawid, 1976). A transcription map of *Drosophila* mtDNA has been constructed by R-loop and RNA transfer hybridization experiments (Merten and Pardue, 1981). Most polyadenylated transcripts and the small rRNA species map in a densely packed, but not overlapping, fashion to cover almost the entire length of the genome outside the AT-rich region. The transcript map is consistent with a model of transcription of the two mtDNA strands in form of polycistronic molecules destined to be processed to the mature products according to the tRNA punctuation model proposed for mammalian mitochondria (Ojala *et al.,* 1980, 1981a).

A different type of evolutionary rearrangement of mtDNA has been observed in sea urchin, in particular in *Strongylocentrotus franciscanus* (Roberts *et al.,* 1983). Here, the striking difference from the mammalian mtDNA pattern is represented by the ''insertion'' of URF1 and URF2 between the 12 S and the 16 S rRNA genes; as a result of this rearrangement, the 16 S rRNA and COI genes have become close to each other. In spite of the difference in gene order, the polarity of the 16 S rRNA gene relative to the COI gene is conserved in sea urchin, thus ruling out a simple inversion of the region containing URF1, URF2, and the 16 S rRNA gene. The gene organization in the remainder of sea urchin mtDNA appears to be colinear with that of human mtDNA. Since, in *Drosophila,* the 12 S and 16 S rRNA genes are close to each other, as in mammals, the rearrangement event responsible for the different gene organization of sea urchin mtDNA probably occurred in the sea urchin lineage after sea urchins and mammals last shared a common ancestor, more than 500 million years ago (Roberts *et al.,* 1983).

VIII. Evolution of Animal Mitochondrial DNA

A. Rapid Nucleotide Sequence Flux in a Frozen Gene Organization

The study of the structure, gene organization, and expression of the mtDNA from a large number of animal species at different evolutionary levels has revealed in the past 15 years two striking, apparently contrasting features of the evolution of this genome. On one hand, there is a conservation of size, structure, informational content, and basic gene organization of mtDNA throughout the animal kingdom. On the other hand, there is an elevated rate of nucleotide sequence evolution of animal mtDNA, which is one order of magnitude higher than that of single-copy nuclear DNA (Dawid, 1972; Brown *et al.,* 1979).

The constancy in size and structure of animal mtDNA was recognized by early investigators (Borst, 1972), and suspected to reflect the conservation of informational content. There is good evidence that the limited variability in size of mtDNA results mainly from differences in the noncoding portion of the genome, which surrounds the origin of replication and corresponds to the D-loop and adjacent region in mtDNA from vertebrate cells (Upholt and Dawid, 1977; Bibb *et al.,* 1981; Anderson *et al.,* 1982) and to the AT-rich segment in mtDNA from

Drosophila species (Wolstenholme *et al.*, 1979). The constancy in gene content and basic pattern of gene organization of animal cell mtDNA has become apparent only recently as a result of the sequencing and gene mapping analysis of mtDNA from several mammalian species, *X. laevis, Drosophila* species, and sea urchin. The extreme compactness of gene organization and the interspersion of tRNA genes with the rRNA and structural genes have been found to be common features of the animal cell mtDNAs so far analyzed, which cover divergence times up to 600 million years. The differences in gene order observed in *Drosophila* and sea urchin mtDNAs relative to the mammalian mtDNAs appear to be theme variations, which do not alter those fundamental features. It seems likely that the various existing subtypes of animal mtDNA evolved in a parallel fashion from an ancestral looser type of gene organization.

From the studies mentioned above, animal cell mtDNA has emerged as a genome evolutionarily frozen in terms of gene content and organization. The evolutionary drive toward simplification of animal mtDNA, which is apparent in the almost complete absence of "unnecessary" intergenic noncoding sequences, in the reduction in size of the rRNA and tRNA genes, and in the simpler modification pattern of their products, has produced a streamlined, extremely tight gene organization; in turn, the compactness of gene organization has decreased the probability of evolutionary rearrangements. The simultaneous development of simplified mechanisms of RNA synthesis and processing, including the proposed role of the tRNA sequences as recognition signals for the processing enzymes, has placed further constraints on the evolution of this genome. Therefore, the present-day animal cell mitochondrial genome appears to be "locked-in" in a variety of forms which differ in gene order, but which all share the basic features of economy and compactness of gene organization and interspersed tRNA gene arrangement.

In striking contrast to the "frozen" state of animal cell gene organization, the nucleotide sequence of animal cell mtDNA evolves very rapidly. The first indication of this rapid sequence evolution was provided by a comparison by heteroduplex analysis of the mtDNAs from two different *Xenopus* species (Dawid, 1972). This analysis revealed that the rate of sequence evolution of mtDNA is at least as high, and possibly higher than that of nuclear DNA; furthermore, it showed that, among the mitochondrial genes, the rRNA and tRNA genes are the most conserved, and that the mitochondrial rRNA genes evolve more rapidly than the nuclear rRNA genes. These conclusions have been confirmed and extended by investigations carried out by restriction endonuclease cleavage site mapping on a variety of mammalian mtDNAs, mostly primates (Brown *et al.*, 1979). This study has revealed that, for divergence times up to 15 million years, the rate of mtDNA change is 10 times higher than that of the single-copy nuclear DNA. Under the assumption that nucleotide substitutions play a major role in the evolution of mammalian mtDNA, an assumption which has been confirmed by the recent DNA sequence analyses, a value of 0.02 substitution per base pair per million years could be estimated for mtDNA. For divergence times greater than 25 million years, the rate of change of mtDNA is similar or slower than that of the single-copy nuclear DNA. Finally, only 25 to 30% of the mitochondrial genome appears to be affected by the fast rate of evolutionary change.

The sequence analysis of the total mtDNA from several mammalian mtDNAs and that of selected fragments of mtDNA from several primate species (Brown *et al.*, 1982) or from several human individuals (Greenberg *et al.*, 1983; Aquadro and Greenberg, 1983), besides verifying the predominance of nucleotide substitutions in the evolution of animal mtDNA, has revealed the occurrence of small deletions or additions (1 to 4 bp) in the rRNA and tRNA genes. Furthermore, it has shown, unexpectedly, that the rate of nucleotide substitution is strongly biased, by a factor of ~ 25, toward transitions, presumably due to a higher rate of transition-causing mutations. The ratio of transitions to transversions, which is ~ 25 in intraspecific comparisons (Greenberg *et al.*, 1983), decreases with increasing divergence times, approaching the value of ~ 0.5, expected in the absence of any bias, for divergence times of 80 million years or greater (Brown, 1983).

Two main explanations have been advanced to account for the high rate of nucleotide sequence evolution of mtDNA. One explanation involves a high rate of mtDNA mutation, resulting from exposure of the DNA to endogenous DNA-damaging agents, such as free radicals or high oxidant levels, or poor fidelity of replication or editing capacity of the mtDNA polymerase, or deficiency in DNA repair (Brown, 1983). The other explanation assumes a high probability of fixation of mildly deleterious mutations in mtDNA: the existence of multiple copies of mtDNA in each cell and the relaxation of translational constraints expected in a small genome coding for a limited set of proteins would indeed be expected to increase the probability of fixation of mutations (Cann *et al.*, 1984; Brown, 1983). It seems likely that both factors, high rate of mtDNA mutation and high probability of fixation of mutations, contribute to the production of the observed high rate of mtDNA sequence evolution.

B. Evolution of Genes and Intergenic Regions

1. *Transfer RNA Genes*

Homologous tRNA genes from different mammalian mtDNAs exhibit a high degree of homology (Table V), with variations depending upon the tRNA species and upon the pair of organisms compared. Most of the differences are due to nucleotide substitutions, although small deletions or additions (1 or 2 bp) have been observed. The changes occur in all portions of the molecule, however, with significant differences in frequency in different structural domains. The most conserved segments appear to be the anticodon loop, the anticodon stem, the

TABLE V

DEGREE OF HOMOLOGY BETWEEN MITOCHONDRIAL GENES OF MAN (H), COW (C), MOUSE (M), RAT (R), AND *Drosophila yacuba* (D)[a]

	% nucleotide homology				% amino acid homology			
	H/M	H/C	H/R	M/D	H/M	H/C	H/R	M/D
12 S rRNA	75	77	76					
16 S rRNA	74	76	70					
tRNAs	68–90	69–96	66–97					
COI	77	79	78	71	90	91	93	75
COII	71	70	72	64	71	73	71	56
COIII	76	79	78	67	86	87	87	64
Cyt. *b*	74	73			78	79		
ATPase 6	72	73		49	75	78		36
URF1	71	73		57	78	78		48
URF2	61	66	61	47	57	63	54	35
URF3	66	70	65		65	74	65	
URF4	68	72	68		67	74	52	
URF4L	69	75	68		67	74	64	
URF5	65	71			63	70		
URF6	63	65			52	63		
URFA6L	65	63	65	48	46	52	48	30

[a]Taken from Bibb *et al.* (1981; H/M), Anderson *et al.* (1982; H/C), Saccone *et al.* (1982; H/R), Pepe *et al.* (1983; H/R), Clary *et al.* (1982; D/M), Clary and Wolstenholme (1983a; D/M), and Clary *et al.* (1983; D/M) The homology of mouse and *Drosophila* URFA6Ls was determined on the basis of the alignment shown in Fig. 6, excluding the gaps from the calculation.

"DHU" stem, and the amino-acyl acceptor stem, while the "TΨC" and "DHU" loops and the extra arm are the most rapidly diverging segments.

The nucleotide substitution rate in the mitochondrial tRNA genes has been estimated to be about 50% of that observed in the protein coding genes, but at least 100 times higher than in the nuclear tRNA genes, on the basis of pairwise sequence comparisons of three tRNA genes from closely related primate species (Brown *et al.*, 1982). This high substitution rate is in part explained by the overall higher mutation rate in mtDNA, in part, however, it must reflect a higher rate of fixation: the latter is probably due to the lower constraints which exist in the tRNA–ribosome and tRNA–synthetase interactions as a result of the change in the genetic code and codon recognition patterns in mitochondria (Brown, 1983).

2. *Ribosomal RNA Genes*

Pairwise sequence comparisons of the rRNA genes from human, mouse, cow, and rat mtDNA have revealed a high degree of homology, with the 12 S rRNA gene being somewhat more conserved than the 16 S rRNA gene (Table V).

Again, most of the changes are represented by nucleotide substitutions, with some small deletions or additions. In the 12 S rRNA molecule, differences occur in both the single-stranded and the base-paired regions of the proposed secondary structure (Anderson *et al.*, 1982). In both the 12 S rRNA and 16 S rRNA, patches of high homology are interspersed with less conserved stretches. In the 16 S rRNA molecule, the most conserved region among mammalian species, and one which exhibits the highest homology to the *E. coli* 23 S rRNA and to the yeast large mitochondrial rRNA species, is near its 3′ end (Blanc *et al.*, 1981a,b; Bibb *et al.*, 1981; Saccone *et al.*, 1982a,b; Anderson *et al.*, 1982). It is interesting that the positions of the nucleotide changes conferring resistance to chloramphenicol in both yeast (Dujon, 1980) and mammalian mtDNA (Blanc *et al.*, 1981a,b; Kearsey and Craig, 1981) are within this conserved region. This portion of the large mitochondrial rRNA molecule from several mammalian species and from yeast can be folded into a common secondary structure, which is perfectly conserved in the *E. coli* 23 S rRNA (Baer and Dubin, 1981). Within this structure, there is the universal UmGmU hairpin, which also contains the puromycin-reactive site in the *E. coli* ribosome. The proposed extensive hairpin formation in this region would bring this putative puromycin-reactive site close to the chloramphenicol resistance sites in the yeast and mammalian large mitochondrial rRNA. Since chloramphenicol interferes with the binding of the 3′ end of the amino-acyl-tRNAs (Nierhaus and Nierhaus, 1973), it has been proposed that this region of the large mitochondrial rRNA molecule is involved in this binding (Baer and Dubin, 1981); probably, this region lies close to the interface of the two ribosomal subunits, since it is protected from kethoxal modification in the 70 S ribosome (Herr and Noller, 1979).

Although no sequence data are available for the mitochondrial rRNA genes from species pairs with short divergence times, from thermal stability measurements and cleavage map analysis, it has been concluded that the evolution rate of the mitochondrial rRNA genes is somewhat less than that of the protein coding genes, but at least 10-fold higher than that of the nuclear rRNA genes (Dawid, 1972; Brown *et al.*, 1979; Ferris *et al.*, 1981a, 1983). Here again, the overall higher mutation rate of mtDNA and a higher rate of fixation, resulting from the lower translational constraints characteristic of mitochondria, have been invoked (Brown, 1983).

3. *Protein Coding Genes*

The availability of the complete sequence of several mammalian mtDNAs and of the almost complete sequence of *Drosophila* mtDNA has allowed a direct analysis of the nucleotide sequence and amino acid sequence divergence of homologous genes among mammals and between mammals and *Drosophila*. As shown in Table V, the protein coding genes of different mammalian species exhibit, in pairwise comparisons, a nucleotide sequence homology varying between 61 and 79% and an amino sequence homology varying between 46 and 93%. With the exception of URF1, the URFs exhibit a lower degree of nucleotide and sequence conservation than the known protein coding genes. Among the known genes, that for COI reveals the highest degree of conservation, while URFA6L is least conserved among the URFs. There is in general a parallelism in the evolution of amino acid and nucleotide sequences. However, in the known genes, with the exception of that for COII, the nucleotide sequence conservation is less than the amino sequence conservation, as expected from the code degeneracy. By contrast, in the URFs, with the exception of URF1, the conservation of amino acid sequence is very similar to, or even lower than the conservation of nucleotide sequence, URFA6L being an extreme example of this tendency. The latter pattern points to the existence of some constraints in the evolution of the nucleotide sequence.

The *Drosophila* mitochondrial protein genes so far analyzed exhibit, not surprisingly, a greater divergence of both amino acid sequence and nucleotide sequence from the homologous mammalian genes than the genes of different mammalian species between each other, as illustrated in Table V for the pair *D. yacuba*/mouse. However, the tendency of the known genes to be in general more conserved than the URFs, with that for COI being the most conserved and URFA6L the least conserved, is observed also here. Also, one notices a general trend for all genes toward a somewhat greater conservation of the nucleotide sequence than of the amino acid sequence.

In the mammalian mitochondrial protein coding genes so far compared, all interspecific nucleotide sequence changes appear to be due to nucleotide substitutions or to small (1 to 3 codons) additions or deletions at the 5′ end or 3′ end of the reading frames. By contrast, small internal additions or deletions corresponding to 1 or 2 codons have been detected in the *Drosophila* mitochondrial protein coding genes, when compared to their mammalian counterparts. In both mammalian and *Drosophila* mtDNAs, nucleotide substitutions occur most frequently in the third positions of the codons. The fraction of base changes which are silent, i.e., do not lead to amino acid substitution, shows a loose positive correlation with the extent of amino acid conservation. The reported proportions of silent changes among the total base changes, in pairwise comparisons between homologous mitochondrial genes of several mammalian species, vary between 51 and 95% for different genes in the pair rat/mouse, and between 14 and 86% in the pairs rat/cow, rat/man, and mouse/man (Bibb *et al.*, 1981; Brown and Simpson, 1982; Pepe *et al.*, 1983). In all cases, the URFs exhibit a significantly lower proportion of silent changes than the known genes, with URFA6L providing an extreme case of this trend. However, it should be pointed out that in mtDNA (Brown and Simpson, 1982; Brown, 1983), as in nuclear genes (Perler *et al.*, 1980), the rate of accumulation of silent substitutions is several times higher than can be estimated from the measured frequency of silent changes in a comparison between mammalian species within divergence times of 80–100 million years, like man, mouse, rat, and cow. The reason for this is that the much higher rate of nucleotide substitution at the silent positions, which are free of functional constraints, as compared to that at the replacement positions, which are subject to stringent functional constraints, results in a saturation of the silent sites within divergence times of 80–100 million years, and consequently in multiple changes at the same site. By comparing mtDNAs from species whose divergence times are considerably less than the saturation time for silent sites, in particular mtDNAs from different primates, it has been possible to estimate that the "true" silent to replacement substitution ratio for divergence times within 15 million years is about five (Brown *et al.*, 1982). This estimate is similar to that reported for nuclear genes (Perler *et al.*, 1980), suggesting that the mitochondrial protein coding genes, though evolving more rapidly than the nuclear genes, are subjected to similar evolutionary constraints. This observation implies that the rapid sequence evolution of animal mtDNA cannot be due only to silent substitutions and, furthermore, that an increased rate of mutation must play an important role in determining the high rate of nucleotide sequence evolution of mtDNA (Brown, 1983).

4. *Regions of the Origins of Replication*

The portion of vertebrate mtDNA surrounding the origin of H-strand replication (O_H) and the homologous AT-rich segment of *Drosophila* mtDNA represent the most rapidly diverging region of these mtDNAs (Upholt and Dawid, 1977; Bibb *et al.*, 1981; Anderson *et al.*, 1982; Greenberg *et al.*, 1983). In vertebrate mtDNAs, this region includes the segment corresponding to the D-loop and extending from O_H to the $tRNA^{Pro}$ gene, and the adjacent segment comprised between the $tRNA^{Phe}$ gene and O_H (Fig. 2). As described above (Section V,D,5), this region contains also two H-strand and one L-strand transcription initiation sites (Montoya *et al.*, 1982). Although, in HeLa cell mtDNA, both strands are completely transcribed in this region (Murphy *et al.*, 1975), the only relatively stable discrete transcript mapping in this mtDNA segment is the 7 S RNA, the polyadenylated L-strand transcript encoded in the ~ 220 bp stretch upstream of O_H (Ojala *et al.*, 1981). An H-strand coded 4 S RNA mapping in the same region has been detected in *Xenopus* mitochondria (Rastl and Dawid, 1979).

The D-loop region and adjacent segments vary greatly in sequence in different animal species, even related ones. Thus, a comparison by heteroduplex and restriction enzyme mapping analysis of the D-loop region of sheep and goat mtDNAs has revealed segments of high sequence divergence at both ends of the D-loop, in particular in the segment upstream of O_H and in the segment overlapping and extending beyond the other end of the D-loop; at the latter end, a 150 bp duplication appears to have occurred in sheep mtDNA (Upholt and Dawid, 1977). Also in other mammalian species, the D-loop region is the least conserved, due to large deletions or addition events and to a high rate of nucleotide substitution, especially marked at the 5′ end and the 3′ end. The D-loop itself is ~ 250 bp shorter in mouse mtDNA than in human mtDNA (Bibb *et al.*, 1981), while the region between the tRNA^{Phe} and O_H is ~ 200 bp shorter in bovine mtDNA than in human mtDNA (Anderson *et al.*, 1982), these differences accounting for most of the size variation of the mtDNAs of these species. In a comparison of the mtDNAs of six primate species, a 95 bp deletion has been observed in gorilla mtDNA in the region between the tRNA^{Phe} gene and O_H (Ferris *et al.*, 1981a).

The observation that the region of mtDNA which contains the origin of mtDNA replication and the initiation sites for mtDNA transcription, and therefore, presumably, the regulatory sequences expected to interact with specific proteins, is the most rapidly diverging region of animal mtDNA, is very intriguing. A sequence comparison of the segment between tRNA^{Phe} and O_H of several mammalian mtDNAs, including human mtDNA, has revealed a conserved array of characteristic stretches [$(GT)_3$, AT-rich-sequence, oligo(G)]; this observation has prompted the suggestion that the arrangement of sequence types rather than the primary sequence may play the main role in the protein–DNA interactions important in mtDNA replication and transcription (Greenberg *et al.*, 1983).

In contrast to the O_H region, the region around the L-strand replication origin (O_L, Fig. 2) appears to be slowly diverging in its primary sequence and highly conserved in its potential secondary structure. In particular, the sizes of the stem and loop of the characteristic hairpin which can be drawn from the DNA sequence in this region appear to be preserved in different mammalian mtDNAs.

A comparison by heteroduplex and restriction mapping analysis of the AT-rich region of the mtDNA from several closely related species of *Drosophila* has revealed that also this region is the site of extensive deletion or addition events and of rapid primary sequence evolution (Fauron and Wolstenholme, 1980).

C. Intraspecific Variation of Mitochondrial DNA

The rapid evolution of mtDNA, as revealed by the interspecific comparisons discussed above, manifests itself also in the extensive intraspecific variation of this genome. Substantial intraspecific heterogeneity of mtDNA has in fact been detected in many mammalian species by restriction enzyme cleavage analysis and, in some cases, by DNA sequencing (Potter *et al.*, 1975; Upholt and Dawid, 1977; Francisco and Simpson, 1977; Avise *et al.*, 1979a,b; Brown, 1980; Castora *et al.*, 1980; Goddard *et al.*, 1981; Ferris *et al.*, 1981b; Denaro *et al.*, 1981; Hauswirth and Laipis, 1982). These studies have provided a wealth of information concerning the rate and mechanisms of mtDNA evolution, possible mechanisms of mtDNA transmission, and several questions pertaining to population genetics.

Particularly extensive have been the investigations on mtDNA polymorphism within human populations (Brown, 1980; Giles *et al.*, 1980; Denaro *et al.*, 1981; Greenberg *et al.*, 1983; Blanc *et al.*, 1983; Cann *et al.*, 1984). In the most comprehensive of these investigations, involving 112 individuals (Cann *et al.*, 1984), about 165 polymorphisms have been identified; these are presumably due, in their great majority, to single base substitutions. The nature and genomic distribution of these polymorphic sites have confirmed in general the conclusions of the analysis of interspecific variation of mtDNA, as concerns the relative rates of nucleotide substitution in noncoding sequences, rRNA genes, tRNA genes, and protein coding genes, the ratio of silent to replacement mutations in protein coding genes, and the bias in favor of transitional over transversional nucleotide changes. Among 19 amino acid replacements inferred from the restriction site gains, nine were radical in terms of size, structure, and polarity of the amino acid involved; several protein products, in particular, COII, COIII, cytochrome *b*, and the products of URF1, URF3, and URF6 exhibited multiple amino acid replacements. In the same set of mtDNA samples, in addition to base substitutions, 14 length variations (from about 6 to 14 bp) have been detected in the D-loop region and in other sites in the genome, probably within noncoding sequences and at junctions between coding sequences (Cann and Wilson, 1983).

The analysis of mtDNA polymorphism in human populations has also yielded valuable data concerning the origin of the human species and the relationship among different human groups (Brown, 1980; Denaro *et al.*, 1981; Ferris *et al.*, 1981a; Blanc *et al.*, 1983). On the basis of restriction endonuclease site polymorphisms, the level of sequence heterogeneity among human individuals has been estimated to be about 0.0018 substitutions per bp per individual, corresponding to a pairwise nucleotide divergence in humans of 0.0036 substitutions per bp (Brown, 1980). A somewhat lower value for pairwise sequence divergence (0.0022) has been derived from a comparative sequence analysis of the *Hin*dIII-C fragment of two human mtDNAs (Brown *et al.*, 1982). These estimates of sequence divergence are significantly lower than those reported for rodents, sheep, goats, and apes (0.005 to 0.05) (Upholt and Dawid, 1977; Avise *et al.*, 1979a,b; Ferris *et al.*, 1981b).On the basis of the estimated rate of base substitution in mammalian mtDNAs of 1% per lineage per 10^6 years (Brown *et al.*, 1979), it has been inferred that the present-day human individuals could have evolved from a small mitochondrially monomorphic population which existed 1.8 to 3.5 $\times$ 10^5 years ago, the differentiation into races having taken place subsequently (Brown, 1980).

D. Apparent Homogeneity of mtDNA within One Individual and the Mechanism of mtDNA Inheritance

A considerable amount of evidence suggests that the mtDNA molecules from different cells of an individual are essentially identical. The apparent mtDNA homogeneity within an individual stands in striking contrast to the fairly extensive intraspecific variation of this genome, and indicates that a rectification mechanism must exist, as postulated for the rRNA gene and other multigene families. It is generally accepted that, in the case of mtDNA, this rectification mechanism is associated with the mode of segregation of the mtDNA molecules during gametogenesis. That mitochondrial DNA in vertebrates is maternally inherited has been established in several systems, by taking advantage of the difference in restriction pattern or of more extensive sequence divergence of the two parental DNAs (Dawid and Blacker, 1972; Hutchinson *et al.*, 1974; Kroon *et al.*, 1977, 1978; Avise *et al.*, 1979a; Giles *et al.*, 1980). However, until now, a small paternal contribution has not been excluded.

The clonal theory of mtDNA rectification (Dawid and Blacker, 1972) assumes that any heterogeneity existing among the mtDNA molecules in the mature oocyte can be resolved into pure genotypes as a result of the partitioning of mitochondria in different fetal tissues during the early cleavage of the egg and of the segregation in the germ cells of only a minor portion of the original mtDNA complement of the oocyte. This reduction mechanism is formally equivalent to a recloning of the mtDNA population at each generation. It has been calculated that, if an immature female germ cell contains 100 to 1000 mtDNA molecules and there are 50 germ cell generations per animal generation, it would take 20 to 200 animal generations to obtain a pure mtDNA population from a mixed population, assuming a random segregation of the molecules at each cell division. This rate could be sufficiently high to assure an almost complete homogeneity of mtDNA in an individual animal (Upholt and Dawid, 1977). However, the total random segregation model would not account for the rapid shifts in mitochondrial genotype which have recently been observed within one Holstein cow maternal lineage (Hauswirth and Laipis, 1982). It has been suggested by the above cited authors that either a selective advantage of one genotype in the segregation process or differential amplification of mtDNA molecules during oocyte development (Piko and Matsumoto, 1976) could explain rapid genotypic shifts. In any case, the various models mentioned above all imply that the germline mtDNA population within one individual is polymorphic. The pedigree analysis of cow mtDNA indeed strongly suggests the existence of heterogeneity in germline mammalian mtDNA (Hauswirth and Laipis, 1982). Direct evidence for polymorphism within the mtDNA from germline cells of individual sea urchins has been recently obtained (Brown, 1983).

IX. Conclusions and Perspectives

This review has touched briefly upon the manifold aspects of the structure, function, and evolution of animal mtDNA, pointing to the more pressing questions raised by the recent unravelling of the mitochondrial genetic code and of the primary structure and gene organization of this remarkable genome. The determination of the complete sequence of human and other mammalian mtDNAs and the parallel detailed description of the structural, mapping, and metabolic properties of the mtDNA transcripts have provided a large amount of detailed information on the structure and function of this genome, which has no parallel in other genetic systems. This information has created the framework for attacking the most crucial problems concerning the mechanisms and regulation of expression of this genome and its evolution.

A great challenge lying ahead, concerning the functional role of the mitochondrial genome in animal cells, will be the elucidation of the physiological significance of the URFs, which represent almost 60% of the informational content of animal mtDNA. Such an achievement will have profound implications from the point of view of both the function of this DNA and its origin and evolution.

It can be expected that, in the next few years, the main features of the mechanisms of mtDNA transcription and RNA processing will be elucidated. However, a possibly more formidable task for the future will be the understanding of the mechanisms involved in the coordination and reciprocal control of expression of the mitochondrial genes and of the nuclear genes involved in mitochondriogenesis, and in the modulation of mitochondrial and nuclear gene activity in relationship to the respiratory demands of the cell.

Although a relatively few animal mtDNAs have so far been analyzed in their structure, they cover a large evolutionary distance (>600 million years), and thus they allow some tentative generalizations concerning their evolution. It seems likely that the mtDNAs from all vertebrates exhibit the same basic pattern of gene organization and expression, and possibly the same detailed gene order as that found in mammals and *Xenopus*. Other types of vertebrates, especially the more primitive, like cyclostomes and fishes, will have to be examined, however, before any conclusion can be reached on how far back the "mammalian type" goes, and on whether any gene order variants exist among early vertebrates. A comparison of the available data concerning the mtDNA gene organization in the two types of invertebrates so far examined, *Drosophila* and sea urchin, indicates that an extensive reshuffling of mitochondrial genes has occurred during invertebrate evolution, still preserving or leading to an organization of mtDNA with the same basic characteristics of compactness and economy. Here too, the analysis of other forms of invertebrates, particularly the primitive ones, will be essential to understand the rules which have governed this gene reshuffling and the mechanisms involved, and to try to establish an evolutionary hierarchy among the various types of gene order. This analysis will also be invaluable to ascertain whether the simplified mode of gene expression which involves transcription from one main promoter on each strand, with processing of polycistronic transcripts signaled by tRNA sequences, appeared very early in animal evolution, and whether in fact it exerted some constraints on the structural evolution. A more extensive analysis of mtDNA from various forms of invertebrates is likely to reveal other variants of the mitochondrial genetic code and codon recognition pattern, which will be invaluable for the elucidation of the mechanisms underlying the evolution of the genetic code under the more relaxed constraints operating in mitochondria (Cann *et al.*, 1984).

ACKNOWLEDGMENTS

The work from my laboratory described in this paper was supported by the National Institutes of Health Grants GM-11726 and T32 GM-07616. I am very grateful to Drs. D. Wolstenholme, B. Roe, and D. Dubin for communicating results prior to publication.

REFERENCES

Aloni, Y., and Attardi, G. (1971a). *Proc. Natl. Acad. Sci. U.S.A.* **68,** 1957–1961.

Aloni, Y., and Attardi, G. (1971b). *J. Mol. Biol.* **55,** 251–270.

Amalric, F., Merkel, C., Gelfand, R., and Attardi, G. (1978). *J. Mol. Biol.* **118,** 1–25.

Anderson, S., Bankier, A. T., Barrell, B. G., de Bruijn, M. H. L., Coulson, A. R., Drouin, J., Eperon, I. C., Nierlich, D. P., Roe, B. A., Sanger, F., Schreier, P. H., Smith, A. J. H., Staden, R., and Young, I. G. (1981). *Nature (London)* **290,** 457–465.

Anderson, S., de Bruijn, M. H. L., Coulson, A. R., Eperon, I. C., Sanger, F., and Young, I. G. (1982). *J. Mol. Biol.* **156,** 683–717.

Aquadro, C. F., and Greenberg, B. D. (1983). *Genetics* **103,** 287–312.

Arcari, P., and Brownlee, G. G. (1980). *Nucleic Acids Res.* **8,** 5207–5212.

Arnberg, A., van Bruggen, E. F. J., ter Schegget, J., and Borst, P. (1971). *Biochim. Biophys. Acta* **246,** 353–357.

Attardi, B., and Attardi, G. (1971). *J. Mol. Biol.* **55,** 231–249.

Attardi, G., Cantatore, P., Chomyn, A., Crews, S., Gelfand, R., Merkel, C., Montoya, J., and Ojala, D. (1982). *In* "Mitochondrial Genes" (P. Slonimski, P. Borst, and G. Attardi, eds.), pp. 51–71. Cold Spring Harbor Laboratory, Cold Spring Harbor, New York.

Aujame, L., and Freeman, K. B. (1979). *Nucleic Acids Res.* **6,** 455–470.

Avise, J. C., Lansman, R. A., and Shade, R. O. (1979a). *Genetics* **92,** 279–295.

Avise, J. C., Giblin-Davidson, C., Laerm, J., Patton, J. C., and Lansman, R. A. (1979b). *Proc. Natl. Acad. Sci. U.S.A.* **76,** 6694–6698.

Baer, R. J., and Dubin, D. T. (1981). *Nucleic Acids Res.* **9,** 323–337.

Barrell, B. G., Anderson, S., Bankier, A. T., de Bruijn, M. H. L., Chen, E., Coulson, A. R., Drouin, J., Eperon, I. C., Nierlich, D. P., Roe, B. A., Sanger, F., Schreier, P. H., Smith, A. J. H., Staden, R., and Young, I. G. (1980). *Proc. Natl. Acad. Sci. U.S.A.* **77,** 3164–3166.

Bianchetti, R., Lucchini, G., and Sartirana, M. L. (1971). *Biochem Biophys. Res. Commun.* **42,** 97–102.

Bibb, M. J., Van Etten, R. A., Wright, C. T., Walberg, M. W., and Clayton, D. A. (1981). *Cell* **26,** 167–180.

Blanc, H., Wright, C. T., Bibb, M. J., Wallace, D. C., and Clayton, D. A. (1981a). *Proc. Natl. Acad. Sci. U.S.A.* **78,** 3789–3793.

Blanc, H., Adams, C. W., and Wallace, D. C. (1981b). *Nucleic Acids Res.* **9,** 5785–5795

Blanc, H., Chen, K.-H., D'Amore, M. A., and Wallace, D. C. (1983). *Am. J. Hum. Genet.* **35,** 167–176.
Bonitz, S. G., Berlani, R., Coruzzi, G., Li, M., Macino, G., Nobrega, F. G., Nobrega, M. P., Thalenfeld, B. E., and Tzagoloff, A. (1980). *Proc. Natl. Acad. Sci. U.S.A.* **77,** 3167–3170.
Borst, P. (1972). *Annu. Rev. Biochem.* **41,** 333–376.
Bothwell, A. L. M., Stark, B. C., and Altman, S. (1976). *Proc. Natl. Acad. Sci. U.S.A.* **73,** 1912–1916.
Brennicke, A., and Clayton, D. A. (1981). *J. Biol. Chem.* **256,** 10613–1O617.
Brown, G. G., and Simpson, M. V. (1982). *Proc. Natl. Acad. Sci. U.S.A.* **79,** 3246–3250.
Brown, W. M. (1980). *Proc. Natl. Acad. Sci. U.S.A.* **77,** 3605–3609.
Brown, W. M. (1983). *In* "Evolution of Genes and Proteins" (M. Nei and R. K. Koehn, eds.), pp. 62–88. Sinauer, Sunderland, Massachusetts.
Brown, W. M., George, M., and Wilson, A. C. (1979). *Proc. Natl. Acad. Sci. U.S.A.* **76,** 1967–1971.
Brown, W. M., Prager, E. M., Wang, A., and Wilson, A. C. (1982). *J. Mol. Evol.* **18,** 225–239.
Brown, T. A., Davies, R. W., Ray, J. A., Waring, R. B., and Scazzocchio, C. (1983). *EMBO J.* **2,** 427–435.
Browning, K. S., and RajBhandary, U. L. (1982). *J. Biol. Chem.* **257,** 5253–5256.
Cann, R. L., and Wilson, A. C. (1984). *Genetics* **104,** 699–711.
Cann, R. L., Brown, W. M., and Wilson, A. C. (1983). *Genetics* **106,** 479–499.
Cantatore, P., and Attardi, G. (1980). *Nucleic Acids Res.* **8,** 2605–2624.
Cantatore, P., De Benedetto, C., Gadaleta, G., Gallerani, R., Kroon, A. M., Holtrop, M., Lanave, C., Pepe, G., Quagliariello, C., Saccone, C., and Sbisà, E. (1982). *Nucleic Acids Res.* **10,** 3279–3289.
Castora, F. J., Arnheim, N., and Simpson, M. V. (1980). *Proc. Natl. Acad. Sci. U.S.A.* **77,** 6415–6419.
Ching, E., and Attardi, G. (1982). *Biochemistry* **21,** 3188–3195.
Chomyn, A., Hunkapiller, M. W., and Attardi, G. (1981). *Nucleic Acids Res.* **9,** 867–877.
Chomyn, A., Mariottini, P., Attardi, G., Strong, D. D., Trovato, D., Riley, M., and Doolittle, R. (1983). *Proc. Natl. Acad. Sci. U.S.A.* **80,** 5535–5539.
Clary, D. O., and Wolstenholme, D. R. (1983a). *Nucleic Acids Res.* **11,** 4211–4227.
Clary, D. O., and Wolstenholme, D. R. (1983b). *Nucleic Acids Res.* **11,** 6859–6872.
Clary, D. O., and Wolstenholme, D. R. (1983c). *Nucleic Acids Res.* **12,** 2367–2379.
Clary, D. O., Goddard, J. M., Martin, S. C., Fauron, C. M.-R., and Wolstenholme, D. R. (1982). *Nucleic Acids Res.* **10,** 6619–6637.
Clary, D. O., Wahleithner, J. A., and Wolstenholme, D. R. (1983). *Nucleic Acids Res.* **11,** 2411–2425.
Clayton, D. A. (1982). *Cell* **28,** 693–705.
Crews, S., and Attardi, G. (1980). *Cell* **19,** 775–784.
Crews, S., Ojala, D., Posakony, J., Nishiguchi, J., and Attardi, G. (1979). *Nature (London)* **277,** 192–198.
Crick, F. H. C. (1966). *J. Mol. Biol.* **19,** 548–555.
Davenport, L. W., Taylor, R. H., and Dubin, D. T. (1976). *Biochim Biophys. Acta* **447,** 285–293.
Dawid, I. B. (1972). *Dev. Biol.* **29,** 139–151.
Dawid, I. B., and Blacker, A. W. (1972). *Dev. Biol.* **29,** 152–161.
de Bruijn, M. H. L. (1983). *Nature (London)* **304,** 234–241.
de Bruijn, M. H. L., Schreier, P. H., Eperon, I. C., and Barrell, B. G. (1980). *Nucleic Acids Res.* **8,** 5213–5222.
Denaro, M., Blanc, H., Johnson, M. J., Chen, K. H., Wilmsen, E., Cavalli-Sforza, L. L., and Wallace, D. C. (1981). *Proc. Natl. Acad. Sci. U.S.A.* **78,** 5768–5772.

Dubin, D. T., and HsuChen, C.-C. (1983). *Plasmid* **9,** 307–320.
Dubin, D. T., and Taylor, R. H. (1978). *J. Mol. Biol.* **121,** 523–540.
Dubin, D. T., Taylor, R. H., and Davenport, L. W. (1978). *Nucleic Acids Res.* **5,** 4385–4397.
Dubin, D. T., Timko, K. D., and Baer, R. J. (1981). *Cell* **23,** 271–278.
Dubin, D. T., Montoya, J., Timko, K. D., and Attardi, G. (1982a). *J. Mol. Biol.* **157,** 1–19.
Dubin, D. T., *et al.* (1982b). *In* "Mitochondrial Genes" (P. Slonimski, P. Borst, and G. Attardi, eds.), pp. 89–98. Cold Spring Harbor Lab., Cold Spring Harbor, New York.
Dubin, D. T., HsuChen, C.-C., Cleaves, G. R., and Timko, K. D. (1984). *J. Mol. Biol.* **176,** 251–260.
Dujon, B. (1980). *Cell* **20,** 185–197.
Eaton, B. T., and Randlett, D. J. (1978). *Nucleic Acids. Res.* **5,** 1301–1314.
England, J., Costantino, P., and Attardi, G. (1978). *J. Mol. Biol.* **119,** 455–462.
Eperon, I. C., Anderson, S., and Nierlich, D. P. (1980). *Nature (London)* **286,** 460–467.
Ephrussi, B., Hottinguer, H., and Tavlitzki, J. (1949). *Ann. Inst. Pasteur* **76,** 419–451.
Fauron, C. M.-R., and Wolstenholme, D. R. (1980). *Nucleic Acids Res.* **8,** 2439–2452.
Ferris, S. D., Wilson, A. C., and Brown, W. M. (1981a). *Proc. Natl. Acad. Sci. U.S.A.* **78,** 2432–2436.
Ferris, S. D., Brown, W. M., Davidson, W. S., and Wilson, A. C. (1981b). *Proc. Natl. Acad. Sci. U.S.A.* **78,** 6319–6323.
Ferris, S. D., Sage, R. D., Prager, E. M., Ritte, U., and Wilson, A. C. (1983). *Genetics* **105,** 681–721.
Fox, T. D., and Weiss-Brummer, B. (1980). *Nature (London)* **288,** 60–63.
Francisco, J. F., and Simpson, M. V. (1977). *FEBS Lett.* **79,** 291–294.
Gaines, G., and Attardi, G. (1984). *J. Mol. Biol.,* **172,** 451–466.
Gelfand, R., and Attardi, G. (1981). *Mol. Cell. Biol.* **1,** 497–511.
Giles, R. E., Blanc, H., Cann, H. M., and Wallace, D. C. (1980). *Proc. Natl. Acad. Sci. U.S.A.* **77,** 6715–6719.
Goddard, J. M., and Wolstenholme, D. R. (1978). *Proc. Natl. Acad. Sci. U.S.A.* **75,** 3886–3890.
Goddard, J. M., and Wolstenholme, D. R. (1980). *Nucleic Acids Res.* **8,** 741–757.
Goddard, J. M., Masters, J. N., Jones, S. S., Ashworth, W. D., and Wolstenholme, D. R. (1981). *Chromosoma* **82,** 595–609.
Godson, G. N., Barrell, B. G., Staden, R., and Fiddes, J. C. (1978). *Nature (London)* **276,** 236–247.
Greenberg, B. D., Newbold, J. E., and Sugino, A. (1983). *Gene* **21,** 33–49.
Grisi, E., Brown, T. A., Waring, R. B., Scazzocchio, C., and Davies, R. W. (1982). *Nucleic Acids Res.* **10,** 3531–3539.
Grohmann, K., Amalric, F., Crews, S., and Attardi, G. (1978). *Nucleic Acids Res.* **5,** 637–651.
Grosskopf, R., and Feldmann, H. (1981a). *Curr. Genet.* **4,** 151–158.
Grosskopf, R., and Feldmann, H. (1981b). *Curr. Genet.* **4,** 191–196.
Hare, J. F., Ching, E., and Attardi, G. (1980). *Biochemistry* **19,** 2023–2030.
Hauswirth, W. H., and Laipis, P. J. (1982). *Proc. Natl. Acad. Sci. U.S.A.* **79,** 4686–4690.
Heckman, J. E., Sarnoff, J., Alzner-DeWeerd, B., Yyn, S., and RajBhandary, U. L. (1980). *Proc. Natl. Acad. Sci. U.S.A.* **77,** 3159–3163.
Herr, W., and Noller, H. F. (1979). *J. Mol. Biol.* **130,** 421–432.
Hirsch, M., and Penman, S. (1973). *J. Mol. Biol.* **80,** 379–391.
Hoeijmakers, J. H. J., Snijders, A., Janssen, J. W. G., and Borst, P. (1981). *Plasmid* **5,** 329–350.
HsuChen, C.-C., Cleaves, G. R., and Dubin, D. T. (1983). *Nucleic Acids Res.,* **11,** 8659–8662.
Hudspeth, M. E. S., Ainley, W. M., Shumard, D. S., Butow, R. A., and Grossman, L. I. (1982). *Cell* **30,** 617–626.

Hutchinson, C. A., III, Newbold, J. E., Potter, S. S., and Edgell, M. H. (1974). *Nature (London)* **251,** 536–538.
Kasamatsu, H., Robberson, D. L., and Vinograd, J. (1971). *Proc. Natl. Acad. Sci. U.S.A.* **68,** 2252–2257.
Kastelein, R. A., Remaut, E., Fiers, W., and van Duin, J. (1982). *Nature (London)* **295,** 35–41.
Kearsey, S. E., and Craig, I. W. (1981). *Nature (London)* **290,** 607–608.
Klukas, C. K., and Dawid, I. B. (1976). *Cell* **9,** 615–625.
Kobayashi, M., Seki, T., Yaginuma, K., and Koike, K. (1981). *Gene* **16,** 297–307.
Kotin, R. M., and Dubin, D. T. (1984). *Biochim. Biophys. Acta* **782,** 106–108.
Kozak, M. (1981). *Curr. Top. Microbiol. Immunol.* **93,** 81–123.
Kroon, A. M., Pepe, G., Bakker, H., Holtrop, M., Bollen, J. E., Vanbruggen, E. F., Cantatore, P., Terpstra, P., and Saccone, C. (1977). *Biochim. Biophys. Acta* **478,** 128–145.
Kroon, A. M., deVos, W. M., and Bakker, H. (1978). *Biochim. Biophys. Acta* **519,** 269–273.
Kyte, J., and Doolittle, R. F. (1982). *J. Mol. Biol.* **157,** 105–132.
Lagerkvist, U. (1978). *Proc. Natl. Acad. Sci. U.S.A.* **75,** 1759–1762.
Lansman, R. A., and Clayton, D. A. (1975). *J. Mol. Biol.* **99,** 777–793.
Lynch, D., and Attardi, G. (1976). *J. Mol. Biol.* **102,** 125–141.
Macreadie, I. G., Novitski, C. E., Maxwell, R. J., John, U., Ooi, B.-G., McMullen, G. L., Lukins, H. B., Linnane, A. W., and Nagley, P. (1983). *Nucleic Acids Res.* **11,** 4435–4451.
Mariottini, P., Chomyn, A., Attardi, G., Trovato, D., Strong, D. D., and Doolittle, R. F. (1983). *Cell* **32,** 1269–1277.
Merten, S. H., and Pardue, M. L. (1981). *J. Mol. Biol.* **153,** 1–21.
Mitchell, M. B., and Mitchell, H. K. (1952). *Proc. Natl. Acad. Sci. U.S.A.* **38,** 442–449.
Montoya, J., Ojala, D., and Attardi, G. (1981). *Nature (London)* **290,** 465–470.
Montoya, J., Christianson, T., Levens, D., Rabinowitz, M., and Attardi, G. (1982). *Proc. Natl. Acad. Sci. U.S.A.* **79,** 7195–7199.
Montoya, J., Gaines, G., and Attardi, G. (1983). *Cell* **34,** 151–159.
Murphy, W., Attardi, B., Tu, C., and Attardi, G. (1975). *J. Mol. Biol.* **99,** 809–814.
Netzker, R., Köchel, H. G., Basak, N., and Küntzel, H. (1982). *Nucleic Acids Res.* **10,** 4783–4794.
Nierhaus, D., and Nierhaus, K. H. (1973). *Proc. Natl. Acad. Sci. U.S.A.* **70,** 2224–2228.
Nierlich, D. P. (1982). *Mol. Cell. Biol.* **2,** 207–209.
Ohi, S., Ramirez, J. L., Upholt, W. B., and Dawid, I. B. (1978). *J. Mol. Biol.* **121,** 299–310.
Ojala, D., and Attardi, G. (1974). *J. Mol. Biol.* **82,** 151–174.
Ojala, D., Merkel, C., Gelfand, R., and Attardi, G. (1980). *Cell* **22,** 393–403.
Ojala, D., Montoya, J., and Attardi, G. (1981a). *Nature (London)* **290,** 470–474.
Ojala, D., Crews, S., Montoya, J., Gelfand, R., and Attardi, G. (1981b). *J. Mol. Biol.* **150,** 303–314.
Oliver, N. A., Greenberg, B. D., and Wallace, D. C. (1983). *J. Biol. Chem.* **258,** 5834–5839.
Pepe, G., Holtrop, M., Gadaleta, G., Kroon, A. M., Cantatore, P., Gallerani, R., De Benedetto, C., Quagliariello, C., Sbisà, E., and Saccone, C. (1983). *Biochem. Int.* **6,** 553–563.
Perler, F., Efstradiatis, A., Lomedico, P., Gilbert, W., Kolodner, R., and Dodgson, J. (1980). *Cell* **20,** 555–566.
Piko, L., and Matsumoto, L. (1976). *Dev. Biol.* **49,** 1–10.
Potter, S. S., Newbold, J. E., Hutchinson, C. A., III, and Edgell, M. H. (1975). *Proc. Natl. Acad. Sci. U.S.A.* **72,** 4496–4500.
Prunell, A., and Bernardi, G. (1974). *J. Mol. Biol.* **86,** 825–841.
Prunell, A., and Bernardi, G. (1977). *J. Mol. Biol.* **110,** 53–74.
Ramirez, J. L., and Dawid, I. B. (1978). *J. Mol. Biol.* **119,** 133–146.
Randerath, E., Agrawal, H. P., and Randerath, K. (1981). *Biochem. Biophys. Res. Commun.* **103,** 739–744.

Rastl, E., and Dawid, I. B. (1979). *Cell* **18,** 501–510.
Roberts, J. W., Grula, J. W., Posakony, J. W., Hudspeth, R., Davidson, E. H., and Britten, R. J. (1983). *Proc. Natl. Acad. Sci. U.S.A.* **80,** 4614–4618.
Robberson, D. L., Kasamatsu, H., and Vinograd, J. (1972). *Proc. Natl. Acad. Sci. U.S.A.* **69,** 737–741.
Robertson, M. D., Altman, S., and Smith, J. D. (1972). *J. Biol. Chem.* **247,** 5243–5251.
Saccone, C., Cantatore, P., Gadaleta, G., Gallerani, R., Lanave, C., Pepe, G., and Kroon, A. M. (1981). *Nucleic Acids Res.* **9,** 4139–4148.
Saccone, C., De Benedetto, C., Gadaleta, G., Lanave, C., Pepe, G., Rainaldi, G., Sbisà, E., Cantatore, P., Gallerani, R., Quagliariello, C., Holtrop, M., and Kroon, A. M. (1982a). *In* "Mitochondrial Genes" (P. Slonimski, P. Borst, and G. Attardi, eds.), pp. 121–128. Cold Spring Harbor Laboratory, Cold Spring Harbor, New York.
Saccone, C., Cantatore, P., De Benedetto, C., Gadaleta, G., Gallerani, R., Holtrop, M., Kroon, A. M., Lanave, C., Pepe, G., Quagliariello, C., and Sbisà, E. (1982b). *In* "Cell Function and Differentiation," 30th Ed., Part B, pp. 367–374. Liss, New York.
Sanger, F., Air, G. M., Barrell, B. G., Brown, N. L., Coulson, A. R., Fiddes, J. C., Hutchinson, C. A., III, Slocombe, P. M., and Smith, M. (1977). *Nature* (London) **265,** 687–695.
Sekiya, T., Kobayashi, M., Seki, T., and Koike, K. (1980). *Gene* **11,** 53–62.
Sibler, A.-P., Bordonné, R., Dirheimer, G., and Martin, R. (1980). *C. R. Acad. Sci. Paris, Ser. D* **290,** 695–698.
Simpson, L., and Simpson, A. (1978). *Cell* **14,** 169–178.
Sprinzl, M., Grueter, F., Spelzhaus, A., and Gauss, D. H. (1980). *Nucleic Acids. Res.* **8,** r1–r22.
Steffens, G. J., and Buse, G. (1979). *Hoppe-Seyler's Z. Physiol. Chem.* **360,** 613–619.
Steitz, J. A., and Jakes, K. (1975). *Proc. Natl. Acad. Sci. U.S.A.* **72,** 4734–4738.
Tabak, H. F., Grivell, L. A., and Borst, P. (1983). *CRC Crit. Rev. Biochem.* **14,** 297–317.
Upholt, W. B., and Dawid, I. B. (1977). *Cell* **11,** 571–583.
Van Etten, R. A., Michael, N. L., Bibb, M. J., Brennicke, A., and Clayton, D. A. (1982). *In* "Mitochondrial Genes" (P. Slonomski, P. Borst, and G. Attardi, eds.), pp. 73–88. Cold Spring Harbor Laboratory, Cold Spring Harbor, New York.
Van Etten, R. A., Bird, J. W., and Clayton, D. A. (1983). *J. Biol. Chem.* **258,** 10104–10110.
Wolstenholme, D. R., Goddard, J. M., and Fauron, M.-R. (1979). *In* "Extrachromosomal DNA" (D. J. Cummings, P. Borst, I. B. Dawid, S. M. Weissman, and C. F. Fox, eds.), pp. 409–425. Academic Press, New York.

Note Added in Proof

A specific mitochondrial RNA polymerase has recently been partially purified from human cells and shown to be able to start transcription on a exogenus human mtDNA template at the rRNA-specific H-strand *in vivo* initiation site and at the L-strand *in vivo* initiation site (D. D. Chang and D. A. Clayton, *Cell* **36,** 635–643, 1984; D. F. Bogenhagen, E. F. Appelgate, and B. K. Yoza, *Cell* **36,** 1105–1113, 1984; D. Shuey and G. Attardi, *J. Biol. Chem.*, in press). The sequences required for initiation of transcription at the two sites have been defined by deletion mutagenesis and found to correspond in each case to a stretch encompassing the transcription start site.

An RNase P-like activity has been partially purified from HeLa cell mitochondria and shown to be different in some of its properties from the nuclear enzyme (C. Doersen, C. Guerrier-Takada, S. Altman, and G. Attardi, *J. Biol. Chem.*, in press).

Section III

Split Genes and the Genetic Evolution of Eukaryotes

INTERNATIONAL REVIEW OF CYTOLOGY, VOL. 93

Exons and the Evolution of Proteins

C. C. F. BLAKE

Laboratory of Molecular Biophysics, University of Oxford, Oxford, England

I. Introduction

The splitting of most eukaryotic structural genes into coding sequences (exons) and noncoding sequences (introns) has been widely interpreted in terms of evolutionary processes concerned with the gene products (Breathnach and Chambon, 1981). The most wide-ranging and influential hypothesis of the value of the exon/intron mosaic in increasing the scope and rate of protein evolution is that proposed by Gilbert (Gilbert, 1978; Tonegawa *et al.*, 1978). He noted two distinct aspects of the gene mosaic that could assist with the production of novel gene products: (1) if exonic regions correspond to individual protein functions, recombination within the intronic regions could reassort these functions into novel protein products (the reading of the gene over a larger region of DNA would considerably enhance the probabilities of recombination); and (2) point mutations at the exon/intron splice junctions could result in the deletion or addition of whole blocks of amino acid sequences, while variable splicing could allow transcription of both the original and new gene products simultaneously.

ISBN 0-12-364493-3

Blake (1978) pointed out in relation to (1) that if exons also corresponded to folded protein units the probability of the new protein achieving the structural integrity it requires for expressing its function will be much improved by allowing integrally folded units to be added to one another. It may thus be possible to identify such units and their functions in contemporary proteins and trace their evolution. So great are the potential evolutionary advantages of a mosaic gene structure that a number of writers (Doolittle, 1978; Darnell, 1978; Reanney, 1979) have suggested that despite its absence in prokaryotic genes, the exon/intron mosaic is likely to be representative of early gene structures.

This article is not intended to be a comprehensive review of the literature on gene structure and protein structure, but in compliance with the requests of the editors, will attempt a critical examination of Gilbert's hypothesis in the light of experimental evidence and seek to set the hypothesis in a more comprehensive framework.

II. Protein Structure

In order to be able to consider the role of exons in protein evolution, it is necessary to give an outline of the principles of protein structure: a more detailed consideration can be found in Schulz and Schirmer (1979). In three-dimensional terms protein molecules, or more precisely polypeptide chains, can be most conveniently described in terms of a structural hierarchy having probably four levels; it should be noted that while the top and bottom levels of this hierarchy can be rigorously defined, the middle two levels are rather more elusive to absolute definition. Considering the structural hierarchy in terms of increasing complexity, the lowest level is represented by the elements of secondary structure—the α-helix, the β-sheet, and the collagen helix; the second level envisages these secondary structures organized into local units of structure, sometimes called supersecondary structures, or sectors; the third level is represented by domains which have many of the properties of a folded protein molecule but are linked covalently through the polypeptide chain to one or more other domains; the final level, the tertiary structure, is the complete folding of the polypeptide chain to form the protein molecule, or the subunit of polymeric proteins.

A. Secondary Structures

The number and types of stable secondary structure in proteins were established definitively by Ramachandran and his colleagues (Ramachandran and Sasisekharan, 1968), by analyzing the configurational energy in terms of the two dihedral angles at the α-carbon atom. As shown in Fig. 1, this plot demonstrates that only three regular extended structures are possible: the α-helix, the β-sheet,

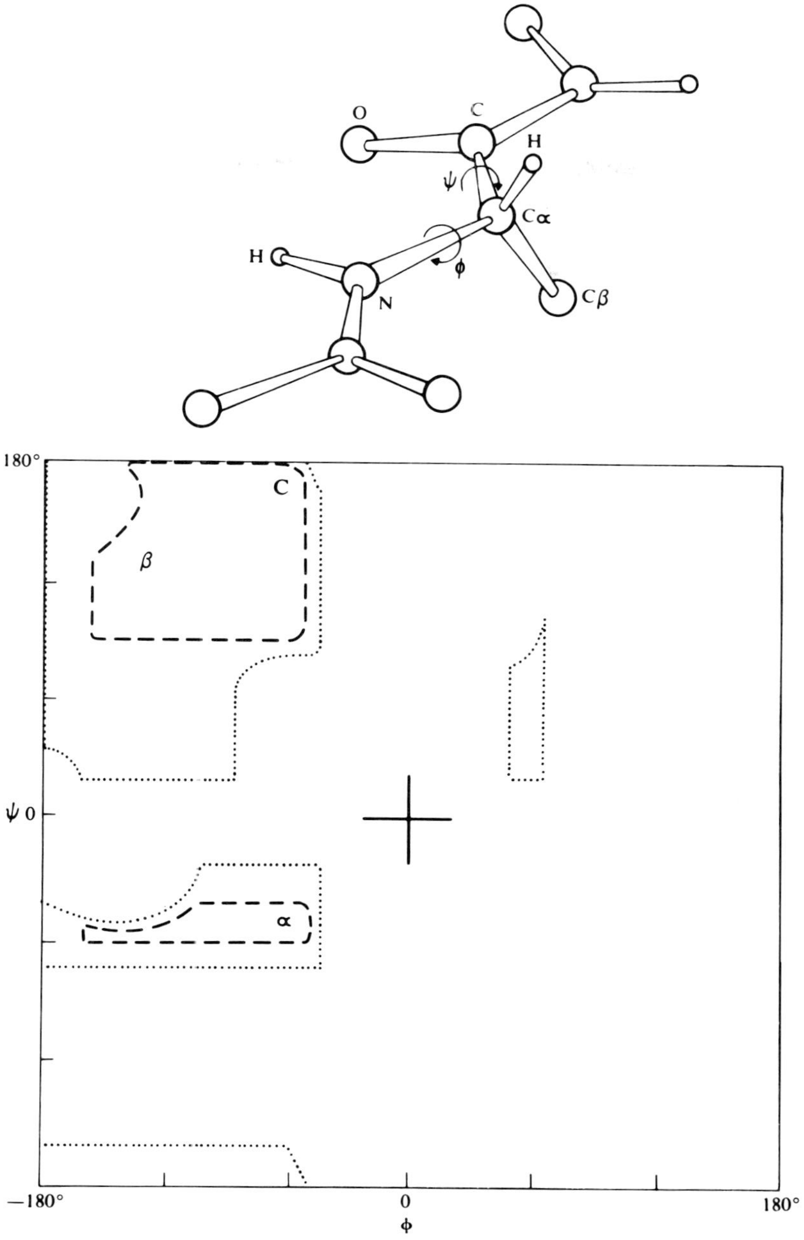

FIG. 1. A Ramachandran Plot of the torsion angles ϕ and ψ, defined at top. The dotted lines enclose areas of allowed conformation, and the broken lines enclose areas of the most favourable conformations. α, β, and C represent the optimum conformations of the α-helix, β-sheet, and collagen helix, respectively.

and the collagen helix. Each of these structures not only occupies a minimum in the energy plot, but is also capable of forming sequential hydrogen bonds between the carbonyl oxygen and the imino hydrogen atoms of the constituent peptide groups. The hydrogen bonds in the α-helix occur between residues four positions apart along the helix; β-strands hydrogen bond to other β-strands running either parallel or antiparallel with respect to one another; the collagen helix hydrogen bonds one polypeptide chain to two others all running in the same direction to form a triple stranded structure (Fig. 2). The collagen helix is the only one of the secondary structure that is absolutely dependent on the sequences of amino acids: it requires a repeated Gly-x-y sequence in each of the three chains.

Subsequent X-ray analysis of a wide range of proteins, both fibrous and globular, has shown that these three secondary structures are the only extended elements of structure present in proteins. With fibrous proteins usually only one type of secondary structure is present and tends to extend throughout the molecule (the tertiary structure is thus the same as the secondary structure) to give the characteristic rod-like or sheet-like molecules. In globular proteins, the secondary structures may be mixed, and are usually short to enable the ensemble of structures to fold together in a compact globular form. Possibly because collagen helix requires three polypeptide chains it is only very rarely found in globular proteins.

B. Tertiary Structures

It will be more convenient to discuss tertiary structures here, and then to consider supersecondary structures and domains in the light of this section. Protein tertiary structures are organized from a folded ensemble of secondary structures. To enable the rigid secondary elements to fold into a globular form loops of chain lacking *repeated* regular structure link the secondary elements and allow the whole chain to articulate. These loops may on occasion be quite lengthy, but the reverse turn, a strictly local regular structure which allows the polypeptide chain to reverse direction in three peptide units, is quite common. Although it is difficult to generalize, the current sample of known globular protein structure suggests that in most proteins between 50 and 75% of the chain is involved in secondary structures, and the remainder in the loops, and possibly N- and C-terminal extensions.

FIG. 2. Drawings of the three regular elements of secondary structure. Top left, the α-helix; top right, the three-stranded collagen helix, one of whose chains is picked out in dark shading. Bottom, a segment of parallel β-sheet, the "pleating" of which is shown on the left. Interpeptide hydrogen bonds are shown in dotted line.

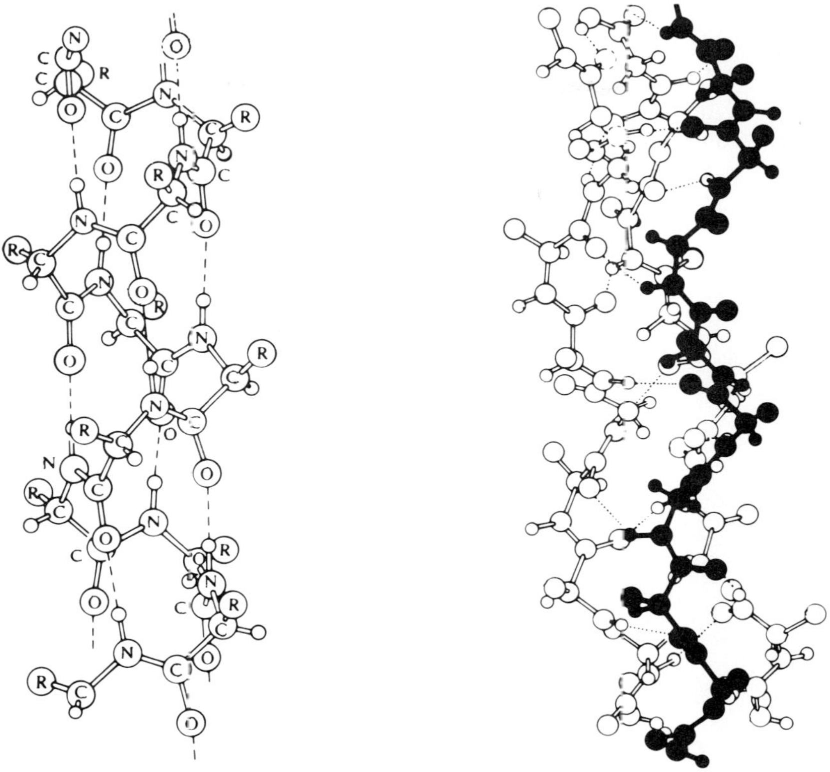

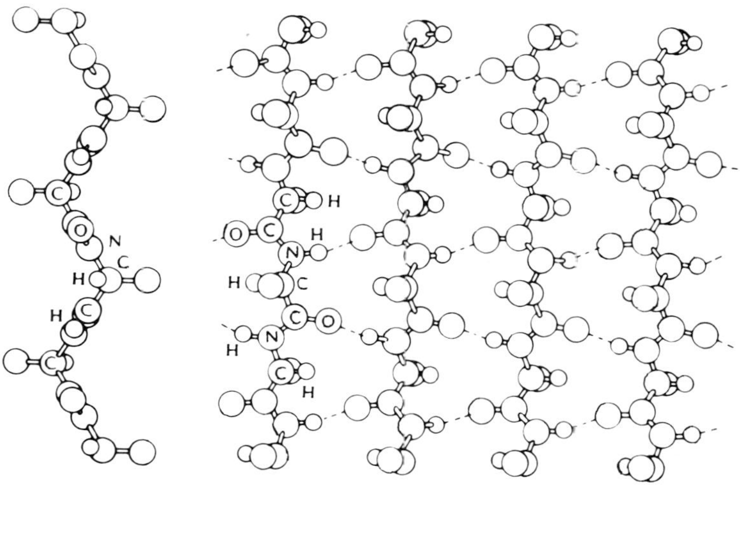

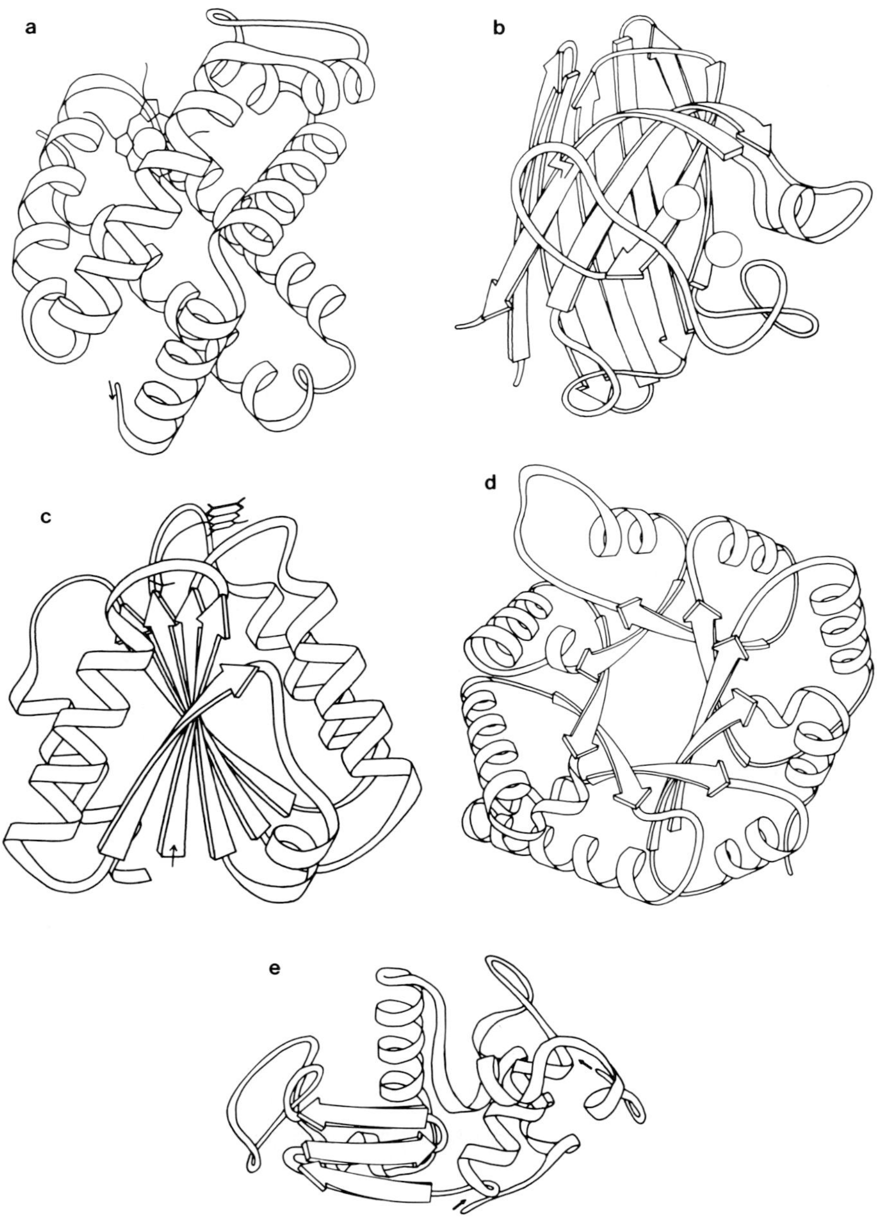

FIG. 3. Schematic drawings by Jane Richardson of the main-chain conformations of examples of the major structural classifications of protein molecules. (a) Hemoglobin β-subunit—an all-α protein; (b) Cu/Zn superoxide dismutase—an all-β protein; (c) flavodoxin—an α/β protein; (d) a subunit of triose phosphate isomerase—an α/β barrel protein; (e) hen egg-white lysozyme—an α+β protein. Reproduced from Phillips *et al.* (1983).

Although short segments of α- and β-secondary structures can be mixed in globular proteins, they do so in a relatively well-organized manner to form four identifiable classes of tertiary structure (Levitt and Chothia, 1976; Richardson, 1981). These four classes are illustrated in Fig. 3. In the "all-α" class, only α-helices are present as secondary structures and the chain folds to pack the helices together in certain well-defined arrangements. In the "all-β" class, all the secondary structures are β-strands which hydrogen bond together to form twisted sheets. There is a tendency for this type of β-sheet to be formed largely, or even exclusively, of antiparallel strands. The completed molecule (or domain) tends to be formed of two β-sheets packed face-to-face, or organized into a barrel-like structure.

In the third class, the "α+β" proteins, both α- and β-secondary structures are present, but are *segregated* in the tertiary structure some parts of which are like the "all-α" class, and other regions are like the "all-β" class. The final class is the "α/β" class in which α-helices and β-strands *alternate* along the polypeptide chain. The tertiary structures of these molecules have a central β-sheet, in which the β-strands tend to be parallel, flanked on either side by α-helices running antiparallel to the local β-strands. It is also possible in this class of protein for the β-sheet to form a barrel, with the associated α-helices forming an outer layer to the barrel structure. Apart from a number of rather small proteins that tend to be rich in disulfide bridges which form somewhat irregular structures, nearly all of the larger protein molecules clearly belong to one of the four classes described above.

C. Domains

A domain may be simply described as a part of a molecule whose structure is organized like a complete globular protein. It differs from a single molecule in that it is connected by the polypeptide chain to one or more other domains of similar characteristics; the majority of proteins with polypeptide chains of more than 200 residues contain domains, and the longer the chain the larger the number of domains. The most characteristic example of domain structure is the immunoglobulins, whose heavy chain in the IgG variants contains four domains of similar size and structure connected by segments of extended polypeptide chain (see Fig. 6), but many enzyme molecules have two or three domains in their polypeptide chains (Fig. 4). Since a domain is defined in terms of being like an individual protein, the structure of domains can be classified in one of the four structural classes, as in (B) above, and different domains in the same polypeptide chain can belong to the same class or to different classes. A nice example of the latter case is pyruvate kinase, which has three domains: one β-class, one α/β in the form of an eight-stranded barrel; and one α/β with a central five stranded β-sheet (Fig. 5).

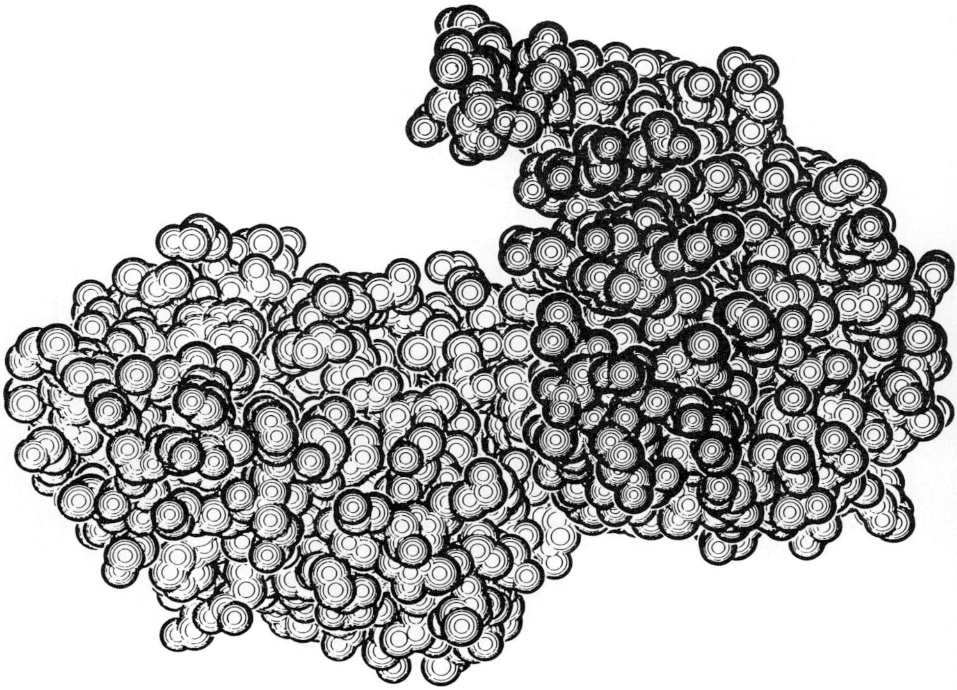

FIG. 4. A space-filling drawing of the two-domain yeast phosphoglycerate kinase molecule. Reproduced by courtesy of H. C. Watson and N. P. C. Walker.

Domains in proteins are often associated with individual functions. For example in domain enzymes, individual substrates or effectors are often bound to individual domains, but the active site is usually located at the interface between two domains. (In phosphorylase, which has three domains, the active site is situated at the point where all three domains come together.) With the immunoglobulins, the antigen binding function is associated with the N-terminal variable domain, but the various cell binding, complement fixation affector functions are associated with the two C-terminal domains. There are many other examples of this kind of behavior that are too numerous to mention in detail.

One particularly intriguing aspect of domains when considering evolutionary properties is the observation that domains with similar functions in different molecules sometimes have similar tertiary structures. The most well known example is the NAD-binding domains in lactate, glyceraldehyde-3-phosphate, and alcohol dehydrogenase (Rossmann *et al.*, 1974), which, in each enzyme, belong to the α/β class and have a core of six parallel β-strands. This structure is also present as the ATP binding domain of phosphoglycerate kinase, and also possibly with the same functions in phosphorylase and the regulatory subunit of aspartate transcarbamylase. In each of these molecules the other domains are quite different, and furthermore the NAD binding domain is N-terminal in LDH and GAPDH but C-terminal in ADH and PGK. Another example is that the central α/β barrel domain of pyruvate kinase is structurally similar to the complete triose phosphate isomerase subunit (Fig. 5), with the active sites of both enzymes, whose catalytic mechanisms involves proton abstraction, is situated in the same place relative to the common structure.

There has been much controversy on the significance of these observations: whether they indicate that particular elements of structure with specific functions had evolved from a common ancestor and become fused with other structures to produce the present-day proteins; or whether these are just examples of convergent evolution in which elements of a particular structural type had been used for specific functions. These discoveries and arguments occurred before split genes were suspected and it seems reasonable now to wait for the particular gene structures to see whether they shed any light on possible evolutionary relationships.

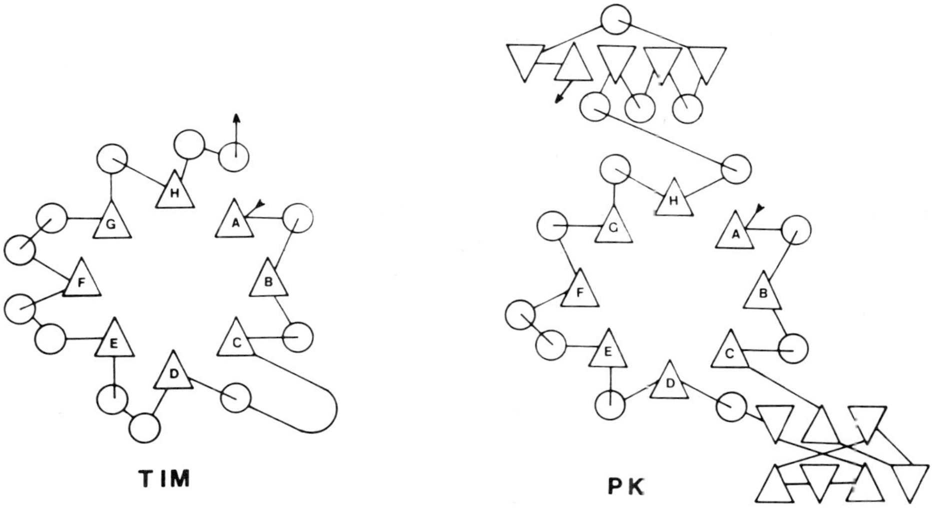

FIG. 5. A schematic drawing of the chain topology of triosephosphate isomerase (TIM) and pyruvate kinase (PK) showing the TIM fold forming the central domain of PK. Helices are indicated by circles and β-strands by triangles. Reproduced from Phillips *et al.* (1983).

D. Supersecondary Structures

There is not much agreement on the exact definition of the elements of structure that lie between secondary structures and domains. There are broadly two approaches to the problem: one is in terms of defining a small number of sequential secondary structures that seem to form local clusters; the other tries to define a small group of amino acid residues, not necessarily sequential, that seem to form a compact unit separated from other similar units.

The first idea of supersecondary structures of the sequential type came from Rao and Rossmann (1973) who noted that the NAD-binding domain in lactate dehydrogenase could be decomposed into two smaller βαβαβ units, each of which was associated with the binding of a mononucleotide. This has been extended (see Schulz and Schirmer, 1979) to include smaller units including the βXβ unit where X can be either a loop or an α-helical segment. It can be seen in Fig. 3 that in the all-α proteins an αα structure often forms compact "hairpin"-like structures, as do ββ units in the all-β proteins, and βα units in the α/β proteins. Many larger proteins have rather simple repetitive structures in these terms: for example the structures of the glycolytic enzymes triose-phosphate isomerase (Banner *et al.*, 1975), and phosphoglycerate kinase (Banks *et al.*, 1979) can be fairly accurately described as an 8-fold and 12-fold repeat, respectively, of the βα unit. The difficulty with definition of this type of supersecondary structure is to know how many secondary structures make up the larger unit.

The problem of attempting to define small groups of amino acid residues that form compact units has been attacked by Rose (1979) and Crippen (1978). They have developed different computer algorithms, working on the three-dimensional coordinate sets of proteins, that rigorously subdivide the total structure into a hierarchical tree, whose smallest units may be compact structures. Unfortunately, the elements of the tree structure that emerge appear to be dependent on the nature of the definitions used to produce the algorithms, and the results of the two methods do not show much internal agreement. A simpler and more useful approach has been reported by Gō (1981) which is based on the two dimensional diagonal plot, a plot of the distances between the *i*th and *j*th residues for all residue pairs in the protein. Groups of residues that are close to one another, but relatively distant from other residue clusters, stand out very clearly on the diagonal plot (see Fig. 15). This analysis is particularly relevant because it has been carried out with intention of relating exons to these compact units of structure, or modules (Gō, 1983). It will be further discussed in the appropriate sections.

III. Correlation of Exons with Protein Structure

Almost inevitably the majority of the gene structures that have been determined experimentally do not correspond with the relatively small number of protein molecules whose three-dimensional structures are known in detail. It is therefore only possible to discuss those gene structures where either the corresponding protein structure is known, or where some structural feature can be reasonably deduced from the amino acid sequence. The latter information is almost entirely concerned with the strong inference that repeated blocks of sequence and/or repeated patterns of disulfide bridges, correspond to repeated three-dimensional structures, probably domains.

A. The Immunoglobulins and Related Molecules

The immunoglobulin molecules are composed of two identical pairs of chains, the light chains and the heavy chains, linked together by disulfide bridges. There are two different types of light chain, λ and κ, and five types of heavy chain, α, γ, δ, ε, and μ, which are differentiated by sequences in the C-terminal parts of the chains. The N-terminal portions of both heavy and light chains are variable in sequence. In structural terms each chain is composed of domains of about 110 amino acid residues—the homology units discovered from the sequence studies—each having the characteristic β-structured "immunoglobulin fold" with a disulfide bridge linking β-strands on opposite sides of the domain, shown in Fig. 6. In the α, γ, and δ classes of the heavy chains there are four domains, V_H, C_{H1}, C_{H2}, C_{H3} with a small "hinge" peptide between the C_{H1} and C_{H2} domains: the ε and μ chains differ only in the replacement of the hinge peptide by an additional immunoglobulin domain. Both λ and κ light chains have two of the characteristic domains, V_L and C_L. In their membrane bound forms the immunoglobins have an additional transmembrane/cytoplasmic, domain, or domains, at the C-terminus of the heavy chain (see Fig. 7). In functional terms, the V_H and V_L domains pair to provide the antigen binding function, while the $C_{H1}C_L$, $C_{H2}C_{H2}$, $C_{H3}C_{H3}$ domain pairs provide the various effector functioning such as complement fixation, and β cell activation.

The studies of the organisation and structure of the immunoglobulin genes have produced a picture of remarkable complexity, but one which goes a long way, perhaps all the way, in explaining the origins of the genetic and somatic generation of antibody diversity. This aspect cannot be discussed in detail here (see reviews by Tonegawa, 1983; Leder, 1982), where the considerations must center on the evolutionary implications of the gene structure. It has been found that the λ light chain genes, the κ light chain genes and the heavy chain genes each reside on different chromosomes in the mouse. In the germ-line, each of these three gene families has the coding sequences for the variable domains some distance away from the constant-domain coding sequences. For the light chains the C_L domain is encoded by a single exon, but the V_L domains are encoded by three exons; an L-exon that encodes the leader sequence; a V-exon that encodes the first ~95 residues of the variable domain; and a J-exon encoding the remain-

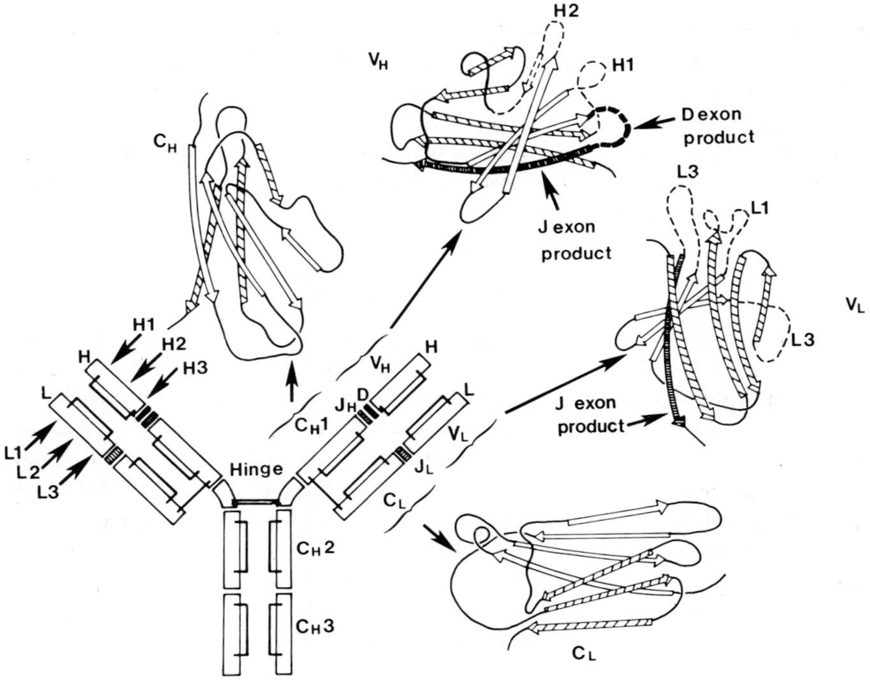

FIG. 6. A schematic drawing of the immunoglobulin (class G) molecule. The complete IgG molecule in the bottom left is shown divided into the major structural domains of its heavy chains (V_H, C_{H1}, Hinge, C_{H2}, C_{H3}) and light chains (V_L, C_L). Intradomain and interchain disulfide bridges are shown in heavy line. The three-dimensional folds of the various domains are shown around the periphery, and include the locations of the J- and D-exon products. Reproduced from Phillips *et al.* (1983).

ing ~15 residues of this domain (Brack *et al.*, 1978). For the heavy chain each of the separate structural elements in the constant part of the chain is encoded by a separate exon. For example, in the γ-chain the C_{H1} domain, the hinge, the C_{H2} and C_{H3} domains are each encoded by single exons, separated by intervening sequences (Sakano *et al.*, 1979). This is also true for the other types of heavy chain, including the transmembrane and cytoplasmic portions of the chains. The V_H domains are encoded in a similar manner to the V_L domains, except that an additional small intron (D) encodes a portion of chain located between the V- and J-exon encoded regions of the V_L domain.

The correlation of the individual structural domains with separate exons seems to provide a clear basis for understanding the role of exons in the evolution of the immunoglobulins. It seems almost certain that the constant domains of the heavy and light chains derive from a single ancestral exon encoding a protein having the immunoglobulin fold, and very likely that the variable domains also derive from the ancestral exon. Multiple tandem duplications of this exon, together with parts of the noncoding sequences that may be expected to have flanked it, could readily account for the contemporary structure of the heavy chain constant region gene. Indeed, Sakano *et al.* (1979) have suggested that the "hinge" exon in the α, γ, and δ chains may have evolved from an extra C region domain corresponding to the C_{H2} domain which is found in its place in the ϵ and μ chains. Duplication of the ancestral exon, and subsequent dispersal is a plausible origin for the light chains. The presence of multiple copies of the V_L and V_H genes, each with its own leader sequence, is almost certainly due to further multiple tandem duplications following the dispersal of the V exon. The J- and D-exons may have originated in one of three ways: they may have been introduced from elsewhere in the genome; they could have been split off from the V-exons by the introduction of an intron late in the evolution of the immunoglobins; or represent an early structure of the primordial exon which was lost by the deletion of one or more introns in the C-domain line prior to the duplication in this line. Overall, the gene structures of the immunoglobulin chains demonstrates the independence of the exon in evolution, and the correlation of V- and C-domains, the leader sequence, the hinge region, and the transmembrane regions with individual exons strongly supports the proposition that exons correspond to protein functions that can be recombined to provide proteins of novel function (Gilbert, 1978).

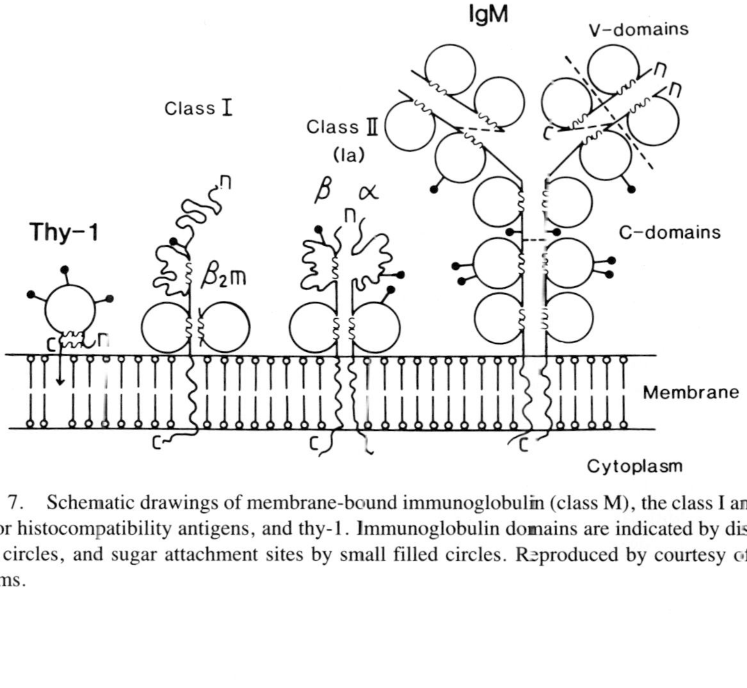

FIG. 7. Schematic drawings of membrane-bound immunoglobulin (class M), the class I and class II major histocompatibility antigens, and thy-1. Immunoglobulin domains are indicated by disulfide-linked circles, and sugar attachment sites by small filled circles. Reproduced by courtesy of A. F. Williams.

Further examples of molecules containing the immunoglobulin fold are the Class I and Class II antigens of the major histocompatibility complex. The structural relationships between the two classes of MHC antigens and the immunoglobulins are shown in Fig. 7. The polypeptide chains of the Class I antigens are organized into three external domains, a transmembrane domain, and a cytoplasmic domain. The membrane-proximal domain has the immunoglobulin fold, and forms a noncovalent association with β_2-microglobulin which itself has the immunoglobulin fold. The gene sequences of the H-2K^d (Steinmetz *et al.*, 1981; Kvist *et al.*, 1983) and H-2L^d (Moore *et al.*, 1982) Class I antigens show that the leader, the three external and the transmembrane domains are each encoded by separate exons. However, the cytoplasmic domain, for which no structural or functional subdivisions are known, is encoded by three small exons. The gene for the associated β_2-microglobulin (Parnes and Seidman, 1982) is composed of three coding sequences, corresponding to the leader sequence, residues 3–95 of the immunoglobulin fold, and the final four residues. Thus the gene organisation of β_2-microglobulin is similar to the L-V-J organization of the immunoglobulin V_L domains. The Class II antigens are also membrane bound proteins in which the functional unit is an $\alpha\beta$ dimer. The polypeptide chains are organized into three domains, an external pair, the membrane-proximal of which has an immunoglobulin fold, and what is probably a single transmembrane/cytoplasmic domain. In the gene structure of the HLA-DR α-chain (Lee *et al.*, 1982), the leader sequence and each of the three structural domains are encoded by separate exons. The similarities of the gene structures of the various immunoglobulin domains in these antigens suggest that they are also derived from the ancestral immunoglobulin exon, and may have been combined with other exons that have no counterparts in the immunoglobulins.

B. α-Fetoprotein/Serum Albumin

Mammalian α-fetoprotein and serum albumin are homologous proteins related by a duplication of an ancestral gene 300–500 Myr ago. Comparison of the internal pattern of disulfide bridges in the amino acid sequences of human and bovine serum albumins led Brown (1977) to propose that the protein had three similar structural domains which must have arisen by a triplication of an ancestral gene encoding a single domain. The binding sites in mammalian α-fetoproteins are not well characterized, but those in serum albumin are well defined and related to the putative three domain structure. Domain I contains the indole binding site, domain II the bilirubin binding site, and domain III the high-affinity fatty acid binding site (the lower affinity fatty acid binding sites on domains I and II are thought to be relics of high-affinity site on domain III). Further consideration of the amino acid sequence of serum albumin by Brown (1977), and McLachlan and Walker (1977), has suggested that the primordial domain was itself produced by the triplication of a basic 73 amino acid sequence.

The studies of the organization of the mouse albumin gene by Kioussis *et al.* (1981) and of the mouse α-fetoprotein gene by Eiferman *et al.* (1981) have confirmed and extended this idea of the evolution of the gene. The coding region of the α-fetoprotein gene contains 14 exons: a leader-sequence encoding first exon, a second short exon, followed by a threefold repeat of four exons (A–D) shown in Fig. 8. The superposition of the exon pattern on the amino acid sequence correlates precisely the threefold repeat of the exons A, B, C, and D with the three putative structural domains. Comparisons of the nucleotide sequences of similar exons in the three domains revealed sequence homologies that confirm the hypothesis of a gene triplication but also shows that exon D in domains I and II is a fusion of exon D in domain III and exon 2. This suggests that the gene for the proposed single domain ancestor was composed of five exons (plus the leader exon) rather than four, giving the model for the gene triplication events shown in Fig. 8. The sequence comparisons between the exons also indicated significant homologies amongst the five exons that encode a single domain, suggesting that these exons themselves had derived from a primordial single exon. However in order to derive the primordial domain com-

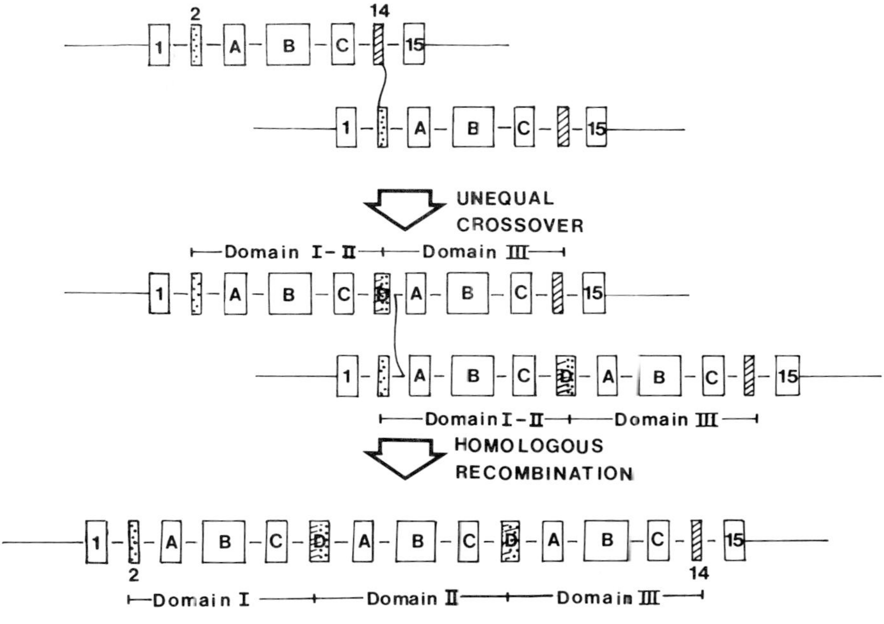

FIG. 8. The scheme for the domain evolution of α-fetoprotein proposed by Eiferman *et al.* (1981). The rectangles represented exons in the α-fetoprotein gene. Exons A–D encode a single α-FP domain, exon 1 encodes the leader sequence, and exon 15 is located in the 3′-untranslated region of the gene. Exons encoding for the early single domain version of the protein (2, A, B, C, and 14) appear to be derived from a single "primordial" exon. Redrawn from Eiferman *et al.* (1981).

posed of exons of different sizes, the amplication of the primordial exon must have involved unequal crossovers during recombination and/or removal of intervening sequences.

C. Collagen

Although collagen is a fibrous rather than globular protein, and thus has no differentiated functional/structural domains, it is in some respects a key example in protein/exon evolution. The collagen molecule is composed of three polypeptide chains, each about 1000 residues long, which combine to form the structurally unique collagen helix. This structure has an absolute requirement that every third residue in the three chains is a glycine (see Fig. 2). Sequence studies have confirmed that collagen molecules contain the mandatory Gly-x-y triplets repeated consecutively 338 times. Because the substitution of a few, perhaps even one, of these internal glycines would destroy the integrity of this essential structural protein, the evolution of the collagen molecule poses problems which seemed insuperable until the gene structure was determined.

The α2-pro-collagen gene in chicken is composed of at least 52 exons, as shown in Fig. 9 (Wozney *et al.*, 1981). The sizes of the exons form a very clear

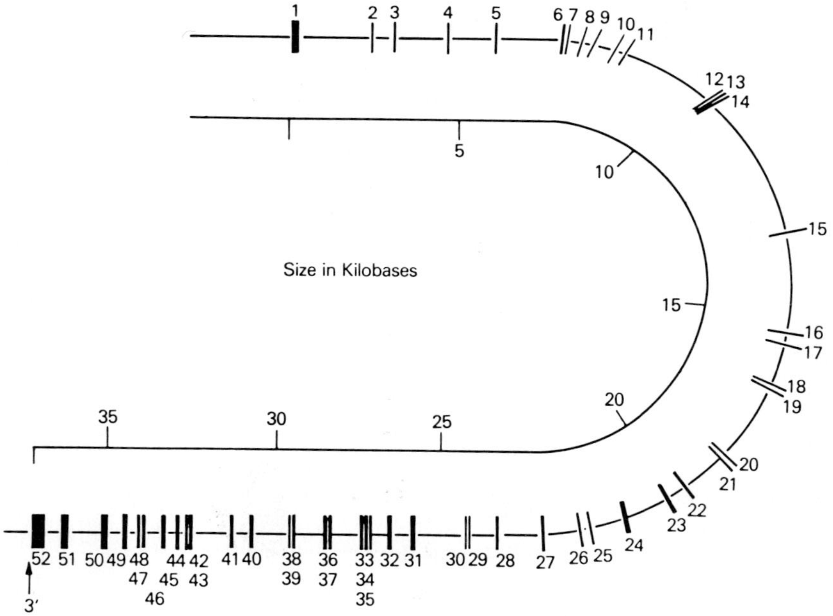

Fig. 9. A drawing of the chicken α2 (type 1) procollagen gene. The short, numbered lines represent the sizes and locations of the 52 exons in the gene. (Reproduced from de Crombrugghe and Pastan (1982).

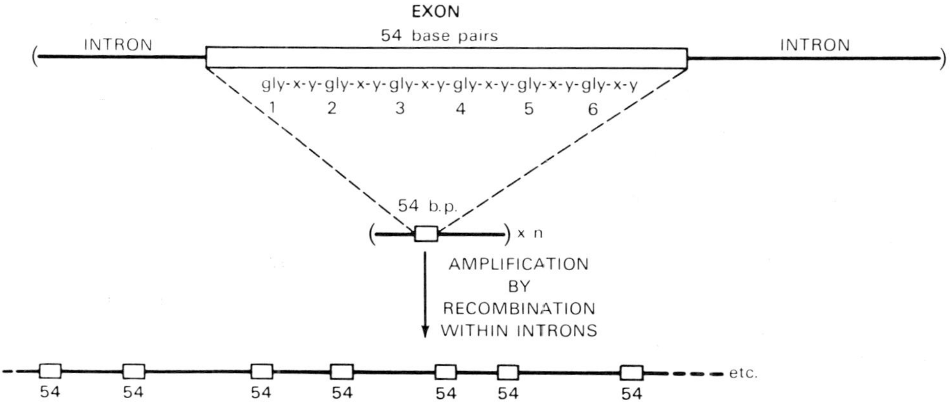

FIG. 10. A scheme for the evolution of the collagen gene. Reproduced from de Crombrugghe and Pastan (1982).

pattern: the majority are 54 bp long, others are exactly twice as long (104 bp) and a few are 9 bp shorter than these (45 and 99 bp). The 54 bp exons are not only the right length to encode a sixfold repeat of the Gly-X-Y repeat, but are also in register with this structure. The 108 bp exons clearly encoded a double length copy of the $(\text{Gly-X-Y})_6$, and the 45 bp and 99 bp exons have one Gly-X-Y sequence deleted with respect to their full length versions. This size distribution of exons leads compellingly to a model for the evolution of the collagen molecule from a single 54 bp exon by amplication to create multiple copies separated by intervening sequences (de Crombrugghe and Pastan, 1982). Precise deletion of intervening sequences between adjacent 54 bp units would lead to the observed double length exons, and deletion of 9 bp from the 54 and 108 bp exons could account for the 45 and 99 bp exons. The overall scheme derived in this way (Fig. 10) accounts in most satisfying manner for the evolution of the collagen molecule.

D. OVOMUCOID

On the basis of sequence homologies and the pattern of disulfide bridges, the polypeptide chains of avian ovomucoids are organized into three structural domains. Each of these domains is capable of binding to one molecule of trypsin, or other member of the serine proteinases. The chicken ovomucoid gene (Stein *et al.*, 1980) is composed of eight exons: an exon encoding the leader sequence, a very short N-terminal exon, and six exons that in similar pairs encode the three structural domains. The relationship between these six exons and the domains is shown in Fig. 11. This correlation shows that the molecule has evolved by tandem duplications from two ancestral exons that encoded a single ovomucoid

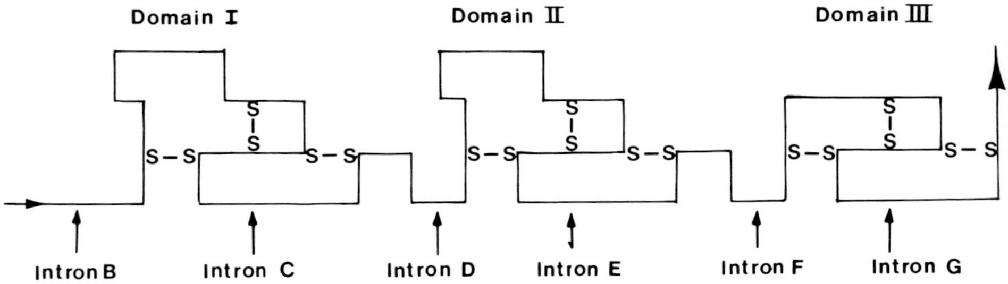

FIG. 11. A schematic drawing of the three structural/functional domains of the ovomucoid molecule. The locations of the six introns (B–G) in this region of the ovumucoid gene are shown. Redrawn from Stein *et al.* (1980).

domain. Stein *et al.*, have suggested that a timescale can be applied to the exon duplications from the finding that dog ovomucoid contains only domains II and III. This suggests that the first duplication took place before the divergence of the birds and the mammals (~ 300 Myr bp), and that the second took place in the birds after divergence.

E. Growth Hormone

Growth hormone is a member of a family of structurally related peptide hormones that includes prolactin and placental lactogen. Their amino sequences suggest that the precursor of this family was formed by duplication of a smaller ancestral gene. The physiological functions of growth hormone can be divided into growth-promoting and antiinsulin activities, and a peptide can be derived from the growth hormone molecule that has anti-insulin but not growth-promoting activity. The sequence of the rat growth hormone gene has been determined by Barta *et al.* (1981). The gene is composed of five exons, in which the leader sequence is encoded by the first exon and the first half of the second exon, regions of sequence homology are found in the second, fourth and twice in the fifth exon but not in the third exon, while the peptide that shows antiinsulin but not growth-promoting activity maps in the third exon. Putting these pieces of information together Barta *et al.* have proposed that the evolution of growth hormone, and the other related hormones, involves tandem duplications of an ancestral exon, the incorporation of coding sequences for the anti-insulin activity from elsehwere, and the precise deletion of the noncoding sequences from between two of the exons, as shown in Fig. 12.

F. β-Crystallin

The crystallins are a family of proteins located in the lens of the vertebrate eye. Sequence determinations of the β- and γ-crystallins suggest that they have

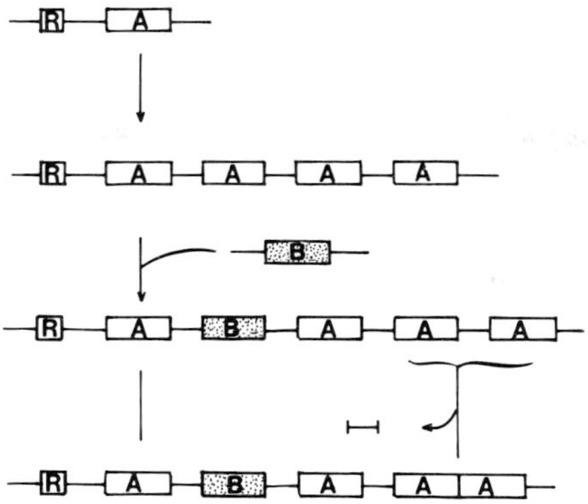

FIG. 12. A scheme for the evolution of the growth hormone gene. The rectangles represent exons; R the regulatory exon; A the "primordial" exon; and B an exon encoding sequences having antiinsulin activity. Redrawn from Barta *et al.* (1981).

evolved from a common ancestral protein. X-Ray analysis of γ-crystallin (Blundell *et al.*, 1981) has shown that molecule has two similar domains each of which is composed of two similar "Greek key" structural motifs. This structure was used as a basis for building a model of β-crystallin which was then shown to be structurally homologous to the γ-crystallin. The gene structure of a mouse β-crystallin has been determined (Inana *et al.*, 1983) and shown to contain four exons. Each of the exons corresponds to one of the "Greek key" structural motifs in the molecule as shown in Fig. 13. It is possible to propose the evolution of the β- and γ-crystallins from a single ancestral exon encoding a "Greek key" structural motif, involving first a tandem duplication to produce a single domain, then a second tandem duplication to produce the contemporary two domain molecule.

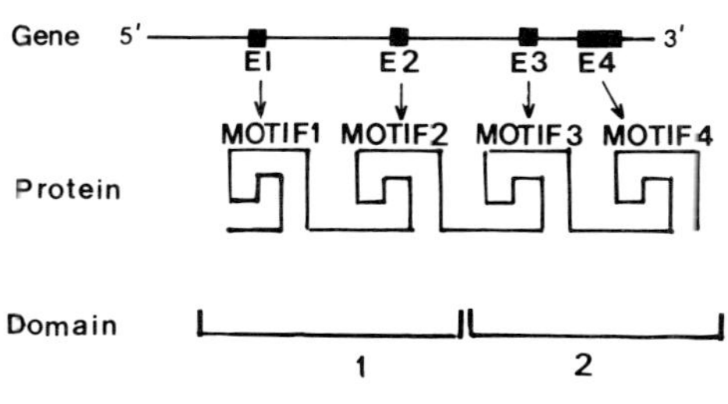

FIG. 13. A drawing of the relationship between exons and the "Greek key" and domain structure of β-crystallin. Redrawn from Inana *et al.* (1983).

G. Hemoglobin

Unlike the other proteins discussed previously, the very similar α- and β-polypeptide chains of hemoglobin have no discernible structural domains, nor any known repeated substructures or sequences. In this respect they are characteristic of a range of other proteins. Nonetheless the functional globin genes contain three or four coding sequences. The analysis of the possible role of exons in the evolution of hemoglobin is more difficult and has less security than for those proteins where intragenic amplication is the major process, and can only be carried out when the detailed three-dimensional structure is known.

The mouse α- and β-globin genes contain three exons (Nishioka and Leder, 1979; Konkel *et al.*, 1979) whose boundaries in sequence terms are slightly different in the two chains, but which coincide precisely in structural terms.

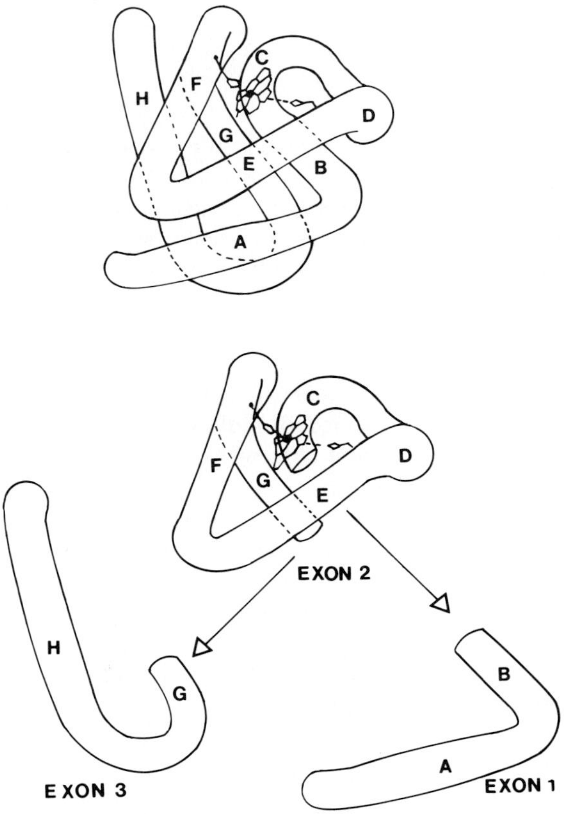

Fig. 14. Schematic drawings of top the hemoglobin subunit, and bottom its three exon products, α-helices are labeled A–H and the location of the heme group is shown.

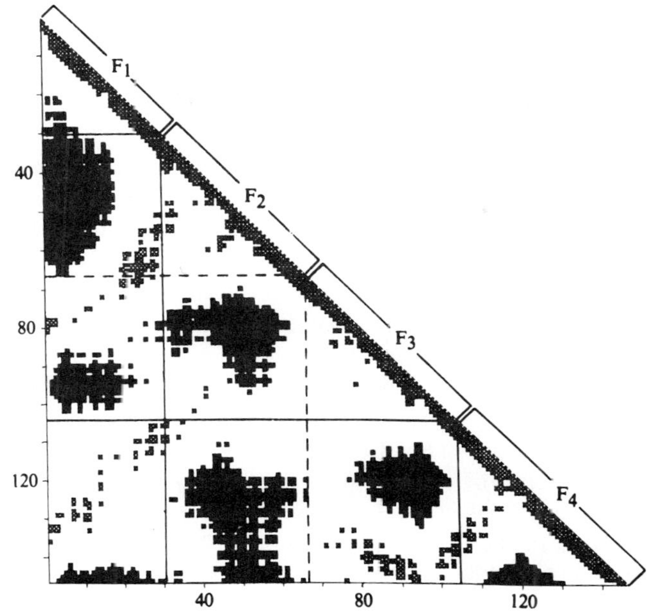

FIG. 15. The Gō plot for the hemoglobin β-subunit. The abscissa and ordinate represent residue numbers and the black areas those interresidue distances greater than 27 Å. F_1, F_2 + F_3, and F_4 represent the exon products in hemoglobin, and the broken line the proposed division of the plot that led to the successful prediction of an intron dividing the central exon into F_2 and F_3 found in leghemoglobin. Reproduced from Gō (1981).

These three exons have been found to be also present in the γ-, δ-, ε-, and ζ-globin genes in a wide range of species from *Xenopus* to man. At first these exons, shown in Fig. 14, seemed to be related in an arbitrary manner to the protein structure and to its function. However, subsequent analysis has shown that there are demonstable, if subtle, relationships between the exons and the protein's structure and function, which reveal intriguing possibilities on its evolutionary history.

The breakthrough in relating the gene structure of the globins to their protein structure was made by Gō (1981). Using a diagonal plot, a two-dimensional representation of interresidue distances in a molecule, she demonstrated (Fig. 15) a number of blocks of residues that were more than 27 Å apart, which were divided off from one another by the exon junctions. This revealed a hitherto unsuspected feature of the globin-fold, namely that there were a number of "compact" units of structure that were relatively distant from one another, and that the exons seemed to correspond to these "compact" structural units. However, Gō noted that to fully divide off the compact units from one another ''dividing'' line was needed, somewhere between residues 66 and 71 in the β-chain. On this basis she proposed that the large central exons in the globin consisted of two fused exons that had been joined between residues 66 and 71. Remarkably this prediction of the location of an intron was quickly verified by the determination of the gene sequence of soybean leghemoglobin by Jensen *et al.* (1981). This gene has an identical first and last exon to the other globin, but the central exon shows an additional intron that splits it between residues 68 and 69, in precise agreement with Gō's prediction. This successful prediction of the location of an intervening sequence in a gene from an examination of the structure of the corresponding protein gives a powerful impetus to the idea that gene structure and protein structure are intimately linked.

The key to the correlation of hemoglobin function with its exon pattern was the observation made independently by Blake (1979) and Gilbert (1979) that the central exon (see Fig. 14) seems to correspond to the hemebinding function of the protein. These observations led Eaton (1980) to analyze the range of hemoglobin functions in terms of its three exons as shown in Fig. 16. As can be seen, this analysis not only confirms that the heme contacts are concentrated in the central exon, but also shows that residues involved in the cooperative and regulatory aspects of hemoglobin oxygenation also map within particular exons. The $\alpha_1\beta_2$ contacts that are responsible for the formation of the cooperative dimer

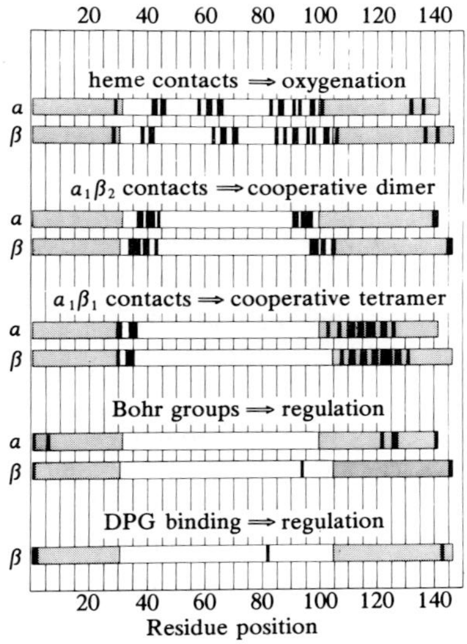

FIG. 16. The correlation of hemoglobin functions with the exon products. Both α- and β-subunits are shown, with exon 1 shaded, exon 2 unshaded, and exon 3 shaded again. The heavy vertical bars indicate residues involved in the particular functions. Reproduced from Eaton (1980).

are concentrated in exon 2; the $\alpha_1\beta_2$ contacts that are responsible for the formation of the cooperative tetramer map largely in exon 3; and the groups giving rise to the Bohr effect and those that bind the effector 2,3-diphosphoglycerate, both involved in the regulation of hemoglobin, are located in exons 1 and 3. On the basis of this analysis Eaton has suggested an evolution of the hemoglobin molecule beginning with the three coding sequences followed by point mutations to give a cooperative dimer, and completed by the replacement of the first and third exons by recombination to achieve the fully active, allosteric tetramer.

The suggestion that the central exon of hemoglobin is the heme binding fragment has been experimentally verified by Craik *et al.* (1981). By clostripain treatment, they excised a protein fragment from hemoglobin that corresponded to the product of the central exon. This fragment was shown to be able to bind heme tightly and specifically, but it was unable to stabilize the heme–oxygen complex without the presence of the side-exon products and the complementary subunit. Of particular interest was the observation that certain spectroscopic properties of the fragment–heme complex were similar to those of the *b*-type cytochromes, because Argos and Rossmann (1979) had already shown that the β-globin chains, cytochrome b_5 and *Pseudomonas* cytochrome c_{551} all share a contiguous helical structure responsible for heme binding, and that the residues of the β-globin chain making up this structural segment correspond almost exactly with the central exon product. This suggests that the globins and some of the cytochromes may be derived from a common, primordial heme binding protein coded by the central exon.

H. Lysozyme

The gene for hen egg-white lysozyme (Jung *et al.*, 1980) is composed of four exons. As in hemoglobin these correspond to compact units of structure revealed by the Gō plot and suggest that the large second exon now represents a fused pair of exons, implying that the ancestral lysozyme gene had five exons (Phillips *et al.*, 1983; Gō, 1983) but no example of this gene has yet been found. In considering the possible relationship between the exons and the activity of lysozyme, Jung *et al.* noted that exon 2 codes for a region that includes the two catalytic residues and part of the substrate binding site, and suggested that a protein fragment coded by this exon might be able to function as a primitive, relatively nonspecific glycosidase. The putative recombination of exon 2 and exon 3 would then result in a protein fragment with a complete active site that could give additional substrate specificity and catalytic efficiency to the primative enzyme. The products of exons 1 and 4 do not contain residues directly involved in the catalytic function but may have had a stabilizing function, or some as yet unknown biochemical function. Gō (1983) has pointed out the proposed missing intron would divide exon 2 into two parts, one of which contained the catalytic residues, and the other the major part of the binding site. With this model the catalytic and binding functions are located on a number of separate exons, as are those of the serine proteinases and carboxypeptidase, possibly representing a general feature of enzymes.

Two other, apparently quite different lysozymes have had their three-dimensional structures determined, T_4 phage lysozyme (Matthews *et al.*, 1981) and goose lysozyme (Grütter *et al.*, 1983), which show a certain, but complex structural similarity to the hen enzyme, which has been diagrammatically represented by Pain (1983) as in Fig. 17. Artymiuk *et al.* (1981) have proposed that the region of structural correspondence between hen and T_4 lysozyme corresponds closely to the exon 2 and 3 products of hen lysozyme, that exon 1 is absent from T_4 lysozyme and that exon 4 of hen lysozyme has been replaced by a complete domain in T_4 lysozyme which may be involved in a cell-binding function not required by hen lysozyme. Neither the phage gene for T_4 lysozyme nor the goose lysozyme gene has been sequenced, so no direct comparison of their exon patterns with that of the hen lysozyme gene can be made.

A totally new light on the possible relationship involving hen and T_4 phage lysozymes has been shed by the development of a new method of structural comparison by Levine *et al.* (1983) that makes use of a three-dimensional Ramachandran plot (see Fig. 1) in which the third dimension is represented by the residue number of the protein. Applying this method to a range of proteins, they have shown that 55 residues of hen egg-white lysozyme (129 residues) and tuna cytochrome *c* (104 residues) can be superimposed with a root mean square difference of 3.75 Å. In relation to the exons of the hen lysozyme gene (Jung *et al.*, 1980) nearly all the equivalenced lysozyme residues are located in exons 2 and 3, and the largest break in the list of equivalenced residues corresponds to the junction between these two exons. Furthermore the location of the single intron in cytochrome *c*, known from the gene sequence of the rat protein (Soarpulla *et al.*, 1981) is located in terms of the equivalenced residues only 3 residues away from the break between exons 2 and 3 in hen lysozyme. A further structural comparison demonstrated by this technique is between phage T_4 lysozyme and bovine carboxypeptidase, where 70 of the 164 residues in the lysozyme and the 307 residues in the carboxypeptidase can be equivalenced with an root mean

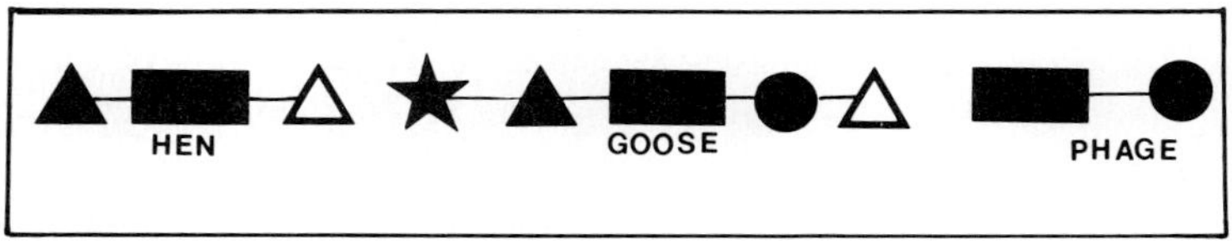

FIG. 17. A diagramatic representation of the structures of hen, goose, and T_4 phage lysozymes. Each symbol represents a particular chain fold: the rectangle representing the fold in the region of the catalytic site. Redrawn from Pain (1983).

square difference of 3.83 Å. This comparison superimposes most of the N-terminal part of T_4 lysozyme chain with five different segments of the carboxypeptidase chain. In relation to exons known in the rat carboxypeptidase gene (Craik *et al.*, 1982) every break in the list of equivalences occurs within a splice junction.

These studies are suggestive of hitherto unsuspected structural relationships between proteins of quite different function. The observations that the parts of the molecules that can be superimposed seem to be correlated with exonencoded fragments, and that splice junctions seem to occur at breaks in the structural overlap, are suggestive of the presence of exons encoding similar structures, and therefore perhaps of common origin, in an otherwise different setting in different genes. If this suggestion can be verified, and at present it is very tentative, we would have the first verification of the shuffling of exons as a mechanism of building genes that was proposed by Gilbert (1978). As Phillips *et al.* (1983) have pointed out, because protein tertiary structures appear to outlast changes in both their chemical structures and in their gene mosaic structures, the three-dimensional structures of proteins may provide the best intimations of evolution that are accessible to us.

IV. Temporal Stability of the Exon/Intron Pattern

The attempt to define the evolutionary history of proteins from the pattern of exons and introns in their genes is based on the assumption that the exon/intron pattern is stable over relatively long periods of the evolutionary time scale. It would also be valuable to know if the pattern changes primarily through the removal of preexisting introns, or the introduction of new ones. The evidence that is available at present seems to give provisional answers to both these questions, but more will be needed to give the necessary degree of certainty.

Evidence is available from a number of gene lines suggesting that the exon/intron pattern is relatively stable over quite long periods of the evolutionary time scale, but there are also suggestions of some variability as between different gene lines. In the human, rabbit, and mouse globin chains the introns are located within the codon for residue 31 and between the codons for residues 99 and 100 in the α-chains, and within residue 30 and between residues 104 and 105 in the β-chains. However the small differences in structure between the α- and β-chains are such that in three-dimensional terms residue 31 of the α-chain is equivalent to residue 30 in the β-chain, and that residues 99 and 100 in the α-chain are equivalent to residues 104 and 105 in the β-chain. This seems unequivocal evidence that the two introns were present in the ancestral gene common to the α- and β-chains and prior to the dispersal of the genes on different chromosomes, estimated at 300–350 Myr ago. Although the soybean leghemoglobin gene has one more intron than the animal hemoglobins (Jensen *et al.*, 1981), the first and third introns appear to be precisely homologous in position to all the other known functional hemoglobin genes. As functional monomeric proteins leghemoglobin and myoglobin are assumed to have evolved earlier from the common ancestral gene than the α- and β-globins. This suggests that the homologous intron/exon splice junctions have been maintained in the globin genes for at least 1000 Myr. Another example of exon stability can be seen in the α-fetoprotein gene (Eiferman *et al.*, 1981) where the splice junctions between the triplet pattern of four exons that make up a single domain must have been maintained since before the divergence of AFP and serum albumin (~300–500 Myr) and probably for much longer to account for the gene triplication. Over longer time scales, determined by comparing homologous bacterial and mammalian proteins with equivalent eukaryotic gene sequences, Craik *et al.* (1983) have noted that splice junctions often coincide with variable length loops in the protein structures, and have proposed that sliding of splice junctions could be a mechanism for generating chain length polymorphism in proteins.

There are now a considerable number of examples ofgene sequences of homologous proteins that differ in the number and location of their introns. It is evident therefore that although the exon/intron pattern is relatively stable, changes do take place that correspond to the introduction, or precise deletion, of introns. But does introduction or deletion of introns dominate in evolution? Plausible mechanisms for each process are possible: three processes for the introduction of introns have been discussed by Crick (1979), while possible processes for their precise deletion may be indicated by the discovery of a number of intronless pseudogenes terminating in poly(A) tracts that implies reverse transcription from mRNA.

Evidence for intron deletion is available in a number of gene lines. The actin genes, of which there are at least six types of mammals, seem particularly variable in the number and location of introns (see Table I). The finding of an intron at residue 150 in soybean actin in an homologous position to an intron in chicken and rat skeletal muscle actin is seen as good evidence that the absence of an intron at this location in the sea urchin J-gene and rat β-actin is due to its deletion (Zakut *et al.*, 1982). On these grounds, they see the data given in Table I as compatible with an ancestral actin gene split at all the positions shown in Table I that subsequently lost different introns in different gene lines. Similarly the presence of both 54 and 108 bp exons in the collagen gene can be more readily understood in terms of deletion of introns between adjacent 54 bp exons, rather than the introduction of introns into the 108 bp exons because this would require insertion to take place precisely in the middle of the larger units. The presence of three introns in plant hemoglobin as compared with two in animal hemoglobin can also be argued as due to deletion of an intron because of the unlikelihood that it could be introduced exactly at the boundary of the compact

TABLE I
INTRONS IN ACTIN GENES[a]

Species	Intron location							
Dictyostelium	—	—	—	—	—	—	—	—
Yeast	—	—	—	—	—	—	—	—
Soybean	20	—	—	150	—	—	—	355
Sea urchin J gene	—	41	121	—	203	267	—	—
Chicken skeletal muscle	—	41	—	150	204	267	327	—
Rat skeletal muscle	—	41	—	150	204	267	327	—
Rat β-cytoplasmic	—	41	121	—	—	267	327	—
Human cardiac muscle	—	41	—	150	204	267	327	—

[a]Data from Zakut *et al.* (1982) and Hamada *et al.* (1982).

units of structure described by Gō (1981). Another clear example of intron deletion comes from analysis of the two tandemly arranged preproinsulin genes in rat: preproinsulin gene I has one intron but gene II has two, one of which is common to gene I (Lomedico *et al.*, 1979). The fact that the single preproinsulin genes of man (Bell, *et al.*, 1980) and chicken (Perler *et al.*, 1980) have two introns exactly homologous to those in rate gene II, suggests that the possession of two introns represents the ancestral state, and that the second intron was deleted during or after duplication in the ancestral rat line. In contrast to these examples there seems to be no clear case of the introduction of an intron into a gene, but this may be due to the greater difficulty of detecting introduction as compared with deletion. We should not therefore rule out intron introduction completely at present.

V. A Concluding Hypothesis

Although it is now quite clear that the presence of separate blocks of coding sequences presents a major stimulus to the evolution of proteins, apparently by the mechanisms proposed by Gilbert (1978), there are still two outstanding gaps in our understanding of exons. These gaps are concerned first with the origin of exons, and second with their relationship with the expressed protein molecule.

Since split genes seem to be characteristic of the organisation of the eukaryotic genome, and continuous genes characteristic of the organization of the prokaryotic genome, there seems to be a *prima facie* case for assuming that introns were introduced into the eukaryotic genome following the divergence of the two cell types to increase the evolutionary potential of the eukaryotes. However, Doolittle (1978) has pointed out that the cells which inserted noncoding sequences into their genes would enjoy no immediate selective advantage, and that the idea that they do so to improve the future evolutionary prospects is not Darwinian because it imputes purpose. To counter these problems, Doolittle proposed that the last common ancestors of the eukaryotes and prokaryotes had unfaithful, and hence rapidly evolving, transcription and translation systems where a split-gene organization could be able to compensate for the unfaithfulness, in addition to having an evolutionary role. Subsequently, the split-genes organization was lost in the prokaryotes in a ''streamlining'' process that traded the efficiency of a compact genome against the loss of evolutionary potential, while the eukaryotes have retained the original genomic structure in following the opposite path. Darnell (1978) has taken a similar view that the split-genes organization is an ancient form, but has sought to argue for its absence in prokaryotes in terms of the independent evolution of prokaryotes and eukaryotes. However, unless there has been widespread ''sideways'' transfer of genes between the two cell lines, there are too many homologous prokaryotic and eukaryotic protein molecules to sustain this point of view.

One obvious difficulty with the proposal that split-genes organization is representative of an early state of cell evolution would seem to be the need for a complex splicing system, involving enzymes, energy sources, and possibly ancillary factors, similar perhaps to the U-type RNA (Lerner *et al.*, 1980), to remove the noncoding sequences from the mRNA. A possible solution to this difficulty may be available from the unexpected discovery by Cech and his colleagues (Kruger *et al.*, 1982; Zaug *et al.*, 1983) that the intron in the rRNA from *Tetrahymena* is excised by the intron acting as its own cleavage-ligation ''enzyme,'' and requires no energy supply or protein involvement. It is possible that the contemporary splicing of the major classes of RNA may involve different mechanisms. For example, mitochondrial RNA contains introns that form complex secondary structures that bring the ends of the introns together (Davies *et al.*, 1982; Michel and Dujon, 1983) but require enzymes, maturases, encoded by the introns to carry out the excision (Lazowska *et al.*, 1980). The removal of introns from mRNA may require yet another, and more general, splicing system. It seems probable that these different splicing mechanisms represent an evolutionary process for adding and refining regulatory control to the splicing process itself. If this is correct, the self-removal of the rRNA intron is probably a primitive splicing mechanism that may be representative of the early splicing process depending only on the sequence of the intron. Thus, the evolutionary arguments and reasonable deductions on the splicing mechanism both point to the split-genes organization being an early form, and this view is supported by the conclusion, albeit tentative, that introns tend to be lost slowly during the course of evolution which, of course, suggests that they were more, rather than less, common in earlier gene lines.

Turning now to the relationship between exons and protein structure it has become evident that exons do not generally correspond to protein domains (Blake, 1978). Although there is a correspondence in certain proteins, for example the immunoglobulins, most domains and single-domain polypeptide chains are encoded on a number of exons, as indicated in the earlier sections in this article. In a similar way, the suggestion by Gō (1981, 1983) that exons delineate compact units of protein structure, or modules, despite its remarkable success in predicting the third intron in the hemoglobin genes, does not appear to be of universal application (Phillips *et al.*, 1983). The partial success of both these approaches suggests a need to seek a more comprehensive definition of the exon/protein structure relationship. One possibility that has not yet apparently been considered in detail (see Blake, 1978) is the relationship between exons and supersecondary structures of the sequential type (see Section II,D).

Surprisingly, no proper analysis of the size distribution of exon-encoded protein fragments has yet been published. An analysis of exon and intron size has been reported by Naora and Deacon (1982), but they do not relate their results to protein structure, and have included multiple copies of some protein genes, and genes with multiple repeats of the same exon, without consideration of the bias this introduces. Figures 18 and 19 show the size distribution of exons in 20 eukaryotic genes, each of which codes for a different protein family, but which does not include the repetitive collagen (Wozney *et al.*, 1981) or vitellogenin (Wahli *et al.*, 1980) gene sequences. Unlike the Naora and Deacon Analysis, Fig. 18 shows a unimodal distribution, centred about 45 residues (corresponding to the 140 bp exon of Naora and Deacon) and suggests that few exons are less than 10 or greater than 100 amino acids in length. Since protein domains tend to lie in the size range 100–300 amino acids, this result is consistent with the observations that most protein domains are encoded by several exons. Taking the mean size of exon products as 45 amino acids (which may to too high as a ''primordial'' size if the proteins analyzed have suffered significant intron deletion), this seems to correlate reasonably well with certain supersecondary structures. As shown in Table II the mean length of a supersecondary structure composed of two α-helices is about 45 amino acids, while the supersecondary αβ and ββ units are in the range of 30–35 amino acids. The values in Table II have been produced by subdividing known protein molecules in the appropriate structural class into a discrete number of supersecondary structures to allow for the loops between the structures and thus produce a practical size for the structural unit.

There is at present an insufficient number of proteins whose tertiary structure and gene structure is known to allow this hypothesis that exons may correspond to sequential supersecondary structures to be tested at all rigorously. However Fig. 20 and Table III show an examination of the nature of the exon products in one example of an all-β protein (chymotrypsin) and an α/β protein (carboxypeptidase), neither of which yields to the Gō analysis. As can be seen a number of

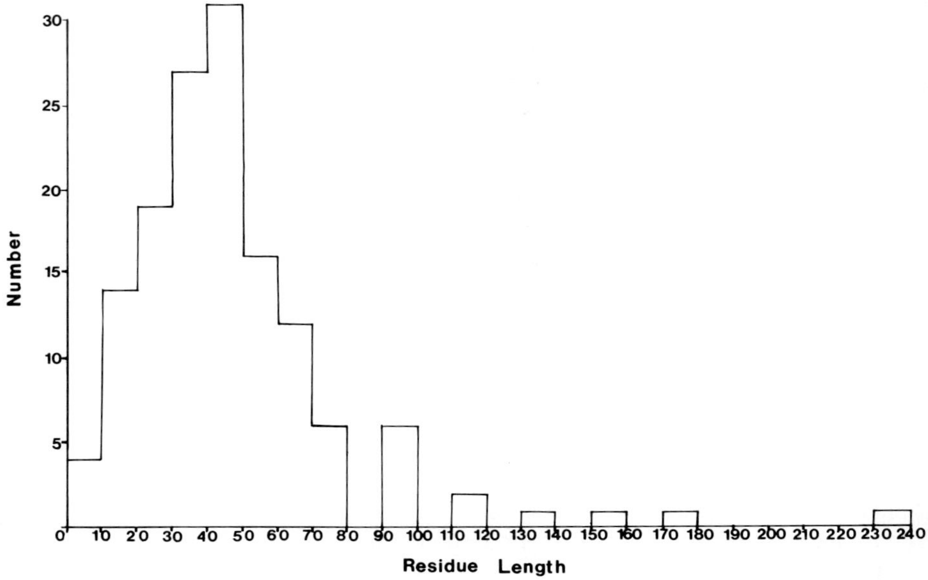

FIG. 18. A histogram of the distribution of exon size from the following protein genes: conalbumin (Cochet *et al.*, 1979); tryptophan oxygenase (Schmid *et al.*, 1982); chymotrypsin (Craik *et al.*, 1982); carboxypeptidase (Craik *et al.*, 1982); dihydrofolate reductase (Craik *et al.*, 1982); hen lysozyme (Jung *et al.*, 1980); γ-interferon (Gray and Goeddel, 1982); human cardiac muscle actin (Hamada *et al.*, 1982); α-fetoprotein (Eiferman *et al.*, 1981); growth hormone (Barta *et al.*, 1981); H-2K^d antigen (Kvist *et al.*, 1983); metallothionine (Durnam *et al.*, 1980); β_2-microglobulin (Parnes and Seidman, 1982); ovalbumin (Heilig *et al.*, 1980); cytochrome oxidase (subunit 1) (Bonitz *et al.*, 1980), δ-crystallin (Jones *et al.*, 1980); ovomucoid (Stein *et al.*, 1980); α-globin (Nishioka and Leder, 1979); complement factor Bb (Campbell and Porter, 1983); chicken preproinsulin (Perler *et al.*, 1980).

TABLE II
SIZES OF SUPERSECONDARY STRUCTURE

Protein	Number of residues	Number and type of structure	Size of structure		References
Hemoglobin	146	7 α	42	αα	Perutz (1969)
Citrate Synthase	437	20 α	44		Remington *et al.* (1982)
Prealbumin	127	8 β	32	ββ	Blake *et al.* (1974)
Chymotrypsin	245	12 β	40		Birktoft *et al.* (1970)
Phosphoglycerate kinase	417	12 βα	35	βα	Banks *et al.* (1979)
Triosephosphate isomerase	249	8 βα	31		Banner *et al.* (1975)

exons seem to be correlated with ββ units, and other structures that seem to fold back on themselves to form compact structural units. It is interesting to note that a trace of the inverted tandem repeat which probably gave rise to the two domain chymotrypsin molecule is discernible, and in complement factor Bb an intron occurs at the domain junction (Campbell and Porter, 1983). In carboxypeptidase the exon products also sometimes correspond to βα units, or their longer variants, αββ and αβα units, which again tend to fold back on themselves.

These simple analyses, while not proving anything, do suggest that it may be worthwhile to consider the consequences of the proposal that exons code for a range of small supersecondary structural units. The immediate implication is, of

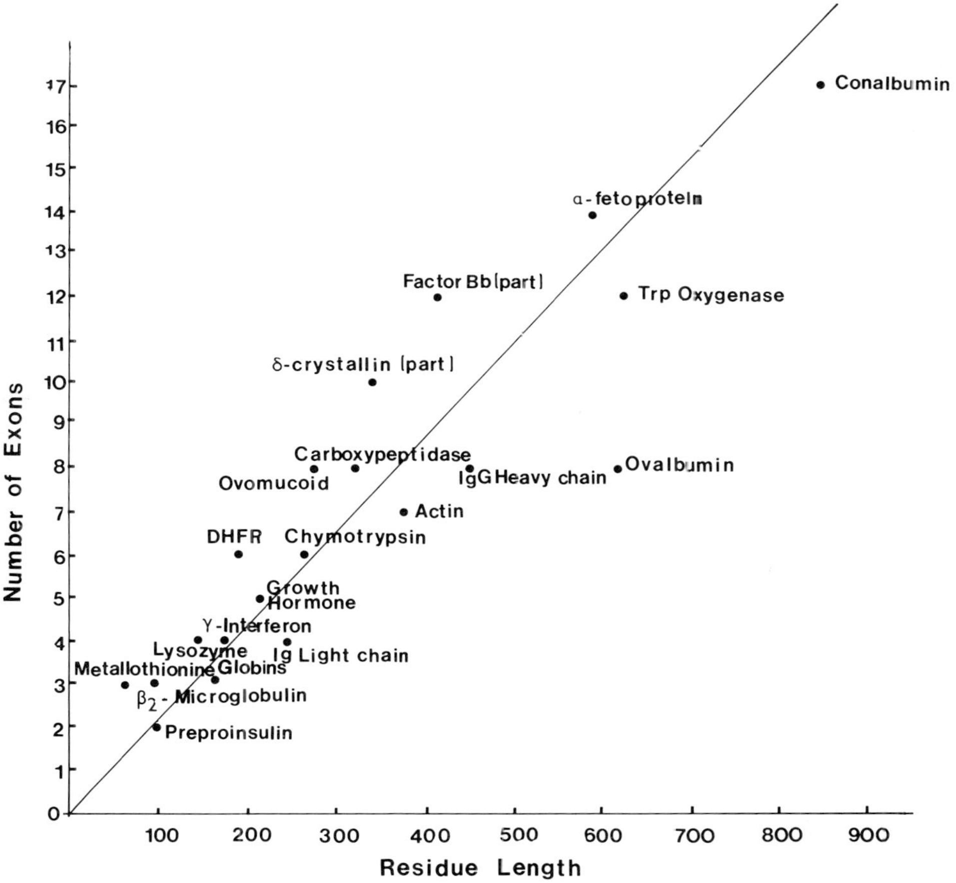

FIG. 19. A plot of the number of exons against the length of the polypeptide chains for the proteins listed in the legend to Fig. 18.

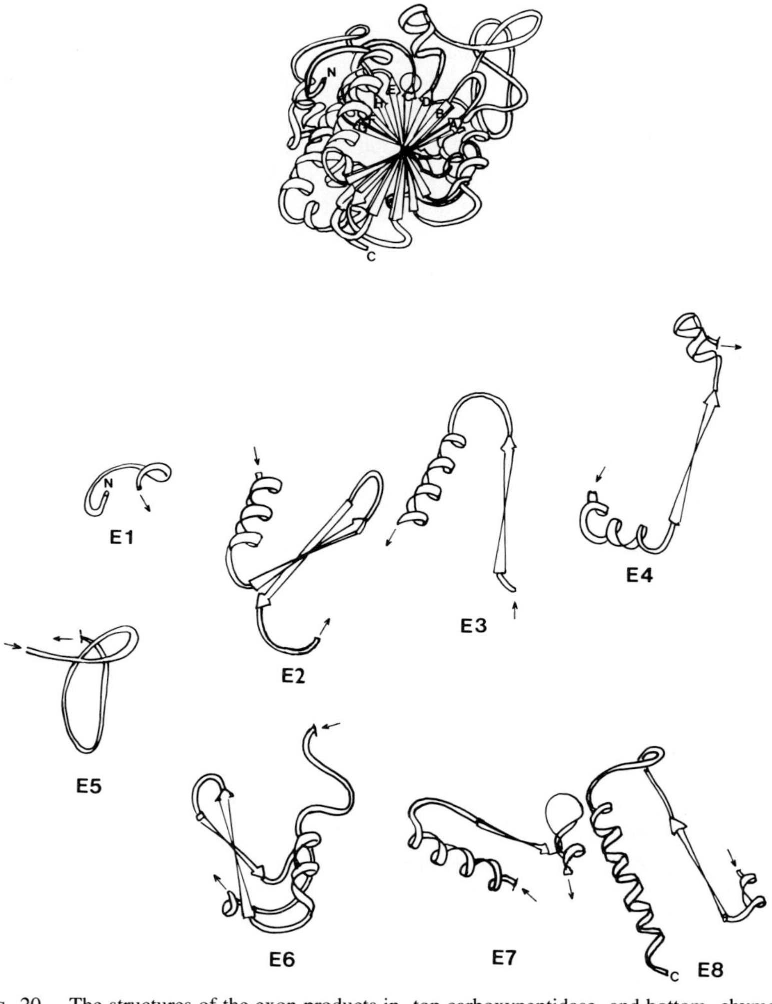

FIG. 20. The structures of the exon products in, top carboxypeptidase, and bottom, chymotrypsin based on drawings made by Jane Richardson.

course, that exons were involved in the very early stages of protein evolution with transcription units being increased in size by the accumulation of exons, either by tandem duplication or exon shuffling, to code for larger, more stable and more efficient proteins. This would generalize the deductions that have already been made for collagen (de Crombrugghe and Pastan, 1982) and α-fetoprotein (Eiferman *et al.,* 1981) molecules, in which it seems very probable

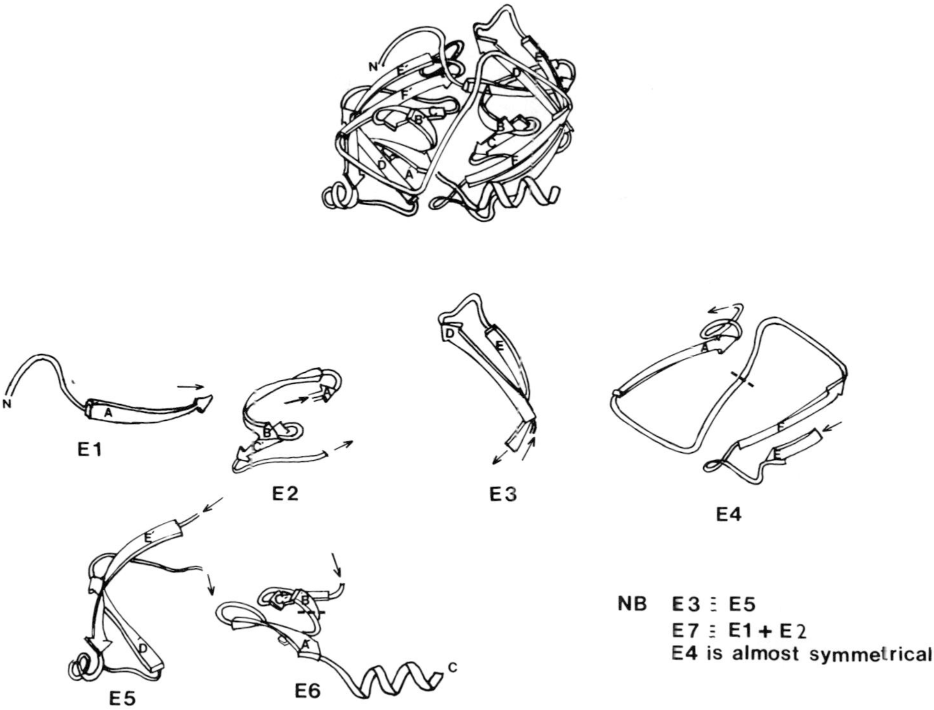

that the large contemporary proteins were derived ultimately from a single primordial exon of quite small size. Such ideas have been discussed previously by Ohno (Ohno, 1981; Ohno *et al.*, 1982). The observations of Craik *et al.* (1982) that splice junctions tend to map at the surfaces of proteins also fit into this proposal, because the link with successive supersecondary structures would tend to lie in surface loops between the structures.

In conclusion, it is now possible to attempt to produce a comprehensive view of the origin and role of exons in protein evolution. We begin with early coding

TABLE III
EXONS AND SUPERSECONDARY STRUCTURES

Protein	Structures of exon products							
	1	2	3	4	5	6	7	8
Chymotrypsin	β	ββ	ββ	βββ	βαβ	βββα		
Carboxypeptidase	Loop	αββ	βα	αβα	Loop	αββ	αβα	αβα

units that expressed small oligopeptides including some potential supersecondary structures. These oligopeptides may not, in isolation, have had stable structures in solution but expressed their structural potential only in association with other similar units, but some like contemporary avian polypeptide (Glover *et al.*, 1983) may have had stable hairpin structures. The flanking noncoding sequences would be such as to allow their potential selfexision as in *Tetrahymena* rRNA. Tandem duplications, and possibly shuffling, of these primordial coding sequences together with their flanking sequences, could lead to early transcription units with exon/intron mosaics that coded for stable protein molecules made up of individual supersecondary structures. Of course, it is impossible to say how many different primordial coding sequences there were, but it is possible that the number was quite small, and that through recombination they were amplified and spread throughout the genome with each copy becoming the potential locus for a different protein molecule. These events would have taken place (if they took place at all) in the ancestors to the eukaryotes and prokaryotes. We must therefore suppose that the split-genes structure was established before the divergence of the two cell types, and that it was retained in the eukaryotic line, but lost from the prokaryotic line by the wholesale deletion of introns, for the reasons proposed by Doolittle (1978). With the establishment of the exon/intron mosaic, the evolutionary mechanisms outlined by Gilbert (1978) would come into play, exploiting the already existing gene structure to build larger, multifunctional, multidomain protein molecules. In general protein function (and domains) would correspond to blocks of exons, but intron deletion could precede amplification or shuffling in some cases to give a functional unit encoded by a single exon. Regulation of protein synthesis could also occur with the development of enzyme-regulated splicing mechanisms. Of course, the record of these possible events in contemporary genes is likely to be mutilated to a greater or lesser extent by subsequent events, such as mutations at splice junctions, intron deletion (or addition), gene conversion, and so on, making it difficult to interpret. Nevertheless, the record does not seem to be completely obliterated, and in some cases seems surprisingly intact, so the study of further gene structures holds the promise of defining the whole of protein evolution in terms of a single type of process.

References

Argos, P., and Rossmann, M. G. (1979). *Biochemistry* **18,** 4951–4957.

Artymiuk, P. J., Blake, C. C. F., and Sippel, A. E. (1981). *Nature (London)* **290,** 287.

Banks, R. D., Blake, C. C. F., Evans, P. R., Haser, R., Rice, D. W., Hardy, G. W., Merrett, M., and Phillips, A. W. (1979). *Nature (London)* **279,** 773–777.

Banner, D. W., Blommer, A. C., Petsko, G. A., Phillips, D. C., Pogson, C. I., and Wilson, I. A. (1975). *Nature (London)* **255,** 609–614.

Barta, A., Richards, R. I., Baxter, J. D., and Shine, J. (1981). *Proc. Natl. Acad. Sci. U.S.A.* **78,** 4867–4871.
Bell, G. T., Pictet, R. L., Rutter, W. J., Cordell, B., Tischer, E., and Goodman, H. M. (1980). *Nature (London)* **284,** 26–32.
Birktoff, J. J., Blow, D. M., Henderson, R., and Steitz, T. A. (1970).*Philos. Trans. R. Soc. Ser. B.* **257,** 67–73.
Blake, C. C. F. (1978). *Nature (London)* **273,** 267.
Blake, C. C. F. (1979). *Nature (London)* **277,** 598.
Blake, C. C. F., Geisow, M. J., Swan, I. D. A., Rerat, C., and Rerat, B. (1974). *J. Mol. Biol.* **88,** 1–12.
Blundell, T. L., Lindley, P., Miller, L., Moss, D., Slingsby, C., Tickle, I. J., Burnell, B., and Wistow, G. (1981). *Nature (London)* **289,** 771–777.
Bonitz, S. G., Coruzzi, G., Thalenfeld, B. E., and Tzagaloff (1980). *J. Biol. Cehm.* **255,** 11927–11941.
Brack, C., Hirama, M., Lenhard-Schuller, R., and Tonegawa, S. (1978). *Cell* **15,** 1–14.
Breathnach, R., and Chambon, P. (1981). *Annu. Rev. Biochem.* **50,** 349–383.
Brown, J. R. (1977). *Proc. FEBS 11th Meet.* **50,** 1–10.
Campbell, R. D., and Porter, R. R. (1983). *Proc. Natl. Acad. Sci. U.S.A.* **80,** 4464–4468.
Cochet, M., Gannon, F., Hen, R., Maroteaux, L., Perrin, F., and Chambon, P. (1979). *Nature (London)* **282,** 567–574.
Craik, C. S., Buchman, S. R., and Beychok, S. (1981). *Nature (London)* **291,** 87–89.
Craik, C. S., Sprang, S., Fletterick, R., and Rutter, W. J. (1982). *Nature (London)* **299,** 180–182.
Craik, C. S., Rutter, W. J., and Fletterick, R. (1983). *Science* **220,** 1125–1129.
Crick, F. (1979). *Science* **204,** 264–271.
Crippen, G. M. (1978). *J. Mol. Biol.* **126,** 315–332.
Crouse, G. F., Simonsen, C. C., McEwan, R. N., and Schimke, R. T. (1982). *J. Biol. Chem.* **257,** 7887–7891.
Darnell, J. E. (1978). *Science* **202,** 1257–1260.
Davies, R. W., Waring, R. B., Ray, J. A., Brown, T. A., and Scazzocchio, C. (1982). *Nature (London)* **300,** 719–724.
de Crombrugghe, B., and Pastan, I. (1982). *Trends Biochem. Sci.* **7,** 11–13.
Doolittle, W. F. (1978). *Nature (London),* **272,** 581.
Durnam, D. M., Perrin, F., Gannon, F., and Palmiter, R. D. (1980). *Proc. Natl. Acad. Sci. U.S.A.* **77,** 6511–6515.
Eaton, W. A. (1980). *Nature (London)* **284,** 183–184.
Eiferman, F. A., Young, P. R., Scott, R. W., and Tilghman, S. M. (1981). *Nature (London)* **294,** 713–718.
Gilbert, W. (1978). *Nature (London)* **271,** 501.
Gilbert, W. (1979). *ICN–UCLA Sympo. Mol. Cell Biol.* **14,** 1–10.
Glover, I., Haneef, I., Pitts, J., Wood, S., Moss, D., Tickle, I., and Blundell, T. (1983). *Biopolymers* **22,** 293–304.
Gō, M. (1981). *Nature (London)* **291,** 90–92.
Gō, M. (1983). *Proc. Natl. Acad. Sci. U.S.A.* **80,** 1964–1968.
Gray, P. W., and Goeddel, D. V. (1982). *Nature (London)* **298,** 859–863.
Grutter, M. G., Weaver, L. H., and Matthews, B. W. (1983). *Nature (London)* **303,** 828–830.
Hamada, H., Petrino, M. G., and Kakunaga, T. (1982). *Proc. Natl. Acad. Sci. U.S.A.* **79,** 5901–5905.
Heilig, R., Perrin, F., Gannon, F., Mandel, J. L., and Chambon, P. (1980). *Cell* **20,** 625–637.
Inana, G., Piatigorsky, J., Norman, B., Slingsby, C., and Blundell, T. (1983). *Nature (London)* **302,** 310–315.

Jensen, E. O., Paludan, K., Hyldig-Nielsen, J. J., Jorgensen, P., and Marcker, K. A. (1981). *Nature (London)* **291,** 677.
Jones, R. E., Bhat, S. P., Sullivan, M. A., and Piatigorsky, J. (1980). *Proc. Natl. Acad. Sci. U.S.A.* **77,** 5878–5883.
Jung, A., Sippel, A. E., Grez, M., and Schutz, G. (1980). *Proc. Natl. Acad. Sci. U.S.A.* **77,** 5759–5763.
Kioussis, D., Eiferman, F., van der RIjin, P., Gorin, M. B., Ingram, R. S., and Tilghman, S. (1981). *J. Biol. Chem.* **256,** 1960–1967.
Konkel, D. A., Maizel, J. V., and Leder, P. (1979). *Cell* **18,** 865–873.
Kruger, K., Grabowski, P. J., Zaug, A. J., Sands, J., Gottschling, D. E., and Cech, T. R. (1982). *Cell* **31,** 147–157.
Kvist, S., Roberts, L., and Dobberstein, B. (1983). *EMBO J.* **2,** 245–254.
Lazowska, J., Jacq, C., and Slonimsky, P. P. (1980). *Cell* **22,** 333–348.
Leder, P. (1982). *Sci. Am.* **246,** 72–83.
Lee, J. S., Trowsdale, S., Travers, P. J., Carey, J., Grosveld, F., Jenkins, J., and Bodmer, W. F. (1982). *Nature (London)* **299,** 750–752.
Lerner, M. R., Boyle, J. A., Mount, S. M., Wolin, S. L., and Steitz, J. A. (1980). *Nature (London)* **283,** 220–224.
Levine, M., Stuart, D. I., and Williams, J. (1983). *EMBO J.* (in press).
Levitt, M., and Chothia, C. (1976). *Nature (London)* **261,** 552–558.
Lomedico, P., Rosenthal, N., Efstratiadis, A., Gilbert, W., Kolodner, R., and Tizard, R. (1979). *Cell* **18,** 545–558.
McLachlan, A. D., and Walker, J. E. (1977). *J. Mol. Biol.* **112,** 543–558.
Matthews, B. W., Grutter, M. G., Anderson, W. F., and Remington, S. J. (1981). *Nature (London)* **290,** 334–335.
Michel, F., and Dujon, B. (1983). *EMBO J.* **2,** 33–38.
Moore, K. W., Sher, B. T., Sun, Y. H., Eakle, K. A., and Hood, L. (1982). *Science* **215,** 679–682.
Naora, H., and Deacon, N. J. (1982). *Proc. Natl. Acad. Sci. U.S.A.* **79,** 6196–6200.
Nishioka, Y., and Leder, P. (1979). *Cell* **18,** 875–897.
Ohno, S. (1981). *Proc. Natl. Acad. Sci. U.S.A.* **78,** 7657–7661.
Ohno, S., Kuniki, K., Hozumi, T., and Matsunaga, T. (1982). *Proc. Natl. Acad. Sci. U.S.A.* **79,** 132–136.
Pain, R. (1983). *Nature (London)* **303,** 750.
Parnes, J., and Seidman, J. G. (1982). *Cell* **29,** 661–669.
Perler, F., Efstratiadis, A., Lomedico, P., Gilbert, W., Kolodner, R., and Dodgeson, J. (1980). *Cell* **20,** 555–566.
Perutz, M. F. (1969). *Proc. R. Soc. Ser. B.* **173,** 113–140.
Phillips, D. C., Sternberg, M. J. E., and Sutton, B. J. (1983). *In* "Evolution from Molecules to Men" (D. S. Bendall, ed.), pp. 145–173. Cambridge Univ. Press, London and New York.
Ramachandran, G. N., and Sasisekharan, V. (1968). *Adv. Protein Chem.* **23,** 283–437.
Rao, S. T., and Rossmann, M. G. (1973). *J. Mol. Biol.* **76,** 241–256.
Reanney, D. (1979). *Nature (London)* **277,** 598–600.
Remington, S., Weigand, G., and Huber, R. (1982). *J. Mol. Biol.* **158,** 111–152.
Richardson, J. S. (1981). *Adv. Protein Chem.* **34,** 167–339.
Rose, G. D. (1979). *J. Mol. Biol.* **134,** 447–470.
Rossmann, M. G., Moras, D., and Olsen, K. W. (1974). *Nature (London)* **250,** 194–197.
Sakano, H., Rogers, J. H., Hüppi, K., Brack, C., Traunecker, A., Maki, R., Wall, R., and Tonegawa, S. (1979). *Nature (London)* **277,** 627–633.
Scarpulla, R. C., Agne, K. M., and Wu, R. (1981). *J. Biol. Chem.* **256,** 6480–6486.
Schmid, W., Scherer, G., Danesch, U., Zentgraf, H., Matthias, P., Strange, C. M., Röwekamp, P., and Schütz, G. (1982). *EMBO J.* **1,** 1287–1293.

Schulz, G. E., and Schirmer, R. H. (1979). "Principles of Protein Structure." Springer-Verlag, Berlin and New York.

Stein, J. P., Caterall, J. F., Kristo, P., Means, A. R., and O'Malley, B. W. (1980). *Cell* **21,** 681–687.

Steinmetz, M., Moore, K. W., Frelinger, J. G., Sher, B. T., Shen, F. W., Boyse, E. A., and Hood, L. (1981). *Cell* **25,** 683–692.

Tonegawa, S. (1983). *Nature (London)* **302,** 575–581.

Tonegawa, S., Maxam, A. M., Tizard, R., Bernard, O., and Gilbert, W. (1978). *Proc. Natl. Acad. Sci. U.S.A.* **75,** 1485–1489.

Wahli, W., David, I. B., Wyler, T., Weber, R., and Ryffel, G. (1980). *Cell* **20,** 107–117.

Wozney, J., Hanahan, D., Tate, V., Boedtker, H., and Doty, P. (1981). *Nature (London)* **294,** 129–135.

Zakut, R., Shani, M., Givol, D., Neuman, S., Yaffe, D., and Nudel, U. (1982). *Nature (London)* **298,** 857–859.

Zaug, A. J., Grabowski, P. J., and Cech, T. R. (1983). *Nature (London* **301,** 578–583.

The Origin and Evolution of Retroposons

JOHN H. ROGERS

MRC Laboratory of Molecular Biology, Cambridge, England

ISBN 0-12-364493-3

Part 1: MECHANISMS OF RNA SPLICING

I. Prologue

Within less than a decade, new molecular techniques have revealed a variety of remarkable and previously unsuspected mechanisms for rearranging the sequence information in eukaryotic genomes. This review deals with two of them which operate on very different timescales: RNA splicing (particularly messenger RNA splicing), which is the excision of transcribed noncoding sequences from RNA during normal gene expression; and "retroposition," which is the insertion of reverse-transcribed sequences from RNA back into the genome during evolution.

RNA splicing was the first and most surprising discovery of eukaryotic molecular genetics. It is now known to occur in all kingdoms of eukaryotes. Although a specific mechanism involving a small nuclear RNA was proposed several years ago, only very recently has the development of *in vitro* splicing made experimental studies possible and lent support to the model.

Retroposons are a class of dispersed sequences in DNA, which appear to have arisen during evolution by a particular mechanism of inserting RNA sequences into chromosomal DNA ("retroposition"). They include processed mRNA pseudogenes, snRNA pseudogenes, the highly repeated *Alu* sequences, and—as this review will show—a variety of other sequences. They are entirely distinct from transposons and retroviruses.

There are three reasons why retroposons should be of general interest. The first, which is central to the theme of this volume, is that they constitute at least several percent of a typical mammalian genome, and show remarkable patterns of mobility and sequence evolution. Second, the mechanism that creates them may well have an important but so far unknown role in normal physiology. And third, the retroposons may include copies of various functional RNAs which have not previously been described. I will present significant new findings, which arise from a review of the existing data, in each of these three areas.

Both for mRNA splicing and for retroposition, most of the data on the probable mechanism come from analysis of the apparently dispensible parts of genomic DNA sequences. For this reason, and for consistency, all sequences will be quoted in DNA form throughout this review.

II. Splicing of Transfer RNA

The recent discovery of introns in two tRNA genes of an archaebacterium (Kaine *et al.* 1983) suggests that tRNA splicing is as old as life itself. Split tRNA genes also occur in yeast, slime moulds, flies, and frogs, and a corresponding enzyme activity has been detected in plants. [Split tRNA genes also occur in chloroplasts, but these are long introns of the mitochondrial type (Section III), not always in exactly the same position as in the nuclear tRNA genes, and may not have been in the ancestral chloroplasts. They will not be considered further here.]

The most detailed studies on the mechanism of tRNA splicing have used the organism in which split tRNA genes were first discovered—yeast. This and other aspects of the yeast tRNA genes have been reviewed by Guthrie and Abelson (1982).

A. Sequence Requirements

All the eukaryotic tRNA introns lie in the anticodon loop of the tRNA, and although there is often a redundant adenine flanking the intron, the intron can always be defined as lying one base 3′ to the anticodon itself (Table I). All pre-

TABLE I
Splice Sites in tRNA Genes

Saccharomyces	* * *	
Tyr[a]	CTGT A A/T T T . . . A C G A A/A T C	Goodman *et al.* (1977)
Phe[a]	CTGA A G/A A A . . . A A G T T/A T C	Valenzuela *et al.* (1978)
Ser-UCG	CTCG A A/T G G . . . C G G C T/A T C	Etcheverry *et al.* (1979)
Leu-3[a]	TTCA A G/A A A . . . G A A T A/C T C	Kang *et al.* (1979); Venegas *et al.* (1979)
Trp	CTCC A A/T T A . . . T T G C A/A T C	Ogden *et al.* (1979)
Ile	CT TA T A/G C A . . . A A G C A/A C G	Peebles *et al.* (1983)
Schizosaccharomyces		
Ser	CTCG A A/T T C . . . G T G G C/A T C	Mao *et al.* (1980)
Neurospora		
Phe	CTGA A G/T C C . . . C A T A A/A T C	Selker and Yanofsky (1980)
Dictyostelium		
Trp	CTCC A A/A A C . . . A A G T T/A T C	Peffley and Sogin (1981)
Xenopus		
Tyr	CTGT A G/G T G . . . G A G C A/A T C	Müller and Clarkson (1980)
Drosophila		
Leu[a]	CTCA A G/A T T . . . G A G C G/T T C	Robinson and Davidson (1981)
Sulfolobus		
Ser	CTCG A G/A C C . . . G G G C G/A C C	Kaine *et al.* (1983)
Leu	CTCA G G/A C C . . . G A G A C/C C C	Kaine *et al.* (1983)
Consensus (13 sequences)	C T C N A $^{G}_{A}$ / $^{A}_{T}$ NN . . . N A GNN / A T C * * * 12 13 8 . 11 $^{7}_{7}$ $^{6}_{5}$ . . . 8 11 . . 10 9 12	

[a]Multiple genes were sequenced for these species; there are a few nucleotide differences but not within the regions shown. *, Anticodon.

tRNAs tested are cleaved in exactly this position during splicing (Guthrie and Abelson, 1982). The introns are all very short (13–60 nucleotides) and have entirely different sequences in different tRNA genes. The 32-bp[1] intron of yeast tRNA$_3^{\mathrm{Leu}}$ can accommodate insertions of 10–30 bp or deletions of 8–13 bp and still be spliced correctly (Johnson *et al.*, 1980; Raymond and Johnson, 1983), although the larger insertions are spliced less efficiently, and a pre-tRNA with a 103-bp insertion is not spliced at all (nor trimmed at the 5′ and 3′ ends). Deletion of 20 bp from the intron, so as to make the last nucleotide a C, does prevent splicing (Raymond and Johnson, 1983).

Each intron in yeast tRNA includes a sequence complementary to the adjacent anticodon, which is presumed to basepair with it (O'Farrell *et al.*, 1978; Etcheverry *et al.*, 1979; Ogden *et al.*, 1979). These secondary structures are supported by studies of nuclease susceptibility around the anticodons (O'Farrell *et al.*, 1978; Etcheverry *et al.*, 1979). However, these structures are not essential for splicing, since they are absent from the introns in *Xenopus* and *Dictyostelium* (Muller and Clarkson, 1980; Peffley and Sogin, 1981) and from the suppressor forms of yeast tRNAs (Goodman *et al.*, 1977).

The precise sequence requirements for splicing are few. A consensus from Table I is

$$\begin{array}{l} \quad 32 \qquad\quad 37 \qquad\qquad\qquad\qquad\qquad\qquad 40 \\ \qquad * * * \\ \mathrm{CTCN}\underline{\mathrm{A}}{}^{\mathrm{A}}_{\mathrm{G}}\ /\ {}^{\mathrm{A}}_{\mathrm{T}}\mathrm{N}\ .\ .\ .\ .\ \mathrm{A}\ \underline{\mathrm{G}}\ \mathrm{N}\ \mathrm{N}\ /\ \mathrm{A}\ \mathrm{TC} \end{array}$$

The anticodon is indicated by asterisks. The consensus sequence apparently owes more to the largely invariant sequence of the tRNA, and to the (A,T)-rich character of the intron, than to specific requirements for splicing. Only the two nucleotides underlined seem to be specifically associated with tRNAs containing introns. Robinson and Davidson (1981) pointed out that most tRNA introns contain a sequence similar to AAATCTTG (or ACTTCG), but this is not a universal rule.

Several mutations have been identified which affect splicing of yeast SUP4 tRNA^{Tyr} when the gene is injected into *Xenopus* oocyte nuclei (Nishikura *et al.*, 1982). They occur at the three nucleotides numbered above. A → G37, at the upstream splice site, only reduces the level of splicing, while C → G32 and C → A40 block splicing completely. A → G37 also slows splicing in yeast (Colby *et al.*, 1981), but there is no loss of splicing accuracy.

[1]Abbreviations: bp, basepairs; kb, kilobases; R, purine; Y, pyrimidine; mrDNA, moderately repeated DNA; hnRNA, heterogeneous nuclear RNA; SV40, simian virus 40; IAP, intracisternal A particle; POMC, proopiomelanocortin; DHFR, dihydrofolate reductase; LINE, long interspersed element; TR, terminal repeat.

The yeast splicing machinery works only on pre-tRNAs, and there are several reasons for believing that it recognizes principally the tertiary structure of the whole tRNA, as do the 5′ and 3′ trimming nucleases. First, the local sequence alone, as described above, does not seem to provide adequate specificity. Second, mature tRNA inhibits splicing *in vitro* (Peebles *et al.*, 1979). Third, the yeast enzymes cannot ligate nonmatching pairs of half-tRNAs (Guthrie and Abelson, 1982). Fourth, of the $tRNA^{Tyr}$ mutants studied by Nishikura *et al.* (1982) in *Xenopus* oocytes, several had mutations in the aminoacyl stem which blocked trimming of the precursor, and in most of these the precursor accumulated without being spliced. [Untrimmed precursors of a different tRNA can be spliced in *Xenopus* extracts (Otsuka *et al.*, 1981), and an untrimmed precursor of one mutant (U6) may have been seen in the experiments of Nishikura *et al* (1982).] Mutation of the D stem also blocks splicing (Baldi *et al.*, 1983).

B. Enzymatic Mechanisms

1. *The Cleavage Reaction*

The study of tRNA splicing *in vitro* (Valenzuela *et al.*, 1978; Knapp *et al.*, 1978; O'Farrell *et al.*, 1978) has been facilitated by the *ts*136 mutant of yeast, which accumulates unspliced pre-tRNAs (see below). Abelson's group have now characterised the yeast reaction in detail, with separated and partially purified endonuclease and ligase components. tRNAs are also correctly spliced in extracts of animal cells. In *Xenopus* oocytes, microdissection indicates that all the enzymes required are in the nucleus (De Robertis *et al.*, 1981).

Endoribonuclease activities involved in tRNA splicing have been identified in yeast and in animal systems, and the yeast enzyme has been partially purified. All these enzymes require Mg^{2+} or some other divalent cation. They all leave a phosphate at the 3′ end and a hydroxyl at the 5′ end; where it has been investigated in detail, the 3′ phosphate is converted to a 2′,3′-cyclic phosphate. Thus the splicing ribonucleases resemble degradative ribonucleases such as RNase T1 and RNase A, which produce 5′ hydroxyl and 3′ phosphate or (as intermediates) 2′,3′-cyclic phosphate termini (Boyer, 1971). Conversely, all other processing ribonucleases leave 3′ hydroxyl and 5′ phosphate termini (Abelson, 1979; Boyer, 1982).

The cleavage enzyme of the yeast tRNA splicing system has been partially (5000-fold) purified (Peebles *et al.*, 1983). It is apparently an integral membrane protein. Nonionic detergent and spermidine enhance the activity of the enzyme; ATP is not required. A single fraction correctly cleaves all 10 yeast pre-tRNAs which are known to contain introns. This involves two precise cuts; the terminal redundancy of some introns is not reflected in the cleavage pattern. (In some cases one cut can be made without the other.) The intron RNA is released intact, as a linear molecule with a 5′ hydroxyl and a phosphate (exact linkage not characterized) at the 3′ end (Knapp *et al.,* 1979). The two half-tRNA molecules carry, respectively, a 2′,3′-cyclic phosphate and a 5′ hydroxyl. The cyclic phosphodiester terminus can be opened to 2′ phosphate by an enzyme in the same preparation.

A cleavage activity has also been obtained from *Xenopus* oocytes, separate from other ribonucleases and from the ligase activity (Otsuka *et al.,* 1981). This enzyme, "XlaI," similarly cleaves yeast pre-tRNAs into half-tRNA molecules and releases the intron in one piece; the half-tRNAs have, respectively, a 5′ hydroxyl and a phosphate at the 3′ end (exact linkage not characterized). A human HeLa cell extract (Filipowicz and Shatkin, 1983), like the yeast enzyme, produces termini with 5′ hydroxyl (for yeast $tRNA^{Ser}_{UCG}$) and 2′,3′-cyclic phosphate (for *Xenopus* $tRNA^{Tyr}$), although for the other half-tRNAs this extract gave partial 3′ degradation of the 5′ half-tRNA or 5′ phosphorylation of the 3′ half-tRNA.

2. *The Ligation Reaction*

The splicing ligation reaction is separable from the cleavage reaction, and yeast half-tRNA molecules can be correctly spliced by ligases from yeast, *Chlamydomonas,* wheat germ, *Xenopus* oocytes, human HeLa cells, and even phage T4-infected bacteria. The ligases are not membrane-bound. Each ligase system can use the 5′ hydroxyl and 2′,3′-cyclic phosphate termini of the yeast half-tRNAs as substrate. However, the reaction mechanisms are different, as summarized in Fig. 1.

a. *Phage T4.* It is remarkable that yeast half-tRNAs can be correctly spliced by a prokaryotic system—extracts of phage T4-infected *E. coli* (Greer *et al.,* 1983b). (Uninfected *E. coli* can also "splice" but the product is a 2′,5′ phosphodiester bond, not the normal 3′,5′ phosphodiester bond.) Phage T4 specifies phosphatase which removes the cyclic phosphate, polynucleotide kinase which adds a 5′ phosphate, and RNA ligase which completes the correct ligation. T4 also specifies several enzymes which cut specific tRNAs, in some cases cutting beside the anticodon (David *et al.,* 1982), but so far attempts to achieve the cleavage step with intron-containing pre-tRNAs have been unsuccessful. Instead, it has been proposed that the T4 tRNA "splicing" enzymes may be used to generate base hypermodifications by cleavage and rejoining (Kaufman and Kallenbach, 1975; Greer *et al.,* 1983b).

b. *Fungi.* The yeast enzyme is soluble and has been partially purified and extensively studied (Greer *et al.,* 1983a).

The immediate substrate at the 3′ end may be either the 2′,3′-cyclic phosphate, or a 2′P, 3′OH structure produced by a cyclic phosphodiesterase. The latter activity is present in yeast extracts, and copurifies both with the endonuclease and with the ligase. However, its role is unclear, since half-tRNAs with 2′ phosphates produced by it cannot be ligated *in vitro,* whereas those with 2′,3′-

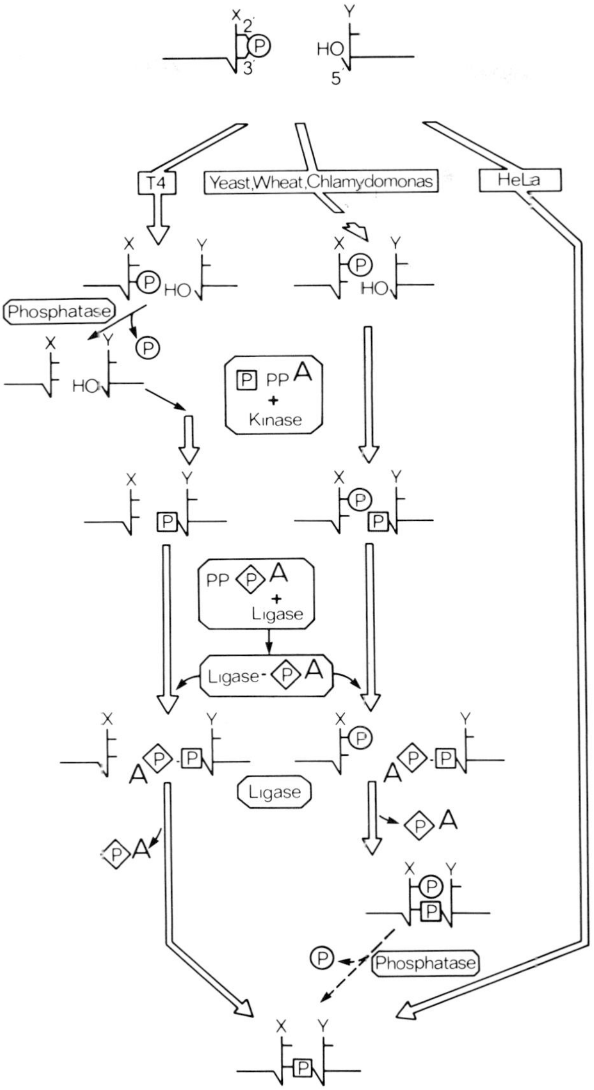

FIG. 1. Pathways of tRNA splicing. Adapted from Greer *et al.* (1983b), by kind permission of the authors and the MIT Press.

cyclic phosphates can (Greer *et al.*, 1983a); also, it is not tRNA-specific (Schwartz *et al.*, 1983). But the phosphate from this terminus ends up in the 2′ position following ligation, so a cyclic phosphodiesterase activity must operate during ligation.

The immediate substrate at the 5′ end is a 5′ phosphate, generated by an ATP-dependent kinase which cannot be separated from the ligase (Greer *et al.*, 1983a). Thus, the final ligation step probably couples 3′OH to 5′P just as do T4 and other ligases. In fact, the detailed mechanism is identical to that of other ligases (Boyer, 1982); AMP is transferred from ATP to a covalent linkage with the ligase, and thence to an adenosine(5′)P-P(5′)RNA structure; this reacts with the 3′ position of the other half-tRNA. This produces a standard 3′,5′ phosphodiester linkage (derived ultimately from ATP) plus an extra 2′ phosphate (derived from the donor splice site in the pre-tRNA):

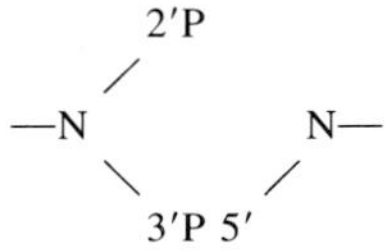

A specific enzyme is probably required to remove the 2′ phosphate, which is not present in mature tRNA, but this enzyme has still to be identified.

c. *Plants*. Both wheat germ (Konarska *et al.*, 1981, 1982) and *Chlamydomonas* (Kikuchi *et al.*, 1982) contain ligases which reproduce the activity of the yeast splicing ligase, joining a 2′,3′-cyclic phosphate terminus to a 5′ hydroxyl or 5′ phosphate terminus to produce the structure shown in the preceding paragraph.

The RNA ligase from wheat germ was first characterized by Konarska *et al.* (1981), when its natural substrate was not known. These authors assayed it using a model substrate, "omega RNA," which is a phosphatased 73-nucleotide RNase T1 fragment from tobacco mosaic virus RNA, bearing a 5′OH and 2′,3′-cyclic phosphate. It becomes circularized with the linkage shown above. This crude wheat germ extract also contains enzymes which convert the cyclic terminus to a 2′ phosphate, and which phosphorylate the 5′ hydroxyl (Konarska *et al.*, 1982; Schwartz *et al.*, 1983). As in yeast, these are likely to be components of the splicing pathway. Subsequent analysis with this extract (Konarska *et al.*, 1982) or with partially purified enzymes (Furneaux *et al.*, 1983; Gegenheimer *et al.*, 1983; Schwartz *et al.*, 1983), and with a variety of substrates, has shown that the 2′P left at the spliced junction comes from the terminal cyclic phosphate and the 3′,5′ phosphodiester group comes from the γ position of ATP. Thus, the overall reaction appears to be identical to that of yeast (Fig. 1).

The wheat germ system has a much wider range of substrates than the yeast system. It can circularize or dimerize omega RNA, potato spindle tuber viroid RNA, poly(A), U_nG, and C_nG, as well as ligating half-tRNAs from all 10 yeast pre-tRNAs tested (Branch *et al.*, 1982; Furneaux *et al.*, 1983; Gegenheimer *et al.*, 1983; Schwartz *et al.*, 1983). With yeast pre-$tRNA_3^{Leu}$ in the presence of the yeast endonuclease, the wheat germ ligase will even circularize the excised intron, by the same mechanism and with no loss of nucleotides; this is the only reported case of circularization of an excised intron in tRNA or mRNA splicing. Apparently this enzyme can ligate any RNAs with the correct termini.

The natural substrate for this ligase is still unknown. Split tRNA genes have not yet been discovered in plant nuclei; however, half-tRNAs can be found both in wheat germ and in *Chlamydomonas,* and can be ligated by the *Chlamydomonas* enzyme (Tyc *et al.*, 1983). Viroids may make use of it in the generation of progeny RNA from long concatamers, since it can circularize natural linear viroid RNA into forms resembling the normal circular form (Branch *et al.*, 1982).

d. *Animals.* Soon after the discovery of split yeast tRNA precursors, it was shown that they could be spliced in *Xenopus* oocyte nuclei (De Robertis and Olson, 1979) or an extract thereof (Ogden *et al.*, 1979) or in a HeLa cell extract (Standring *et al.*, 1979). Subsequently, split tRNA genes have been found in *Xenopus* itself (Table I).

Winicov and Button (1982) reported that mouse L cell nuclei incorporate γ label from ATP into internal phosphodiester bonds in tRNA-sized products, suggesting a splicing mechanism similar to that in yeast and wheat. However, *Xenopus* oocytes (Nishikura and De Robertis, 1981) and HeLa cell extracts (Filipowicz and Shatkin, 1983) show a different, simpler, tRNA splicing mechanism; they derive the phosphate at the spliced junction not from ATP, but from the pre-tRNA.

In the HeLa cell extract (Filipowicz and Shatkin, 1983), yeast and *Xenopus* tRNA genes can be transcribed, cleaved, and spliced in the same extract. The spliced junction is a normal 3′,5′ phosphodiester formed from the 3′ phosphate of the 5′ half-tRNA, i.e., from the upstream splice site. No ATP is required and no 2′ phosphate is formed, although a 2′,3′-cyclic phosphate may still be a substrate since half-tRNA carrying it is formed during the reaction. This reaction thus looks like the reverse of the RNAse T1 reaction. Also, omega RNA bearing a 2′,3′-cyclic phosphate and 5′OH can be circularized by the HeLa cell or *Xenopus* oocyte extracts and, again, this phosphate is the one incorporated into the spliced junction (Filipowicz *et al.*, 1983). A 3′ phosphate can be used but only when it is first cyclized by an ATP-dependent activity in the extracts; if the 5′ end is phosphorylated it is not ligated (Filipowicz *et al.*, 1983).

C. Coupling to Other Steps in tRNA Maturation

The maturation of tRNAs is a complicated process, involving trimming of the 5′ and 3′ ends, addition of 3′-terminal CCA, and many base modifications. Most of these processes normally precede splicing and occur also in unsplit tRNAs, which are the majority. Splicing is apparently required only for one base modification (see below). Whether splicing itself depends on the performance of the preceding modifications is more difficult to answer, because it is difficult to generate a specific defect in any one of these processes. The studies which have been made involve either the yeast mutant *ts*136 (Section II,D), in which the accumulated unspliced pre-tRNAs already carry most of their modifications, or cloned yeast tRNA genes injected into *Xenopus* oocytes (Melton *et al.*, 1980), which may not be a strictly physiological system.

The trimming of the 5′ and 3′ ends of the precursors normally precedes splicing (Guthrie and Abelson, 1982; Melton *et al.*, 1980) but it is not essential for splicing in extracts of *Xenopus* oocytes (Otsuka *et al.*, 1981) or of HeLa cells (Standring *et al.*, 1981; Filipowicz and Shatkin, 1983). Most base modifications also precede splicing, in animal and yeast systems (Kang *et al.*, 1979; Melton *et al.*, 1980; Nishikura and De Robertis, 1981; Otsuka *et al.*, 1981; Filipowicz and Shatkin, 1983). However, *ts*136 intermediates usually lack modifications in the vicinity of the intron in the anticodon arm—i^6A in $tRNA^{Tyr}$, Y in $tRNA^{Phe}$, and 2′-O-methylations (Kang *et al.*, 1979). These defects could be due to the different conformation of the anticodon loop prior to splicing; i^6A is produced in a $tRNA^{Tyr}$ from a gene which lacks the intron (Johnson and Abelson, 1983).

One modification that seems to depend on the presence of the intron in $tRNA^{Tyr}$ is the anticodon pseudouridine, a modification that is not present in any other yeast tRNAs. This pseudouridine is present in the *ts*136 accumulated pre-$tRNA^{Tyr}$, but when a suppressor $tRNA^{Tyr}$ gene (*SUP6*) was reconstructed to delete the intron (Wallace *et al.*, 1980; Johnson and Abelson, 1983) the modification did not take place, resulting in reduced suppressor activity of the tRNAs. Three other modifications did take place correctly. Mutational analysis of a similar $tRNA^{Tyr}$ gene (*SUP4*) (Kurjan *et al.*, 1980; Koski *et al.*, 1980) showed that, of 26 different mutations which were selected for loss of suppressor activity, only one was within the intron, and its fault was that it produced premature termination of transcription. Therefore, no general role for splicing in tRNA maturation has yet been identified.

D. Yeast Mutations Affecting tRNA Splicing

The best-known mutant affecting tRNA splicing, *ts*136 in the gene *rna1*, was isolated by Hutchinson *et al.* (1969), who showed that at the nonpermissive temperature production of most mRNAs ceases immediately. The cells accumulate nuclear mRNA or pre-mRNA (Shiokawa and Pogo, 1974) and also precursors to rRNAs and to a subset of tRNAs. The latter are all tRNA precursors which contain introns (Hopper *et al.*, 1978; Knapp *et al.*, 1978; Kang *et al.*, 1979); they have mature 5′ and 3′ termini and most of the base modifications. Pre-tRNAs without introns do not accumulate (Guthrie and Abelson, 1982). The defect may be in some aspect of nuclear RNA transport (Hutchison *et al.*, 1969); if the tRNA cleavage enzyme is attached to the nuclear membrane, accumulation of intron-containing pre-tRNAs may be secondary to the transport defect (Peebles *et al.*, 1983).

A second temperature-sensitive mutant, *los1*-1, accumulates the same set of intron-containing pre-tRNAs as *ts*136, but does not seem to affect rRNA or mRNA synthesis (Hopper *et al.*, 1980). The *los*1-1 and *rna*1 mutations complement each other and segregate independently in meiosis. Clearly the *los*1-1 gene product is a prime candidate for a component of the splicing machinery, but no temperature-sensitive biochemical defect has yet been reported.

III. Splicing of Ribosomal RNA, Mitochondrial RNA, and Chloroplast RNA

Splicing is widespread in rRNA and mtRNA of fungi and protozoa, and in ctRNA (chloroplast RNA) of plants. In at least one case it can be catalyzed by the RNA structure alone, with no proteins present. This suggests that it may be a very ancient reaction. But one cannot prove that it existed in the ancestors of mitochondria, because it is likely that the original distribution of these introns has been altered; there is now evidence for abundant genetic exchange between mitochondria, chloroplasts, and nuclei, and for the insertion of homologous mitochondrial introns at different locations.

The mechanisms are reviewed in detail in the chapter by T. Cech in this volume. The main features may be summarized here as follows. For the major group of introns (Group I), splicing requires a specific, elaborate secondary structure within the intron; the only consistent features at the actual splice site are a T preceding the intron and a G ending it. Splicing of *Tetrahymena* nuclear rRNA can occur *in vitro* without any proteins. However, splicing of some mtRNAs *in vivo* depends on the existence of open reading frames within some introns, implying that these splicing reactions require intron-encoded proteins. The self-catalyzed *Tetrahymena* reaction does not resemble the tRNA reactions; the products of cleavage are 3′OH and 5′ guanylate, and the product of ligation is a normal 3′,5′ phosphodiester bond derived from the phosphate at the downstream splice point.

Other mtRNAs and ctRNAs (Group II) have a different intron structure and longer splice site consensus sequences (Michel and Dujon, 1983; Cech, 1983). One ctRNA has even stranger introns consisting almost entirely of A and T (Montandon and Stutz, 1983). Clearly there are novel splicing mechanisms still to be discovered.

IV. Splicing of Messenger RNA

Messenger RNA splicing is by far the most widespread type of splicing. Almost all protein-coding genes in vertebrates contain introns, as do many in plants and in lower animals, and some in fungi. Such genes are transcribed into large nuclear RNA precursors—the "heterogeneous nuclear RNA" (hnRNA or pre-mRNA), which was studied devotedly by some and dismissed as an artifact by others, until the revolution in techniques and the discovery of RNA splicing generated a new paradigm of gene expression in which hnRNA took an unquestioned place. The 5′ and 3′ ends of the hnRNA are the same as those of the mature mRNA, with the same posttranscriptional modifications—the 5′ cap and the 3′ poly(A). The large size of hnRNA is due to introns which may be many times the length of the exons. For reviews, see Perry *et al.* (1976), Darnell (1979), Abelson (1979), and Breathnach and Chambon (1981).

Unfortunately, in contrast to the detailed biochemical analysis of tRNA and rRNA splicing, the evidence on the mechanism of mRNA splicing is still largely circumstantial, because until recently it could not be reproduced *in vitro*. The major clue was the existence of extensive sequence requirements around the splice points. It was proposed that this reflected basepairing with the U1 species of small nuclear RNA. Recent *in vitro* experiments support this hypothesis. Thus, while tRNA splicing may depend on extensive structure in the exons, and rRNA and mtRNA splicing on extensive structure in the intron, mRNA splicing may depend on specific duplex formation at the splice sites themselves.

Most work on mRNA splicing has used animal systems (Sections IV, A–E). Fungal and plant genes have very similar splicing signals but their mechanisms may differ in detail; they will therefore be discussed separately (Sections IV, F–H). Yeast, in particular, offers both simpler splicing patterns and the possibility of genetic analysis.

A. Introns in Animals: Sequence Requirements

The first splicing rule to be discovered, and the most invariant, is the "GT-AG rule" (Breathnach *et al.*, 1978): virtually all introns begin with GT and end with AG. These dinucleotides are the cores of longer splicing signals (Seif *et al.*, 1979). The most recent compilation (Mount, 1982), from over a hundred published intron sequences, leads to the following consensus sequence. Essentially the same consensus is obtained from insects (Snyder *et al.*, 1982) as from vertebrates. [Bases present in $> 36\%$ of sequences are shown, with the actual percentage frequency below. A few plant and fungal introns were included by Mount (1982) but do not affect the consensus.]

Upstream site (donor site):

−3				+1						+7
(C) (A)	A	G	/	G	T	A	A	G	T	(A)
43 40	64	73		100	100	62	68	84	63	39

Downstream site (acceptor site):

−15										−5						+2
T	(T)	T	T	T	(T	T	T	T)	T	T	(N)	C	A	G	/ G	(T)
51	44	50	53	60	49	49	45	45	57	58	–	65	100	100	52	37

At the upstream site, all but two of the compiled sequences agree in at least 6/9 positions with $^{C}_{A}$AG/GTRAGT (R = A or G) (Mount, 1982), and all but four agree in at least 7/10 positions with $^{C}_{A}$RG/GTRRGTR (a consensus which allows for G.U basepairs with U1 snRNA; see below). Curiously, most of the exceptions are in chickens, although chickens show the same consensus as other animals. Although the GT–AG rule was inviolate in this compilation, several recently sequenced upstream splice sites have shown GC instead of GT. These are AAG/GCAAGCG in the α^{D}-globin gene of chicken and duck (Dodgson and Engel, 1983; Erbil and Niessing, 1983), AA/GCTGAGT at a cryptic site used in a mutant β-globin gene of rabbit (Wieringa *et al.*, 1983), and AAG/GCAAGTT at an optional site in the αA-crystallin gene of mouse (King and Piatigorsky, 1983). (The rabbit cryptic site fits the consensus better if the offending C is looped out.)

At the downstream splice site, the GT–AG rule is still inviolate. There is $^{C}_{T}$AG/ in all but five cases and usually AG/G in the remainder; again, the exceptions are mostly from chickens. The −4 position shows no base preference. The preceding 11 nucleotides are mostly pyrimidines; each position registers > 43% T and 19–36% C. This region never contains the dinucleotide AG. [However, AG has recently been found at −5, in the sequence TCCCAGCAG/GT (Van het Schip *et al.*, 1983), and in another such junction both AG sites can be used (Maurer *et al.*, 1981).]

There is usually at least one nucleotide of terminal redundancy flanking introns, but the exact boundaries can always be defined by the GT–AG rule. The redundancy is attributable mainly to the overlap in upstream and downstream consensus sequences, and some introns do not show any redundancy (Mount, 1982).

Studies on natural and artificial mutants have confirmed that these sequences at the splice sites are necessary and sufficient for splicing.

1. *Splice site sequences are sufficient: deletions of flanking sequences.* Outside the consensus sequences described, there is no universal resemblence between different introns, nor sequence conservation within homologous ones. Experiments on simian virus 40 (SV40), and on globin genes cloned within it, have shown that an upstream splice site is still functional if all the 5′ exon sequences are deleted, or if the intron sequences are deleted to within 29 or 6 bp of the splice point. A downstream splice site is still functional if all the 3′ exon sequences are deleted, or if the intron sequences are deleted to within 23 or 11 bp of the splice point (Khoury *et al.*, 1979; Piatak *et al.*, 1981; Ghosh *et al.*, 1981;

Kühne *et al.*, 1983; Wieringa *et al.*, 1984; and citations in Elder *et al.*, 1981).

Admittedly, many apparently good matches to the consensus sequences can be found which are not used. However, many of them can be used given sequence changes elsewhere that affect the normal pattern of splicing. This topic will be discussed in Section IV,E,6.

2. *Splice site sequences are necessary: local mutations.* The importance of the upstream splice site has been proved by a variety of mutations that affect it. In particular, natural splice site mutants in the globin genes of thalassaemia patients, and artificially constructed globin splice site mutants, have been tested for proper splicing on introduction into human or monkey cells. The results are shown in Table II.

Changes in the invariant GT always prevent the use of the site. The other changes which reduce splicing all decrease the match between the mutant sequence and the consensus (although some of them still retain a better match than some natural splice sites). Wieringa *et al.* (1983) produced five separate G ↔ A

TABLE II
MUTATIONS AFFECTING UPSTREAM SPLICE SITES[a]

Consensus	CAG/GTAAGTA		CAG/GTAAGTA	
Down mutations				
Yeast	CTG/GTATGTT	to	CTG/GGGCTTG	Gallwitz (1982)
Yeast	ATG/GTATGTN	to	ATG/GTATACN	Pikielny *et al.* (1983)
SV40	AAG/GTACCTA	to	AAG/GTTGACG	Ghosh *et al.* (1982)
Adenovirus	AGG/GTGAGGA	to	AGG/GTGAATA	Solnick (1981)
Adenovirus	AGG/GTGAGGA	to	AGG/GGGAGGA	Montell *et al.* (1982)
Drosophila	AAG/GTAACTA	to	AAG/GCGATCT	Benyajati *et al.* (1982)
Mammals				
Collagen	GTG/GTAGGTT	to	GTG/GTAGCGA	Esumi *et al.* (1983)
Immunoglobulin	TAG/GTGAGTG	to	TAC/CTGAGTG	Hozumi *et al.* (1981)
Globin	AGG/GTGAGTC	to	AGG/ATGAGTC	Treisman *et al.* (1982)
Globin	AGG/GTGAGTT	to	AGG/ATGAGTT	Wieringa *et al.* (1983)[b]
Globin	GAG/GTGAGGC	to	GAG/GCTCCCT	Felber *et al.* (1982)
Globin	CAG/GTTGGTA	to	CAG/ATTGGTA	Treisman *et al.* (1983)
Globin[c]	CAG/GTTGGTA	to	CAG/GTTGCTA	Treisman *et al.* (1983)
Globin[c]	CAG/GTTGGTA	to	CAG/GTTGGCA	Treisman *et al.* (1983)
Up mutations				
Globin	GTG/GTGAGGC	to	GAG/GTGAGGC	Goldsmith *et al.* (1983)
Globin	GAT/GTAAGAG	to	GAG/GTAAGAG	Dobkin *et al.* (1983)
Globin	CAG/CTACCAT	to	CAG/GTACCAT	Treisman *et al.* (1983)
Globin	GTG/GTGAGGC	to	GTG/GTAAGGC	Orkin *et al.* (1982)

[a]Adapted from Mount (1983).

[b]Separate A–G transitions at five other locations in this sequence (AGG/GTGA) had no effect (Wieringa *et al.*, 1983).

[c]These two reduce use of mutant site; all other down mutations eliminate it.

transitions in rabbit globin splice sites which all left good matches to the consensus, and these did not affect splicing.

Mutations which create new functional upstream splice sites (Table II) all involve increased match to the consensus.

No single-site mutations of downstream sites have been reported. However, splicing is abolished by deletion of 25 bp of the human β-globin gene (Orkin *et al.*, 1983), or of 17 bp of the *Drosophila* alcohol dehydrogenase gene (Benyajati *et al.*, 1982), eliminating the invariant AG.

B. U1 snRNA and Splicing

1. *The snRNPs*

All metazoan and protozoan cells contain discrete, stable, small nuclear RNAs (snRNAs). In mammals the major ones are named U1 to U6. They have unusual 5′ caps and many base modifications, and they have all been sequenced by Harris Busch's group and others (Reddy and Busch, 1981; Busch *et al.*, 1982). Secondary structures have been predicted for all these snRNAs in solution, and they show some common features (Reddy and Busch, 1981; Krol *et al.*, 1981; Branlant *et al.*, 1982; Nojima, 1983). In the nucleus, they exist in distinct ribonucleoprotein particles (snRNPs). For reviews on the general properties of the snRNAs and snRNPs, see Zieve (1981), Lerner and Steitz (1981), and Busch *et al.* (1982). Here I will concentrate on the possible involvement of U1 snRNA in mRNA splicing.

U3 snRNA has different properties from the others and resides in the nucleolus. U1, U2, U4, U5, and U6 are found in the nucleoplasm, in 10 S ribonucleoprotein particles (snRNPs), which share five or six polypeptides including the "Sm" antigens (Lerner and Steitz, 1979, 1981; MacGillivray *et al.*, 1982). This set of antigens is recognized by anti-Sm sera from many human patients with systemic lupus erythematosus, a discovery which has greatly facilitated the study of the snRNPs. Monoclonal antibodies against "Sm" have also been produced (Lerner *et al.*, 1981). The stoichiometry is probably one snRNA per 10 S particle (Lerner and Steitz, 1979). U1- and U2-containing particles can be distinguished because each has a characteristic protein in addition to the "Sm" set (67–68 kDa in the case of U1), and some lupus patients have antibodies specific for the U1 snRNP. Thus the snRNPs can be separated both biochemically and immunologically (Lerner and Steitz, 1979, 1981; Takano *et al.*, 1981; Sri-Widada *et al.*, 1982; Wooley *et al.*, 1983; Hinterberger *et al.*, 1983; Kinlaw *et al.*, 1983; Bringmann *et al.*, 1983).

2. *Proposed Role of U1 snRNA*

The demonstration of a universal consensus sequence for mRNA splice sites (Seif *et al.*, 1979) prompted a search for matching consensus sequences in the snRNAs. The most abundant of them, U1, had been sequenced (Reddy *et al.*, 1974; minor revisions by Ro-Choi and Henning, 1977, and Busch *et al.*, 1982) and was found to be exactly complementary to the combined consensus sequences for upstream and downstream splice sites. This immediately suggested the model for basepairing shown in Fig. 2, in which U1 snRNA basepairs with both ends of an intron in an hnRNA molecule, bringing them together in exact register for cutting and splicing (Lerner *et al.*, 1980; Rogers and Wall, 1980).

The proposed basepairing involves the first 22 nucleotides of mammalian U1 snRNA, and should not be affected by the 2′-O-methyl or pseudouridine modifications of the snRNA. Since most introns have some terminal redundancy, one could envisage that branch migration in the duplex in Fig. 2 might permit variability in the exact position of splicing. Alternatively, basepairing of upstream and downstream sites with U1 could be sequential rather than simultaneous, so that each site would basepair maximally, and the splice would always be made at the same point.

At position −3 of the upstream splice site, the A often found instead of C could perhaps form a splaying A.G interaction as in tRNA^{Phe} (Lerner *et al.*, 1980). At positions +3 and +4, where A is most abundant, G is the second most abundant (Mount, 1982); this means that G.U basepairs could replace A.U basepairs, consistent with observations on natural and mutant splice sites (see Table II). At the downstream site, the pyrimidine-rich stretch in the pre-mRNA is proposed to basepair with the purine-rich stretch in the snRNA. This complementarity is variable; to improve it in individual cases, it seems reasonable to allow looping-out of a few bases four nucleotides before the downstream splice point, at a position where there is no consensus nucleotide (Rogers and Wall, 1980). However, the purine-rich stretch of U1 is not conserved in *Drosophila* (Mount and Steitz, 1981), and does not bind to a downstream splice site *in vitro* (Mount *et al.*, 1983; see below); therefore Mount *et al.* suggest that the pyrimidine-rich segment of the downstream splice site may be recognized by some mechanism other than U1 basepairing. Even if only the YAG/ at the downstream site anneals with U1, three basepairs should be sufficient to provide precise alignment of the hnRNA, as it is in the codon–anticodon and Shine-Dalgarno

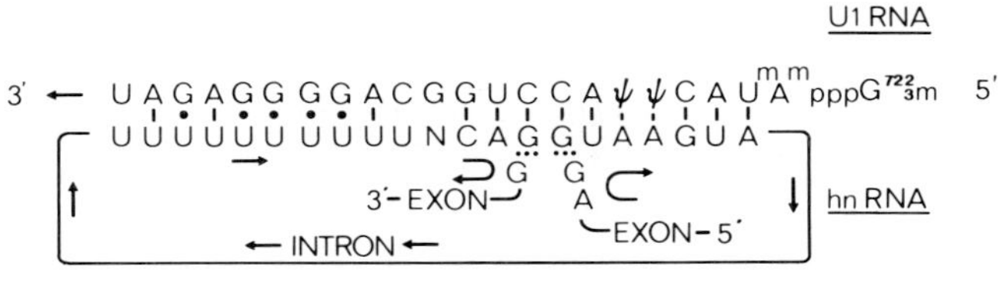

FIG. 2. Model for alignment of splice sites by basepairing to U1 RNA. Dots indicate the predicted points of cutting and splicing. Adapted from Rogers and Wall (1980), with updated U1 sequence from Busch *et al.* (1982).

(1974) interactions of mRNA. Other factors would be expected to provide stability when hnRNA and U1 snRNA are appropriately basepaired. Molecular modelling shows that the upstream and downstream splice sites could indeed basepair simultaneously with U1 RNA, putting the RNA chains in a plausible conformation for a simple splicing reaction (MacCumber and Ornstein, 1984).

3. *Relevant Properties of U1 snRNP*

A number of properties of U1 snRNPs are consistent with their being part of the RNA splicing complex. U1 (or Species D) is the most abundant snRNA in mammalian cells, at ~ 10^6 copies per cell, and is as stable as ribosomal RNA (Weinberg and Penman, 1968, 1969; Ro-Choi and Busch, 1974). Kano *et al.* (1978) found as much snRNP in chicken erythrocytes as in erythroblasts and liver, but Lerner *et al.* (1980) found much less in erythrocytes, consistent with a role in gene expression. U1 snRNPs are physically associated with the large RNP particles which contain hnRNA in 30 S hnRNP subunits (Zieve and Penman, 1976, 1981; Deimel *et al.*, 1977; Northemann *et al.*, 1977; Guimont-Ducamp *et al.*, 1977; Howard, 1978; Gallinaro and Jacob, 1979; Barque *et al.*, 1981; Maxwell *et al.*, 1981; Zieve, 1981). U1 RNA and hnRNA can be coisolated from these complexes and dissociated by 70% formamide (Flytzanis *et al.*, 1978) or 2 *M* urea (Northemann *et al.*, 1979), which suggests that U1 RNA is bound to hnRNA by short regions of basepairing. The best evidence for this basepairing *in vivo* is that U1 RNA can be crosslinked to hnRNA in cells, using an intercalating photochemical reagent (Calvet and Pederson, 1981).

The snRNAs and their associated antigens are virtually identical in different mammals (Lerner *et al.*, 1980; Busch *et al.*, 1982). The rat U1a RNA sequence (Busch *et al.*, 1982) is very similar to those of human and chicken U1 (Branlant *et al.*, 1980), and *Drosophila* U1 is 72% identical to human U1 (Mount and Steitz, 1981). Similar snRNA species, not yet sequenced, occur throughout the animal kingdom (Rein and Penman, 1969; Hellung-Larsen and Frederiksen, 1977; Lerner *et al.*, 1980), in plants (Krol *et al.*, 1983), and even in amoeba (Goldstein, 1976). U1 hybridizes to genomic DNA of sea urchin, fly, and slime mould (Blin *et al.*, 1983). The "Sm" and U1 "RNP" antigens are conserved not only throughout the mammals, but also in insects (Lerner *et al.*, 1980; Wooley *et al.*, 1982). So are the interactions between U1 and the proteins of the snRNP, in that U1 RNAs and U1-binding proteins from human and *Drosophila* are interchangeable (Wieben *et al.*, 1983).

Secondary structures for vertebrate U1 snRNA can be predicted from the sequence (e.g., Ohshima *et al.*, 1981), and three groups who have used various methods to choose between the alternatives have arrived at very similar structures. Epstein *et al.* (1981) used limited ribonuclease digestion of the snRNA and snRNP. Branlant *et al.* (1981) did a much more extensive study of U1 RNA in solution, observing accessibility of the RNA to nucleases and to hybridization.

Mount and Steitz (1981) used comparative modeling of the vertebrate and *Drosophila* sequences. The three groups agree on the locations of the four main duplex regions, and differ only in the fact that limited sliding is possible within some of them between pyrimidine-rich and purine-rich strands. Within this common structure, the sequences of *Drosophila* and human U1 show many compensating base changes, providing strong support for the biological significance of the structure (Mount and Steitz, 1981). The single-stranded regions are the most conserved, including both the "connector" sequence YAAU$_n$G which is common to U1, U2, U4, and U5 RNAs (Krol *et al.,* 1981; Branlant *et al.,* 1982; Nojima, 1983), and also including the first 11 nucleotides, which are supposed to basepair with splice sites. These 11 nucleotides are identical in *Drosophila* and human, and the structure leaves them unpaired within U1; this is implied by the fact that they are available in solution for nuclease digestion (Branlant *et al.,* 1981) and for basepairing with a complementary oligodeoxyribonucleotide (Lazar and Jacob, 1982). This latter basepairing mimics that proposed for splice sites. The first few nucleotides of U1 are also susceptible to nicking (possibly during isolation) which removes the U1 cap and first six nucleotides, producing the U1a* variant (Lerner *et al.,* 1980).

Mount *et al.* (1983) have shown that purified U1 snRNP can indeed bind specifically to an upstream splice site in naked RNA, within a 508-nucleotide *in vitro* transcript. The binding protected 15–17 nucleotides covering the upstream splice site, and there was 5-fold less binding to a transcript with no upstream splice site. The binding depended on protein, in that U1 RNA alone did not compete, and binding was not seen with partially Proteinase K-digested U1 snRNP; conversely, nuclease-digested U1 snRNP bound RNAs at high levels but entirely nonspecifically. Thus it seems that the protein component has a general affinity for RNA, and that U1 RNA confers specificity for the U1-complementary region at the upstream splice site. Although the downstream splice site was not specifically bound nor required, and there was no cleavage or splicing in the reaction, this is the most direct evidence yet for the proposed role of U1 snRNP.

4. *Are Other snRNAs Involved in RNA Processing?*

In view of the similar properties of the U-snRNAs, there has been speculation that they might all be involved in various steps of RNA processing. Indeed, U2, like U1, appears to be basepaired to hnRNA *in vivo* (Calvet *et al.,* 1982). Therefore, various other models involving snRNA basepairing have been proposed.

The first was the proposal of Murray and Holliday (1979) that the adenovirus-encoded VA RNAs could align both the exons and introns of adenovirus pre-mRNA. Unfortunately, there are multiple possible (G,C)-rich interactions involving these sequences, and the proposed splice site interaction is only possible for a minor VA RNA species. The VA RNAs do indeed hybridize to adenovirus mRNAs (Mathews, 1980), but are not associated with adenovirus pre-mRNA (Gallinaro *et al.*, 1980; Blanchard, 1980). The major VA RNA seems rather to be required for efficient translation of late viral mRNAs (Thimmappaya *et al.*, 1982).

Other alignments of snRNAs with RNA processing sites have been proposed by Harada and Kato (1980), Krayev *et al.* (1980), Kato and Harada (1981), Ohshima *et al.* (1981), and Gojobori and Nei (1981).

None of these interactions meets the criteria that make the U1 hypothesis so attractive: that it applies to all introns in pre-mRNAs, that it aligns the cleavage sites precisely, and that the basepairing between the snRNA and the consensus splice sites is perfect. So, while U2, U4, etc. may well have related functions, no convincing models for these functions have yet been proposed.

The only other proposed interaction with comparable credentials is that of U3 RNA with pre-rRNA (Bachellerie *et al.*, 1983). U3 snRNA is located in the nucleolus where pre-rRNA is transcribed, and is basepaired to pre-rRNA but not to mature rRNA (Prestayko *et al.*, 1970; Zieve and Penman 1976; Calvet and Pederson, 1981). It contains an 8-base sequence which is exactly complementary to a conserved 8-base sequence in an otherwise nonconserved spacer, adjacent to where cleavage of the pre-rRNA takes place.

C. mRNA Splicing *in Vitro*

RNA splicing has been reported in isolated nuclei with endogenous viral RNAs (Blanchard *et al.*, 1978; Yang *et al.*, 1981), but without detailed characterization of the products, and such splicing usually occurs only at very low efficiency (Manley *et al.*, 1982). *Xenopus* oocytes can splice an injected *in vitro* transcript (Green *et al.*, 1983). However, for detailed biochemical analysis, cell-free splicing is needed.

Extracts of mammalian cells will transcribe genes accurately *in vitro* (Manley *et al.*, 1980), with addition of 5′ caps though not of 3′ poly(A), and substantial efforts have been devoted to obtaining splicing in similar systems. Most investigators have used adenovirus templates. Early results were suggestive (Weingärtner and Keller, 1981; Goldenberg and Raskas, 1981), but others could not detect splicing in similar extracts (Handa *et al.*, 1981). It is only recently that *in vitro* splicing has been obtained reproducibly and the products characterized convincingly. Whole-cell extracts, similar to those of Manley *et al.* (1980) and Handa *et al.* (1981), show coupled transcription, capping, and splicing of exogenous globin exons (Kole and Weissman, 1982) or adenovirus exons (Padgett *et al.*, 1983). The products were shown to be spliced by partial sequencing. However, the efficiency was no more than 2–10%.

Extra RNA inhibits these reactions (Padgett *et al.*, 1983), and other inhibitors could be present. More efficient splicing has been obtained using purified pre-mRNA as substrate, and also with extracts enriched in nuclear components. Hernandez and Keller (1983) made RNA *in vitro* from a template encoding the adenovirus late leader, then purified the RNA and obtained splicing of it in a HeLa cell extract, with an efficiency of up to 14%. The reaction with nuclear extract plus postnuclear S100 fraction (lacking ribosomes and other cytoplasmic particles) was more efficient than with a whole cell extract. The activity was sensitive to heating or to Proteinase K, so must involve a protein component. Admittedly, this late leader splice is one of the fastest known in vivo (Manley *et al.*, 1982), and other introns may be less efficient substrates (Hernandez and Keller, 1983).

The highest efficiency has been reported for nuclear extracts with pure exogenous RNA. Goldenberg and Hauser (1983) obtained 80% efficiency with adenovirus E2 transcript from infected cells, purified by successive sucrose gradients, and Krainer *et al.* (1984) obtained up to 90% efficiency with globin transcript from a phage promoter, made and capped *in vitro*. Primed sequencing showed these splices to be accurate. Uncapped RNA was spliced less efficiently, with prominent cuts at cryptic sites, but absence of poly(A) made no difference (Krainer *et al.*, 1984).

Is U1 snRNP required for *in vitro* splicing? All the extracts used contain snRNPs, but in preparation of nuclear extracts many of the snRNPs are lost due to leakage into the cytoplasm (Hernandez and Keller, 1983). These authors found that splicing by nuclear extract was substantially stimulated by a cytoplasmic subfraction which was 10% snRNPs. The cytoplasmic fractions alone did not splice.

The effect of human anti-(U1) snRNP antiserum has been studied in live cells, where it inhibited some split-gene expression, although not all, perhaps for technical reasons (Lenk *et al.*, 1982). When anti-(U1)snRNP and anti-Sm antibodies are used in isolated nuclei (Yang *et al.*, 1981), and in the more reliable *in vitro* systems (Goldenberg *et al.*, 1983; Hernandez and Keller, 1983; Padgett *et al.*, 1983b), they completely inhibit splicing, as predicted, whereas control sera do not. Thus the experiments done so far support the involvement of U1 snRNP in splicing. Given pure RNAs made *in vitro,* and efficient splicing systems, we can hope for rapid elucidation of the steps in the reaction. Recent reports reinforce this hope (Krainer *et al.*, 1984; Grabowski *et al.*, 1984).

D. Coupling to Other Steps in mRNA Maturation

1. *Transcription, Capping, and Polyadenylation*

Splicing is not directly coupled to transcription, as shown by the existence of complete primary transcripts *in vivo* (Section IV,E,5), and by the splicing of isolated pre-mRNAs *in vitro* (see above) and in *Xenopus* oocytes injected with a capped *in vitro* transcript (Green *et al.*, 1983). It is not known whether capping is essential, since caps are added very early *in vivo* (Salditt-Georgieff *et al.*, 1980; Babich *et al.*, 1980) and *in vitro,* and are necessary for stability in the *Xenopus* oocytes (Green *et al.*, 1983).

Splicing normally follows polyadenylation, but this is not obligatory either, since splicing can occur *in vivo* if polyadenylation is inhibited (Zeevi *et al.*, 1981), as well as on poly(A)$^-$ transcripts in the *in vitro* systems and in injected oocytes. The adenovirus late leader exons can even be spliced before the transcript is complete (Section IV,E,5).

The nucleoside analog DRB (5,6-dichloro-1-β-D-ribofuranosylbenzimidazole) blocks synthesis of most mRNAs, including those of split genes in adenovirus, SV40, and globin, by enhancing premature termination after several hundred nucleotides (Fraser *et al.*, 1978; Tweeten and Molloy, 1981; Skolnick-David *et al.*, 1982). However, it has no effect on expression of unsplit genes for interferon or adenovirus pIX (Sehgal *et al.*, 1978; Vennström *et al.*, 1979). This has prompted suggestions that it interferes with an intron-dependent process, possibly recognition of the splice sites, which would imply a previously unsuspected relation between transcription and splicing.

Intercalating drugs such as proflavine inhibit hnRNA turnover and mRNA production (Snyder *et al.*, 1971; Brinker *et al.*, 1973; Yannarell *et al.*, 1977; Chiu *et al.*, 1980). Since processing of rRNA precursors is also inhibited, these drugs may affect processes other than RNA splicing.

2. *Is Splicing Essential?*

Splicing is not essential for all genes, since cellular genes for interferons and histones (Nagata *et al.*, 1980; Hentschel and Birnstiel, 1981) and some viral genes (Aleström *et al.*, 1980; Wagner *et al.*, 1981; Treisman *et al.*, 1981) do not contain introns. In SV40 recombinants, it is possible to produce and translate unspliced mRNA from the herpesvirus thymidine kinase gene (Hanahan *et al.*, 1980), from an SV40-insulin fusion gene (Gruss *et al.*, 1981), and from the SV40 late (19 S) region itself (Piatak *et al.*, 1981; Ghosh *et al.*, 1981; White *et al.*, 1982).

However, two other genes studied in SV40-derived vectors do require splicing for stable expression (reviewed by Elder *et al.*, 1981). One is the SV40 late (16 S) VP1 gene. When this is replaced by a cDNA copy, the late region is transcribed but produces little or no stable RNA and no VP1 (Gruss *et al.*, 1979). Insertion of a complete globin intron into this gene rescues expression of a stable RNA with the correct globin splice (Gruss and Khoury, 1980). The second example is the β-globin gene. A transcription unit containing an SV40 promoter and globin cDNA does not produce stable RNA, but it can be rescued by incorporating more SV40 sequence containing the "19 S" intron (Hamer *et al.*, 1979), or by incorporating a globin intron (Hamer and Leder, 1979). Even though the intron does not have to be in the globin part of the transcription unit, the problem is with the globin cDNA sequence, since if globin cDNA is inserted into an unsplit herpesvirus thymidine kinase gene, the gene can no longer be expressed (H. Pelham, personal communication). So apparently transcripts of β-globin have to pass through the splicing machinery before they can be transported and stabilized. Pederson (1983) has proposed a model whereby split and unsplit gene transcripts follow different processing pathways in the nucleus.

E. Organization of Splicing Pathways

1. *The Problem*

No one understands how the correct pairs of splice sites are brought together. Virtually all pulse-labeled primary transcripts are correctly processed into mRNA (see Section IV,E,5). And yet a typical transcription unit contains several normal introns, plus many other sequences with apparently adequate matches to the consensus splicing signals, both in exons and in introns (Staden, 1984).

There is probably a minimum distance that must separate the splice sites of an individual intron. Vertebrate genes include many introns around 100 bp in length, but none shorter than 80 or 79 bp (Jung *et al.*, 1980; Bothwell *et al.*, 1981), although a few viral introns are as short as 74 bp or 66 bp (Sugisaki *et al.*, 1980; Ziff, 1980). [A 31-bp excision occurs in some SV40 RNA, but could represent splicing from concatameric transcripts (Ghosh *et al.*, 1978).] A minimum length of 70–80 nucleotides has recently been demonstrated by experiments on a globin intron (Wieringa *et al.*, 1984). The minimum lengths are shorter in insects (54 bp) and nematodes (47 bp) (Henikoff *et al.*, 1983; Kramer *et al.*, 1982).

Over a short range, one might envisage that the pair of sites best matching the consensus, and more than the minimum distance apart, would be spliced. However, apparently adequate alternatives often exist (Staden, 1984), and can be used in mutant genes (see Section IV,E,6). Thus, even the short introns of globin genes present problems of specificity.

Over a longer range, there is so far no upper limit on the length of an intron, and increasing the length of one from 5 to 13 kb did not prevent gene expression (Cullen *et al.*, 1982). Even simple transcription units, making a single mRNA, can be as long as 48 kb (Hirano *et al.*, 1983; Van Ommen *et al.*, 1983), with as many as 51 exons (Tolstoshev and Solomon, 1982), and individual introns can be as long as 17 kb (Van Ommen *et al.*, 1983; Shimizu *et al.*, 1983; McGrath *et al.*, 1983). The rat thyroglobulin gene may be over 150 kb long with a 30-kb intron (Avvedimento *et al.*, 1984). In complex transcription units, making multiple mRNAs, the splices can reach even further (Ziff, 1980; Garber *et al.*, 1983). The longest may be a splice across 180 kb of the immunoglobulin heavy chain locus, joining the V-J_H exon to the C_ϵ exons (Yaoita *et al.*, 1982). However this event has not been demonstrated in a cloned cell line, and the fact that the intervening DNA segments are more commonly deleted (Honjo *et al.*, 1981) may indicate that such a gigantic transcription unit is not efficient.

Two classes of model have been widely considered to explain how the correct splicing patterns are chosen. One invokes secondary structure in the RNA to bring correct splice sites close together. The other proposes that the splicing machinery scans in one direction along the RNA, recognizing one end of an intron first and searching for a matching end. The latter class includes models for either 5′→3′ or 3′→5′ scanning, and for scanning either along the whole RNA or separately within each intron.

We have already seen that individual splice sites can function despite deletions of their surrounding sequences. More extensive data on the normal and abnormal patterns of splicing, reviewed in the following sections, clearly show that there is no universal rule for selecting correct pairs of sites, and in particular that none of the strict secondary-structure or scanning models can be correct. In Section IV,E,7 I will speculate on more elaborate models of the same two classes that may still be viable.

2. *The Role of Secondary Structure*

Secondary structures can of course be devised for many pairs of splice sites, and they may well facilitate correct splicing in individual cases. One type would align the regions around the two splice points by basepairing between them. Zain *et al.* (1979) devised structures of this type for adenovirus splice sites, and Trapnell *et al.* (1980) did so for 17 pairs of splice sites; some of these structures should be stable (although others are no better than in random sequences). Another type of structure would align the ends of both exons with a third stretch of RNA; these can be predicted from mRNA sequences, as was done by Altenburger *et al.* (1980), Naora *et al.* (1980), Akusjärvi and Petterson (1979), and Lomedico *et al.* (1979). Nussinov and Tinoco (1981, 1982) predicted a secondary structure for the entire SV40 late transcript. It puts the splice points some distance apart, but remains to be tested experimentally.

The only case for which there is not only a theoretical prediction for the complete intron structure (Trapnell *et al.*, 1980), but also detailed experimental data concerning the flanking exon structures, is the short intron of the α- and β-globin genes. Pavlakis *et al.* (1980) found that α mRNAs of mouse and rabbit showed very similar nuclease protection patterns and would fit into similar secondary structures, in which the spliced point was in a weakly basepaired stem. The β mRNAs of mouse and rabbit have the spliced point in a similar position in a conserved short stem, but the remainder of the β secondary structure is not conserved.

Even if such structures are real, they cannot be universal, in view of the deletion experiments reviewed above, the examples of exon shuffling reviewed below, and the fact that exons can be as short as 11 bp (Tate *et al.*, 1983).

Nor is secondary structure within the intron likely to be important, in view of the lack of conservation of intron sequences. Dispersed repeated sequences presumably contribute to the secondary structure of large introns within which they lie, since they produce basepairing within and/or between hnRNAs (Jelinek *et al.*, 1974, 1978). But they cannot be of great significance, since they are present in certain rodent genes but absent from the human homologues (see Section VI,C).

One case in which distant sequences clearly do affect the ratio of alternatively spliced RNAs was reported by Khoury *et al.* (1979). In the SV40 early region, RNA resulting from splicing of one intron was greatly diminished by various deletions in an exon region 100–250 nucleotides upstream. Similar modulation of alternative splicing pathways by deletions can be found in the SV40 late region (Piatak *et al.*, 1981; Ghosh *et al.*, 1981, 1982). It may be relevant that both sets of deletions diminish splicing of exceptionally short introns downstream of them. But one caveat is that these deletions could alter the relative stability of the RNAs following splicing, rather than affecting splicing directly.

3. *Artificial Exon Shuffling*

A number of researchers have reconstructed transcription units with heterologous exons and introns, and typically find that they are spliced together properly in cells, so long as the natural splice sites are provided in pairs. Heterologous spliced pairs of exons include SV40-globin (Chu and Sharp, 1981), adenovirus-immunoglobulin (Kaufman and Sharp, 1982), and α-fetoprotein exons (Young *et al.*, 1982). Adenovirus and SV40 exons can be spliced together in hybrid viruses, although with use of a cryptic downstream splice site (Khoury *et al.*, 1980). And the immunoglobulin V-J_H exon can be spliced not only to any of the homologous C_H gene segments during the normal immune response, but also to nonhomologous hinge or C-terminal exons (Dunnick *et al.*, 1980; Neuberger, 1983). Many natural immunoglobulin heavy chain mutants (Franklin and Frangione, 1975) probably also result from illegitimate splicing.

4. *Natural Exon Shuffling: Complex Transcription Units*

''Complex transcription units'' are those which can produce multiple mRNAs encoding different gene products; they often involve alternative patterns of RNA splicing. Sometimes the alternatives are regulated, but the splicing pattern is rarely if ever regulated directly. Often the splicing pattern is dependent on alternative promoters or poly(A) sites. In such situations, complex transcription units often use the most 5′ available upstream site, and the most 3′ available downstream site or set of sites. However, there are also more complicated patterns.

The best studied complex transcription units are those of *adenoviruses and papovaviruses,* in which RNA splicing was first discovered. These viruses still display the most baroque patterns of splicing known (reviewed by Ziff, 1980, and this volume; Tooze, 1981; Flint, 1982). In various transcription units, they show optional exons, optional introns, and introns with multiple downstream sites, or multiple upstream sites, or both. In SV40, the splicing patterns may depend on the extent of sequences further upstream (see Section IV,E,2). In the 26-kb major late transcription unit of adenovirus, which has five alternative poly(A) sites, the consequent splicing events can stretch over some 20 kb; the set of downstream sites closest to the actual poly(A) site is used, but each set itself includes alternative splicing patterns.

In most cases, there is no evidence for direct regulation of splicing. One case in the adenovirus E1b region, in which the ratio of two differently spliced mRNAs changes during infection (Spector *et al.*, 1978), is mainly due to regulation of the relative stability of the two mRNAs (Wilson and Darnell, 1981), although differential splicing may also be involved (Montell *et al.*, 1984). However, the adenovirus major late transcription unit exhibits two prima facie cases of regulated splicing. One concerns the fourth and fifth sets of 3′-coterminal RNAs, which are normally distinct sets; in monkey cells, downstream splice sites of the fourth set are used with the poly(A) site of the fifth set (Klessig and Chow, 1980). The other case concerns the first set of 3′-coterminal late RNAs. This set is also represented early in infection, by a single species which is differently spliced from those seen later (Kitchingman and Westphal, 1980; Akusjärvi and Persson, 1981). The absence of the "late" alternatives is not due to deficient transport or stability, so this appears to represent direct regulation of splicing (Nevins and Wilson, 1981). Both of these cases seem likely to involve some virus-coded factor which modulates splicing, but the mechanism is unknown.

The only cellular genes reported to approach adenovirus in complexity are the *immunoglobulin genes* (reviewed by Honjo *et al.*, 1981; Rogers and Wall, 1984). There are multiple upstream splice sites in the cluster of J gene segments, but only the most 5′ J segment (joined to a V segment) is actually used (Schibler *et al.*, 1978; Gilmore-Hebert and Wall, 1979; Perry *et al.*, 1980: see Section IV,E,5). In the immunoglobulin heavy chain genes, there are also alternative downstream sites, so that the V-J exon can be spliced to the C_μ and C_δ gene segments (Moore *et al.*, 1981; Maki *et al.*, 1981) and possibly to other C_H gene segments in the same transcription unit (Yaoita *et al.*, 1982). There are also alternative 3′ splice patterns within each C_H gene segment, producing membrane and secreted forms of the heavy chains (Early *et al.*, 1980; Rogers and Wall, 1984); as with adenovirus, these appear to depend on the use of alternative poly(A) sites.

Several *other cellular genes* show alternative splicing patterns which allow them to produce more than one polypeptide.

In some, as in adenovirus and immunoglobulin heavy chain genes, there is alternative 3′ end splicing depending on alternative poly(A) sites. The *Drosophila* glycinamide ribotide transformylase gene (Henikoff *et al.*, 1983) and the bovine kininogen gene (Kitamura *et al.*, 1983) both have a second poly(A) site which allows splicing of an additional exon. The rat calcitonin gene (Amara *et al.*, 1982) also has two poly(A) sites, which are tissue-specific, and are associated with splicing to two alternative 3′ exons. The *Drosophila* myosin heavy chain gene (Rozek and Davidson, 1983) also appears to have developmentally regulated alternative 3′ exons.

Similarly, developmentally regulated promoters can produce alternative 5′ exons. The *Drosophila* alcohol dehydrogenase gene (Benyajati *et al.*, 1983) merely makes one additional splice for the more 5′ of its promoters, whereas the mouse α-amylase gene (Young *et al.*, 1981) has a separate 5′ exon associated with each promoter. A chicken myosin light chain gene (Nabeshima *et al.*, 1984) has a remarkable pattern of interspersed alternative exons: two separate promoters produce mRNAs containing either exons 1 and 4 or exons 2 and 3, followed by shared exons.

There are also genes which produce alternative polypeptides which differ only by the insertion of a few amino acids internally, and are not regulated. These can result from closely spaced alternative upstream splice sites (chicken ovomucoid: Stein *et al.*, 1980), alternative downstream sites (rat prolactin: Maurer *et al.*, 1981; human growth hormone: Wallis, 1980), an optional intron (rat γ-fibrinogen: Crabtree and Kant, 1983), or an optional exon (mouse αA-crystallin: King and Piatigorsky, 1983; human c-Ki-*ras:* Shimizu *et al.*, 1983; McGrath *et al.*, 1983). And the mouse histocompatibility genes show a variety of these patterns leading to minor C-terminal and N-terminal variations in closely related genes (Steinmetz *et al.*, 1981; Reyes *et al.*, 1982; Kvist *et al.*, 1983; Lalanne *et al.*, 1983).

5. *Natural Pathways of Splicing*

Transcripts containing introns have been observed for many genes, by pulse-labeling, electron microscopy, RNA gel blots, or S1 nuclease mapping. They are normally polyadenylated and restricted to the nucleus, and are mostly likely to be pre-mRNAs.

Candidates for unspliced, polyadenylated primary transcripts have been identified for many viral transcription units (Nevins and Darnell, 1978; Lai *et al.*, 1978; Horowitz *et al.*, 1978; Berget and Sharp, 1979; Goldenberg and Raskas, 1979; Weber *et al.*, 1980) and cellular genes (Curtis *et al.*, 1977; Bastos and Aviv, 1977; Kinniburgh *et al.*, 1978; Roop *et al.*, 1978; Schibler *et al.*, 1978; Gilmore-Hebert and Wall, 1978, 1979; Choi *et al.*, 1979, 1980; Perry *et al.*, 1980; Nordstrom *et al.*, 1979; Harpold *et al.*, 1979; Nelson *et al.*, 1983), although only S1 nuclease mapping can prove that no splicing has occurred (Lai *et al.*, 1978; Kinniburgh and Ross, 1979; Grosveld *et al.*, 1981; Mariman *et al.*, 1983). On pulse-labeling cells with [^{3}H]uridine, these primary transcripts are labeled with no lag, and begin to be processed within less than 5 minutes. Intermediate-sized molecules, followed by mRNA, appear within ~2–30 minutes. Most if not all of the primary transcript molecules are chased into cytoplasmic mRNA (Ross, 1976; Bastos and Aviv, 1977; Curtis *et al.*, 1977; Nevins and Darnell, 1978; Schibler *et al.*, 1978; Gilmore-Hebert and Wall, 1979; Tsai *et al.*, 1980; Weber *et al.*, 1980; Maquat *et al.*, 1980).

Pulse-chase studies have usually not been done with sufficient resolution to resolve multiple intermediates, and have never demonstrated an obligate order of splicing. However, a combination of methods indicates that, in some transcription units, some introns are preferentially removed first.

In the *adenovirus major late leader,* the first intron is rapidly removed before other introns, beginning even before transcription is complete; but cleavages also occur equally early within both the first and the second introns, and may represent alternative processing pathways (Manley *et al.*, 1982; Keohavong *et al.*, 1982; Mariman *et al.*, 1983). In the *adenovirus E2* transcription unit, the longer, more 5′ intron is removed before the shorter, more 3′ intron (Goldenberg and Raskas, 1979; Weber *et al.*, 1980).

For *chicken ovomucoid* (Tsai *et al.*, 1980), a pulse-chase showed that introns near the middle are excised first, the most 3′ intron later, and the most 5′ intron last of all. Gel blots and electron microscopy showed that the major species in steady-state nuclear RNA are indeed the intermediates in this pathway. However, the pathway is not followed by all ovomucoid transcripts.

For β-*globin* in the mouse (Kinniburgh and Ross, 1979; Donaldson *et al.*, 1982), several RNA species have incompletely spliced introns, and the pulse-chase data cited above imply that most if not all of them are true intermediates. Most commonly, the first splice removes only the middle of the larger (3′) intron, probably between a pair of "cryptic" sites that obey the normal sequence rules; the next splice (which in some molecules happens first) removes the smaller (5′) intron; and a final splice removes the remaining 280 nucleotides of the larger intron. Minor species show splicing at cryptic sites in the smaller intron, but it seems likely that these would be dead-end products. In the rabbit β-globin transcript (Grosveld *et al.*, 1981), there are similar species using cryptic splice sites, but here the more 5′ intron is usually removed before any splicing has occurred elsewhere.

The globin transcripts are the only well-documented cases of stepwise excision of introns. No such intermediates can be detected in the immunoglobulin κ transcripts, although these include a prominent pre-mRNA containing a single large intron with spare upstream (J_κ) sites (Schibler *et al.*, 1978; Gilmore-Hebert and Wall, 1978, 1979; Perry *et al.*, 1980; Choi and Wall, 1979, 1980). Cutting at a cryptic upstream site in a collagen transcript has been reported (Avvedimento *et al.*, 1980), but it has not been shown to be part of the normal splicing pathway. (Two other bands seen, which were supposed to represent cutting at cryptic sites lacking GT–AG, may have been degradation products.)

Thus those preferred pathways which can be demonstrated appear to be idiosyncratic. In other genes, it has not been possible to show any preferred pathway; for example, the *Xenopus* vitellogenin RNAs (at steady state) have a haphazard distribution of introns (Ryffel *et al.*, 1980). Such analyses may well be confused by slow or dead-end branchings of the normal pathway; their products could theoretically accumulate to dominate the steady-state nuclear RNA.

One class of these might be RNAs which had had some introns removed but not in the normal order. If these were difficult to process further, they would represent not the normal pathway but the opposite. Examples of genes in which a mutation in one intron slows the processing of other introns (Section IV,E,6) show that this possibility is not too far-fetched.

A second class might result from cutting at one splice site without another, as has been shown for at least one adenovirus RNA cut at a downstream site (Mariman *et al.*, 1983), and suggested for several immunoglobulin RNA species. One of these is an abundant small poly(A)$^+$ RNA containing the last two exons of the μ_m transcription unit (M1, M2) spliced together (Kemp *et al.*, 1983). Some of this RNA has its 5′ end at the 5′ end of the M1 exon; however, most of it begins ~ 20 nucleotides downstream, for unknown reasons. Other candidates for "sliced" immunoglobulin RNAs are also abundant nuclear species, but have been mapped only at low resolution on RNA gel blots. One may be cut at the 3′ end of the V-J_κ exon (Perry *et al.*, 1980), and another at the 3′ end of the μ M1 exon (Rogers, 1981).

An additional poly(A)$^+$ "amputated transcript" from the μ and γ2b transcription units probably represents a molecule cut at both of the alternative poly(A) sites, and contains intron sequences followed by the spliced M1 and M2 exons (Rogers, 1981; Nelson *et al.*, 1983). Such molecules are not produced at all frequently from the adenovirus alternative poly(A) sites (Nevins and Darnell, 1978).

Excised introns alone have never been clearly detected, and must be rapidly degraded, in keeping with the rapid turnover of hnRNA sequences in general.

6. *Mutant Genes and Cryptic Splice Sites*

There has been much analysis of mutant genes, especially globin genes, of which many natural splicing mutants have been found in thalassaemia patients (reviewed by Mount and Steitz, 1983). Although some of these patients' bone marrow cells show abnormal splicing *in vivo* (Kantor *et al.*, 1980; Ley *et al.*, 1982), the aberrant RNAs are often unstable, so most studies have been done with cloned genes in expression vectors in human or monkey cells. Other mutants have been constructed or selected *in vitro*. The results have been informative as regards pathology, but as regards the mechanism of RNA splice site selection, they have ruled out all simple models.

Table II lists the local mutations which inactivate upstream splice sites. In addition, some larger deletions remove splice sites and create alternative patterns. Inactivation of an upstream site never simply prevents splicing. Instead, the normal downstream site may be joined to the normal upstream site of the next intron upstream (with removal of the affected exon). This has been observed for an immunoglobulin gene (Choi *et al.*, 1980; Seidman and Leder, 1980) and for a globin gene (Treisman *et al.*, 1982). In other cases, the normal downstream site is joined to one or more cryptic upstream sites—usually 5′ to the defective site, although occasionally shortly 3′. This has been seen in one immunoglobulin gene (Hozumi *et al.*, 1981), in adenovirus (Montell *et al.*, 1984), and in several globin genes (Orkin *et al.*, 1981; Felber *et al.*, 1982; Treisman *et al.*, 1982, 1983; Wieringa *et al.*, 1983, 1984). The cryptic sites are used even though many of them still have a poorer match to the consensus, and to U1 RNA, than has the mutant site. In the splicing of the globin large intron studied by Treisman *et al.* (1982) and Wieringa *et al.* (1983), use of the cryptic sites even requires cutting at a GT or GC which is not in the normal position in the consensus sequence (Mount and Steitz, 1983). Thus there may be a bias in favor of the more 5′ of a set of upstream sites, offsetting the bias in favor of a well-matched sequence.

Inactivation of a downstream site is much less common. In one thalassemia gene with the downstream site deleted, the remainder of the intron is left unspliced (Orkin *et al.*, 1983). The same happens in the adenovirus E1a region (Carlock and Jones, 1981).

Creation of a new upstream site can have various effects. (1) Two independent mutations that "improve" a cryptic upstream site in a globin exon cause it to be used in addition to the normal upstream site (Orkin *et al.*, 1982; Goldsmith *et al.*, 1983; Table II). [This is the same as one of the cryptic sites used in mutants studied by Treisman *et al.* (1983)]. (2) When the upstream site of a globin intron is duplicated, the more 5′ copy is always used (Lang and Spritz, 1983; Kühne *et al.*, 1983). (3) Two independent mutations which generate different upstream sites in a human globin intron are both used in conjunction with the normal downstream site, and induce the use of a cryptic downstream site in conjunction with the normal upstream site; thus a new exon is created which splits the original intron into two parts (Spence *et al.*, 1982; Dobkin *et al.*, 1983; Treisman *et al.*, 1983). Remarkably, one of these mutant genes produces some RNA in which only the 5′ "half-intron"—between unmutated normal and cryptic sites—has been spliced out, as though the mutant site sequesters the normal downstream site even before cutting at it.

Creation of a new downstream site can also have various effects. In one natural mutant due to a single base change in the intron (Spritz *et al.*, 1981; Westaway and Williamson, 1981; Busslinger *et al.*, 1981; Fukumaki *et al.*,

1982), and in one experimental duplication of a downstream site (Lang and Spritz, 1983), the more 5′ copy is used exclusively. The natural mutation is only 20 bp 5′ to the normal splice point so may perturb the normal site. In three other experimental duplications (Kühne *et al.*, 1983), the more 3′ copy is always used. Insertion of a downstream site into a different part of the gene (Kühne *et al.*, 1983) can also activate a cryptic upstream site—the same one that is activated when the normal upstream site is deleted.

No universal rule emerges from these patterns, but in a clear majority of cases, the outermost available sites are used. In particular, almost all of them have the effect of using all the available normal downstream sites; supernumerary ones are either spliced out or find cryptic upstream sites.

Mutations in complex transcription units of adenovirus (E1a) and SV40 (early and late) have more complicated effects (Montell *et al.*, 1982; Solnick, 1981; Lai and Khoury, 1979; Khoury *et al.*, 1979; Ghosh *et al.*, 1982). Some produce no abnormally spliced RNA, probably because an alternative normal splicing pathway is used instead, although the choice of pathway is not readily predictable. When all normal splicing patterns are barred, one may find unspliced RNA (Ghosh *et al.*, 1981, 1982; Carlock and Jones, 1981; White *et al.*, 1982; Villareal and White, 1983) and/or RNA from novel promoters with their own splicing patterns (White *et al.*, 1982) and/or no stable RNA at all (Lai and Khoury, 1979). In some of these cases, mutant transcripts may be lost because these are RNAs which require splicing in order to be stable (see Section IV,D,2).

The fate of abnormally spliced RNA. Frequently, in splice site mutants, unspliced RNA accumulates, indicating that the abnormal splicing pathway, if possible at all, is slower than a normal one. Such RNA can take a variety of forms. For adenovirus Ela (Carlock and Jones, 1981), as for SV40 late RNA (Piatak *et al.*, 1981; Ghosh *et al.*, 1981, 1982), the unspliced mutant RNA is used as messenger. More commonly, in globin mutants, specific precursors accumulate in the nucleus (Kantor *et al.*, 1980; Maquat *et al.*, 1980; Ley *et al.*, 1982). These RNAs may include introns other than the one directly affected, in globin and in *Drosophila* alcohol dehydrogenase (Busslinger *et al.*, 1981; Ley *et al.*, 1982; Treisman *et al.*, 1982; Benyajati *et al.*, 1982). In mutants of a rat albumin gene (Esumi *et al.*, 1983) and of the SV40 late region (Villareal and White, 1983), the accumulated unspliced RNA consists of very heterogenerous nuclear molecules; for SV40, these are mostly poly(A)$^-$, and are both larger and smaller than the normal primary transcript.

In globin mutants using cryptic upstream splice sites in the first exon, no intron sequences are left unspliced, and precursors do not accumulate. But these mutants produce much less stable RNA than the wild-type genes (Felber *et al.*, 1982; Treisman *et al.*, 1983; Goldsmith *et al.*, 1983), so the abnormal RNA is probably unstable. This instability is probably a general phenomenon; in many cases, aberrantly spliced RNAs which are detected in cells transformed with a cloned mutant gene are less abundant, or absent, in the human patient (Orkin *et al.*, 1981, 1982; Fukumaki *et al.*, 1982; Felber *et al.*, 1982). Therefore, the abnormal RNA is probably unstable in the normal cell. This need not result directly from its spliced structure, but could be a consequence of defective translation; even a simple amber mutation results in the very low RNA levels seen in one β-thalassemia (Moschonas *et al.*, 1981). However, it remains unclear how nuclear processes distinguish between aberrantly spliced RNAs, which are transported and/or degraded, and unspliced RNAs, which accumulate.

7. *Possible Mechanisms*

In conclusion, for individual short introns, the mutant patterns make it clear that splicing does not simply select the pair of splice sites which best match the consensus, and a range of evidence rules out any simple rule of 5′→3′ or 3′→5′ scanning. This leaves secondary structure, and there is some indirect evidence in its favor. However, the exon shuffling experiments show that there can be no universal structure. What secondary structure may be able to do is to sequester the cryptic splice sites so that they are less accessible (Rogers *et al.*, 1979; Khoury *et al.*, 1979; Ghosh *et al.*, 1982). A plausible scenario has been offered by Kühne *et al.* (1983).

For long introns and long transcripts, secondary structure seems a perilously flimsy framework for directing splicing patterns. And in complex transcription units, the correct use of arrays of alternative splice sites, such as the immunoglobulin J segments or the adenovirus 3′-coterminal gene sets, seems to call for some more systematic mechanism. It is notable that in the complex transcription units, as in the individual mutant introns, it is usually the outermost available splice sites which are preferentially used.

A modified scanning model could account for much of the data, so I will explain it briefly as the minimum possible systematic mechanism for directing splicing patterns, although, as I will explain below, there are serious problems even with this model. The required process is convergent scanning, from 5′→3′ and from 3′→5′ simultaneously. This requires that the splicing enzyme — or, more precisely, the enzyme which scans and recognizes splice sites — should operate in two modes. In one mode, it would start from the 5′ end of the RNA, bind to the first upstream splice site (with an efficiency dependent on the match with U1 RNA), and travel 3′ with it. When it encountered the first downstream site, the two sites would be spliced (or tagged for subsequent splicing), and the enzyme would continue 3′ to repeat the process with the next intron. Meanwhile, a similar splicing enzyme operating in the second mode would start from the 3′ end of the RNA, bind to the first downstream splice site, and travel 5′ with it searching out upstream sites. Eventually the two enzymes would meet somewhere near the middle, bringing about the final pairing of splice sites if they met in an intron, and would then dissociate.

This model accounts for the preferential use of the outermost alternative splice sites in most complex transcription units, and also the great lengths of intron which separate them from the rest of the transcription unit; long introns are necessary to make the direction of scanning unambiguous. However, it involves a number of ad hoc assumptions (for example, some transcription units would require faster scanning from $3' \rightarrow 5'$ than from $5' \rightarrow 3'$). The major problem is that the preference for outermost sites is not universal; in conflict with it are the positions of some apparently good sites in introns (Staden, 1984), a few mutant splicing patterns (Choi *et al.,* 1980; Lang and Spritz, 1983), the pattern of myosin light chain splicing (Nabeshima *et al.,* 1984), and, most seriously, the kinetic data on stepwise and nondirectional splicing (Section IV,E,5). Thus, even a fairly elaborate scanning model cannot be comprehensive.

The time now seems ripe for new experimental approaches to the mechanism of splice site selection, given improved techniques and a greater appreciation of the conceptual problems. Purified poly(A)$^-$ precursors can now undergo splicing *in vitro* or in oocytes (Section IV,C); it is important to know whether such precursors with multiple introns will be spliced in the correct pattern (as predicted from the secondary structure model), or whether this requires directionality conferred by the transcription process or by the poly(A) tail (as predicted by the convergent scanning model). Further studies of splicing patterns with reconstituted transcription units introduced into cells would also be useful. To test the convergent scanning hypothesis, a natural array of alternative sites (such as the immunoglobulin J segments) could be placed for splicing near the "wrong" end of a transcription unit, where the usual order of preference should be reversed. To test the secondary structure hypothesis, a splice site could be introduced into a transcription unit with a synthetic stem-and-loop structure in various positions with respect to it.

It may also be worth considering possible structural constraints within the nucleus. Phasing of the subunit structure of the hnRNP could be relevant, since the protein component is nonrandomly distributed on hnRNP (Calvet and Pederson, 1978; Beyer *et al.,* 1980; Steitz and Kamen, 1981; Van Eekelen *et al.,* 1982; Ohlsson *et al.,* 1982). The hnRNP appears to be associated with nuclear matrix structures during splicing (Miller *et al.,* 1978; Mariman *et al.,* 1982; Ben-Ze'ev *et al.,* 1982; Gallinaro *et al.,* 1983). These structures could be relevant to several observations already mentioned, which suggest that, in contrast to unsplit gene transcripts, transcripts destined for splicing must follow a rather elaborate trajectory through the nuclear processing machinery. These observations are (1) the selective inhibition of split-gene transcription by DRB (Section IV,D,1); (2) the requirement for splicing to stabilize certain transcripts (Section IV,D,2); (3) the possible selective accumulation of mRNAs from unsplit genes in heatshock (Pederson, 1983); (4) the effects that mutation in one pair of splice sites can have on splicing of other pairs (Section IV,E,6); (5) the nuclear accumulation of unspliced pre-mRNAs in some splice site mutants, in contrast to the transport or instability of aberrantly spliced RNAs (Section IV,E,6); (6) the evidence that some illegitimate splices may be prevented *in vivo* by RNA transport (Svensson *et al.*, 1983). While these observations do not yet add up to a coherent picture, they hint that the biochemical pathway of splicing may correspond to a physical pathway of some complexity.

F. Introns in Plants: Sequence Requirements

Enough plant intron sequences have now been published for an independent consensus sequence to be derivable. Slightom *et al.* (1983), from 20 sequences, derived the consensus:

$$\mathrm{TG/\underline{GT}AAGTAT\ldots\ldots \overset{(A)}{(T)}_{11}\overset{A}{G}T\underline{AG}/GT.}$$

However, with 30 sequences, weighted so as to reduce the predominance of multiple related genes, the consensus becomes effectively the same as in animals (Table III):

$$\mathrm{\underline{AG}/GTAAGTA\ldots\ldots(T)_{n}NY\underline{AG}/GTA.}$$

The GT–AG rule always holds. The downstream site is YAG/ in all but four cases, where it is YGAAG/GT. The consensus also shows T residues extending further into the intron, reflecting a general (T)-richness of many of the introns.

The fit of the individual sequences to the plant consensus is poor. At the upstream site, two introns (Shah *et al.*, 1983; Slightom *et al.*, 1983) have only a 4/9 match to the consensus, and the nonmatching portions have transversions, not transitions. At the downstream site, many plant introns show no (T)-rich character at all immediately preceding the YAG/ signal. What is most embarrassing for the U1 basepairing hypothesis is that, where A appears in the consensus, T is usually the second most abundant base, and vice versa. This is not the case in animal sequences.

This suggested that there might be an internal control sequence like the one in yeast (see below). Oligo(T) or oligo(A) are seen in many of the introns. However, a search through the 3′ parts of the introns (from the penultimate YAG to the actual splice site: analysis not shown) revealed not even a tetranucleotide which was present in all of them. It seems incredible that the often poor splice site sequences alone would be adequate signals. Perhaps there could be various internal signals.

Several plant introns are less than 90 bp long and the shortest so far is 72 bp (Slightom *et al.*, 1983).

TABLE III

CONSENSUS SPLICE SITES IN PLANTS[a]

Upstream site

	−4				+1									+10
A	10	4	16	1	0	0	23	18	5	7	17	9	8	9
G	7	9	0	23	30	0	1	1	21	2	0	3	2	1
C	10	12	0	4	0	0	3	3	0	5	3	5	6	6
T	3	5	14	2	0	30	3	8	4	16	10	13	14	14
	.	.	A/T	G /	G	T	A	A	G	T	A	T	T	T
	.	.	A	G /	G	T	A	A	G	T	A	.	T	T
	.	.	67	72	100	100	71	58	63	50	60	.	47	58

Downstream site

	−20																			−1				+4
A	10	6	8	7	6	7	4	2	6	8	10	13	13	8	7	6	11	4	30	0	5	5	13	6
G	4	3	4	4	10	4	3	3	8	2	8	5	4	1	6	1	13	0	0	30	24	4	9	10
C	6	5	3	4	2	7	3	4	4	4	2	2	1	2	1	2	2	11	0	0	1	5	4	5
T	10	16	15	15	12	12	20	21	12	15	10	10	11	19	16	21	4	15	0	0	0	16	4	9
N										1			1											
	.	T	T	T	T	.	T	T	.	T	.	A	A	T	T	T	G	T	A	G /	G	T	A	.
	.	T	T	.	T	.	T	T	.	T	.	.	.	T	T	T	.	C/T	A	G /	G	T	A	.
	.	47	46	.	44	.	61	68	.	54	.	.	.	60	50	72	.	46/43	100	100	78	44	64	.

[a]This consensus was derived from 30 different intron sequences: soybean leghemoglobin (Hyldig-Nielsen *et al.*, 1982; Brisson and Verma, 1982), soybean and maize actin (Shah *et al.*, 1982, 1983), French bean phaseolin (Slightom *et al.*, 1983), soybean ribulose bisphosphate carboxylase (Berry-Lowe *et al.*, 1982), soybean 7 S seed storage protein (Schuler *et al.*, 1982), and maize sucrose synthetase (Sheldon *et al.*, 1983). The actual frequencies of each nucleotide are given. In the bottom row, a weighted consensus is shown, in which only single weight was given to the set of three closely related leghemoglobin genes, and to exons in the set of three actin genes. In this row, percentage frequencies (> 43%) are given.

G. INTRONS IN FUNGI: SEQUENCE REQUIREMENTS

Most genes in yeast are unsplit, but several introns have now been found. They can be as short as 54 bp (Miller, 1984).

Comparison of six sequences (Langford and Gallwitz, 1983; Pikielny *et al.*, 1983) leads to the following preliminary consensus (with actual base frequencies shown below):

$$\text{AAG / GTATGT}\genfrac{}{}{0pt}{}{\text{A}}{\text{T}} \;\ldots\ldots\; \text{TACTAACA (N)}_{5\text{-}52}\genfrac{}{}{0pt}{}{\text{T}}{\text{C}}\text{AG / TT}$$

$$3\,3\,5 \quad 6\,6\,6\,6\,6\,6\,\genfrac{}{}{0pt}{}{3}{3} \qquad 6\,6\,6\,6\,6\,6\,6\,5 \qquad \genfrac{}{}{0pt}{}{3}{3}\,6\,6\;\;3\;\;3$$

This generally resembles the animal consensus but differs significantly in that the pyrimidine-rich stretch at the animal downstream site is replaced by an "internal conserved sequence" (ICS) in yeast. All the introns conform to the consensus remarkably closely.

At the upstream site, the sequence /GTATGT may be essential, since splicing is abolished by deletions of part of it (Table II). Unlike mutant animal genes, these two mutants do not use any cryptic splice sites, but merely accumulate large amounts of unspliced precursor RNA. Similarly, a transcription unit containing two complete downstream sites uses only the single accurate upstream site (Langford and Gallwitz, 1983).

The novel structure of the downstream site and ICS was suggested by deletion mutants in two different genes, which localized the signal not to the 3′ end of the intron but to a short internal region, which contained the sequence TACTAAC in each gene. This ICS is found once, and only once, in all yeast introns sequenced, 9–55 bp before the downstream splice point. Introns from other organisms, which lack this ICS, cannot be spliced in yeast (Beggs *et al.*, 1980; Langford *et al.*, 1983). Using experimental gene constructs, Langford and Gallwitz (1983) showed that, in conjunction with a good upstream site, the ICS is necessary and sufficient to specify downstream splicing, usually at the next AG. [The next-but-one AG was used instead in constructs where the first AG would be cut only 5 or 8 bp from the ICS, but this may be because YAG is the optimal signal and GAG is not used. One intron in the *MAT*al gene ends only 9 bp following the ICS, with YAG (Miller, 1984).]

Studies of one split gene showed that many RNA molecules have a cut within the ICS, thus: TACTAA/CA (Pikielny *et al.*, 1983). It was proposed that this cut occurs during normal splicing.

Pikielny *et al.* (1983) suggest that the ICS substitutes for U1 RNA and forms basepairs with the GT at the upstream splice site; the cuts in the ICS and in the upstream splice point could then be made symmetrically. Alternatively, one can predict that both the upstream splice site and the ICS will basepair with the (so far unidentified) yeast U1 RNA, by analogy with the model of Fig. 2. Note that the variable space in the yeast downstream consensus coincides with the undefined nucleotide at −4 in the animal consensus, consistent with looping-out of RNA in both cases.

In *Neurospora,* the intron sequences are more divergent than in yeast, and there have been no experiments to locate the required signals. A consensus sequence from 11 introns, including a possible equivalent of the yeast ICS, is as follows (Kinnaird and Fincham, 1984):

```
G / G  T  A  A  G  T . . . . . R  C  T  R  A  C . . . . Y  A  G /
6   11 11 10 6  11 10          9  11 10 11 11 10        11 11 11
```

The signals in *Aspergillus* introns conform to a similar consensus (Boel *et al.*, 1984).

H. Yeast Mutations Affecting mRNA Splicing

While the *rna*1 mutation apparently affects RNA transport, and the *los*1 mutation affects tRNA splicing (Section II,D), the *rna*2 mutation in yeast seems to affect mRNA splicing. This temperature-sensitive mutation was isolated by Hartwell *et al.* (1970) and mainly leads to depletion of the mRNAs encoding ribosomal proteins (Warner and Gorenstein, 1977). Concomitantly, longer transcripts of the r-protein genes accumulate (Rosbash *et al.,* 1981; Fried *et al.,* 1981; Leer *et al.,* 1982; Bromley *et al.,* 1982). For at least one gene (rp51), the accumulated transcript has an intron which fails to be excised (Rosbash *et al.,* 1981). It seems that *rna*2 blocks mRNA splicing generally, since intron-containing transcripts also accumulate rather than being spliced for the actin and *MAT*a1 genes (Teem *et al.,* 1983; Miller, 1984), and the expression of a chimaeric split gene is also blocked (Teem and Rosbash, 1983). With the power of modern yeast genetics, it should soon be possible to identify the *RNA*2 gene product, which is likely to play a central role in mRNA splicing.

References to Part 1

Abelson, J. (1979). *Annu. Rev. Biochem.* **48,** 1035–1069.

Akusjärvi, G., and Pettersson, U. (1979). *Cell* **16,** 841–850.

Akusjärvi, G., and Persson, H. (1981). *Nature (London)* **292,** 420–426.

Aleström, P., Akusjärvi, G., Perricaudet, M., Matthews, M. B., Klessig, D. F., and Pettersson, U. (1980). *Cell* **19,** 671–681.

Altenburger, W., Steinmetz, M., and Zachau, H. G. (1980). *Nature (London)* **287,** 603–607.

Amara, S. G., Jonas, V., Rosenfeld, M. G., Ong, E. S., and Evans, R. M. (1982). *Nature (London)* **298,** 240–244.

Avvedimento, V. E., Vogeli, G., Yamada, Y., Maizel, J. V., Jr., Pastan, I., and de Crombrugghe, B. (1980). *Cell* **21,** 689–696.

Avvedimento, V. E., Musti, A. M., Obici, S., Cocozza, S., and DiLauro, R. (1984). *Nucleic Acids Res.* **12,** 3461–3472.

Babich, A., Nevins, J. R., and Darnell, J. E. (1980). *Nature (London)* **287,** 246–248.

Bachellerie, J.-P., Michot, B., and Raynal, F. (1983). *Mol. Biol. Rep.* **9,** 79–86.

Baldi, M. I., Mattoccia, E., and Tocchini-Valentini, G. P. (1983). *Cell* **35,** 109–115.

Barque, J. P., Yeni, P., Peraudeau, L., Danon, F., and Larsen, C. J. (1981). *Biochem. Biophys. Res. Commun.* **99,** 284–291.

Bastos, R. N., and Aviv, H. (1977). *Cell* **11,** 641–650.

Beggs, J. D., van den Berg, J., van Ooyen, A., and Weissmann, C. (1980). *Nature (London)* **283,** 835–840.

Benyajati, C., Place, A. R., Wang, N., Pentz, E., and Sofer, W. (1982). *Nucleic Acids Res.* **10,** 7261–7272.

Benyajati, C., Spoerel, N., Haymerle, H., and Ashburner, M. (1983). *Cell* **33,** 125–133.

Ben-Ze'ev, A., Abulafia, R., and Aloni, Y. (1982). *EMBO J.* **1,** 1225–1231.

Berget, S. M., and Sharp, P. A. (1979). *J. Mol. Biol.* **129,** 547–565.

Berry-Lowe, S. L., McKnight, T. D., Shah, D. M., and Meagher, R. B. (1982). *J. Mol. Appl. Genet.* **1,** 483–498.

Beyer, A. L., Miller, O. L., and McKnight, S. L. (1980). *Cell* **20,** 75–84.
Blanchard, J.-M. (1980). *Biochem. Biophys. Res. Commun.* **97,** 524–529.
Blanchard, J.-M., Weber, J., Jelinek, W., and Darnell, J. E. (1978). *Proc. Natl. Acad. Sci. U.S.A.* **75,** 5344–5348.
Blin, N., Weber, T., and Alonso, A. (1983). *Nucleic Acids Res.* **11,** 1375–1387.
Boel, E., Hansen, M. T., Hjort, I., Hoegh, I., and Fiil, N. P. (1984). *EMBO J.* **3,** 1581–1585.
Bothwell, A. L. M., Paskind, M., Reth, M., Imanishi-Kari, T., Rajewsky, K., and Baltimore, D. (1981). *Cell* **24,** 625–637.
Boyer, P. D. (ed.) (1971). *The Enzymes 3rd Ed.* **4.**
Boyer, P. D. (ed.) (1982). *The Enzymes 3rd Ed.* **15.**
Branch, A. D., Robertson, H. D. Greer, C., Gegenheimer, P., Peebles, C., and Abelson, J. (1982). *Science* **217,** 1147–1149.
Branlant, C., Krol, A., Ebel, J.-P., Lazar, E., Gallinaro, H., Jacob, M., Sri-Widada, J., and Jeanteur, P. (1980). *Nucleic Acids Res.* **8,** 4143–4154.
Branlant, C., Krol, A., and Ebel, J.-P. (1981). *Nucleic Acids Res.* **9,** 841–858.
Branlant, C., Krol, A., Ebel, J.-P., Lazar, E., Haendler, B., and Jacob, M. (1982). *EMBO J.* **1,** 1259–1265.
Breathnach, R., and Chambon, P. (1981). *Annu. Rev. Biochem.* **50,** 349–383.
Breathnach, R., Benoist, C., O'Hare, K., Gannon, F., and Chambon, P. (1978). *Proc. Natl. Acad. Sci. U.S.A.* **75,** 4853–4857.
Bringmann, P., Rinke, J., Appel, B., Reuter, R., and Lührmann, R. (1983). *EMBO J.* **2,** 1129–1135.
Brinker, J. M., Madore, H. P., and Bello, L. J. (1973). *Biochem. Biophys. Res. Commun.* **52,** 928–934.
Brisson, N., and Verma, D. P. S. (1982). *Proc. Natl. Acad. Sci. U.S.A.* **79,** 4055–4059.
Bromley, S., Hereford, L., and Rosbash, M. (1982). *Mol. Cell Biol.* **2,** 1205–1211.
Brunel, C., Sri Widada, J., Lelay, M.-N., Jeanteur, P., and Liautard, J.-P. (1981). *Nucleic Acids Res.* **9,** 815–830.
Busch, H., Reddy, R., Rothblum, L., and Choi, Y. C. (1982). *Annu. Rev. Biochem.* **51,** 617–654.
Busslinger, M., Moschonas, N., and Flavell, R. A. (1981). *Cell* **27,** 289–298.
Calvet, J. P., and Pederson, T. (1978). *J. Mol. Biol.* **122,** 361–378.
Calvet, J. P., and Pederson, T. (1981). *Cell* **26,** 363–370.
Calvet, J. P., Meyer, L. M., and Pederson, T. (1982). *Science* **217,** 456–458.
Carlock, L., and Jones, N. C. (1981). *Nature (London)* **294,** 572–574.
Cech, T. R. (1983). *Cell* **34,** 713–716.
Chiu, N. H., Bruszewski, W. B., and Salzman, N. P. (1980). *Nucleic Acids Res.* **8,** 153–168.
Choi, E., Kuehl, M., and Wall, R. (1979). *In* "Eucaryotic Gene Regulation" (R. Axel, T. Maniatis, and C. F. Fox, eds.), pp. 421–431. Academic Press, New York.
Choi, E., Kuehl, M., and Wall, R. (1980). *Nature (London)* **286,** 776–779.
Chu, G., and Sharp, P. (1981). *Nature (London)* **289,** 378–382.
Colby, D., Leboy, P. S., and Guthrie, C. (1981). *Proc. Natl. Acad. Sci. U.S.A.* **78,** 415–419.
Crabtree, G. R., and Kant, J. A. (1982). *Cell* **31,** 159–166.
Cullen, B. R., Kopchick, J. J., and Stacey, D. W. (1982). *Nucleic Acids Res.* **10,** 6177–6190.
Curtis, P. J., Mantei, N., van den Berg, J., and Weissmann, C. (1977). *Proc. Natl. Acad. Sci. U.S.A.* **74,** 3184–3188.
Darnell, J. E. (1979). *Prog. Nucleic Acid Res. Mol. Biol.* **22,** 327.
David, M., Borasio, G. D., and Kaufmann, G. (1982). *Proc. Natl. Acad. Sci. U.S.A.* **79,** 7097–7101.
Deimel, B., Louis, C., and Sekeris, C. E. (1977). *FEBS Lett.* **73,** 80–84.
De Robertis, E. M., and Olson, M. V. (1979). *Nature (London)* **278,** 137–143.

De Robertis, E. M., Black, P., and Nishikura, K. (1981). *Cell* **23,** 89–93.
Dobkin, C., Pergolizzi, R. G., Bahre, P., and Bank, A. (1983). *Proc. Natl. Acad. Sci. U.S.A.* **80,** 1184–1188.
Dodgson, J. B., and Engel, J. D. (1983). *J. Biol. Chem.* **258,** 4623–4629.
Donaldson, D. S., McNab, A. R., Rovera, G., and Curtis, P. J. (1982). *J. Biol. Chem.* **257,** 8655–8660.
Dunnick, W., Rabbitts, T. H., and Milstein, C. (1980). *Nature (London)* **286,** 669–675.
Early, P., Rogers, J., Davis, M., Calame, K., Bond, M., Wall, R., and Hod, L. (1980). *Cell* **20,** 313–319.
Elder, J. T., Spritz, R. A., and Weissman, S. M. (1981). *Annu. Rev. Genet.* **15,** 295–340.
Epstein, P., Reddy, R., and Busch, H. (1981). *Proc. Natl. Acad. Sci. U.S.A.* **78,** 1562–1566.
Erbil, C., and Niessing, J. (1983). *EMBO J.* **2,** 1339–1343.
Esumi, H., Takahashi, Y., Sato, S., Nagase, S., and Sugimura, T. (1983). *Proc. Natl. Acad. Sci. U.S.A.* **80,** 95–99.
Etcheverry, T., Colby, D., and Guthrie, C. (1979). *Cell* **18,** 11–26.
Felber, B. K., Orkin, S. H., and Hamer, D. H. (1982). *Cell* **29,** 895–902.
Filipowicz, W., and Shatkin, A. J. (1983). *Cell* **32,** 547–557.
Filipowicz, W., Konarska, M., Gross, H. J., and Shatkin, A. J. (1983). *Nucleic Acids Res.* **11,** 1405–1418.
Flint, S. J. (1982). *Biochim. Biophys. Acta* **651,** 175–208.
Flytzanis, C., Alonso, A., Louis, C., Krieg, L., and Sekeris, C. E. (1978). *FEBS Lett.* **96,** 201–205.
Franklin, E. C., and Frangione, B. (1975). *Contemp. Top. Mol. Immunol.* **4,** 89–126.
Fraser, N. W., Sehgal, P. B., and Darnell, J. E. (1978). *Nature (London)* **272,** 590–593.
Fried, H. M., Pearson, N. J., Kim, C. H., and Warner, J. R. (1981). *J. Biol. Chem.* **256,** 10176–10183.
Fukumaki, Y., Ghosh, P. K., Benz, E. J., Reddy, V. B., Lebowitz, P., Forget, B. G., and Weissman, S. M. (1982). *Cell* **28,** 585–593.
Furneaux, H., Pick, L., and Hurwitz, J. (1983). *Proc. Natl. Acad. Sci. U.S.A.* **80,** 3933–3937.
Gallinaro, H., and Jacob, M. (1979). *FEBS Lett.* **104,** 176–182.
Gallinaro, H., Gattoni, R., Stevenin, J., and Jacob, M. (1980). *Biochem. Biophys. Res. Commun.* **95,** 20–26.
Gallinaro, H., Puvion, E., Kister, L., and Jacob, M. (1983). *EMBO J.* **2,** 953–960.
Gallwitz, D. (1982). *Proc. Natl. Acad. Sci. U.S.A.* **79,** 3493–3497.
Garber, R. L., Kuroira, A., and Gehring, W. J. (1983). *EMBO J.* **2,** 2027–2036.
Gegenheimer, P., Gabius, H.-J., Peebles, C. L., and Abelson, J. (1983). *J. Biol. Chem.* **258,** 8365–8373.
Ghosh, P. K., Reddy, V. B., Swinscoe, J., Lebowitz, P., and Weissman, S. M. (1978). *J. Mol. Biol.* **126,** 813–846.
Ghosh, P. K., Ray, P., Barkan, A., Mertz, J. E., Weissman, S. M., and Lebowitz, P. (1981). *Proc. Natl. Acad. Sci. U.S.A.* **78,** 1386–1390.
Ghosh, P. K., Piatak, M., Mertz, J. E., Weissman, S. M., and Lebowitz, P. (1982). *J. Virol.* **44,** 610–624.
Gilmore-Hebert, M., and Wall, R. (1978). *Proc. Natl. Acad. Sci. U.S.A.* **75,** 342–345.
Gilmore-Hebert, M., and Wall, R. (1979). *J. Mol. Biol.* **135,** 879–891.
Gojobori, T., and Nei, M. (1981). *J. Mol. Evol.* **17,** 245–250.
Goldenberg, C. J., and Hauser, S. D. (1983). *Nucleic Acids Res.* **11,** 1337–1348.
Goldenberg, C. J., and Raskas, H. J. (1979). *Cell* **16,** 131–138.
Goldenberg, C. J., and Raskas, H. J. (1981). *Proc. Natl. Acad. Sci. U.S.A.* **78,** 5430–5434.
Goldenberg, C. J., DiMaria, P., and Hauser, S. D. (1983). *UCLA Symp. Mol. Cell. Biol.* **8** (in press).

Goldsmith, M. E., Humphries, R. K., Ley, T., Cline, A., Kantor, J. A., and Nienhuis, A. W. (1983). *Proc. Natl. Acad. Sci. U.S.A.* **80,** 2318–2322.

Goldstein, L. (1976). *Nature (London)* **261,** 519–521.

Goodman, H. M., Olson, M. V., and Hall B. D. (1977). *Proc. Natl. Acad. Sci. U.S.A.* **74,** 5453–5457.

Grabowski, P. J., Padgett, R. A., and Sharp, P. A. (1984). *Cell* **37,** 415–427.

Green, M. R., Maniatis, T., and Melton, D. A. (1983). *Cell* **32,** 681–694.

Greer, C. L., Peebles, C. L., Gegenheimer, P., and Abelson, J. (1983a). *Cell* **32,** 537–546.

Greer, C. L., Javor, B., and Abelson, J. (1983b). *Cell* **33,** 899–906.

Grosveld, G. C., Koster, A., and Flavell, R. (1981). *Cell* **23,** 573–584.

Gruss, P., and Khoury, G. (1980). *Nature (London)* **286,** 634–637.

Gruss, P., Lai, C.-J., Dhar, R., and Khoury, G. (1979). *Proc. Natl. Acad. Sci. U.S.A.* **76,** 4317–4321.

Gruss, P., Efstratiadis, A., Karanthanasis, S., König, M., and Khoury, G. (1981). *Proc. Natl. Acad. Sci. U.S.A.* **78,** 6091–6095.

Guimont-Ducamp, C., Sri-Widada, J., and Jeanteur, P. (1977). *Biochimie* **59,** 755–758.

Guthrie, C., and Abelson, J. (1982). *In* The Molecular Biology of the Yeast Saccharomyces, (J. N. Strathern, E. W. Jones, and J. R. Broach, eds.), pp. 487–528. Cold Spring Harbor, New York.

Hamer, D. H., and Leder, P. (1979). *Cell* **18,** 1299–1302.

Hamer, D. H., Smith, K. D., Boyer, S. H., and Leder, P. (1979). *Cell* **17,** 725–735.

Hanahan, D., Lane, D., Lipsich, L., Wigler, M., and Botchan, M. (1980). *Cell* **21,** 127–139.

Handa, H., Kaufman, R. J., Manley, J. L., Gefter, M. L., and Sharp, P. A. (1981). *J. Biol. Chem.* **256,** 478–482.

Harada, F., and Kato, N. (1980). *Nucleic Acids Res.* **8,** 1273–1285.

Harpold, M. M., Dobner, P. R., Evans, R., Bancroft, F. C., and Darnell, J. E. (1979). *Nucleic Acids Res.* **6,** 3133–3144.

Hartwell, L. H., McLaughlin, C. S., and Warner, J. R. (1970). *Mol. Gen. Genet.* **109,** 42–56.

Hellung-Larsen, P., and Frederiksen, S. (1977). *Comp. Biochem. Physiol.* **58B,** 273–281.

Henikoff, S., Sloan, J. S., and Kelly, D. (1983). *Cell* **34,** 405–414.

Hentschel, C. C., and Birnstiel, M. L. (1981). *Cell* **25,** 301–313.

Hernandez, N., and Keller, W. (1983). *Cell* **35,** 89–99.

Hinterberger, M., Pettersson, I., and Steitz, J. A. (1983). *J. Biol. Chem.* **258,** 2604–2613.

Hirano, H., Yamada, Y., Sullivan, M., de Crombrugghe, B. D., Pastan, I., and Yamada, K. M. (1983). *Proc. Natl. Acad. Sci. U.S.A.* **80,** 46–50.

Honjo, T., Nakai, S., Nishida, Y., Kataoka, T., Yamawaki-Kataoka, Y., Takahashi, N., Obata, M., Shimizu, A., Yaoita, Y., Nikaido., and Ishida N. (1981). *Immunol. Rev.* **59,** 33–67.

Hopper, A. K., Banks, F., and Evangelidis, V. (1978). *Cell* **14,** 211–219.

Hopper, A. K., Schultz, L. D., and Shapiro, R. A. (1980). *Cell* **19,** 741–751.

Horowitz, M., Laub, O., Bratosin, S., and Aloni, Y. (1978). *Nature (London)* **275,** 558–559.

Howard, E. F. (1978). *Biochemistry* **17,** 3228–3236.

Hozumi, N., Wu, G. E., Murialdo, H., Roberts, L., Vetter, D., Fife, W. L., Whitely, M., and Sadowski, P. (1981). *Proc. Natl. Acad. Sci. U.S.A.* **78,** 7019–7023.

Hutchison, H. T., Hartwell, L. H., and McLaughlin, C. S. (1969). *J. Bacteriol.* **99,** 807–814.

Hyldig-Nielsen, J. J., Jensen, E. Ø., Paludan, K., Wiborg, O., Garrett, R., Jorgensen, Ø. P., and Marcker, K. A. (1982). *Nucleic Acids Res.* **10,** 689–701.

Jelinek, W., Molloy, G., Fernandez-Munoz, R., Salditt, M., and Darnell, J. E. (1974). *J. Mol. Biol.* **82,** 361–370.

Jelinek, W., Evans, R., Wilson, M., Salditt-Georgieff, M., and Darnell, J. E. (1978). *Biochemistry* **17,** 2776–2783.

Johnson, P. F., and Abelson, J. (1983). *Nature (London)* **302,** 681–687.

Johnson, J. D., Ogden, R., Johnson, P., Abelson, J., Dembeck, P., and Itakura, K. (1980). *Proc. Natl. Acad. Sci. U.S.A.* **77,** 2564–2568.

Jung, A., Sippel, A. E., Grez, M., and Schütz, G. (1980). *Proc. Natl. Acad. Sci. U.S.A.* **77,** 5759–5763.

Kaine, B. P., Gupta, R., and Woese, C. R. (1983). *Proc. Natl. Acad. Sci. U.S.A.* **80,** 3309–3312.

Kang, H. S., Ogden, R. C., Knapp, G., Peebles, C. L., and Abelson, J. (1979). *In* "Eucaryotic Gene Regulation" (A. Axel, T. Maniatis, and C. F. Fox, eds.), pp. 69–84. Academic Press, New York.

Kano, Y., Komatsu, H., Nakanoin, K., and Fujiwara, Y. (1978). *Exp. Cell Res.* **115,** 444–447.

Kantor, J. A., Turner, P. H., and Nienhuis, A. W. (1980). *Cell* **21,** 149–157.

Kato, N., and Harada, F. (1981). *Biochem. Biophys. Res. Commun.* **99,** 1477–1485.

Kaufmann, G., and Kallenbach, N. R. (1975). *Nature (London)* **254,** 452–454.

Kaufman, R. J., and Sharp, P. A. (1982). *Mol. Cell Biol.* **2,** 1304–1319.

Kemp, D. J., Morahan, G., Cowman, A. F., and Harris, A. W. (1983). *Nature (London)* **301,** 84–86.

Keohavong, P., Gatoni, R., LeMoullec, J. M., Jacob, M., and Stevenin, J. (1982). *Nucleic Acids Res.* **10,** 1215–1229.

Khoury, G., Gruss, P., Dhar, R., and Lai, C.-J. (1979). *Cell* **18,** 85–92.

Khoury, G., Alwine, J., Goldman, N., Gruss, P., and Jay, G. (1980). *J. Virol.* **36,** 143–151.

Kikuchi, Y., Tyc, K., Filipowicz, W., Sänger, H. L., and Gross, H. J. (1982). *Nucleic Acids Res.* **10,** 7521–7529.

King, C. R., and Piatigorsky, J. (1983). *Cell* **32,** 707–712.

Kinlaw, C. S., Robberson, B. L., and Berget, S. M. (1983). *J. Biol. Chem.* **258,** 7181–7189.

Kinnaird, J. A., and Fincham, J. R. S. (1984). *Gene* **26,** 253–260.

Kinniburgh, A. J., and Ross, J. (1979). *Cell* **17,** 915–921.

Kinniburgh, A. J., Mertz, J. E., and Ross, J. (1978). *Cell* **14,** 681–693.

Kitamura, N., Takagaki, Y., Furuto, S., Tanaka, T., Nawa, H., and Nakanishi, S. (1983). *Nature (London)* **305,** 545–549.

Kitchingman, G. R., and Westphal, H. (1980). *J. Mol. Biol.* **137,** 23–48.

Klessig, D. F., and Chow, L. T. (1980). *J. Mol. Biol.* **139,** 221–242.

Knapp, G., Beckmann, J. S., Johnson, P. F., Fuhrman, S. A., and Abelson, J. (1978). *Cell* **14,** 221–236.

Knapp, G., Ogden, R. C., Peebles, C. L., and Abelson, J. (1979). *Cell* **18,** 37–45.

Kole, R., and Weissman, S. M. (1982). *Nucleic Acids Res.* **10,** 5429–5445.

Konarska, M., Filipowicz, W., Domdey, H., and Gross, H. J. (1981). *Nature (London)* **293,** 112–116.

Konarska, M., Filipowicz, W., and Gross, H. J. (1982). *Proc. Natl. Acad. Sci. U.S.A.* **79,** 1474–1478.

Koski, R. A., Clarkson, S. G., Kurjan, J., Hall, B. D., and Smith, M. (1980). *Cell* **22,** 415–425.

Krainer, A. R., Maniatis, T., Ruskin, B., and Green, M. R. (1984). *Cell* **36,** 993–1005.

Kramer, J. M., Cox, G. N., and Hirsh, D. (1982). *Cell* **30,** 599–606.

Krayev, A. S., Kramerov, D. A., Skryabin, K. G., Ryskov, A. P., Bayev, A. A., and Georgiev, G. P. (1980). *Nucleic Acids Res.* **8,** 1201–1208.

Krol, A., Branlant, C., Lazar, E., Gallinaro, H., and Jacob, M. (1981). *Nucleic Acids Res.* **9,** 2699–2716.

Krol, A., Ebel, J.-P., Rinke, J., and Luhrmann, R. (1983). *Nucleic Acids Res.* **11,** 8583–8594.

Kühne, T., Wieringa, B., Reiser, J., and Weissmann, C. (1983). *EMBO J.* **2,** 727–733.

Kurjan, J., Hall, B. D., Gillam, S., and Smith, M. (1980). *Cell* **20,** 701–709.

Kvist, S., Roberts, L., and Dobberstein, B. (1983). *EMBO J.* **2,** 245–254.

Lai, C.-J., and Khoury, G. (1979). *Proc. Natl. Acad. Sci. U.S.A.* **76,** 71–75.

Lai, C.-J., Dhar, R., and Khoury, G. (1978). *Cell* **14,** 971–982.

Lalanne, J.-L., Cochet, M., Kummer, A.-M., Gachelin, G., and Kourilsky, P. (1983). *Proc. Natl. Acad. Sci. U.S.A.* **80,** 7561–7565.

Lang, K. M., and Spritz, R. A. (1983). *Science* **220,** 1351–1355.

Langford, C. J., and Gallwitz, D. (1983). *Cell* **33,** 519–527.

Langford, C., Nellen, W., Niessing, J., and Gallwitz, D. (1983). *Proc. Natl. Acad. Sci. U.S.A.* **80,** 1496–1500.

Laub, O., Jakobowits, B., and Aloni, Y. (1980). *Proc. Natl. Acad. Sci. U.S.A.* **77,** 3297–3301.

Lazar, E., and Jacob, M. (1982). *Nucleic Acids Res.* **10,** 1193–1201.

Leer, R. J., van Ramsdonk-Duin, M. M. C., Molenaar, C. M. T., Cohen, L. H., Mager, W. H., and Planta, R. J. (1982). *Nucleic Acids Res.* **10,** 5869–5878.

Lenk, R. P., Maizel, J. V., and Crouch, R. J. (1982). *Eur. J. Biochem.* **121,** 475–482.

Lerner, M. R., and Steitz, J. A. (1979). *Proc. Natl. Acad. Sci. U.S.A.* **76,** 5495–5499.

Lerner, M. R., and Steitz, J. A. (1981). *Cell* **25,** 298–300.

Lerner, M. R., Boyle, J. A., Mount, S. M., Wolin, S. L., and Steitz, J. A. (1980). *Nature (London)* **283,** 220–224.

Lerner, E. A., Lerner, M. R., Janeway, C. A., and Steitz, J. A. (1981). *Proc. Natl. Acad. Sci. U.S.A.* **78,** 2737–2741.

Ley, T. J., Arragnou, N. P., Pepe, G., and Nienhuis, A. W. (1982). *Proc. Natl. Acad. Sci. U.S.A.* **79,** 4775–4779.

Lomedico, P., Rosenthal, N., Efstratiadis, A., Gilbert, W., Kolodner, R., and Tizard, R. (1979). *Cell* **18,** 545–557.

MacCumber, M., and Ornstein, R. L. (1984). *Science* **224,** 402–405.

MacGillivray, A. J., Carroll, A. R., Dahi, S., Naxakis, G., Sadaie, M. R., Wallis, C. M., and Jing, T. (1982). *FEBS Lett.* **141,** 139–147.

McGrath, J. P., Capon, D. J., Smith, D. H., Chen, E. Y., Seeburg, P. H., Goeddel, D. V., and Levinson, A. D. (1983). *Nature (London)* **304,** 501–506.

Maki, R., Roeder, W., Traunecker, A., Sidman, C., Wabl, M., Raschke, W., and Tonegawa, S. (1981). *Cell* **24,** 353–365.

Manley, J. L., Fire, A., Cano, A., Sharp, P. A., and Gefter, M. L. (1980). *Proc. Natl. Acad. Sci. U.S.A.* **77,** 3855–3859.

Manley, J. L., Sharp, P. A., and Gefter, M. L. (1982). *J. Mol. Biol.* **159,** 581–599.

Mao, J., Schmidt, O., and Söll, D. (1980). *Cell* **21,** 509–516.

Maquat, L. E., Kinniburgh, A. J., Beach, L. R., Honig, G. R., Lazerson, J., Ershler, W. B., and Ross, J. (1980). *Proc. Natl. Acad. Sci. U.S.A.* **77,** 4287–4291.

Mariman, E. C. M., van Eekelen, C. A. G., Reinders, R. J., Berns, A. J. M., and van Vendrooij, W. J. (1982). *J. Mol. Biol.* **154,** 103–119.

Mariman, E. C. M., van Beek-Reinders, R. J., and van Venrooij, W. J. (1983). *J. Mol. Biol.* **163,** 239–256.

Mathews, M. B. (1980). *Nature (London)* **285,** 575–577.

Maurer, R. A., Erwin, C. R., and Donelson, J. E. (1981). *J. Biol. Chem.* **256,** 10524–10528.

Maxwell, E. S., Maundrell, K., Puvion-Dutilleul, F., and Scherrer, K. (1981). *Eur. J. Biochem.* **113,** 233–247.

Melton, D. A., De Robertis, E. M., and Cortese, R. (1980). *Nature (London)* **284,** 143–148.

Michel, F., and Dujon, B. (1983). *EMBO J.* **2,** 33–38.

Miller, A. (1984). *EMBO J.* **3,** 1061–1065.

Miller, T. E., Huang, C.-Y., and Pogo, A. O. (1978). *J. Cell Biol.* **76,** 675–704.

Montandon, P.-E., and Stutz, E. (1983). *Nucleic Acids Res.* **11,** 5877–5892.

Montell, C., Fisher, E. F., Caruthers, M. H., and Berk, A. J. (1982). *Nature (London)* **295,** 380–384.

Montell, C., Fisher, E. F., Caruthers, M. H., and Berk, A. J. (1984). *Mol. Cell. Biol.* **4,** 966–972.
Moore, K. W., Rogers, J., Hunkapiller, T., Early, P., Nottenburg, C., Weissman, I., Bazin, H., Wall, R., and Hood, L. E. (1981). *Proc. Natl. Acad. Sci. U.S.A.* **78,** 1800–1804.
Moschonas, N., De Boer, E., Grosveld, F. G., Dahl, H. H. M., Wright, S., Shewmaker, C. K., and Flavell, R. A. (1981). *Nucleic Acids Res.* **9,** 4391–4401.
Mount, S. (1982). *Nucleic Acids Res.* **10,** 459–472.
Mount, S. M. (1983). Ph.D. dissertation, Yale University.
Mount, S. M., and Steitz, J. A. (1981). *Nucleic Acids Res.* **9,** 6351–6368.
Mount, S. M., and Steitz, J. A. (1983). *Nature (London)* **303,** 380–381.
Mount, S. M., Pettersson, I., Hinterberger, M., Karmas, A., and Steitz, J. A. (1983). *Cell* **33,** 509–518.
Müller, F., and Clarkson, S. G. (1980). *Cell* **19,** 345–353.
Murray, V., and Holliday, R. (1979). *FEBS Lett.* **106,** 5–7.
Nabeshima, Y., Fujii-Kuriyama, Y., Muramatsu, M., and Ogata, K. (1984). *Nature (London)* **308,** 333–338.
Nagata, S., Mantei, N., and Weissmann, C. (1980). *Nature (London)* **287,** 401–408.
Naora, H., Deacon, N. J., and Buckle, D. (1980). *J. Theor. Biol.* **85,** 83–87.
Nelson, K. J., Haimovich, J., and Perry, R. P. (1983). *Mol. Cell. Biol.* **3,** 1317–1332.
Neuberger, M. S. (1983). *EMBO J.* **2,** 1373–1378.
Nevins, J. R., and Darnell, J. E. (1978). *Cell* **15,** 1477–1493.
Nevins, J. R., and Wilson, M. C. (1981). *Nature (London)* **290,** 113–118.
Nishikura, K., and De Robertis, E. M. (1981). *J. Mol. Biol.* **145,** 405–420.
Nishikura, K., Kurjan, J., Hall, B. D., and De Robertis, E. M. (1982). *EMBO J.* **1,** 263–268.
Nojima, H. (1983). *J. Biochem.* **93,** 1713–1716.
Nordstrom, J. L., Roop, D. R., Tsai, M.-J., and O'Malley, B. W. (1979). *Nature (London)* **278,** 328–331.
Northemann, W., Scheurlen, M., Gross, V., and Heinrich, P. C. (1977). *Biochem. Biophys. Res. Commun.* **76,** 1130–1137.
Northemann, W., Klump, H., and Heinrich, P. C. (1979). *Eur. J. Biochem.* **99,** 447–456.
Nussinov, R., and Tinoco, I. (1981). *J. Mol. Biol.* **151,** 519–533.
Nussinov, R., and Tinoco, I. (1982). *Nucleic Acids Res.* **10,** 351–363.
O'Farrell, P. Z., Cordell, B., Valenzuela, P., Rutter, W. J., and Goodman, H. M. (1978). *Nature (London)* **274,** 438–445.
Ogden, R. C., Beckman, J. S., Abelson, J., Kang, H. S., Söll, D., and Schmidt, O. (1979). *Cell* **17,** 399–406.
Ohlsson, R. I., van Eekelen, C., and Philipson, L. (1982). *Nucleic Acids Res.* **10,** 3053–3068.
Ohshima, Y., Itoh, M., Okada, N., and Miyata, T. (1981). *Proc. Natl. Acad. Sci. U.S.A.* **78,** 4471–4474.
Orkin, S. H., Goff, S. C., and Hechtman, R. L. (1981). *Proc. Natl. Acad. Sci. U.S.A.* **78,** 5041–5045.
Orkin, S. H., Kazazian, H. H., Antonarakis, S. E., Ostrer, H., Goff, S. C., and Sexton, J. P. (1982). *Nature (London)* **300,** 768–769.
Orkin, S. H., Sexton, J. P., Goff, S. C., and Kazazian, H. H. (1983). *J. Biol. Chem.* **258,** 7249–7251.
Otsuka, A., De Paolis, A., and Tocchini-Valentini, G. P. (1981). *Mol. Cell. Biol.* **1,** 269–280.
Padgett, R. A., Hardy, S. F., and Sharp, P. A. (1983a). *Proc. Natl. Acad. Sci. U.S.A.* **80,** 5230–5234.
Padgett, R. A., Mount, S. M., Steitz, J. A., and Sharp, P. A. (1983b). *Cell* **35,** 101–107.
Pavlakis, G. N., Lockard, R. E., Vamvakopoulos, N., Rieser, L., RajBhandary, U. L., and Vournakis, J. N. (1980). *Cell* **19,** 91–102.

Pederson, T. (1983). *J. Cell Biol.* **97,** 1321–1326.

Peebles, C. L., Ogden, R. C., Knapp, G., and Abelson, J. (1979). *Cell* **18,** 27–35.

Peebles, C. L., Gegenheimer, P., and Abelson, J. (1983). *Cell* **32,** 525–536.

Peffley, D. M., and Sogin, M. L. (1981). *Biochemistry* **20,** 4015–4021.

Perry, R. P., Bard, E., Hames, B. D., Kelley, D. E., and Schibler, U. (1976). *Prog. Nucleic Acid Res. Mol. Biol.* **19,** 275–292.

Perry, R. P., Kelley, D. E., Coleclough, C., Seidman, J. G., Leder, P., Tonegawa, S., Mattheyssens, G., and Weigert, M. (1980). *Proc. Natl. Acad. Sci. U.S.A.* **77,** 1937–1941.

Piatak, M., Subramanian, K. N., Roy, P., and Weissman, S. M. (1981). *J. Mol. Biol.* **153,** 589–618.

Pikielny, C. W., Teem, J. L., and Rosbash, M. (1983). *Cell* **34,** 395–403.

Prestayko, A. W., Tonato, M., and Busch, H. (1970). *J. Mol. Biol.* **47,** 505–515.

Raymond, G. J., and Johnson, J. D. (1983). *Nucleic Acids Res.* **11,** 5969–5988.

Reddy, R., and Busch, H. (1981). *Cell Nucl.* **8,** 261–306.

Reddy, R., Ro-Choi, T. S., Henning, D., and Busch, H. (1974). *J. Biol. Chem.* **249,** 6486–6494.

Rein, A., and Penman, S. (1969). *Biochim. Biophys. Acta* **190,** 1–9.

Reyes, A. A., Schöld, M., Itakura, K., and Wallace, R. B. (1982). *Proc. Natl. Acad. Sci. U.S.A.* **79,** 3270–3274.

Robinson, R. R., and Davidson, N. (1981). *Cell* **23,** 251–259.

Ro-Choi, T. S., and Busch, H. (1974). *Cell Nucl.* **5,** 151–208.

Ro-Choi, T. S., and Henning, D. (1977). *J. Biol. Chem.* **252,** 3814–3820.

Rogers, J. H. (1981). Ph.D. dissertation, University of California, Los Angeles.

Rogers, J., and Wall, R. (1980). *Proc. Natl. Acad. Sci. U.S.A.* **77,** 1877–1879.

Rogers, J., and Wall, R. (1984). *Adv. Immunol.* **35,** 39–59.

Rogers, J., Clarke, P., and Salser, W. (1979). *Nucleic Acids Res.* **6,** 3305–3321.

Roop, D. R., Nordstrom, J. L., Tsai, S. Y., Tsai, M.-J., and O'Malley, B. W. (1978). *Cell* **15,** 671–685.

Rosbash, M., Harris, P. K. W., Woolford, J. L., and Teem, J. L. (1981). *Cell* **24,** 679–686.

Ross, J. (1976). *J. Mol. Biol.* **106,** 403–420.

Rozek, C. E., and Davidson, N. (1983). *Cell* **32,** 23–34.

Ryffel, G. U., Wyler, T., Muellener, D. B., and Weber, R. (1980). *Cell* **19,** 53–61.

Salditt-Georgieff, M., Harpold, M., Chen-Kiang, S., and Darnell, J. E. (1980). *Cell* **19,** 69–78.

Schibler, U., Marcu, K. B., and Perry, R. P. (1978). *Cell* **15,** 1495–1509.

Schuler. M. A., Schmitt, E. S., and Beachy, R. N. (1982). *Nucleic Acids Res.* **10,** 8225–8244.

Schwartz, R. C., Greer, C. L., Gegenheimer, P., and Abelson, J. (1983). *J. Biol. Chem.* **258,** 8374–8383.

Sehgal, P., Lyles, D., and Tamm, I. (1978). *Virology* **89,** 186–198.

Seidman, J. G., and Leder, P. (1980). *Nature (London)* **286,** 779–783.

Seif, I., Khoury, G., and Dhar, R. (1979). *Nucleic Acids Res.* **6,** 3387–3398.

Selker, E., and Yanofsky, C. (1980). *Nucleic Acids Res.* **8,** 1033–1042.

Shah, D. M., Hightower, R. C., and Meagher, R. B. (1982). *Proc. Natl. Acad. Sci. U.S.A.* **79,** 1022–1026.

Shah, D. M., Hightower, R. C., and Meagher, R. B. (1983). *J. Mol. Appl. Genet.* **2,** 111–126.

Sheldon, E., Ferl., R. Fedoroff, N., and Hannah, L. C. (1983). *Mol. Gen. Genet.* **190,** 421–426.

Shimizu, K., Birnbaum, D., Ruley, M. A., Fasano, O., Suard, Y., Edlund, L., Taparowsky, E., Goldfarb, M., and Wigler, M. (1983). *Nature (London)* **304,** 497–500.

Shine, J., and Dalgarno, L. (1974). *Proc. Natl. Acad. Sci. U.S.A.* **71,** 1342–1346.

Shiokawa, K., and Pogo, A. O. (1974). *Proc. Natl. Acad. Sci. U.S.A.* **71,** 2658–2662.

Skolnick-David, H., Hay, N., and Aloni, Y. (1982). *Proc. Natl. Acad. Sci. U.S.A.* **79,** 2743–2747.

Slightom, J. L., Sun, S. M., and Hall, T. C. (1983). *Proc. Natl. Acad. Sci. U.S.A.* **80,** 1897–1901.

Snyder, A. L., Kann, H. E., and Kohn, K. W. (1971). *J. Mol. Biol.* **58,** 555–565.
Snyder, M., Hunkupiller, M., Yuen, D., Silvert, D., Fristrom, J., and Davidson, N. (1982). *Cell* **29,** 1027–1040.
Solnick, D. (1981). *Nature (London)* **291,** 508–510.
Spector, D. J., McGrogan, M., and Raskas, H. J. (1978). *J. Mol. Biol.* **126,** 395–414.
Spence, S. E., Pergolizzi, R. G., Donovan-Peluso, M., Kosche, K., Dobkin, C. S., and Bank, A., (1982). *Nucleic Acids Res.* **10,** 1283–1294.
Spritz, R. A., Jagadeeswaran, P., Choudary, P. V., Biro, P. A., Elder, J. T., DeRiel, J. K., Manley, J. L., Gefter, M. L., Forget, B. G., and Weissman, S. M. (1981). *Proc. Natl. Acad. Sci. U.S.A.* **78,** 2455–2459.
Sri-Widada, J., Assens, C., Liautard, J.-P., Jeanteur, P., and Brunel, C. (1982). *Biochem. Biophys. Res. Commun.* **104,** 457–462.
Staden, R. (1984). *Nucleic Acids Res.,* in press.
Standring, D. N., Venegas, A., and Rutter, W. J. (1979). *Proc. Natl. Acad. Sci. U.S.A.* **78,** 5963–5967.
Stein, J. P., Catterall, J. F., Kristo, P., Means, A. R., and O'Malley, B. W. (1980). *Cell* **21,** 681–687.
Steinmetz, M., Moore, K. W., Frelinger, J. G., Sher, B. T., Shen, F.-W., Boyse, E. A., and Hood L. (1981). *Cell* **25,** 683–692.
Steitz, J. A., and Kamen, R. (1981). *Mol. Cell Biol.* **1,** 21–34.
Sugisaki, H., Sugimoto, K., Takanami, M., Shiroki, K., Saito, I., Shimojo, H., Sawada, Y., Uemizu, Y., Uesugi, S., and Fujinaga, K. (1980). *Cell* **20,** 777–786.
Svensson, C., Pettersson, U., and Akusjärvi, G. (1983). *J. Mol. Biol.* **165,** 475–499.
Takano, M., Golden, S., Sharp, G. C., and Agris, P. F. (1981). *Biochemistry* **20,** 5929–5936.
Tate, V. E., Finer, M. H., Boedtker, H., and Doty, P. (1983). *Nucleic Acids Res.* **11,** 91–104.
Teem, J. L., and Rosbash, M. (1983). *Proc. Natl. Acad. Sci. U.S.A.* **80,** 4403–4407.
Teem, J. L., Rodriguez, J. R., Tung, L., and Rosbash, M. (1983). *Mol. Gen. Genet.* **192,** 101–103.
Thimmappaya, B., Weinberger, C., Schneider, R. J., and Shenk, T. (1982). *Cell* **31,** 543–551.
Tolstoshev, P., and Solomon, E. (1982). *Nature (London)* **300,** 581–582.
Tooze, J. (ed.) (1981). "Molecular Biology of Tumor Viruses, Part 2: DNA Tumor Viruses." Cold Spring Harbor Laboratory Press, Cold Spring Harbor, New York.
Trapnell, B. C., Tosltoshev, P., and Crystal, R. G. (1980). *Nucleic Acids Res.* **8,** 3659–3672.
Treisman, R., Novak, U., Favoloro, J., and Kamen, R. (1981). *Nature (London)* **292,** 595–600.
Treisman, R., Proudfoot, N. J., Shander, M., and Maniatis, T. (1982). *Cell* **29,** 903–911.
Treisman, R., Orkin, S. H., and Maniatis, T. (1983). *Nature (London)* **302,** 591–596.
Tsai, M-J., Ting, A. C., Nordstrom, J. L., Zimmer, W., and O'Malley, B. W. (1980). *Cell* **22,** 219–230.
Tweeten, K. A., and Molloy, G. R. (1981). *Nucleic Acids Res.* **9,** 3307–3319.
Tyc, K., Kikuchi, Y., Konarska, M., Filipowicz, W., and Gross, H. J. (1983). *EMBOJ.* **2,** 605–610.
Valenzuela, P., Venegas, A., Weinberg, F., Bishop, R., and Rutter, W. J. (1978). *Proc. Natl. Acad. Sci. U.S.A.* **75,** 190–194.
Van Eekelen, C., Ohlsson, R., Philipson, L., Mariman, E., van Beek, R., and van Venrooij, W. (1982). *Nucleic Acids Res.* **10,** 7115–7131.
Van het Schip, A. D., Meijlink, F. C. P. W., Strijker, R., Gruber, M., van Vliet, A. J., van de Klundert, J. A. M., and AB, G. (1983). *Nucleic Acids Res.* **11,** 2529–2540.
Van Ommen, G.-J.B., Arnberg, A. C., Baas, F., Brocas, H., Sterk, A., Tegelners, W. H. H., Vassart, G., and de Vijlder, J. J. M. (1983). *Nucleic Acids Res.* **11,** 2273–2285.
Venegas, A., Quiroga, M., Zaldivar, J., Rutter, W. J., and Valenzuela, P. (1979). *J. Biol. Chem.* **254,** 12306–12309.
Vennström, B., Persson, H., Pettersson, U., and Philipson, L. (1979). *Nucleic Acids Res.* **7,** 1405–1418.

Villareal, L. P., and White, R. T. (1983). *Mol. Cell. Biol.* **3,** 1381–1388.
Wagner, M. J., Sharp, J. A., and Summers, W. C. (1981). *Proc. Natl. Acad. Sci. U.S.A.* **78,** 1441–1445.
Wallace, R. B., Johnson, P. F., Tanaka, S., Schöld, M., Itakura, K., and Abelson, J. (1980). *Science* **209,** 1396–1400.
Wallis, M. (1980). *Nature (London)* **284,** 512.
Warner, J. R., and Gorenstein, C. (1977). *Cell* **11,** 201–212.
Weber, J., Blanchard, J.-M., Ginsberg, H. S., and Darnell, J. E. (1980). *J. Virol.* **33,** 286–291.
Weinberg, R. A., and Penman, S. (1968). *J. Mol. Biol.* **38,** 289–304.
Weinberg, R. A., and Penman, S. (1969). *Biochim. Biophys. Acta* **190,** 10–29.
Weingärtner, B., and Keller, W. (1981). *Proc. Natl. Acad. Sci. U.S.A.* **78,** 4092–4096.
Westaway, D., and Williamson, R. (1981). *Nucleic Acids Res.* **9,** 1777–1788.
White, R. T., Berg, P., and Villareal, L. P., (1982). *J. Virol.* **42,** 262–274.
Wieben, E. D., Madore, S. J., and Pederson, T. (1983). *Proc. Natl. Acad. Sci. U.S.A.* **80,** 1217–1220.
Wieringa, B., Meyer, F., Reiser, J., and Weissmann, C. (1983). *Nature (London)* **301,** 38–43.
Wieringa, B., Hofer, E., and Weissmann, C. (1984). *Cell* **37,** 915–925.
Wilson, M. C., and Darnell, J. E. (1981). *J. Mol. Biol.* **148,** 231–251.
Winicov, I., and Button, J. D. (1982). *Mol. Cell. Biol.* **2,** 241–249.
Wolley, J. C., Cone, R. D., Tartof, D., and Chung, S.-Y. (1982). *Proc. Natl. Acad. Sci. U.S.A.* **79,** 6762–6766.
Wooley, J. C., Zukerberg, L. R., and Chung, S.-Y. (1983). *Proc. Natl. Acad. Sci. U.S.A.* **80,** 5208–5212.
Yang, V. W., Lerner, M. R., Steitz, J. A., and Flint, S. J. (1981). *Proc. Natl. Acad. Sci. U.S.A.* **78,** 1371–1375.
Yannarell, A., Niemann, M., Schumm, D. E., and Webb, T. E. (1977). *Nucleic Acids Res.* **4,** 603–611.
Yaoita, Y., Kumagai, Y., Okumura, K., and Honjo, T. (1982). *Nature (London)* **297,** 697–699.
Young, P. R., Scott, R. W., Hamer, D. H., and Tilghman, S. M. (1982). *Nucleic Acids Res.* **10,** 3099–3116.
Young, R. A., Hagenbüchle, O., and Schibler, U. (1981). *Cell* **23,** 451–458.
Zain, S., Gingeras, T. R., Bullock, P., Wong, G., and Gelinas, R. E. (1979). *J. Mol. Biol.* **135,** 413–433.
Zeevi, M., Nevins, J. R., and Darnell, J. E. (1981). *Cell* **26,** 39–46.
Zieve, G. W. (1981). *Cell* **25,** 296–297.
Zieve, G., and Penman, S. (1976). *Cell* **8,** 19–31.
Zieve, G., and Penman, S. (1981). *J. Mol. Biol.* **145,** 501–523.
Ziff, E. B. (1980). *Nature (London)* **287,** 491–499.

Part 2: THE STRUCTURE AND EVOLUTION OF RETROPOSONS

V. Introduction and Summary

At least three classes of dispersed sequences in mammalian genomes share common properties which have suggested that all three are generated from RNA intermediates by a common mechanism involving reverse transcription (Van Arsdell *et al.*, 1981; Jagadeeswaran *et al.*, 1981; Hollis *et al.*, 1982; Sharp, 1983). These are the processed pseudogenes, representing coding mRNAs; the small nuclear RNA pseudogenes, representing snRNAs; and the "*Alu*" family and similar sequences, which constitute most of the dispersed, moderately repeated DNA of mammalian genomes. In this review I shall refer to all such sequences as retroposons (Rogers, 1983a), denoting their RNA origin and their dispersed positions.

Their common properties include sequence boundaries exactly corresponding to RNA species, a repetitive (A)-rich tail at the 3′ side, and direct terminal repeats (TR) of 8–19 bp of the flanking sequence at the 5′ and 3′ ends (Fig. 3).

The (A)-rich tails or their complements were first detected as (A,T)-rich segments associated with repeated DNA units (now known to be mostly *Alu*-like units) (Straus and Birnboim, 1974; Flavell *et al.*, 1977) and as (U)-rich segments in the nonmessenger portions of nuclear RNA (now known to be mostly introns) (Molloy *et al.*, 1972). Many such segments in cloned genomic DNA are in fact retroposon tails. For processed pseudogenes the tails represent the poly(A) of the original mRNA, but for other retroposons their origin is unclear.

The TRs are presumed to be analogous to those generated by transposon insertion, and to be generated by a staggered nick in the target site, with the inserted sequence being joined to the separated ends.

Detailed reviews have been published on snRNA pseudogenes (Denison and Weiner, 1982) and *Alu*-like dispersed repeats (Schmid and Jelinek, 1982; Jelinek and Schmid, 1982; Singer, 1982b; Georgiev *et al.*, 1982). Here I will concentrate on recent discoveries and new analyses which reveal a number of new classes and families of retroposons with interesting properties, and also clues to a common mechanism for the dispersal of all classes. These analyses have been helped by recognition of the characteristic (A)-rich tails and TRs of retroposons in many published genomic DNA sequences. Thus several new retroposons (Table IV) have been located for this review, and the exact arrangements of clustered retroposons have been defined.

Recently identified classes include the longest (the 6–7 kb "long interspersed elements" or LINEs) and the shortest [isolated (A)-rich tails]. Within the ubiquitous *Alu*-like class of short dispersed repeats, several new families and sub-

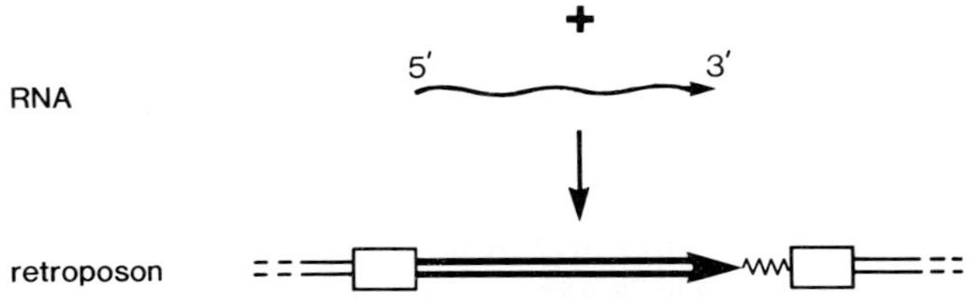

FIG. 3. Diagram of retroposon structure. Zigzag denotes variable (A)-rich tail.

TABLE IV

NEWLY IDENTIFIED SEQUENCES OF RETROPOSONS

Retroposon family	Gene	Location (orientation)		Reference
Alu (3′ part)	Tubulin	Intron (plus)[a]	1244–1486	Lee *et al.* (1983)
Alu	c-*myc*	Intron (minus)	3000–2741	Colby *et al.* (1983)
Alu[b]	ζ-Globin	5′ flanking (plus)	1–110[b]	Proudfoot *et al.* (1982)
Alu[b]	ζ-Globin	3′ flanking (minus)	2441–2411[b]	Proudfoot *et al.* (1982)
Alu[b]	Actin pseudogene 2	5′ flanking (plus)	1–62[b]	Moos and Gallwitz (1983)
B1	H-2 27.1	5′ flanking (plus)	39–172	Steinmetz *et al.* (1981)
B1[b]	H-2L[d]	Intron (minus)	1743–1675[b]	Moore *et al.* (1982)
B1	H-2K[d]	Intron (minus)	2020–1890	Kvist *et al.* (1983)
B2[b]	Growth hormone	3′ flanking (minus)	2158–2137[b]	Barta *et al.* (1981)
B2[b]	H-2I-A_{β}^{b}	3′ flanking (minus)	10000–9823[b]	Larhammar *et al.* (1983)
R.dre.1	SVS IV	3′ flanking (minus)	202–127	Harris *et al.* (1983)
Rat MIF-1[b,c]	SVS IV	3′ flanking (minus)	480–310[b]	Harris *et al.* (1983)
Rat MIF-1[b,c]	Prolactin	Intron (minus)	(190)–1669[b]	Cooke and Baxter (1982)
Mouse MIF-1 (3′ parts)	Kallikrein	Intron (plus)	6355–6785	Mason *et al.* (1983)
Mouse MIF-1 (3′ part)	H-2I-A_{β}^{b}	Intron (minus)	5789–5317	Larhammar *et al.* (1983)
*Kpn*I (3′ end)	$^{G}\gamma$-Globin	5′ flanking (minus)	690–616	Duncan *et al.* (1981)
Alu[d]	Gastrin	Intron (plus)	−263–−142[b]	Kato *et al.* (1983)
B2[d]	α-Lactalbumin	Intron (minus)	2875–2714	Qasba and Safaya (1984)
R.dre.1[d]	rRNA	NTS region 2 (minus)	118–43	Mroczka *et al.* (1984)

[a] 5′ to indicated *Alu*.

[b] 5′ end of retroposon not sequenced.

[c] This mrDNA was located approximately, but not identified, by the authors cited; each gene contains 1.5–2 kb of the rat MIF-1 homolog, including the sequenced 3′ end (. . . TCYTAATA$_{5-6}$). Details of these LINEs and their subfamily relationships are given in Rogers (1985).

[d] These were published too late for inclusion in Table VI.

families have been discovered in rodents and artiodactyls; here I show that some of them have novel relationships to different polymerase III transcription units, including tRNAs.

Analysis of the tail structures in all available retroposon sequences provides clues to the mechanism of retroposon insertion. In all primate and rodent retroposon classes, these tails usually have the structure A_n and/or $(NA_x)_y$, where N is most often C. Artiodactyl retroposon tails have a more complex structure. These structures seem likely to be polymerized anew in many or all insertion events, and I will describe new models for their generation.

Analysis of the insertion sites reveals that they are often (A)-rich themselves, and that retroposons are particularly prone to insert into preexisting retroposon tails. This produces clusters which may generate new retroposon families, and may offer a variety of novel possibilities for evolution.

It is worth emphasizing that the mechanisms used in dispersing retroposons appear to be completely distinct from those used by retroviruses, or by insertion sequences and transposons, which together constitute the second major class of mobile sequence in eukaryotes. The superficial similarities between retroposons and retroviruses apparently represent independent solutions to the same problem. Apart from reverse transcription itself, retroviruses display none of the hallmarks of retroposons, neither in oncogene acquisition nor in their normal replicative cycles. Instead, they employ an elaborate set of encoded repeats and priming sites to produce a DNA provirus which appears to be a transposon (Gilboa *et al.*, 1979; Temin, 1980; Varmus, 1982). Other eukaryotic transposons and insertion sequences include intracisternal A particle (IAP) genes in mice (Ono *et al.*, 1980), *sigma* and TY-1 in yeast, and several families in *Drosophila* (reviewed by Jelinek and Schmid, 1982). In all of them, the transposed sequence ends with inverted repeats, usually terminating 5′-TG...CA-3′; the TR generated on insertion is only 4–6 bp; the transcription unit (if any) begins and ends some distance inside the transposon; and transposition therefore is likely to occur as in bacteria, by a DNA-mediated process (Calos and Miller, 1980; Harshey and Bukhari, 1981). There is no evidence that retroposons can undergo transposition in this fashion. (However, transposons might undergo "retroposition," to form pseudogenes like those from other transcription units described below, which would appear to be slightly shortened transposons or proviruses until their characteristic terminal sequences were analyzed.)

VI. The Varieties of Retroposons

A. Processed Pseudogenes

Pseudogenes are sequences homologous to functional genes which have been altered so as to prevent any normal function. Many pseudogenes retain the original gene organization and evidently represent adjacent duplications of the original genes. Others, with which we are concerned here, correspond to the mRNA sequence alone. They have lost their introns precisely, acquired an (A)-rich tail, and been inserted into unrelated sequences, often on a different chromosome (Hollis *et al.*, 1982).

When the first pseudogene lacking introns, a mouse α-globin sequence, was discovered, its RNA origin was not obvious because it had atypical 5′ and 3′ ends (Nishioka *et al.*, 1980; Vanin *et al.*, 1980). Pseudogenes for human immunoglobulin C_λ and C_ϵ and proopiomelanocortin (POMC) have characteristic 3′ ends but atypical 5′ ends (Hollis *et al.*, 1982; Battey *et al.*, 1982; Ueda *et al.*, 1982; Uhler *et al.*, 1983). However, less specialized genes have revealed several examples with all the characteristics of retroposons: one pseudogene for human metallothionein II (Karin and Richards, 1982), three for human tropomyosin (Macleod and Talbot, 1983; A. R. Macleod, personal communication), three for human dihydrofolate reductase (DHFR) (Chen *et al.*, 1982; Masters *et al.*, 1983), three for rat cytochrome *c* (Scarpulla and Wu, 1983), two and probably more for human actin (Moos and Gallwitz, 1983; Minty *et al.*, 1983), one for rat α-tubulin (Lemischka and Sharp, 1982), and no less than five for human β-tubulin, including three pseudogenes apparently derived independently from a single parental gene (Wilde *et al.*, 1982a,b; Lee *et al.*, 1983). There are also processed genes or pseudogenes for the rat Ha-*ras* and human Ki-*ras* oncogenes (DeFeo *et al.*, 1981; McGrath *et al.*, 1983) whose 3′ ends have not been sequenced.

Most of these specimens show the 5′ and 3′ ends, poly(A), and splicing patterns which imply reverse transcription of normal mRNA. In the two cases where the exact initiation site of the equivalent mRNA is known (Karin and Richards, 1982; Lemischka and Sharp, 1982), the first transcribed (capped) nucleotide is faithfully represented in the pseudogene, either immediately abutting the 5′ TR or only one nucleotide from it.

Most of these "processed genes" are defined as pseudogenes because they have obviously deleterious mutations in the coding sequence. Even those which do not are unlikely to be transcribable, since they cannot carry their promoters with them.

The α-globin pseudogene is an apparent exception, retaining homology to the promoter; but since they must be created in the germline, it is not surprising that the globin, immunoglobulin, and POMC pseudogenes have atypical 5′ ends. These genes are normally expressed only in differentiated tissues, and transcription might well initiate at different sites in germline cells. [However, it is curious that these are the pseudogenes with $(NA_x)_y$ tails, whereas all but one of the others have pure, or nearly pure, oligo(A) tails.] Those genes which have produced multiple accurate processed pseudogenes, sometimes outnumbering the real genes, are those which are expected to be transcribed in all cells.

The retroposon-like features of these pseudogenes make it clear that retroviral transduction is not responsible for them, even though this process can be made to work (Shimotohno and Temin, 1982). [There are IAP genes on each side of the α-globin processed pseudogene, but they could not have transduced it as they are some distance away and wrongly oriented (Lueders *et al.*, 1982).] It is conceivable that retrovirus virions might simply transport mRNAs from other tissues into the germline (Bernstein *et al.*, 1983), but in general there is no evidence that retroviruses are involved.

B. Small Nuclear RNA Pseudogenes

The discrete, stable, small nuclear RNA species, U1 to U6, each hybridize to $\sim 10^2$–10^3 sequences in mammalian genomes. The majority of these appear to be pseudogenes by the criterion of multiple base substitutions (Denison *et al.*, 1981; Hayashi, 1981), and of these many appear to be retroposons, since they have (A)-rich tails and/or TRs and are embedded in unrelated sequences (Van Arsdell *et al.*, 1981; Denison and Weiner, 1982). In contrast, bona fide snRNA genes have a conserved 3′ flanking sequence that is not (A)-rich (Ohshima *et al.*, 1981; Roop *et al.*, 1981; Manser and Gesteland, 1982; Watanabe-Nagasu *et al.*, 1983). Sequences of 15 out of 16 dispersed snRNA pseudogenes include a copy of the first transcribed nucleotide of the RNA, and in seven of them it is immediately adjacent to the TR.

However, Denison and Weiner (1982) have identified multiple classes which imply multiple mechanisms for producing dispersed snRNA pseudogenes. Seven specimens (class III) are typical retroposons—full-length pseudogenes with (A)-rich tails and TRs (Van Arsdell *et al.*, 1981; Hayashi, 1981; Ohshima *et al.*, 1981; Piechaczyk *et al.*, 1982; Monstein *et al.*, 1983; Nojima and Kornberg, 1983). Another seven specimens (class IV) are highly 3′-truncated pseudogenes with TRs (in most cases) but no (A)-rich tails (Van Arsdell *et al.*, 1981; Bernstein *et al.*, 1983; Hammarström *et al.*, 1982; Nojima and Kornberg, 1983). The remaining two specimens (class II), both representing U1 snRNA, are slightly 3′-truncated pseudogenes with no TRs (Denison and Weiner, 1982).

New snRNA pseudogene sequences are now corroborating this classification. Van Arsdell and Weiner (1984) report five more pseudogenes for U2, all of class IV. Murphy *et al.* (1984) report four pseudogenes for 7SK RNA—two full-length ones (class III) cointegrated with *Alu* sequences, and two 3′-truncated ones (class IV) with TRs. Most previous examples of class IV pseudogenes had overlap between the truncated 3′ end and the TR, and these new examples all have such overlap, of 2–6 bp (if a one-base mismatch in a few cases may be ignored). Most class IV pseudogenes also have one or two nucleotides of overlap between the 5′ TR and the 5′ end (Van Arsdell and Weiner, 1984). These overlaps presumably reflect the process by which class IV pseudogenes are created, as discussed in Section VII.

C. Short Interspersed (*Alu*-like) Sequences

1. *General Properties*

Before cloning, analyses of chromosomal DNA renaturation showed that most organisms possess a "short-period interspersion pattern," i.e., short stretches (< 1 kb) of moderately repeated sequences (mrDNA) separated by longer (several kb) sequences of low copy number (reviewed by Britten and Kohne, 1968; Davidson and Britten, 1973; Jelinek and Schmid, 1982; Bouchard, 1982). For mammals, it is now clear that most of the mrDNA consist of retroposons, which belong to different sequence families in different mammalian orders.

The prototypes of these sequences, now called *Alu* in primates and B1 in rodents, were first recognized when it became apparent that the same sequences were present in the mrDNA, in rapidly annealing structures in hnRNA, in discrete small RNAs, and in dispersed polymerase III transcription units (Jelinek *et al.*, 1980). In primates, all these sequences belong to the single *Alu* family; rodents and artiodactyls have multiple families with little mutual homology (Table V).

TABLE V
Families of Short Retroposons[a]

Retroposon family	Retroposon length	Copy number	Homologous RNA	RNA length
Primates				
(*Alu*.A)	120	(—)	7 SL	295
(*Alu*.B)	146	(—)	7 SL	295
Composite				
Alu	282	3–5 × 10^5 (human)		
Rodents				
B1	129	~1 × 10^5 (mouse)	4.5 S	90
B2	180	~0.8 × 10^5 (mouse)	4.5 SI	98
R.dre.1	87	1–1.5 × 10^5 (rat)	BC1	160
(OBY)	63	(—)	?	?
Composite				
B2-OBY	243	Some (mouse)		
Artiodactyls				
(A)	117	Some?[b]	?	?
(C)	82	(—)	tRNA	75
Composites				
A1-A2	261	Many[b]		
C-A3	205	Many[b]		

[a]Copy numbers per haploid genome are from Houck *et al.* (1979) and Rinehart *et al.* (1981) (*Alu*); Kramerov *et al.* (1979) (B1, B2); Milner *et al.* (1984) (R.dre.1).

[b]The total reiteration frequency of bovine A sequences is ~10^5 (Watanabe *et al.*, 1982).

The general properties of *Alu* and "*Alu*-equivalent" sequences, including their similarity to other retroposons, have been reviewed in detail (Jagadeeswaran *et al.,* 1981; Schmid and Jelinek, 1982; Jelinek and Schmid, 1982; Singer, 1982; Georgiev *et al.,* 1982). Here I will concentrate on more recent sequence data and analyses which expand our knowledge of the various families and of their common properties.

a. *Boundary Sequences.* In primates and rodents, all the genomic *Alu*-like sequences carry variable (A)-rich 3′ tails and are flanked by TRs (Bell *et al.,* 1980a; Krayev *et al.,* 1980), except where there is clear evidence of deletion (see footnote to Table VI). Some families (B2, R.dre.1) include an encoded (A)-rich tail, but this is followed by the usual variable one. In several instances the *Alu*-like sequence has inserted into a known target, and the TR is demonstrably a duplication of the target sequence (see Section VII,A).

The 5′ and 3′ ends of individual family members are often one or two nucleotides longer or shorter than the consensus sequence, in contrast to the usually exact termini of mRNA and snRNA pseudogenes. This may reflect the different properties of evolving polymerase III transcription units. Several examples have much more extensive deficiencies at their 5′ ends (Baralle *et al.,* 1980; Tsukada *et al.,* 1982; Comb *et al.,* 1983; A. R. Macleod, personal communication). Since all of these partial inserts are flanked by typical TRs, 3′ fragments are apparently sufficient for insertion. 5′-truncation could presumably arise during reverse transcription.

Alu-like sequences, and retroposons in general, have a strong tendency to insert into each others' (A)-rich tails (see Section VIII). This has apparently generated composites which are themselves propagated as single retroposons. The primate *Alu* is the classic example, being a dimer of homologous sequences; I will present several new examples below.

b. *Sequence Homologies.* From their mutual homologies, all the primate and rodent families fall into two superfamilies, named after the *Alu*/B1 and B2 families. Similarities within and between them are displayed in Fig. 4. I have used a diagonal-matrix computer program, DIAGON (Staden, 1982), to find the best regions of alignment for this figure; they include the major known alignments (Haynes *et al.,* 1981; Krayev *et al.,* 1982; Singer, 1982b) and a few new ones. While B1 and B2 have some conserved elements in common (see below), they are differently arranged, and outside these elements there is no significant similarity. Also the artiodactyl A consensus (Watanabe *et al.,* 1982) and the chicken dispersed repeat CR1 (Stumph *et al.,* 1981) have been reported to align with *Alu*/B1, but no significant homology emerges on diagonal matrices (not shown).

The internal elements which are shared by most *Alu*-like families (with the notable exception of the artiodactyl A family), and are also relatively conserved within each family, are indicated in Fig. 4. They are (1) The polymerase III promoter boxes "A" and "B" (see below). (2) A short symmetrical sequence of

```
Promoter   GTGGCnnAGTGG                                                  GGGTTCGAnn-CC
Alu.A    { GGCtGGGCGTGGTGGCtCACACCTGTAATCCCAGCactTtGGGAGGCcGAGGtgGGtGGATCACcTGAGgTCAG-
         {                                                      -GAGTTCAAGA-CCAGCCTGGcCAACATgGGAAACCCCGTCTCTActAAAAAtAcAAAAAttA-(Alu.B)
Alu.B     -GCCGGGCGTGGTGGCGCGCGCCTGTAATCCCAGCTACTCGGGAGGCTGAGGCAGGAGaATCGCTTGAacCCAGGA*CCAGCCTGGCCAACAGAGGAGACTcCCaTCTC----A_n
7SL      ...GCCGGGCGcGGTGGCGCGtGCCTGTAGTCCCAGCTACTCGGGAGGCTGAGGCtGGAGGATCGCTTGAGTCCAGGA*CCAGCCTGGGCAACATAGGAGACCCCGTCTCT_OH
B1 (CH)  { cCaGGcAtTGGTGGCaCACACCTtTAGTCCCAGC-ACTCAGGAGGCAGAGGCAGGAGGATCACTTGAGTTCCAGAgCCAGCCTGGtCTACAGA-
         {                                                      -TTGAGTTCCAGG-aCAGCtnnnnnnACAGA-GAAACCCTGTCTn----A_n
4.5S     pppGCCGGtAGTGGTGGCGCACGCCgGTACgatttGCTga--AGGAGGCAGAGGCAGagGGATCAC--GAGTTCGAGG-CCAGCCTGGGCTACAcAtTtttt_OH

b
Promoter             GTGGCnnACTGG                               GGGTTCGAnnCC
OBY             GGAcctG-aGCg-AGTGGggttGAcC----------ggagtA-gAGGTCCTaAaTTCAATTCCCAaCAACCcCATt-A_n
R.dre.1   GGGGCTGGgGAttTaGCTCAGTGGT--AGAGCgCTtACctaggaagCgcAAGGcCCTGGGTTCGgTcCCCAGCtCCg------A_7gAAccA_n
B2 (M)    GCGGCTGGAGAGATGGCTCAGTGGTTAAGAGCACTGACTGCTCTTCCA-AAGGTCCTGAGTTCAAATCCCAGCAACCACATGGTGGCTCA-CAACCATCTGTAATGAGATCTG-
B2 (CH)   GaGGCTGGAGAGATGGCTC-GaGGTTAACAGCACcaACTGCTgTTCCA-gAGGTCCTGAGTTCAATTCCCAGCAACCACATGGTGGCTCA-tAACaATCTaTAATGAGATCTG-
4.5S_I    pppGGCTGGAGAGATGGCTCAGccGTTAAagGCTa--GGCTCACAACCAaAAatataaCAGTTCGgTTCCCAGCACCCACg-GcTGtCTCtcCAgCCAcCTtTt(OH)

                                                                                    10
(B1)     ...GaGGCaGGAG-GATcaCTt-GaGtTccAGAGC-CaGcCTGgTC-TaCA-gA----tTGAGTTCcAg...   (Alu)...TGGTGGCTCA-CA-CC---TGTAATcccAgC...
            47
                                                             (B1)...tTGAGTTCcAgagCCAGCc------TGGT...
                                                                    63

(R.dre.1)                GgGgCTGggGAtttAGCT-CAGTG...CTTAccTA....
B2 (M)    -ATGCCCTCTTCTGGTGTGTCTGAAGAC--AGCTACAGTG-TACTTATATATAA(TA_3)_3-TCTTT-A_n
B2 (CH)   -gTGCCCTCTTCTGGTGTGcagatAtAtatgGaagCAGaatgttgTATAcATAA(TA_3)_3ATCTT--A_n
```

FIG. 4. Mutual homologies of *Alu*-like sequences. Regions of homology were identified using the DIAGON program (Staden, 1982). Promoter boxes (Fowlkes and Shenk, 1980) are indicated, and other conserved elements are underlined. Some possible secondary structure elements are overlined with arrows. (A)-rich tails are zigzag-overlined, lower-case letters indicate nonmatching nucleotides (n = not assigned), and hyphens indicate gaps introduced to maximize homology. (a) Alu/Bl superfamily. Human *Alu* consensus (Deininger *et al.*, 1981) is shown in two halves labeled A and B. Chinese hamster B1 consensus is modified from that of Haynes *et al.* (1981) at the 3′ end to exclude a B2 sequence from the set; it is now very similar to the mouse B1 consensus (Kalb *et al.*, 1983). Mouse 4.5 S RNA sequence from Harada and Kato (1980). Human 7 SL RNA sequence from Ullu *et al.* (1982a). Asterisk marks additional sequence in *Alu*.B (39 bp) and 7 SL (191 bp); this includes homology to the first half of the *Alu*.A "B" box, which is shown as a short duplication in *Alu*.A. *Alu*.A and B1 sequences are broken to indicate duplications. (b) B2 superfamily. OBY is from Ou *et al.* (1983), and R.dre.1 from clone pIB224 of Sutcliffe *et al.* (1982). Mouse B2 consensus from Fig. 6. Chinese hamster B2 sequence from clone 19c (Haynes *et al.*, 1981). 4.5 SI RNA sequence from Busch *et al.* (1982). A region of transposed identity is boxed.

unknown function: CCAGCCTGG in *Alu* and B1 (Schmid and Jelinek, 1982) and CCACATGG in B2 (missing from R.dre.1). There is no obvious homologue in artiodactyl C sequences, but CCATGG in the A sequences could be an equivalent. (3) A region which can form at least five base pairs with the 5′ end of the consensus sequence. This overlaps elements (1) and/or (2) (see Fig. 4). It is not a promoter element, since individual transcribed *Alu* sequences can lack it, but it could be a preferred element of RNA secondary structure.

c. *Polymerase III Promoter Structure.* Many *Alu,* B2, and R.dre.1 units, and some B1 units, can function as polymerase III transcription units (see references below). All these sequences contain internal elements closely resembling the consensus "A" and "B" boxes which form the split internal promoter of tRNAs and adenovirus VA RNAs (Fowlkes and Shenk, 1980; Galli *et al.,* 1981; Elder *et al.,* 1981; Duncan *et al.,* 1981; Hall *et al.,* 1982; Ciliberto *et al.,* 1983). The promoter boxes can also be recognized in the other retroposon families in Table V, except for the artiodactyl A family. They are also present in the homologous 4.5 S, 4.5 SI, and OAX RNAs, and weakly in 7 SL RNA (see Sections VI,C,2–4).

The regions necessary for transcription of *Alu* have been mapped by Fuhrmann *et al.* (1981) and Paolella *et al.* (1983). An "A" box is necessary, but the requirements are very loose, and can sometimes be satisfied by randomly substituted sequences. In one case, a naturally 5′-truncated *Alu* sequence had an "A" box contributed by the flanking sequences. The "B" box is more important, as in tRNA genes. Transcription can also be influenced by sequences further 3′.

It has been proposed that the "A" and "B" boxes are recognized not only by their primary sequence but also by their tertiary structure on the DNA "plus" strand, mimicking the D loop and TψCG loop of tRNA (Hall *et al.,* 1982; Goddard *et al.,* 1983; see Fig. 5a). I have searched the *Alu*-like sequences for evidence of similar structures. In the *Alu*/B1 superfamily, neither stem is present, neither in 4.5 S RNA, nor in the *Alu*/B1 consensuses, nor in individual transcribed *Alu* sequences (Duncan *et al.,* 1981; Pan *et al.,* 1981; Paolella *et al.,* 1983; Miyake *et al.,* 1983). However, the one B1 copy known to be transcribed (Young *et al.,* 1982) could form a 6-bp "D stem" and a 3-bp "TψCG stem," and also has a better "A" box than the B1 consensus.

In the B2 superfamily, all members except OBY can form a 4-bp D stem and many can form a 3- to 5-bp TψCG stem. R.dre.1 (Fig. 5c) and OAX (Ackerman, 1983) can form perfect D and TψCG stems in the correct positions relative to the promoter boxes. (They are not tRNA genes as they cannot form anticodon nor aminoacyl stems.) B2 has the same sequences, except for a shorter TψCG stem, but 4.5 SI RNA and OBY have no TψCG stem at all.

This analysis shows that the D and TψCG stems are not essential for promoter activity. Nor are they likely to contribute to RNA structure, at least for R.dre.1,

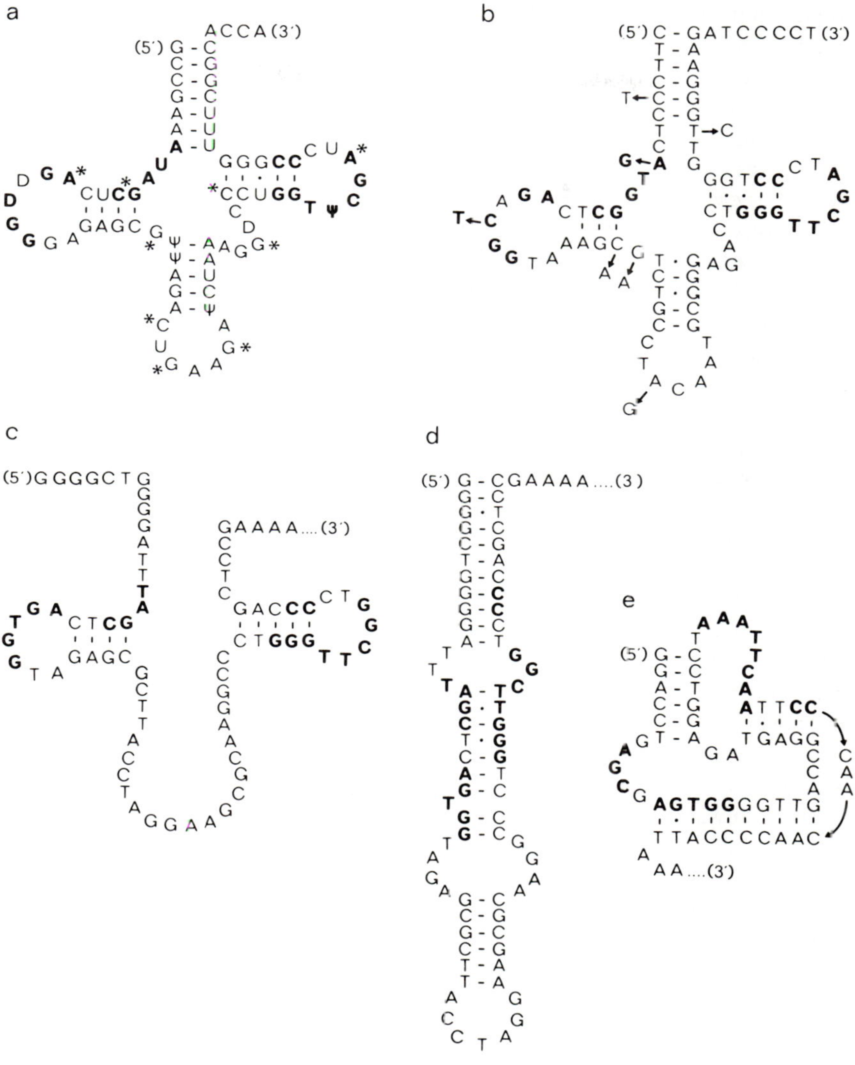

FIG. 5. Secondary structures of some polymerase III transcription units and retroposons. Retroposon sequences are shown as DNA. Conserved polymerase III promoter boxes are shown in heavy type. (a) Mammalian tRNA^{Phe} (Roe *et al.*, 1975). U, D, and ψ correspond to T in the gene; other base modifications are indicated by asterisks. (b) Artiodactyl retroposon sequence C, folded like a tRNA. Arrows indicate alternative nucleotides in the consensus sequence. (c) Rat retroposon sequence R.dre.1, with the promoter boxes folded like tRNA arms. (d) R.dre.1: alternative secondary structure. (e) Mouse retroposon sequence OBY: predicted secondary structure for the specimen sequenced by Ou *et al.* (1983).

where a much stronger structure should exclude them (Fig. 5d); they are also inconsistent with the 5′ end basepairing of other B2 superfamily sequences mentioned above. However, they do exist in the B2 superfamily, and their function remains obscure.

d. *Transcripts and Functions.* There are several types of RNA species which hybridize to *Alu*-like sequences. First, as mentioned above, most *Alu*-like sequences can function as polymerase III transcription units, although with widely varying efficiency. This is true for most *Alu* units (Duncan *et al.*, 1979, 1981; Fritsch *et al.*, 1980; Elder *et al.*, 1981; Shen and Maniatis, 1982; Manley and Colozzo, 1982), B2 units (Haynes and Jelinek, 1981), and R.dre.1 units (Sutcliffe *et al.*, 1984), and for some B1 units (Young *et al.*, 1982). In all these cases, transcripts begin at the 5′ end of the repeat sequence and run beyond the 3′ end, to terminate at fortuitous oligo(T) tracts (Bogenhagen and Brown, 1981) in the flanking sequence. Some transcripts, particularly of B2, may be cleaved and polyadenylated (Georgiev *et al.*, 1982); this has been shown to occur for another polymerase III transcript (Carlson and Ross, 1983).

Second, most *Alu*-like sequences are homologous to discrete polymerase III transcripts (e.g. 4.5 S and 7 SL RNAs; see Sections VI,C,2–4), which are conserved in evolution and are distinct from the heterogeneous transcripts of the bulk of the retroposons.

Third, *Alu*-like sequences are abundant in hnRNA, because they are often present in introns, in either orientation (Jelinek and Schmid, 1982; Ryskov *et al.*, 1983). Fourth, a few also lie in 3′ untranslated segments and thus find their way into mRNAs. A small percentage of mouse mRNA molecules contain B1 or B2 in the 3′ untranslated segment, always in the "plus" orientation (Kinniburgh and Martin, 1976; Tokarskaya *et al.*, 1980; Kramerov *et al.*, 1982; Ryskov *et al.*, 1983). Likewise, *Alu* is present in a few human mRNAs (Jelinek *et al.*, 1978; Calabretta *et al.*, 1981). [R.dre.1 was originally reported in 3′ untranslated segments, but these may actually be introns (Sutcliffe *et al.*, 1982; Milner *et al.*, 1984).] Dispersed repeats are also found in 3′ untranslated segments of mRNAs in other animals—*Xenopus* (Spohr *et al.*, 1981, 1982), sea urchin (Carpenter *et al.*, 1982), and *Drosophila* (Tchurikov *et al.*, 1982). In all these cases, a role in gene regulation has been proposed, but has not yet been demonstrated.

The B2-containing mRNAs, at least, are not coregulated; one is specific to liver (Ryskov *et al.*, 1983), another is a histocompatibility class I gene activated in transformed cells (Steinmetz *et al.*, 1981; Lalanne *et al.*, 1982; Brickell *et al.*, 1983), and others are the ubiquitously expressed H-2D^{b} and H-2L^{d} genes (Reyes *et al.*, 1982; Moore *et al.*, 1982; Evans *et al.*, 1982).

Two very interesting recent findings suggest that some retroposons may indeed be regulated, if not agents of regulation. The first is that all types of R.dre.1 transcripts are brain-specific (Sutcliffe *et al.*, 1982, 1984; see Section VI,C,5). The second is that cytoplasmic poly(A)$^{+}$ RNA of some mouse tumor cell lines, but not of normal liver tissue, contains transcripts of various dispersed repeats, including heterogeneous large RNAs containing MIF-1, IAP, and B1 transcripts (Georgiev *et al.*, 1982) and small RNAs containing B2 transcripts (Kramerov *et al.*, 1982; Bennett *et al.*, 1984). Heterogeneous poly(A)$^+$ B2-containing RNAs are also seen in 9-day embryos and in embryonic or embryonal-carcinoma cell lines, and they vanish upon differentiation (Murphy *et al.*, 1983; Brickell *et al.*, 1983; Bennett *et al.*, 1984). It remains to be seen whether this represents positively regulated expression, or perhaps the cessation of normal regulation and whether it is of physiological importance. It may be relevant that *Alu* sequences near the human ϵ globin gene appear to be coregulated with it (Allan and Paul, 1984). Conceivably, retroposons may become available for transcription merely as a byproduct of neighboring gene activity.

Many other hypotheses have been offered concerning physiological functions for *Alu*-like sequences, for example in DNA replication, gene regulation, or RNA processing. They have been discussed elsewhere (Davidson and Britten, 1973, 1979; Jelinek *et al.*, 1980; Krayev *et al.*, 1980, 1982; Singer, 1982b; Georgiev *et al.*, 1982; Sharp, 1983). None of them has yet been rigorously tested experimentally.

The main reason for doubting any functional role for *Alu*-like sequences is their extreme lability in evolution. Related species can have divergent consensus sequences (e.g., hamster and mouse B2, Fig. 4), and different mammalian orders have entirely different consensuses. Nor are individual locations generally conserved; certain introns which contain *Alu*-like sequences in one species (mouse histocompatibility, rat growth hormone, goat globins; see Figs. 7, 10, and 11) contain no equivalent sequences in the human (DeNoto *et al.*, 1981; Malissen *et al.*, 1982; Shapiro *et al.*, 1983).

An alternative point of view is that they have no physiological function, but are pseudogenes for polymerase III genes, which have been amplified as a form of selfish DNA (Orgel and Crick, 1980; Doolittle and Sapienza, 1980). Whereas the processed pseudogenes for polymerase II transcripts have lost their promoters, the internal promoters of polymerase III transcription units should allow their retroposon copies to be transcribed and reverse-transcribed again and again; thus they could undergo multiple rounds of "retroposition" and be selected only for their efficiency at doing so. Indeed, there is extensive evidence for diversification in the B2 retroposon superfamily (see below). I will refer to this model in describing the *Alu*-like families, and discuss it in more detail in Section IX,A.

The true situation may lie somewhere between the extremes outlined; even if retroposons are amplified as selfish DNA, a minority of them might well influence the genes near which they inserted, or their transcripts might acquire a function.

2. *Primates: The Alu Family*

The predominant short dispersed repeat in primate genomes is *Alu* itself, with 3–5 $\times$ 10^5 copies comprising 3–6% of the human genome, occurring on average every 5–9 kb, and differing on average by 13% from the consensus sequence (Houck *et al.,* 1979; Rinehart *et al.,* 1981; Deininger *et al.,* 1981; Schmid and Jelinek, 1982). *Alu* is a dimer of two homologous sequences, the first (*Alu.*A) carrying the functional polymerase III promoter, and the second (*Alu.*B) having a substantial "insert." They are connected by an (A)-rich linker.

These sequences are homologous to the 7 SL RNA (Weiner, 1980), which has been sequenced and in which the "insert" is larger (Ullu *et al.,* 1982a; Li *et al.,* 1982; Fig. 4). 7 SL is an essential functional RNA, involved in the synthesis of secreted proteins (Walter and Blobel, 1982). It is almost identical in human and rat, and very similar in chickens (Erikson *et al.,* 1973; Balmain *et al.,* 1982). The "Alu" and "insert" portions seem to form separate domains in the protein-bound native state of 7SL (Gundelfinger *et al.,* 1983). 7 SL is a polymerase III transcript (Frederiksen *et al.,* 1974; Zieve *et al.,* 1977).

Thus, *Alu* may be a dimer of 7 SL pseudogenes, which have arisen by internal deletions. One may speculate that 7 SL, like 5 S RNA, uses a specialized mechanism of polymerase III transcription, and that mutations and deletions improved the "A" and "B" boxes and thus enabled *Alu* to use a more general mechanism. However, there are complications in this story which will be discussed in Section IX,A.

Alu sequences include GAGGCNGAGGC, which corresponds to the T antigen-binding sequence of the SV40 replication origin (Jelinek *et al.,* 1980). Its significance is unknown.

There seems to be only one *Alu* family consensus sequence (Rinehart *et al.,* 1981). A few *Alu* units have been found with the central (A)-rich linker replaced by $(CA_x)_y$ or $(CA)_n$ (see Section VI,F). It was suggested that the latter type represents an *Alu* subfamily (Saffer and Lerman, 1983). While this remains to be established, it is likely that multiple *Alu* sequences have been active in retroposition, since the *Alu* sequences are much more divergent than the TRs which they generated on insertion (Fukumaki *et al.,* 1982).

Alu units must have been inserted over a prolonged period, since a long duplicated sequence in the α-globin cluster contained one *Alu* before duplication, and one copy of it has acquired an additional *Alu* pair subsequently (Hess *et al.,* 1983). At least some of the *Alu* inserts in the β-globin cluster antedate the divergence of apes and humans (Fritsch *et al.,* 1980; Barrie *et al.,* 1981; Maeda *et al.,* 1983), and the *Alu* consensus is the same, at least approximately, in human, chimpanzee, and owl monkey. In a prosimian (*Galago*), the consensus has many substitutions, although it is descended from the same ~300-nucleotide heterodimeric structure (Daniels *et al.,* 1983).

The galago also has a second type of *Alu* dimer, in which the *Alu*A half is replaced by a nonhomologous sequence which we may call G (Daniels and Deininger, 1983). G has excellent polymerase III promoter boxes, and also occurs as an isolated dispersed repeat. Thus it may be an unrelated *Alu*-like retroposon which combined with an *Alu* sequence to form the now-ubiquitous G-*Alu*B unit.

3. *Rodents: The Alu/B1 Superfamily*

Mice, rats, and hamsters all contain two abundant "*Alu*-equivalent" families, B1 and B2 (Kramerov *et al.*, 1979; Krayev *et al.*, 1980, 1982; Haynes *et al.*, 1981). B1 is homologous to primate *Alu* but has a large duplication in the 3′ portion (Fig. 4). One B1 specimen carries an active polymerase III promoter (Young *et al.*, 1982), but most are inactive.

There is a 4.5 S RNA in rodent cells which is homologous to B1, but terminates before the end of the B1 sequence (Fig. 4) (Jelinek and Leinwand, 1978; Harada and Kato, 1980; Haynes *et al.*, 1981). It is found in the nucleus and the cytoplasm, basepaired to hnRNA and mRNA, presumably to copies of B1 units. It has not been detected in human cells, but has been conserved within rodents, since it is only 3% different between mouse and Chinese hamster (Harada *et al.*, 1979; Haynes *et al.*, 1981), whereas the consensus B1 sequences of these species differ by 15% (compare Figure 4 with Kalb *et al.*, 1983).

Two subfamilies of mouse B1 sequences have been proposed. However, the only members of the first subfamily are from a sequence tabulation (Krayev *et al.*, 1980) which differs from the corresponding published gels. The latter actually conform to the second consensus (Kalb *et al.*, 1983), as do other B1 sequences.

Rabbits also contain dispersed repeats which are transcribed by polymerase III and have been reported to hybridise weakly to *Alu* (Fritsch *et al.*, 1980); but their major polymerase III retroposon family, called C, is not related to *Alu* or any other known sequence (Cheng *et al.*, 1984).

4. *Rodents: The B2 Superfamily*

The abundant B2 family of rodents is very active both in transcription and in evolution. It is homologous to a wide variety of transcripts and retroposons, which I designate the B2 superfamily. None of these is known to have physiological functions, nor to occur in other mammalian orders. However, a partly homologous *Xenopus* RNA called OAX is transcribed in the frogs' oocytes, from tandemly repeated genes (Ackerman, 1983). This implies that the B2 superfamily does include functional, conserved sequences.

The B2 family sequence was defined in the Chinese hamster by Haynes *et al.* (1981), in the rat by Page *et al.* (1981) and Barta *et al.* (1981), and in the mouse by Krayev *et al.* (1982). An updated compilation is presented in Fig. 6. Mouse specimens differ from the consensus by 8.8% on average.

a. *B2 Transcripts.* B2 transcripts are abundant in cells (Kramerov *et al.*, 1982; Georgiev *et al.*, 1982). Most are in hnRNA, and some are in mRNA (see Section VI,C,1). A substantial number of B2 RNAs probably represent poly-

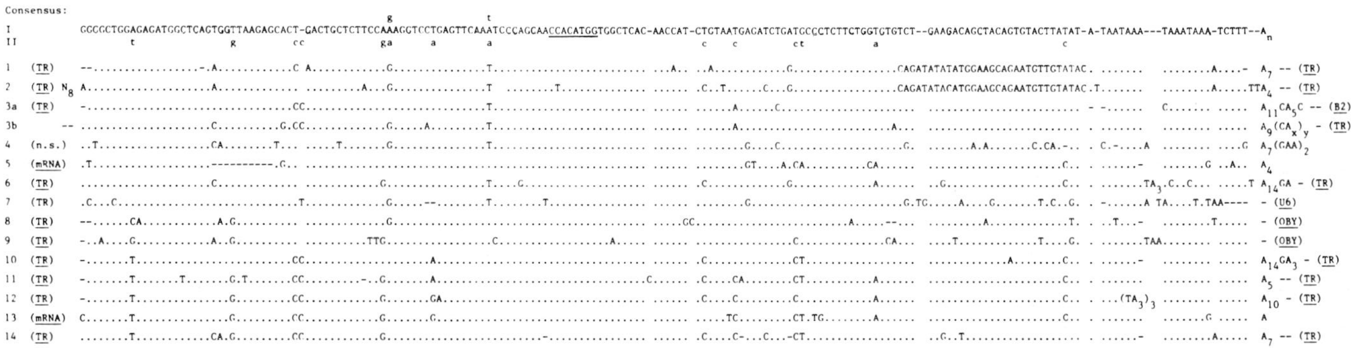

FIG. 6. Compilation of B2 sequences. Consensus sequence is shown on the top line, for mouse subfamily I, which is generally identical to the rat and hamster consensus. Bases in lower case above it are found in some subfamily I and most rat and hamster sequences; bases in lower case below it are the mouse subfamily II consensus. Individual sequences are identical to the subfamily I consensus except where indicated. Hyphens indicate gaps introduced to maximise homology. Chinese hamster (1,2): random clones of Haynes and Jelinek (1981). Rat (3a,b): tandem copies in growth hormone gene (Barta *et al.*, 1981; Page *et al.*, 1981; Fig. 10a). Mouse subfamily I: (4) 3′ to H-2I-A_{β}^{b}GB gene (Table IV); (5) in H-2 mRNA (Steinmetz *et al.*, 1981; Lalanne *et al.*, 1982); (6) random clone Mm14 (Krayev *et al.*, 1982); (7) in cluster (Ohshima *et al.*, 1981; Fig. 10b); (8) in cluster (Ou *et al.*, 1983; Fig. 10g); (9) in H-2 gene 27.1 (Steinmetz *et al.*, 1981; Fig. 10f). Mouse subfamily II: (10) in cluster (Kalb *et al.*, 1983; Fig. 10c); (11) in rDNA sequence (Kominami *et al.*, 1983); (12) random clone Mm61 (Krayev *et al.*, 1982); (13) in liver mRNA (Ryskov *et al.*, 1983); (14) random clone Mm98 (Georgiev *et al.*, 1982).

merase III transcription of individual B2 units (Haynes and Jelinek, 1981). Many of these transcripts are small (6 S–7 S, or 200–600 nucleotides) and many consist almost entirely of B2 sequence with a full-length poly(A) tail (Kinniburgh and Martin, 1976; Kramerov *et al.*, 1982; Bennett *et al.*, 1984). Since B2 units include an encoded $(TA_x)_y$ tail which includes the polyadenylation signal, followed by TCT_{2-4} which can act as a polymerase III termination signal (Haynes and Jelinek, 1981), these may be polyadenylated polymerase III transcripts. These B2 transcripts are more abundant in tumor cells and embryonic cells than in normal cells (see Section VI,C,1).

The rodent 4.5 SI RNA (Ro-Choi *et al.*, 1972; Jelinek and Leinwand, 1978; Busch *et al.*, 1982) is also related to B2, just as 4.5 S RNA is related to B1 (Fig. 4). It binds the La antigen (Reddy *et al.*, 1983), which is characteristic of polymerase III transcripts (Hendrick *et al.*, 1981).

b. *B2 Evolution.* The B2 superfamily is remarkably fissiparous. The 3′ part of the sequence is either truncated or nonhomologous in comparisons of mouse and hamster B2, rat R.dre.1, rat 4.5 SI RNA, frog OAX, and one variant specimen (B2* in Fig. 10c). In the mouse B2, this 3′ part may represent a second segment of homology to the 5′ end of R.dre.1 (Fig. 4), although this is too limited to stand out significantly on diagonal matrices (not shown). Even within the mouse, there appear to be two subfamilies of B2 (Fig. 6), although with considerable sequence exchange between them. Rat B2 and hamster B2 are much closer to subfamily I than to subfamily II.

c. *R.dre.1.* A major dispersed repeat family in the rat was discovered by Sutcliffe *et al.* (1982), who found copies of it in the untranslated segments of several brain-specific RNAs, and proposed that it was a tissue-specific "identifier" (ID) sequence, in line with the models of Davidson and Britten (1973).

It is now clear that this is a major retroposon family, named R.dre.1, which has been amplified recently in the rat (Lemischka and Sharp, 1982). R.dre.1 sequences are homologous to B2 (Barta *et al.*, 1981), and have typical (A)-rich tails and TRs. Copies have been sequenced incidentally in a tubulin processed pseudogene (Lemischka and Sharp, 1982), in Harvey sarcoma virus (Dhar *et al.*, 1982), and in the vicinity of six other rat genes (Barta *et al.*, 1981; Hurst and Parker, 1983; Watanabe-Nagasu *et al.*, 1983; Tani *et al.*, 1983; Schuler *et al.*, 1983; Harris *et al.*, 1983), as well as in more cDNA clones (Milner *et al.*, 1984).

The evidence for recent, or current, amplification is as follows. (1) R.dre.1 has only been found in rat sequences. (2) Two of the six genomic copies sequenced are precisely absent from allelic loci (Barta *et al.*, 1981; Page *et al.*, 1981; Schuler *et al.*, 1983); such allelism has been seen for only one other retroposon (see Fig. 10a,e). (3) The average divergence of eight copies from the consensus shown in Fig. 4 is only 4.8%.

There may be two subfamilies, although they are not completely distinct. The three sequences of Sutcliffe *et al.* (1982) and Barta *et al.* (1981) belong to one (Fig. 4), while most of the other sequences include at least two of the three changes seen in the sequence of Hurst and Parker (1983), and do not have the encoded (A)-rich tail.

R.dre.1 hybridizes to several types of RNA (Sutcliffe *et al.*, 1982, 1984): polymerase II hnRNA with R.dre.1 in introns; heterogeneous small polymerase III RNAs, due to the R.dre.1 promoter; and BC1, which is a 160-nucleotide cytoplasmic poly(A)$^+$ RNA. Remarkably, all these RNAs are made almost exclusively in brain. Thus R.dre.1 may carry a brain-specific signal which enhances polymerase III and adjacent polymerase II transcription (Sutcliffe *et al.*, 1984). If so, R.dre.1 elements should also be required in brain-specific genes of other species. Alternatively, perhaps these retroposons are preferentially inserted into regions which, for other reasons, are only transcribable in brain.

RNA transcribed from R.dre.1 should form a strong hairpin structure (Fig. 5d) (also noted by Milner *et al.*, 1984).

d. *OBY*. A 63-bp retroposon candidate was discovered in the tail of a mouse B2 sequence by Ou *et al.* (1982) (see Fig. 10g). Inspection of this "OBY" sequence reveals an apparent polymerase III promoter; the 5′ half, including the "A" box, is a novel sequence, while the 3′ half, including the "B" box, is closely homologous to the middle of B2 (Fig. 4). If this were transcribed, it could form a strong, "locked" secondary structure (Fig. 5e).

It seems that this B2-OBY composite is a member of a new retroposon family, since a related composite had already been sequenced by Steinmetz *et al.* (1981), and is precisely defined for the first time in Figs. 10f and 11. In this sequence, irregular duplications of the 5′ end of the OBY portion have destroyed its "A" box and its potential secondary structure. It also extends into a longer region of B2 sequence at the 3′ end. It is possible that OBY-(B2) retroposed independently into the tail of a preexisting B2 in both cases, but much more likely that B2-OBY-(B2) was retroposed as a composite, because OBY has not yet been found in isolation, and in each case the OBY sequence begins immediately following the 3′ TCTTT of the B2.

Thus mouse B2-OBY probably represents conglomeration of retroposon sequences in the same way as human *Alu,* and the two specimens sequenced appear to have evolved through multiple rounds of retroposition.

It remains to be seen whether all these members of the B2 superfamily correspond to conserved polymerase III transcription units, or whether they are all variants of the B2 retroposons themselves, undergoing repeated rounds of transcription, reverse transcription, and sequence rearrangement (see Section IX).

5. *Artiodactyls: The A and C Families*

The bovine genome is estimated to contain 8–14 families of dispersed repeats, constituting 6–10% of the genome (Mayfield *et al.*, 1980). DNA sequencing has now revealed some of these families, which are very remarkable, both in their internal structure and in their propensity to cluster. These families were defined independently in the goat globin genes (Schon *et al.*, 1981; Shapiro *et al.*, 1983) and in the cow POMC gene (Watanabe *et al.*, 1982). For the globin specimens, and for one POMC specimen, there are homologous goat or human loci which lack the repeated sequences, so the inserts can be identified precisely and the TRs shown to be duplications of the target. [Other POMC specimens may be parts of larger rearrangements, as they have poor TRs, if any, and lie in regions which are totally nonhomologous to the human gene sequences of Takahashi *et al.* (1983).] With these data, one can now define clearly the sequence families involved (Fig. 7).

The 117-bp repeat defined by Watanabe *et al.* (1982) corresponds to units D and A of Schon *et al.* (1981); here it will be called A. The sequence tabulation of Watanabe *et al.* (1982) clearly shows the subfamilies which are here called A1, A2, and A3. The A sequence has little if any homology to primate *Alu* (Watanabe *et al.*, 1982) or to polymerase III promoters.

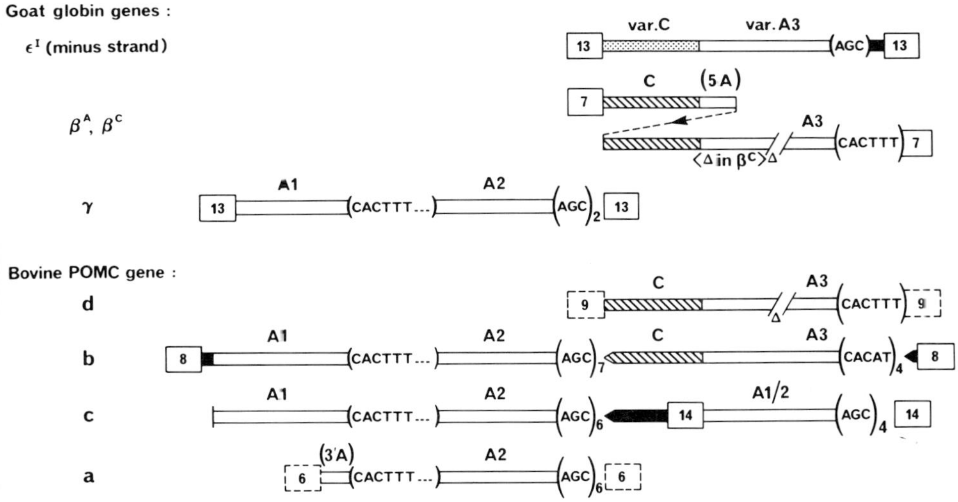

FIG. 7. Artiodactyl retroposons. Dispersed repeats identified in the goat globin genes (Schon *et al.*, 1981; Shapiro *et al.*, 1983) and bovine POMC gene (Watanabe *et al.* 1982) are displayed to indicate the arrangement of the repeat families involved. White, A units; hatched, C units; black, extra sequences. Numbers in boxes indicate lengths of TRs, shown dotted for short, imprecise TRs of doubtful significance. Consensus linker between A1 and A2 is CACTTTCACTTTTCACTT-TCATGCATT. A few single base changes have been ignored in summarizing the repeating tails in the figure. △ marks sites of apparent deletion within A sequences. For A consensus sequences, see Watanabe *et al.* (1982). For C consensus sequence, see Fig. 5B. [Recently, Schimenti and Duncan (1984) have characterized similar repeats in bovine globin genes; the β gene is like that of the goat, while the γ gene contains an A1–A2 insertion in a different position from that of the goat.]

All but one of the A sequences are in multimeric assemblages with other repeated sequences (Fig. 7). The first type of assemblage is analogous to the primate *Alu* dimer; it consists of

$$A1\text{–(linker)–}A2\text{–}(AGC)_n$$

where A1 and A2 have somewhat different consensus sequences. The linker is based on repeats of the nucleotide sequence CACTTT. Two of the four examples are 5′-truncated. There is one case of an isolated insertion of an A sequence [insert A1/2 in Fig. 7 = insert c_3 of Watanabe *et al.* (1982), with a TR equivalent to nucleotides 2062–2049 in the human sequence of Takahashi *et al.* (1983)], but this has an A1-like 5′ half and an A2-like 3′ half, so may have resulted from deletion within an A1-A2 dimer.

The second type of assemblage is

$$C\text{–}A3\text{–}(CACTTT)$$

where A3 is another distinct subfamily of A, and C is a separate sequence, so named by Schon *et al.* (1981). Some C-A3 units have undergone deletions (particularly in the β^C-globin gene, although the homologue in the β^A-gene is less altered), and one in the ϵ^I gene is 26% divergent.

The C consensus includes virtually perfect polymerase III promoter boxes, and is 65% identical to the consensus sequence compiled by Galli *et al.* (1981) and Gauss and Sprinzl (1983) for eukaryotic tRNAs. The score is 47/72 matches, with one insertion in the variable loop; identical scores are reached with three individual mammalian tRNAs, for example $tRNA^{Phe}$ in Fig. 5a. Furthermore, the C sequence can be folded into a tRNA-like structure in which all the invariant and semiinvariant bases are present in the correct positions, except that the aminoacyl stem is not properly formed (Fig. 5b). It therefore seems likely that C was derived from a tRNA gene. The "anticodon" of C corresponds to cysteine, and yeast $tRNA^{Cys}$ matches C at 45/72 positions with single-base insertions in the variable and D loops; no mammalian $tRNA^{Cys}$ has yet been sequenced for comparison.

The A1-A2 and C-A3 families are identified as retroposons in that they have variable repetitive 3′ tails and produce typical TRs on insertion. Since C-A3 is clearly a single retroposon, it seems likely that it uses a polymerase III promoter in the tRNA-like C sequence coupled to a reverse-transcription priming site in the A sequence. This could have arisen by a solo A unit inserting 3′ to a tRNA gene such that they were cotranscribed. [An alternative possibility may be direct RNA–RNA ligation, as suggested by defective interfering variants of Sindbis virus in which a tRNA is attached at the 5′ end of some viral sequences (Monroe and Schlesinger, 1983).] However, the absence of a typical promoter in A, and the complex nature of the tails, make these sequences exceptional. Although they are not retrovirus-like transposable elements, Schon *et al.* (1981) pointed out some similarities, so it is possible that the mobility of the A sequences depends on a somewhat modified mechanism of retroposition.

D. Long Interspersed Sequences (LINES)

In addition to short, *Alu*-like sequences, some mammalian genomes contain 10^4–10^5 copies of sequences 6–7 kb long, called long interspersed elements or LINEs (Singer, 1982) They comprise 1–8% of the genome. It is now clear that there is a single major LINE family in primates (*Kpn*I) (Adams *et al.*, 1980; Singer, 1982), and also in rodents (MIF-1) (Brown and Dover, 1981; Gebhard and Zachau, 1983a; Bennett and Hastie, 1983). Furthermore, *Kpn*I and MIF-1 are 60–70% homologous, except at the 3′ end (Singer *et al.*, 1983a; summarized by Rogers, 1983c). A rat equivalent described here for the first time (Table IV: DIAGON analysis nct shown) is very similar to MIF-1, although again less so towards the 3′ end.

Most of the MIF-1 copies are heterogeneously 5′-truncated (as are a few *Alu*-like sequences), so that the 5′ end is represented $\sim 3 \times 10^4$ times in the mouse genome, while the 3′ end is represented $\sim 10^5$ times (Gebhard and Zachau, 1983a; Fanning, 1983; Bennett and Hastie, 1983). The isolated 3′ fragments of MIF-1 constitute the previously named R and BAM5-R families of dispersed repeats (Fanning, 1982; Gebhard *et al.*, 1982; Lueders and Paterson, 1982; Wilson and Storb, 1983). *Kpn*I also has 5′-truncated copies (Thayer and Singer, 1983; Sun *et al.*, 1984). A previously unrecognised one (Table IV) is only 75 bp long.

A number of genomic LINE 3′ ends have now been sequenced (see Tables IV and VI). All have variable (A)-rich tails which coincide with the 3′ ends of the *in vivo* transcripts (DiGiovanni *et al.*, 1983). And the 5′-truncated specimens which have been totally sequenced are flanked by TRs just like those of other retroposons. In two cases (Gebhard and Zachau, 1983a), the TRs are duplications of MIF-1 sequences into which the second, 5′-truncated MIF-1 (R) sequence has inserted. These associations may be frequent (Bennett and Hastie, 1983). Therefore most, if not all, LINEs are retroposons.

Rearrangements appear to be common. Lerman *et al.* (1983) described a 3′ *Kpn*I fragment attached at the 5′ end to an inverted fragment from further upstream in the consensus, the whole being a single insertion. The same type of rearrangement is seen in a MIF-1 sequence of Gebhard and Zachau (1983a), and in a MIF-1 sequence in a mouse kallikrein gene (Table IV). [In the latter specimen, nucleotides 6355–6583 and 6578–6785 of Mason *et al.* (1983) align with MIF-1 segments E24 (Singer *et al.*, 1983a), reversed, and R (Gebhard *et al.*, 1982), 3′ part.] In at least two of these inversion-deletion variants, there is a short, imprecise overlap between the two parts of the sequence. Thus these variants might arise via foldbacks of the nascent cDNA strand during reverse transcription, as sometimes occurs during cDNA cloning *in vitro* (Land *et al.*, 1981). Potter (1984) has sequenced an even more complicated rearranged specimen.

The origin of the full-length LINEs is unknown, since none has yet been completely sequenced, nor their transcription unit(s) mapped. MIF-1 contains homology to the IAP long terminal repeat (Brown and Huang, 1982). Both MIF-1 and *Kpn*I are indeed transcribed, at a low level (Fanning, 1982; Lerman *et al.*, 1983; DiGiovanni *et al.*, 1983; Kole *et al.*, 1983; Soriano *et al.*, 1983). Most of the transcripts are made by polymerase II, but most are not polyadenylated, and are restricted to the nucleus (Shafit-Zagardo *et al.*, 1983; Sun *et al.*, 1984). These transcripts could reflect unresolved internal complexity in the full-length LINEs, and/or transcription fortuitously promoted from flanking sequences. In fact, it remains possible that the LINE retroposons, although comprising the great majority of copies, are processed pseudogenes derived from some parental copy or copies which have not yet been detected. Recently it was shown that they include an open reading frame, conserved in rodents and primates (Potter, 1984; Martin *et al.*, 1984). The many defective copies must arise frequently from parental copies in which this protein-coding function is maintained.

One apparently full-length LINE, a 6.4-kb *Kpn*I unit in the human β-globin gene cluster (Adams *et al.*, 1980), has been sequenced at both ends. Remarkably, the whole LINE is embedded in the tail of an *Alu* unit (Miyake *et al.*, 1983). A second similar pair has also been identified, and the *Kpn*I 5′ end is the same, although the 3′ end has been lost (Miyake *et al.*, 1983). Since the *Alu* portions of these two pairs are much more divergent than the *Kpn*I portions, they seem likely to represent independent insertions of *Kpn*I into the *Alu* tail (see Section VIII and Fig. 10j). However, they raise the possibility that some, if not all, full-length LINEs may actually be transcribed from *Alu* promoters and dispersed as *Alu*-LINE composites. The first composite sequenced does indeed produce a long transcript from the *Alu* promoter *in vitro,* although it only extends 2.3 kb into the *Kpn*I unit (Manley and Colozzo, 1982).

E. Other Retroposon Candidates

In view of the abundance of pseudogenes for known genes, mammalian genomes probably contain many retroposons with a wide variety of sequences. Some will be moderately populous families, such as the human O family ($\gtrsim 5 \times 10^3$ copies) (Sun *et al.*, 1984). Others will be of low copy number. Candidates can be found in many long noncoding sequences by searching for repetitive (A)-rich segments. For three sequences, additional grounds have been reported for believing them to be novel retroposons.

1. Gebhard *et al.* (1982) and Gebhard and Zachau (1983a) identified a 0.3 kb sequence apparently inserted into the tail of a BAM5-R sequence (Fig. 10d). It

has AATAAA preceding the poly(A) tail, but neither an open reading frame nor a canonical polymerase III promoter.

2. Feagin *et al.* (1983) identified DI-2, a 0.5 kb sequence which corresponds to a 0.6 kb poly(A)$^+$ RNA. It is present in an intron of the mouse DHFR gene and in ~ 20 other genomic sites. I identify it as a retroposon because one sequenced copy has the structure:

$$\underline{A_5TAATGTGTGTTG}\text{—(0.5 kb homology)—}\underline{AATAAA}\ N_{15}\ A_4\ \underline{A_8TGTGTGTTG}$$

(The other sequenced copy must have lost the 3′ end.)

3. J. Trowsdale (personal communication) has identified a 0.5 kb sequence in the human HLA-SB_β gene which corresponds to an abundant 0.65 kb poly(A)$^+$ RNA. Again, it has a typical AATAAA, oligo(A) tail, and TR; there are ~ 30 copies dispersed around the genome.

Surprisingly, these retroposons in the human HLA-SB_β intron and in the mouse DHFR intron (DI-2) are about 80% identical (J. Trowsdale, personal communication), and so define a new retroposon family. Given their comparatively low copy numbers, the independent, unplanned discovery of two members of this family seems to be a remarkable coincidence. Although none of the copies so far sequenced has an open reading frame, nor an obvious polymerase III promoter, the strong primate–rodent sequence conservation implies that these retroposons have originated from a functional gene of some sort.

F. Minimal Inserts

In addition to all these RNA insertions, there is also the "zero option" (Fig. 8). In the large intron of a mouse globin gene (Konkel *et al.*, 1979), there is an insertion which consists of an (A)-rich tail and TR, but no inserted RNA sequence at all!

A sequence in the human c-*myc* intron (minus strand) (Colby *et al.*, 1983) may be another minimal insert:

$$\underline{A_4GTCCAA}\ GAGGGCG_5AAGGAATA_4GA_6\ \underline{A_4GTCCAA}.$$

We have seen that many $(NA_x)_y$ tracts in mammalian DNA are retroposon tails, and a few are retroposon-like minimal inserts. Some others, when compared with homologous sequences, are evidently small insertions or expansions in shorter (A)-rich tracts. $(CA_x)_y$ has been seen to replace both the internal (A)-

β^{min} AATTTTGAT [AAGAATTATTCTATGA] $A_6TA_2TA_3TA_6TA_4$ [AAGAATTATTCTATGA] CACACAAAATTT

β^{maj} AATTaTtAT AAGAAT---cCTATGt -------------------------------- CAaACAgAATTT

FIG. 8. "Zero" insert. An insertion on the minus strand of the large intron of the mouse globin β^{minor} gene, compared with the β^{major} gene. Sequences from Konkel *et al.* (1979). TR sequences are boxed.

rich linker in *Alu* (Tsukada *et al.*, 1982; Krolewski *et al.*, 1983) and an internal oligo(A) sequence in MIF-1 (Brown and Piechaczyk, 1983). These clearly are insertions of $(CA_x)_y$ without significant terminal duplications, and the same may happen in other oligo(A) sequences (e.g., Harrington and Chikaraishi, 1983; Moos and Gallwitz, 1983; Shapiro *et al.*, 1983). These inserts probably have a variety of origins. Of course, they may frequently arise from replication slippage or unequal crossing-over *in situ* (Streisinger *et al.*, 1966; Smith, 1976), which probably affects all classes of simple sequences (Tautz and Renz, 1984). They could also result from promiscuous random insertions, such as the 6- and 13-bp insertions between 5-bp TRs in individual copies of MIF-1 (Brown and Piechaczyk, 1983), or the insertion of 1.5 copies of a tandemly repeated sequence into an opened *Eco*RI site (Shimizu *et al.*, 1983). They could also result from retroposon-style "zero" insertions as in Fig. 8. The absence of a TR does not preclude a retroposon-like origin, as it is possible that retroposons themselves may not create TRs when they insert into pre-existing retroposon tails (see Section VIII).

The alternating copolymer $(CA)_n$ can also replace the internal linker in *Alu* (Hamada *et al.*, 1982; Saffer and Lerman, 1983), and has also been found inserted into a B2 sequence (Gebhard and Zachau, 1983b), with terminal duplication of only 2–3 bp. Therefore, these copolymer sequences can also be inserted en bloc. $(CA)_n \cdot (TG)_n$ is a widespread dispersed repeat in mammalian and other genomes (Miesfeld *et al.*, 1981; Hamada *et al.*, 1982; Hamada and Kakunaga, 1982), and a few such tracts have TRs of 8 bp or more (Miesfeld *et al.*, 1981; Shen and Smithies, 1982; Höchtl and Zachau, 1983; Rogers, 1983b). [A second simple sequence noted by Höchtl and Zachau (1983) may also be independently mobile and is discussed in section VIII.] However, $(CA)_n$ does not generally occur in retroposon tails, so may be dispersed by a somewhat different mechanism.

G. Nonmammalian Retroposons?

It is not yet clear whether retroposons exist in nonmammalian vertebrates. The absence of processed pseudogenes for tubulin and cytochrome *c* in chickens (Lopata *et al.*, 1983; Limbach and Wu, 1983) and for U2 snRNA in *Xenopus* (Mattaj and Zeller, 1983) suggests that retroposons are not common. The presence of mrDNA is no guide in itself, since it could consist of other types of mobile element. In chickens, a dispersed repeat family named CR1 has been identified and some resemblance to *Alu* suggested (Stumph *et al.*, 1981, 1983). However, diagonal matrix analysis (not shown) shows no significant similarity (even if the published 245-bp consensus is abridged to minimise gaps). These repeats have imprecise boundaries, with no sign of TRs or repetitive tails, so one cannot tell what type of mobility they may have. Likewise, two repetitive families in *Xenopus* and one in sea urchin have imprecise borders and no clear retroposon-like characteristics (Spohr *et al.*, 1981, 1982; Carpenter *et al.*, 1982). Other repeated sequences in sea urchin (Posakony *et al.*, 1983; Liebermann *et al.*, 1983) appear to resemble the bacterial insertion sequences. However, there is a sea urchin histone pseudogene with poly(A) tail and 6-bp TRs (Liebermann *et al.*, 1983).

In *Drosophila,* there are several mobile sequences which have characteristics of retroposons. All four candidate families are long (maximum 3.4–5.0 kb) and at least the first three can be variably 5′-truncated like the mammalian LINEs. The Type I and Type II inserts which interrupt some ribosomal RNA genes are the first two candidates, although other origins have been proposed for them (Rae *et al.*, 1980; Rae, 1981; Dawid and Rebbert, 1981; Roiha and Glover, 1981; Glover, 1981). Type I insertion generates a 7–15 bp TR but no (A)-rich tail, while Type II generates an oligo(A) tail but no TR. Unlike mammalian retroposons, each type has been found at only a single site in the rRNA genes, and there is often a deletion of several nucleotides 5′ to the insert. Copies of Type I elsewhere in the genome appear to have arisen by imprecise excision from the rDNA rather than by retroposition (Roiha *et al.*, 1981), and Type I transcripts also include some rRNA sequence (Long and Dawid, 1979). However, the existence of 3.4 kb transcripts colinear with the Type II inserts (Kidd and Glover, 1981) is consistent with their being retroposons.

A third family of *Drosophila* inserts, called F, is more like the mammalian retroposons. F sequences have inserted into a Type I insert and elsewhere, and they generate a poly(A) tail and TR of 10–13 bp (Dawid *et al.*, 1981; DiNocera *et al.*, 1983). A member of a fourth family, called G, has inserted in turn into an F insert within a Type I insert (DiNocera and Dawid, 1983); it also has a poly(A) tail, but with a short deletion at the insertion site. Denison and Weiner (1982) and DiNocera *et al.* (1983) have discussed the similarities of these four families to snRNA pseudogenes and other retroposons.

An apparent retroposon has also been found in a unicellular trypanosome, where some rRNA genes contain an insertion of two tandem copies of a protein-coding, poly(A)-tailed sequence, with a 7-bp TR (Hasan *et al.*, 1984).

VII. Mechanism of Retroposon Insertion

A. Sequences Generated at the Insertion Site

The insertion of a retroposon has never been seen directly, with the possible exception of an *Alu* insertion into SV40 (Dhruva *et al.*, 1980) which, unfortunately, had deletions in target, insert, and junctions. The belief that retroposons are insertions rests not on direct observation, but on their locations in the genome.

Processed pseudogenes are found on different chromosomes from the parent genes, and *Alu* units are scattered throughout the genome. There are many cases where a continuous sequence in one locus is interrupted by a retroposon with TR in a homolog, probably representing insertion rather than deletion, and a few cases where the insertion is allelic; several of these are included in Figs. 7, 10, and 11. Most conclusively, several retroposons have now been found inserted into copies of known repeated sequences: *Drosophila* F and mouse B2 into rDNA-associated repeats (Dawid *et al.*, 1981; Kominami *et al.*, 1983), *Alu* and *Kpn*I into α-satellite (Grimaldi and Singer, 1982; Thayer and Singer, 1983), and MIF-1 into copies of itself (twice) (Gebhard and Zachau, 1983a). Other specimens, of *Alu* (twice) and R.dre.1, have been found within processed pseudogenes (Lemischka and Sharp, 1982; A. R. Macleod, personal communication).

These nine cases show clearly that the TR is generated on insertion of the retroposon. In each case it is a precise duplication of the original target sequence. In eight of these cases it is 13–19 bp long (average 14.4 bp), and the ninth, where it appears to be 5 bp, may have deleted a repeated oligonucleotide within an original 16 bp TR.

Similar TR lengths are found in all classes of retroposon (Table VI). The overall average is 13.1 bp. The true average should probably be slightly higher, because there is evidence that most of the few TRs now less than 9 bp long have undergone partial deletions. Most of the 15 retroposons bounded by them have extra sequences at the 5′ or 3′ ends (e.g., Bernstein *et al.*, 1983; Tsukada *et al.*, 1982; Grimaldi *et al.*, 1981; Haynes *et al.*, 1981; Degen *et al.*, 1983), which may be relics of longer TRs, as such extra sequences are rare elsewhere. However, there are a few retroposons precisely flanked by 8-bp TRs (e.g., Fig. 10f), and one by 5-bp TRs (Jagadeeswaran *et al.*, 1982). The longest fully accurate TR is 19 bp (Fig. 10d).

The TRs are often, but not always, (A)-rich, and they often overlap the (A)-rich tail. As will be discussed later, some insertions are actually within preexisting (A)-rich tails. Conversely, some are into potential hairpin structures in the target DNA, notably the "zero" insert in Fig. 8. These characteristics are reminiscent of the (A,T)-rich or symmetrical character of many transposon insertion sites, and may reflect a preference for easily denatured targets. However, there is no consistent homology between the TRs and the inserts, nor among the TRs themselves.

The (A)-rich tails often have a complex structure, with some (though not usually all) of the following elements:

$$\text{TR} - \text{RNA sequence} - A_n - (NA_x)_y - \text{(A)-rich sequence and TR}$$

An analysis of these elements, in 103 published retroposon tails from primates and rodents, is presented in Table VI. Clustered retroposons are excluded but inspection of Fig. 10 shows that they have the same general properties. Most specimens are from processed pseudogenes, snRNA pseudogenes, *Alu*/B1 repeats, and "classical" B2 and R.dre.1 repeats, and these tails have remarkably regular properties. Among them, 45 have A_n, 14 have $(NA_x)_y$, and 18 have both, always with the oligo(A) segment first. Ten of them have some less regular sequence before the TR is reached, but most of these cases may be attributable to partial deletion of the TRs as noted above. [In 2 of the 10, where the TR is of normal length, the extra sequence includes $(GA)_n$ (Duncan *et al.*, 1981; Kvist *et al.*, 1983), and one processed pseudogene has a tail consisting almost entirely of $(GA)_n$ (Uhler *et al.*, 1983), so this may be a rare variety of tail.] More common are the 37 tails whose (A)-rich repetitive structure extends one or more nucleotides into the TR.

$(NA_x)_y$ tails are more common in *Alu* and B1 units and in immunoglobulin pseudogenes than in other mRNA and snRNA pseudogenes and in B2 units, but they can be found in all these classes of retroposon, with N most often being C. The cluster in Fig. 10a demonstrates that some B2 units can have $(CA_x)_y$ tails, and the sequence $(CA_3)_{4-7}$, present in several pseudogenes (Hollis *et al.*, 1982; Battey *et al.*, 1982; Chen *et al.*, 1982; Piechaczyk *et al.*, 1982), is close to the average for all such tails. Artiodactyl repeats were not included in Table VI as they have more complex repetitive tails, but it may be significant that these tails also include the CA dinucleotide (Fig. 7).

In summary, the (A)-rich tails of these retroposon classes are variable in sequence from one insertion to another, but have a well-defined repetitive structure, which seems too regular to be due to random processes. They bear no consistent homology either with the target site sequence, even if it is (A)-rich, or with the donor RNA, even though the latter includes 3′ poly(A) for mRNAs and encoded (A)-rich 3′ ends for B2 units. Therefore, much of the tail must be generated anew in many if not all insertion events with these classes of retroposon. The "zero" insert of Fig. 8 may represent synthesis of such a tail in isolation.

Somewhat different properties are shown by the tails of the other retroposon classes in Table VI: the tail of a variant B2 (Kalb *et al.*, 1983), the five tails of R.dre.1 units which lack the encoded (A)-rich segment (Tani *et al.*, 1983; Schuler *et al.*, 1983; Hurst and Parker, 1983; Dhar *et al*, 1982; Harris *et al.*, 1983), and the 19 tails of LINEs. These tails are more irregular, with A tracts being interspersed frequently but nonperiodically with G and T (or T_n), and hardly ever with C. Thus the categorisation of LINE tails in Table VI is largely arbitrary, especially since the full extent of many of the tails is unknown. However, where the TR is identified, it starts with at least two A residues in all but one case, so that overlap with the irregular tails of LINEs and R.dre.1 units is even more frequent than overlap with the regular tails of other retroposon classes. The numbers are still too small to show clearly whether this is a distinct class of tails implying a distinct mechanism of synthesis.

TABLE VI

ANALYSIS OF SEQUENCES GENERATED UPON RETROPOSON INSERTION[a]

		(A)-rich tails[b]							Terminal repeats[c]			
			Composition			N						
Class	Type	Number of specimens	n	x	y	C,CC	G	T	Number of specimens	Overlap	TR length (average)	TR %A
Processed pseudogenes	A_n	13	6–24	—	—	—	—	—	13	8/13	10–15 (13.1)	39%
	$(NA_x)_y$	1	—	4–8	5	1	—	—				
	A_n–$(NA_x)_y$	2	4	3	5–7	2	—	—				
snRNA pseudogenes	A_n	4	14–18	—	—	—	—	—	12	4/6 (+4/6*)	5–23 (13.0)	46%
	$(NA_x)_y$	2	—	3–9	3–5	1	1	—				
	Irregular	1	6	—	—	—	(+)	+				
Alu and B1 repeats	A_n	20	4–38	—	—	—	—	—	35	16/35	5–21 (12.9)	42%
	$(NA_x)_y$	11	—	2–7	3–13	7	2	2				
	A_n–$(NA_x)_y$	13	4–14	3–6	2–(>11)	6	3	4				
B2 and R.dre.1 repeats	A_n	9	8–30	—	—	—	—	—	15	15/15	9–18 (13.9)	49%
	A_n–$(NA_x)_y$	3	7–14	2–3	2	—	3	—				
	Irregular	5	6–14	—	—	—	+	+				
LINEs	A_n	6	5–16	—	—	—	—	—	8	7/8	6–15 (12.3)	63%
	A_n–$(NA_x)_y$	4	5–7	2–5	2–3	—	—	4				
	$(NA_x)_y$	2	—	2–7	3–4	—	1	1				
	Irregular	7	5–13	—	—	(±)	+	+				

[a](A)-rich tails and terminal repeats are analyzed for all published sequences (late 1983) of single chromosomal retroposons with intact 3′ ends. This excludes (1) retroposons in cDNA clones; (2) retroposons in free circles (which lack TRs); (3) retroposons in fused tandem insertions (where the original tails cannot be disentangled; Fig. 10); (4) the following additional sequences, which appear to have suffered deletions, since they are missing the (A)-rich tail, the TR, and the 3′ end of the retroposon: one processed pseudogene (Nishioka *et al.*, 1980; Vanin *et al.*, 1980), two snRNA pseudogenes (Denison and Weiner, 1982; Bernstein *et al.*, 1983; these may alternatively represent a separate insertion mechanism), and three *Alu*/Bl units (Steinmetz *et al.*, 1981; Seeburg, 1982; Lee *et al.*, 1983; see Table IV); and (5) another *Alu* which has lost its 3′ TR, as compared with an intact duplicated copy of the locus (Hess *et al.*, 1983). The 109 sequences included are as follows: 17 processed pseudogenes (references in text, Section VI,A); 13 snRNA pseudogenes (references in text, Section VI.B); 43 *Alu*/Bl sequences (Bell *et al.*, 1980a; Baralle *et al.*, 1980; Krayev *et al.*, 1980; Duncan *et al.*, 1981; Grimaldi *et al.*, 1981; Pan *et al.*, 1981; Haynes *et al.*, 1981; Ohshima *et al.*, 1981; Ottolenghi and Giglioni, 1982; Tsukada *et al.*, 1982; Shen and Maniatis, 1982; Shen and Smithies, 1982; Jagadeeswaran *et al.*, 1982; Coggins *et al.*, 1982; Young *et al.*, 1982; Monson *et al.*, 1982; Grimaldi and Singer, 1982; Lee *et al.*, 1983; Myers *et al.*, 1983; Degen *et al.*, 1983; Hess *et al.*, 1983; Das *et al.*, 1983; Gebhard and Zachau, 1983b; Kalb *et al.*, 1983; A. R. Macleod, personal communication; and Table IV); 10 B2 sequences (Fig. 6 and Table IV); 7 R.dre.1 sequences (references in text, Section VI,C,3); and 19 LINEs (Gebhard *et al.*, 1982; Lueders and Paterson, 1982; Wilson and Storb, 1983; Gebhard and Zachau, 1983a; DiGiovanni *et al.*, 1983; Lerman *et al.*, 1983; Thayer and Singer, 1983; Potter and Jones, 1983; and Table IV).

[b](A)-rich tails. All the 109 sequences have (A)-rich tails except for the six class IV snRNA pseudogenes (see text). One processed pseudogene was excluded because it has an atypical $(GA)_n$ tail (Uhler *et al.*, 1982). One Bl unit (Krayev *et al.*, 1980) apparently has two tails in tandem: $A_5(CA_{3-5})_7CACC-A_{12}GA_3G(GA_{3-4})_{>10}$. Thus there were 103 tails for analysis. The table does not include the encoded tails of B2 and some R.dre.1 units (Fig. 4), nor the C_n tracts following two Bl units (Coggins *et al.*, 1982; Gebhard and Zachau, 1983b); these are all followed by the variable-sequence tails analyzed here. Several tails have short, more complex sequence preceding the TR, as discussed in the text. N denotes the major interspersed nucleotide; occasional base substitutions are ignored. The tail lengths include any overlap with the TR. For "irregular" tails, *n* denotes the length of the longest oligo(A) stretch, and G and T are the major non-A nucleotides.

[c]Terminal repeats. Of the 88/109 cases where both the 5′ and 3′ ends have been completely sequenced, all have TRs except two processed pseudogenes (Ueda *et al.*, 1982; Scarpulla and Wu, 1983) and one poly(A)-terminated snRNA pseudogene (Denison and Weiner, 1982). Two *Alu* units were also excluded whose TRs of 5 and 7 bp may reflect deletion within repetitious 16-bp TRs (Shen and Maniatis, 1982; A. R. Macleod, personal communication). Thus there were 83 TRs for analysis. "Overlap" denotes the number of sequences with one or more nucleotides overlap between the TR and the (A)-rich tail [including oligo(A) in irregular tails], plus (*) overlap with snRNA sequence in the absence of a tail.

B. Models for Synthesis and Insertion

Presumably the reverse transcription and insertion mechanism is a normal cell function. Retroposons obviously cannot encode it themselves, and although the evidence so far only requires these insertions to appear on an evolutionary timescale, they are too abundant and too uniform to be aberrations. The mechanism must also be remarkably faithful to copy the first nucleotide of an RNA, even when it is capped.

One puzzling problem is how cDNA synthesis is primed. For *Alu* units, the transcripts extend past the (A)-rich segment into flanking sequence, and end with oligo(U), the usual RNA polymerase III terminator sequence. Jagadeeswaran *et al.*, (1981) therefore suggested that this could basepair with the (A)-rich segment and thus self-prime cDNA synthesis. For processed pseudogenes, derived from polyadenylated RNAs, the primer would have to be a separate (U)- or (T)-rich molecule. In many cases the (A)-rich insertion site could suffice (Moos and Gallwitz, 1983). For snRNA pseudogenes, which are derived from RNAs with neither a transcribed nor an added (A)-rich sequence, at least two different priming mechanisms seem to be required—one to produce complete or nearly complete pseudogenes with typical (A)-rich tails (classes II and III of Denison and Weiner, 1982), and another to produce 3′-truncated pseudogenes without (A)-rich tails (class IV). Class IV pseudogenes for U3 correspond to a cDNA fragment that can be made by self-primed reverse transcription *in vitro* (Bernstein *et al.*, 1983), and thus may require no other primer. This has not been demonstrated for other snRNAs, and for U2 and U4 (and possibly some U3) pseudogenes it may be significant that the truncated 3′ end of the snRNA overlaps the TR by 1–6 nucleotides (see Section VI,B). In these cases, possibly the nicked target DNA primed directly onto a short (A)-rich homology in the snRNA.

Thus various priming mechanisms can be envisaged, but none of those mentioned can be universal. Models proposed by Van Arsdell *et al.* (1981) and Jagadeeswaran *et al.* (1981) would use some of these priming mechanisms to make a cDNA strand on a free RNA. Nonspecific ligation processes (see Section VI,F) are then invoked to insert the cDNA at a staggered nick in the chromosome. Van Arsdell and Weiner (1984) propose a model in which the insertion is achieved by interrupting a nicking-closing reaction in the chromosome.

These scenarios do not account for the very regular nature of the variable repetitive tails in most classes of retroposon. The synthesis of the tail, particularly of the common $(CA_x)_y$ type, may be analogous to several other processes in which cleavage of RNA or DNA occurs in the vicinity of a CA dinucleotide and is followed by amplification or ligation of the free end (Rogers, 1983b). A particularly intriguing parallel is the generation of new chromosome termini in ciliate protozoa (King and Yao, 1982; Boswell *et al.*, 1982), which involves cleavage and tandem amplification of the sequences C_4A_2 or C_4A_4. Although other species have different telomere sequences (see references in Rogers, 1983b), *Tetrahymena* termini can function as telomeres in yeast, maintaining the same pattern of nicks following C_4A_2 sequences as in the donor, which suggests that they are involved in universal cell processes (Szostak and Blackburn, 1982; Cavalier-Smith, 1983).

One can devise two models for retroposon insertion which include telomere-like roles for CA sequences, based on 3′ terminal addition either of $(CA_x)_y$ as in *Tetrahymena* (King and Yao, 1982), or of its complement as in the ciliate *Oxytricha* (Boswell *et al.*, 1982). In each case the TR is presumed to be created by a staggered nick at the insertion site. Both models include some features that seem arbitrary at present, but that is to be expected if retroposons make use of a system that is adapted for some specific but unknown cell function.

The first model simply adapts those described above which invoke a variety of priming mechanisms, some unknown, to make a cDNA strand on a free RNA. An (A)-rich or (CA)-rich tail would then be added in place of the former 3′ end of the RNA, either onto RNA or onto a second cDNA strand. The (A)-rich tail might confer telomeric properties which would somehow enhance the survival of the molecule, and/or might promote its integration. Evidently, several of these steps will remain ambiguous until they can be detected experimentally.

The second model (Fig. 9) combines the priming, tailing, and insertion mechanisms into a single process which could be looked for experimentally. In this model, the first event is the 3′ tailing of a nick in chromosomal DNA with an oligo(T), (T)-rich, or (TG)-rich sequence. This would act as a primer for cDNA synthesis on the 3′ poly(A) of mRNA or on the transcribed (A)-rich tail of *Alu*-like RNA. For tailless snRNAs, as mentioned above, the nicked DNA strand could prime cDNA synthesis internally without any tail synthesis being required.

In order to ensure a complete copy of the RNA, this model separately proposes a mechanism to create a second nick in the target DNA and ligate the (decapped) 5′ end of the RNA to it (Fig. 9). The RNA cap does in fact resemble the high-energy intermediate, A(5′)pp(5′)N..., used by known DNA and RNA ligases (Lehman, 1974; Ohtsuka *et al.*, 1976; see Section II). The DNA-snRNA linkages reported by Pederson and Bhorjee (1979) may provide a precedent for such a ligation mechanism. Presumably, uncapped RNAs or 3′ fragments can also be inserted. The insertion of the retroposon would be completed by general repair processes.

The atypical nature of some tails is a problem for all these models. Tails of artiodactyl repeats might reflect artiodactyl-specific telomere sequences. The irregular tails of LINEs and divergent R.dre.1 units are also puzzling, but their near-universal overlap with their TRs may suggest frequent direct priming from the target site. The snRNA pseudogenes are particularly diverse, and different mechanisms may be responsible for some of them.

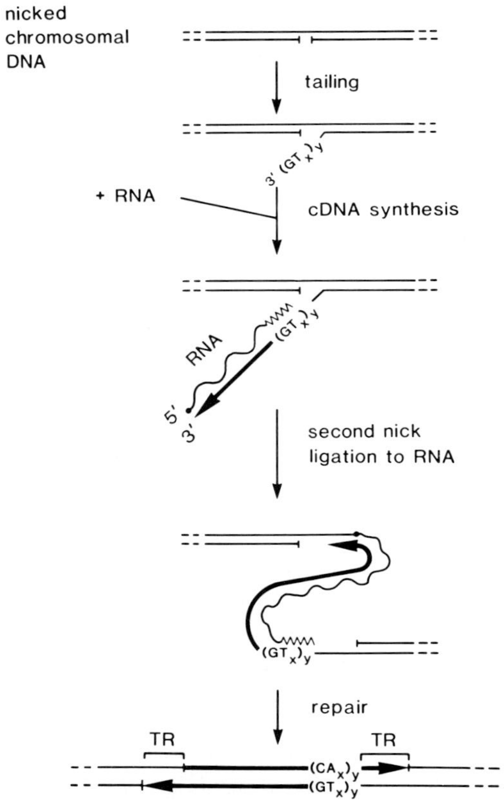

FIG. 9. A model for retroposon insertion. The wavy line represents RNA, and the thick line new DNA. The initial polymerization must be variable: $(GT_x)_y$ as shown here, or T_n, or $(NT_x)_y$. For clarity, the second nick and its ligation to the RNA are shown as occurring after the first nick and cDNA synthesis, but they could occur concurrently.

It is not known at what stage in the life cycle retroposons are created. The early embryo seems a likely candidate, since this is when mouse B2 sequences, which have produced many retroposons, are abundantly transcribed (Murphy *et al.*, 1983; Bennett *et al.*, 1984), while class I histocompatibility genes, which have produced no retroposons (Winoto *et al.*, 1983), are transcribed weakly if at all.

The source of the reverse transcriptase, which must be active in the germline, is unknown. One possible source (Sharp, 1983), at least in mice, is the IAP quasi-proviruses, which produce particles containing reverse transcriptase (Wilson and Kuff, 1972) and are particularly active in early embryos (Kelly and Condamine, 1982). An alternative possibility, implicit in the model of Fig. 9, is that retroposon synthesis is a specialized process performed by a set of enzymes that are used in some unknown but similar process in the normal life cycle.

VIII. The Clustering of Retroposons

From renaturation analyses of rodent genomes, several groups concluded that most short interspersed repeats occur in pairs or small mixed clusters (Bonner *et al.*, 1973; Cech and Hearst, 1976; Moyzis *et al.*, 1981). Indeed, several authors now have pointed out the presence of multiple retroposons in cloned DNA which they have sequenced. Reexamination of these clusters shows that retroposons have a tendency to integrate either adjacent to one another, or directly into each others' tails (Fig. 10).

One of the most obvious of these clusters occurs in an intron of the rat growth hormone gene (Fig. 10a) (Barta *et al.*, 1981; Page *et al.*, 1981). The whole cluster is enclosed by a single TR. In a Hooded rat clone, the cluster is a tandem pair of B2 sequences, each with a tail beginning A_xCA_y; probably the second inserted into the tail of the first. In a Sprague–Dawley rat clone, the sequences are virtually identical except that a R.dre.1 unit, tailed with A_{17}, is inserted into the $(CA_x)_y$ tail of the second B2. Although the $(CA_x)_y$ sequences give the appearance of a TR, they have not actually been duplicated in the R.dre.1 insertion, and the same may be true of the other tail insertion events listed below.

A second cluster (Ohshima *et al.*, 1981) (Fig. 10b) includes a pseudogene for U6 snRNA. The U6 insertion site can now be recognized as the (A)-rich tail of a B2 sequence. A single TR bounds these two retroposons. The pair within their TRs lie three nucleotides upstream from a B1 sequence.

A different style of clustering is shown in Fig. 10c. This cluster (Kalb *et al.*, 1983) consists of a B1 and a variant B2-related sequence with separate but adjacent TRs, and a standard B2 unit with its own 13-bp TRs lying 64 bp downstream.

Figures 10d–j show more tandem pairs in which a second retroposon is inserted into the tail of the first (see Caption for details, and also Fig. 7b). In several cases, the TR is a duplication of a homologous, unoccupied site. One such case (Fig. 10e) is a single insertion in the H-2I-A^b_β gene, which is absent from the A^d_β allele. This one has a spectacular 231-bp double tail, including a long stretch in which the G/A ratio progressively increases, as noted generally in Section VII,A. This stretch may be a third independent element in the composite, since a remarkably similar stretch (which is not flanked by TRs nor by any known retroposon) was sequenced by Höchtl and Zachau (1983):

$$CA_3(GA_3)_6{**}(GA)_{29}{*}(G_3A)_8(GGAA)_{14}$$

[Asterisks indicate individual A-G substitutions within the repeating tracts.]

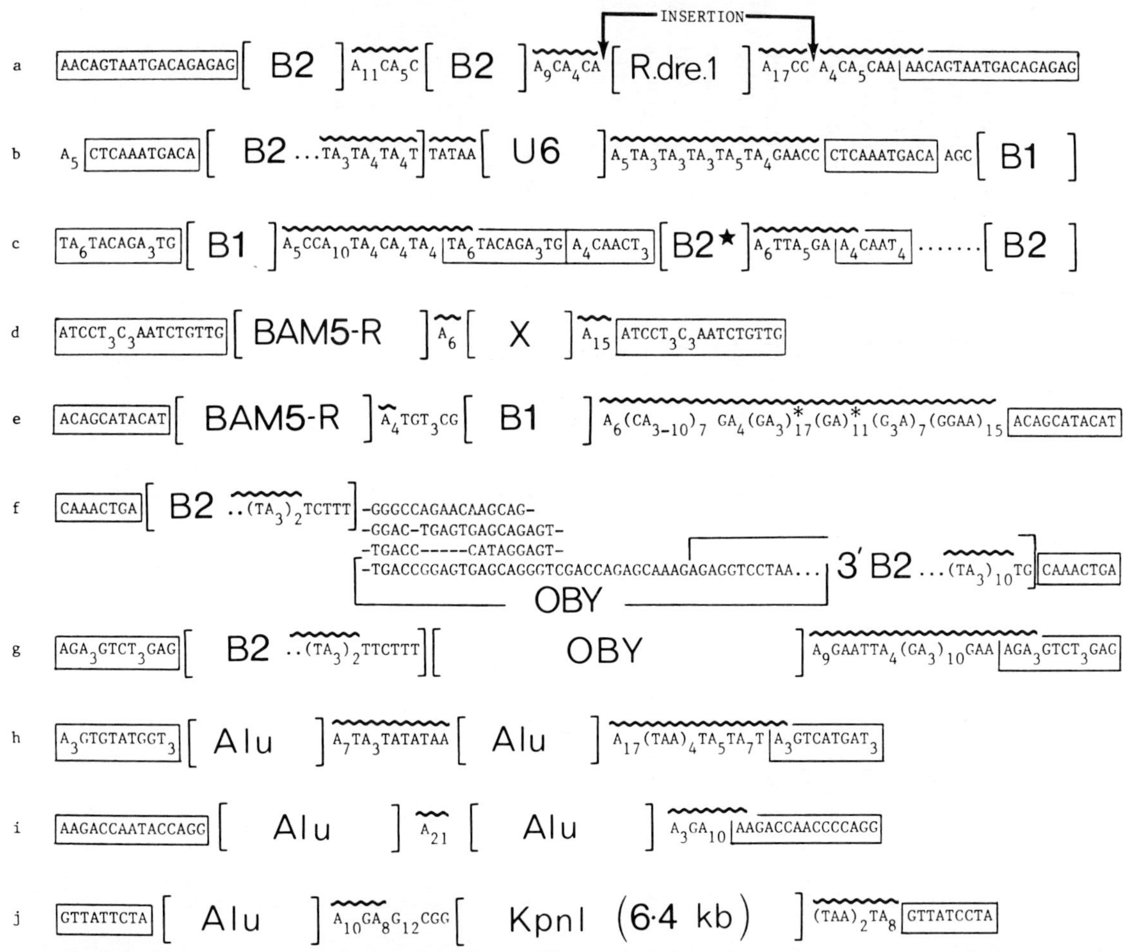

a AACAGTAATGACAGAGAG [B2] A11CA5C [B2] A9CA4CA [R.dre.1] A17CC A4CA5CAA AACAGTAATGACAGAGAG
INSERTION
b A5 CTCAAATGACA [B2 ...TA3TA4TA4T] TATAA [U6] A5TA3TA3TA3TA5TA4GAACC CTCAAATGACA AGC [B1]
c TA6TACAGA3TG [B1] A5CCA10TA4CA4TA4 TA6TACAGA3TG A4CAACT3 [B2★] A6TTA5GA A4CAAT4 ...... [B2]
d ATCCT3C3AATCTGTTG [BAM5-R] A6 [X] A15 ATCCT3C3AATCTGTTG
e ACAGCATACAT [BAM5-R] A4TGT3CG [B1] A6(CA3-10)7 GA4(GA3)*17(GA)*11(G3A)7(GGAA)15 ACAGCATACAT
f CAAACTGA [B2 ..(TA3)2TCTTT]
-GGGCCAGAACAAGCAG-
-GGAC-TGAGTGAGCAGAGT-
-TGACC-----CATAGGAGT-
-TGACCGGAGTGAGCAGGGTCGACCAGAGCAAAGAGAGGTCCTAA... 3' B2 ...(TA3)10TG] CAAACTGA
OBY
g AGA3GTCT3GAG [B2 ..(TA3)2TTCTTT][OBY] A9GAATTA4(GA3)10GAA AGA3GTCT3GAC
h A3GTGTATGGT3 [Alu] A7TA3TATATAA [Alu] A17(TAA)4TA5TA7T A3GTCATGAT3
i AAGACCAATACCAGG [Alu] A21 [Alu] A3GA10 AAGACCAACCCCAGG
j GTTATTCTA [Alu] A10GA8G12CGG [KpnI (6·4 kb)] (TAA)2TA8 GTTATCCTA

Insertion of one retroposon into another's tail is thus very common. As other possible examples, a B1 unit 5′ to the H-2-like 27.1 gene (Table IV) is inserted into a long $(CA_{3-5})_n$ sequence; one R sequence is inserted into $(TA_3)_n$ repeats (Wilson and Storb, 1983); and a mouse B2 and several human *Alu* units are inserted into $(NA_x)_y$-like sequences, some of which could be retroposon tails. A mouse B1 with, apparently, two tandem tails (see footnote to Table VI) may be a relic of a similar event. The processed pseudogene for human Ig C_ϵ, whose abnormal 5′ exon has been mapped by Ueda *et al.* (1982), also may have been inserted into a poly(A) tract, without producing a substantial TR. [An actin processed pseudogene lies 3′ to a 230-bp $(GA_x)_y$ sequence (Moos and Gallwitz, 1983), which is an *Alu* tail, but it ends ~ 60 bp from the pseudogene.]

The fused dimers and multimers in Fig. 10 are all flanked by typical TRs, which, if successive insertions were involved, are presumed to be the TRs generated by the first insertion. TRs can also be defined flanking the second insert, simply because both components have (A)-rich tails; however, these TRs are often short or imprecise. In particular, the polymorphic R.dre.1 insertion into a B2 tail (Fig. 10a) does not seem to have generated any new duplication. This could of course be because TRs decay rapidly within (A)-rich tracts, or because

FIG. 10. Examples of superposed or adjacent retroposons. These are from published sequences, but in most cases the identities and/or boundaries of the elements have been revised or newly defined according to improved knowledge of the retroposon families concerned. (A)-rich tails are zigzag-overlined and TRs are boxed. [Additional TRs could be defined within the (A)-rich tails.] B2 and R.dre.1 units include their encoded (A)-rich tails (Fig. 4), or variants of them where shown. (a) Cluster from the plus strand of an intron of the growth hormone gene of a Sprague–Dawley rat (Barta *et al.*, 1981). In a Hooded rat gene (Page *et al.*, 1981), the sequence between the arrows is absent. (b) Cluster including a U6 pseudogene (Ohshima *et al.*, 1981). The first 7 bp of the B1 unit are missing, and sequences 3′ to the B1 were not published. (c) Cluster from Kalb *et al.* (1983) including a B2 variant (B2*) with a novel 3′ segment, whose 3′ end is here defined arbitrarily. A B2 unit flanked by separate 13-bp TRs lies 64 bp downstream. (d) Tandem pair consisting of BAM5-R (5′-truncated and internally duplicated) with a 300-bp sequence of unknown affinities (X) apparently inserted into its tail (Gebhard *et al.*, 1982). This whole composite is inserted into a MIF-1 sequence (H. Zachau, personal communication); the TR duplicates the known sequence (Brown, 1983). (e) Tandem pair consisting of BAM5-R (5′-truncated) with a B1 sequence (Larhammar *et al.*, 1983; Table IV). This is inserted on the minus strand of an intron in the mouse H-2I-A_β^b; the TR duplicates the unoccupied sequence in the allelic A_β^d gene (Malissen *et al.*, 1983). In the long tail, which may include a third mobile element (see text), asterisks indicate single A–G substitutions. (f) Tandem pair of B2 and OBY. The OBY sequence has irregular duplications at the 5′ end, and is fused to a B2 sequence at the 3′ end, which has an internal deletion (not marked). This composite is inserted on the minus strand of an intron in H-2 gene 27.1 (Steinmetz *et al.*, 1981; see Fig. 11); the TR duplicates the unoccupied site of homologous genes. (g) Tandem pair of B2 and OBY, from Ou *et al.* (1983). (h) Tandem pair of *Alu* units. This is an insertion in one duplicated unit in the human α-globin cluster; the TR represents the site which is unoccupied in the other duplication unit (Hess *et al.*, 1983). (i) Tandem pair of *Alu* units, from the human prothrombin gene (Degen *et al.*, 1983). (j) Tandem pair of *Alu* and full-length *Kpn*I, from the human β-globin gene cluster (Miyake *et al.*, 1983).

most clusters have lost their original TRs in undergoing multiple retropositions as a single unit. However, it seems possible that insertion into an oligo(A) tract does not generate the usual TR. This would increase the similarity between insertion of retroposons and of some minimal inserts (Section VI,F).

The R.dre.1 polymorphism in Fig. 10a is the only case where there is clear evidence that the insertion of the final element occurred at the present location. For all the other tandem pairs, it is possible that the second insertion occurred elsewhere and the pair has been retroposed as a single unit. For the (BAM5-R)-B1 pair in Fig. 10e, this is unlikely, since B1 sequences are not generally associated with MIF-BAM5-R sequences (Bennett and Hastie, 1983). For the B2-OBY units in Fig. 10f,g, the coupling probably did take place elsewhere, as discussed in Section VI,C,4. Some consensus sequences are themselves fused dimers, such as primate *Alu* and artiodactyl A1-A2.

It is interesting to ask whether the human fused dimers, *Alu–Alu* and *Alu–KpnI*, retropose as single units. From the renaturation of genomic DNA, Fox *et al.* (1982) estimated that as much as one-third of all *Alu* members might exist as fused *Alu–Alu* dimers. However, only two have been discovered so far (Fig. 10h,i), and these probably represent separate fusions, as they do not have similar internal linkers, nor similar mutations. The same is likely to be true for the two *Alu–KpnI* dimers sequenced by Miyake *et al.* (1983) (see Section VI,D), but since these are the first two full-length *KpnI* units studied, there may be many more such pairs.

Retroposons also seem to insert in pairs with precisely adjacent TRs, although these examples are less clear. The clearest are the mouse sequences of Fig. 10c, and a pair of human *Alu* sequences inserted into the middle of a tropomyosin processed pseudogene (A. R. Macleod, personal communication). A case in which a U6 pseudogene abuts a B1 unit (Fig. 10b), and two similar cases (Ohshima *et al.*, 1981; Fukumaki *et al.*, 1983) may be comparable, but unfortunately the TR sequences of these adjacent units were not published. Also, there are two *Alu* sequences almost adjacent in an intron (Lee *et al.*, 1983; Table IV), but the more 5′ of these has diverged and lost any TRs which it once had.

These cases imply a degree of site preference in retroposon insertion, possibly because the preexisting TRs and (A)-rich tails are sites particularly prone to random nicking. Such a site preference may incidentally reduce the frequency of deleterious insertions into coding sequences.

There may also be a more general, regional preference in retroposon insertion. This is more difficult to establish, since *Alu*-like units or LINEs occur every few kilobases on average throughout the whole genome, but it is suggested by several cases where homologous genes have different retroposons in similar positions. These cases are the different B1 units in two *H-2* genes (Fig. 11), several species-specific *Alu*-like units in and around the human and bovine POMC genes (Tsukada *et al.*, 1982; Watanabe *et al.*, 1982; Takahashi *et al.*, 1983), and the *Alu* and B2 units just 3′ to the human and rat growth hormone genes (DeNoto *et*

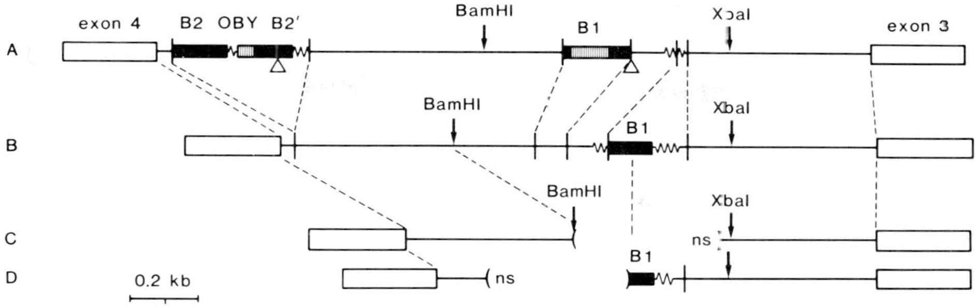

FIG. 11. Retroposons in the third intron of mouse H-2 genes. The histocompatibility class I genes are interspersed with numerous repeated sequences, both in the pig (Singer *et al.*, 1983b) and in the mouse. While the density of repeats may be no higher than in other gene families, the histocompatibility complex offers unusual opportunities for comparing retroposons in homologous sites. An analysis of the long intron of several H-2 genes is presented here. Moore *et al.* (1982) pointed out the presence of various B1 and B2 sequences in this intron, but their arrangement has not previously been properly defined. This figure compares the minus strands of gene 27.1 (A: Steinmetz *et al.*, 1981), the K^d gene (B: Kvist *et al.*, 1983), and the partially sequenced L^d genes (C: Evans *et al.*, 1982; D: Moore *et al.*, 1982) which are very similar to K^d except for a deletion of 0.3–0.4 kb in the unsequenced portion ("ns"). The introns of 27.1 and K^d can be aligned well except for two divergent blocks. One of these is the insertion in Fig. 10f. The other apparently represents independent insertions of B1 into the two genes at points 190 bp apart, with a ~ 90-bp deletion covering the 3′ junction in 27.1, and also with expansion/contraction in the somewhat repetitious flanking regions. In K^d, the B1 is inserted into a preexisting (A)-rich segment which may be an ancient B1 tail. In 27.1, the B1 has an SV40-homologous hexanucleotide (GAGGCA) amplified 17-fold. [Gene 27.1 also has a B1 in the 5′ flanking region (Table IV). The L^d (but not K^d) gene also carries a B2 in the 3′ untranslated segment, as do some other H-2 genes (Steinmetz *et al.*, 1981; Lalanne *et al.*, 1982; Reyes *et al.*, 1982; Brickell *et al.*, 1983).]

al., 1981; Table IV). In each case the retroposons are in the same orientation in the homologous genes, and their insertion sites are within $\lesssim$ 0.1 kb of each other, but do not correspond exactly. There are many other cases where *Alu*-like units have been detected close to, although not adjacent to, each other. These may be regions which for an unknown reason are particularly favorable for accumulation of retroposons, and perhaps of other dispersed repeats such as IAPs (Lueders and Paterson, 1982; Lueders *et al.*, 1982; Cole *et al.*, 1982).

IX. Evolutionary Possibilities

A. Relation to Functional RNA Species

Section VI,C described the hypothesis that *Alu*-like retroposons are polymerase III pseudogenes, selected only for their efficiency at undergoing multiple rounds of retroposition, and propagating as "selfish DNA." The origins proposed for several *Alu*-like retroposons share common features: acquisition by an existing gene or retroposon of an efficient promoter for polymerase III transcription and an efficient priming site for reverse transcription. Thus it was proposed that *Alu* evolved from 7 SL by acquiring an "improved" promoter in the 5′ part, and an (A)-rich priming site exactly at the 3′ end. The artiodactyl C-A3 unit may have originated by fusing a tRNA gene with its promoter to an A retroposon with its 3′ repetitive tail. *Alu–Kpn*I dimers may similarly depend on the *Alu* promoter. Experimental dissection of these composites could test these proposals.

But this picture of one-way evolution, from gene to retroposon to other retroposons, may be too simple. Ullu *et al.* (1982b) showed that the "insert" part of 7 SL RNA is indeed more conserved than the "*Alu*" part in 7 SL RNAs of chickens and frogs. On the other hand, many *Alu*-hybridizing sequences without the 7 SL "insert" are present in mrDNA, not only in mammals but also in chickens, frogs, sea urchins, some insects, and even some plants (Ullu *et al.*, 1982b; Blin *et al.*, 1983). So although 7 SL is the only abundant, widely conserved RNA to which *Alu* hybridizes, and the *Alu* and B1 retroposons do seem to be divergent variants of the much more conserved 7 SL sequence in mammals, it seems that *Alu*-related sequence has been maintained in evolution independently of the 7 SL "insert"—which, therefore, could indeed have been introduced into a preexisting *Alu*-related sequence.

The 4.5 S RNA also suggests two-way evolution between retroposons and discrete polymerase III transcripts. To summarize Section VI,C,3, this RNA is found in rodent cells (but not primate cells), and is homologous to B1 (more so than to 7 SL or *Alu*) but shorter than B1. It therefore must have arisen from a B1 sequence, probably a retroposon. This could easily have occurred by the appearance of a termination signal, oligo(T), within the B1 sequence. However, 4.5 S RNA is much more conserved between mouse and hamster than is the consensus B1 sequence. Therefore, 4.5 S appears to be a conserved (functional?) RNA which evolved from a nonconserved (selfish?) repetitive family.

The distinction between functional and selfish transcription units may thus be blurred. There is no reason why some polymerase III retroposons could not encode functional RNAs; it is their rapid sequence divergence that implies that they mostly do not.

B. Evolution of Retroposon Families

As described in Section VI,C, some of the *Alu*-like superfamilies have divided into several families and subfamilies, sometimes by scattered base substitutions, and sometimes by structural rearrangements. This section will summarize the evolutionary mechanisms that have been proposed and others that may be possible.

In a proliferating family of retroposons, the individual members may diverge rapidly because there is no proofreading in transcription, and perhaps not in reverse transcription. Given such divergence, at least three processes could give rise to subfamilies. One is simply the balance between creation and loss of copies, if most of the supply comes from only a few genes (Sharp, 1983). The second is gene conversion (Munz *et al.*, 1982; Fig. 12a), which can set up a new subfamily if many active copies are preferentially converted (Dover, 1982), or if one very active copy is stochastically converted. (On the other hand, gene conversion acting more democratically can mix up subfamilies.) The third process is tandem insertion followed by homologous recombination, leaving a single hybrid copy (Fig. 12b).

The structural variations occur mostly at the 3′ ends, and could involve any type of insertion or deletion, so long as a priming site is provided. Replacement of the 5′ end, as in OBY, could arise by insertion of a 5′-truncated copy adjacent to any sequence resembling an "A" box, as in the case of one *Alu* unit (Paolella *et al.*, 1983). The second galago Alu family, G-*Alu*.B (Daniels and Deininger, 1983; Section VI,C,2) may have arisen by insertion of a G retroposon into the (A)-rich linker of an *Alu* dimer. The frequency of retroposon clustering also opens up the possibility of nonhomologous fusions between different sequences (Fig. 12c). Tandem insertion is also the likely origin for dimeric retroposons (Section VIII); and two retroposons separated by some host sequence might also be able to form a single unit, transferring the sequence between them (Jagadeeswaran *et al.*, 1981; Haynes *et al.*, 1981).

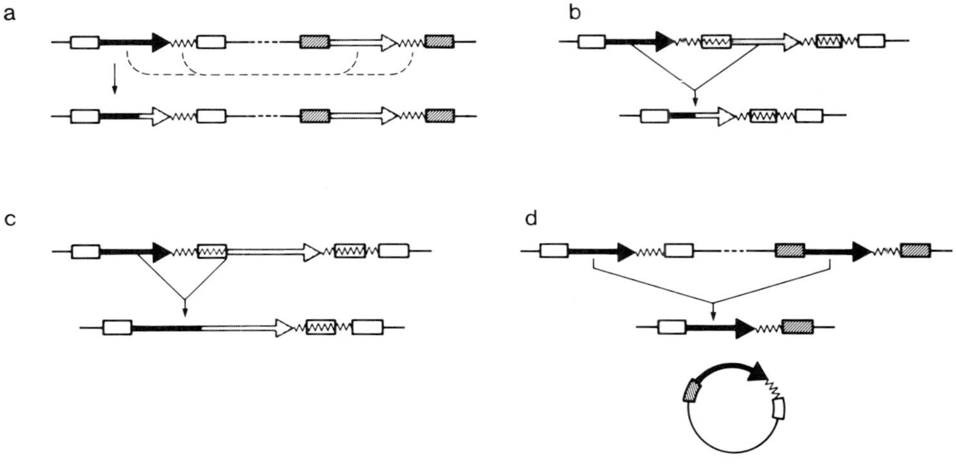

FIG. 12. Possible evolutionary interactions between pairs of retroposons. Symbols are as in Fig. 1. (a) Gene conversion; (b) homologous fusion; (c) nonhomologous fusion; (d) excision of free circles.

C. Evolution in Chromosomal DNA

What is the fate of a retroposon following integration? Retroposons probably cannot transpose as DNA (see Section V). In fact, three inherited deletions in the human globin gene clusters, which are terminated at *Alu* units, have breakpoints not at the ends of the units, but inside them (Orkin and Michelson, 1980; Jagadeeswaran *et al.*, 1982; Ottolenghi and Giglioni, 1982).

The abundant *Alu*-like and LINE sequences might be hotspots for recombinational rearrangements, as has been suggested for the growth hormone gene cluster (Barsh *et al.*, 1983) and for a highly labile *Alu*-rich cluster (Calabretta *et al.*, 1982). It has not been shown that homologous crossing-over occurred in these clusters; indeed, homologous but unequal crossing-over between dispersed repeats could be disastrous. The globin deletions cited above seem to have occurred between *Alu* and nonhomologous sequences. However, homologous recombination may be the origin of free *Alu*-containing circles (Haynes *et al.*, 1981; Fig. 12d).

These *Alu*-containing circles exist in monkey and human cells, and some of them hybridise to an *Alu*-rich chromosomal region (Calabretta *et al.*, 1982). An origin by recombination between chromosomal *Alu* sequences is supported by the absence of TRs from several *Alu* units in circles, and one such circle is indeed a circularly permuted copy of a chromosomal sequence (Krolewski *et al.*, 1982, 1984). Some of the smallest circles are simply circularized *Alu* units (Krolewski and Rush, 1984), which were postulated to be reverse transcription intermediates, but could also arise by recombination between TRs or within tandem inserts. These circles could perhaps reintegrate by recombination with another *Alu* sequence elsewhere.

Gene conversion may act among dispersed repeated sequences, as mentioned above. It is particularly plausible as a mechanism for the concerted evolution of LINE subfamilies, without gross changes in copy number, in closely related species (Brown and Dover, 1981; Dover, 1982; Rogers, 1985). Gene conversion could also operate on the parent genes of snRNA pseudogenes (Bernstein *et al.*, 1983) or of processed mRNA pseudogenes, and this may be the only contribution of processed pseudogenes to evolution. This could explain the removal of one intron from the rat insulin I gene (Lomedico *et al.*, 1979; Bell *et al.*, 1980b), and possibly longer-term removal of introns in the actin genes (Fornwald *et al.*, 1983), a calmodulin gene (Stein *et al.*, 1983), and the ovalbumin and α_1-antitrypsin genes (Leicht *et al.*, 1982). Otherwise, there seems to be nothing that processed pseudogenes can do except lie and rot.

It is not clear whether *Alu*-like retroposons are ever removed, since there is so far no evidence that they existed before the mammalian radiation. However, lower animals contain other mrDNA families which must undergo evolutionary turnover (Bouchard, 1982). If *Alu*-like retroposons are removed, a major mechanism could be deletion between the TRs. A typical repeat of 13 bp, 100–300 bp apart, should lead to a significant deletion frequency (Albertini *et al.*, 1982).

D. Conclusions

We have seen that retroposons are major constituents of mammalian genomes. In rodents and primates, up to 10% of the genome may be composed of them—mostly the self-propagating polymerase III retroposons of the *Alu* type, plus the still-mysterious LINEs, but also smaller families representing hundreds or perhaps thousands of other genes. Most if not all are likely to be nonfunctional or "selfish" DNA. Their insertions and rearrangements are a major factor in the structural evolution of mammalian genomes, which behave like a turbulent sea within which the conserved gene segments are islands of stability.

Curiously, although other eukaryotes have rare retroposon-like elements, it is only in mammals that they have become the predominant dispersed repetitive sequences. Indeed, they may still be proliferating in our genomes. The reason for this is unknown, and there will be great interest in elucidating the mechanism that creates them. Even if retroposons themselves do not have any function for the organism, it seems possible that the retroposition mechanism does, and that reverse transcription will join RNA splicing as an unexpected but vital process in mammalian physiology.

Acknowledgments

I am particularly grateful to all the authors whose DNA sequences have provided the raw material for this article, and to those who have furnished data prior to publication. I also thank Randolph Wall, for encouraging my interest in RNA splicing; John Abelson, Stephen Mount, and Joan Steitz, for useful tabulations; Eric Ackerman, Sandy Macleod, Deborah Wilde, and David Zimmern, for valuable discussions; Rodger Staden and Jon Karn, for instruction in running computer programs; and Sydney Brenner and the Medical Research Council for support.

References to Part 2

Ackerman, E. (1983). *EMBO J.* **2,** 1417–1422.

Adams, J. W., Kaufman, R. E., Kretschmer, P. J., Harrison, M., and Nienhuis, A. W. (1980). *Nucleic Acids Res.* **8,** 6113–6128.

Abertini, A. M., Hofer, M., Calos, M. P., and Miller, J. H. (1982). *Cell* **29,** 319–328.

Allan, A., and Paul, J. (1984). *Nucleic Acids Res.* **12,** 1193–1200.

Balmain, A., Krumlauf, R., Vass, J. K., and Birnie, G. D. (1982). *Nucleic Acids Res.* **10,** 4259–4277.

Baralle, F. E., Shoulders, C. C., Goodbourn, S., Jeffreys, A., and Proudfoot, N. J. (1980). *Nucleic Acids Res.* **8,** 4393–4404.
Barrie, P. A., Jeffreys, A. J., and Scott, A. F. (1981). *J. Mol. Biol.* **149,** 319–336.
Barsh, G. S., Seeburg, P. H., and Gelinas, R. E. (1983). *Nucleic Acids Res.* **11,** 3939–3958.
Barta, A., Richards, R. I., Baxter, J. D., and Shine, J. (1981). *Proc. Natl. Acad. Sci. U.S.A.* **78,** 4867–4871.
Battey, J., Max, E. E., McBride, W. O., Swan, D., and Leder, P. (1982). *Proc. Natl. Acad. Sci. U.S.A.* **79,** 5956–5960.
Bell, G. I., Pictet, R., and Rutter, W. J. (1980a). *Nucleic Acids Res.* **8,** 4091–4109.
Bell, G. I., Pictet, R. L., Rutter, W. J., Cordell, B., Tischer, E., and Goodman, H. M. (1980b). *Nature (London)* **284,** 26–32.
Bennet, K. L., and Hastie, N. D. (1984). *EMBO J.* **3,** 467–472.
Bennett, K. L., Hill, R. E., Pietras, D. F., Woodworth-Gutai, M., Kane-Haas, C., Houston, J. M., Heath, J. K., and Hastie, N. D. (1984). *Mol. Cell. Biol.* (in press).
Bernstein, L. B., Mount, S. M., and Weiner, A. M. (1983). *Cell* **32,** 461–472.
Blin, N., Weber, T., and Alonso, A. (1983). *Nucleic Acids Res.* **11,** 1375–1387.
Bogenhagen, D. F., and Brown, D. D. (1981). *Cell* **24,** 261–270.
Bonner, J., Garrard, J., Gottesfeld, J., Holmes, D. S., Sevall, J. S., and Wilkes, M. (1973). *Cold Spring Harbor Symp. Quant. Biol.* **38,** 303–310.
Boswell, R. E., Klobutcher, L. A., and Prescott, D. M. (1982). *Proc. Natl. Acad. Sci. U.S.A.* **79,** 3255–3259.
Bouchard, R. A. (1982). *Int. Rev. Cytol.* **76,** 113–193.
Brickell, P. M., Latchman, D. S., Murphy, D., Willison, K., and Rigby, P. W. J. (1983). *Nature (London)* **306,** 756–760.
Britten, R. J., and Kohne, D. E. (1968). *Science* **161,** 529–540.
Brown, A., and Huang, R. (1982). *Proc. Natl. Acad. Sci. U.S.A.* **79,** 6123–6127.
Brown, S. D. M. (1983). *Gene* **23,** 95–97.
Brown, S. D. M., and Dover, G. (1981). *J. Mol. Biol.* **150,** 441–466.
Brown, S. D. M., and Piechaczyk, M. (1983). *J. Mol. Biol.* **165,** 249–256.
Busch, H., Reddy, R., Rothblum, L., and Choi, Y. C. (1982). *Annu. Rev. Biochem.* **51,** 617–654.
Calabretta, B., Robberson, D. L., Maizel, A. L., and Saunders, G. F. (1981). *Proc. Natl. Acad. Sci. U.S.A.* **78,** 6003–6007.
Calabretta, B., Robberson, D. L., Barrera-Saldana, H. A., Lambrou, T. P., and Saunders, G. F. (1982). *Nature (London)* **296,** 219–225.
Calos, M. P., and Miller, J. H. (1980). *Cell* **20,** 579–595.
Carlson, D. P., and Ross, J. (1983). *Cell* **34,** 857–864.
Carpenter, C. D., Bruskin, A. M., Spain, L. M., Eldon, E. D., and Klein, W. H. (1982). *Nucleic Acids. Res.* **10,** 7829–7842.
Cavalier-Smith, T. (1983). *Nature (London)* **301,** 112–113.
Cech, T. R., and Hearst, J. E. (1976). *J. Mol. Biol.* **100,** 227–256.
Chen, M.-J., Shimada, T., Moulton, A. D., Harrison, M., and Nienhuis, A. W. (1982). *Proc. Natl. Acad. Sci. U.S.A.* **79,** 7435–7439.
Cheng, J.-F., Printz, R., Callaghan, T., Shuey, D., and Hardison, R. C. (1984). *J. Mol. Biol.* **176,** 1–20.
Ciliberto, G., Raugei, G.. Costanzo, F., Dente, L., and Cortese, R. (1983). *Cell* **32,** 725–733.
Coggins, L. W., Vass, J. K., Stinson, M. A., Lanyon, W. G., and Paul, J. (1982). *Gene* **17,** 113–116.
Colby, W. W., Chen, E. Y., Smith, D. H., and Levinson, A. I. (1983). *Nature (London)* **301,** 722–725.
Cole, M. D., Ono, M., and Huang, R. C. C. (1982). *J. Virol.* **42,** 123–130.

Comb, M., Rosen, H., Seeburg, P., Adelman, J., and Herbert, E. (1983). *DNA* **2,** 213–225.

Cooke, N. E., and Baxter, J. D. (1982). *Nature (London)* **297,** 603–606.

Daniels, G. R., and Deininger, P. L. (1983). *Nucleic Acids Res.* **11,** 7595–7610.

Daniels, G. R., Fox, G. M., Loewensteiner, D., Schmid, C. W., and Deininger, P. L. (1983). *Nucleic Acids Res.* **11,** 7579–7593.

Das, H. K., Lawrance, S. K., and Weissman, S. M. (1983). *Proc. Natl. Acad. Sci. U.S.A.* **80,** 3543–3547.

Davidson, E. H., and Britten, R. J. (1973). *Q. Rev. Biol.* **48,** 565–613.

Davidson, E. H., and Britten, R. J. (1979). *Science* **204,** 1052–1059.

Dawid, I. B., and Rebbert, M. L. (1981). *Nucleic Acids Res.* **9,** 5011–5020.

Dawid, I. B., Long, E. O., DiNocera, P. P., and Pardue, M. L. (1981). *Cell* **25,** 399–408.

DeFeo, D., Gonda, M. A., Young, H. A., Chang, E. H., Lowry, D. R., Scolnick, E. M., and Ellis, R. W. (1981). *Proc. Natl. Acad. Sci. U.S.A.* **78,** 3328–3332.

Degen, S. J. F., MacGillivray, R. T. A., and Davie, E. W. (1983). *Biochemistry* **22,** 2087–2097.

Deininger, P. L., Jolly, D. J., Rubin, C., Friedmann, T., and Schmid, C. W. (1981). *J. Mol. Biol.* **151,** 17–33.

Denison, R. A., and Weiner, A. M. (1982). *Mol. Cell. Biol.* **2,** 815–828.

Denison, R. A., van Arsdell, S. W., Bernstein, L. B., and Weiner, A. W. (1981). *Proc. Natl. Acad. Sci. U.S.A.* **78,** 810–814.

DeNoto, F. M., Moore, D. D., and Goodman, H. M. (1981). *Nucleic Acids Res.* **9,** 3719–3730.

Dhar, R., Ellis, R. W., Shih, T. Y., Oroszlan, S., Shapiro, B., Maizel, J., Lowy, D., and Scolnick, E. (1982). *Science* **217,** 934–937.

Dhruva, B. R., Shenk, T., and Subramanian, K. N. (1980). *Proc. Natl. Acad. Sci. U.S.A.* **77,** 4514–4518.

DiGiovanni, L., Haynes, S. R., Misra, R., and Jelinek, W. R. (1983). *Proc. Natl. Acad. Sci. U.S.A* **80,** 6533–6537.

DiNocera, P. P., and Dawid, I. B. (1983). *Nucleic Acids Res.* **11,** 5475–5482.

DiNocera, P. P., Digan, M. E., and Dawid, I. B. (1983). *J. Mol. Biol.* **168,** 715–727.

Doolittle, W. F., and Sapienza, C. (1980). *Nature (London)* **284,** 601–603.

Dover, G. (1982). *Nature (London)* **299,** 111–117.

Duncan, C., Biro, P. A., Choudary, P. V., Elder, J. T., Wang, R. R. C., Forget, B. G., De Riel, J. K., and Weissman, S. M. (1979). *Proc. Natl. Acad. Sci. U.S.A.* **76,** 5095–5099.

Duncan, C. H., Jagadeeswaran, P., Wang, R. C., and Weissman, S. M. (1981). *Gene* **13,** 185–196.

Elder, J. T., Pan, J., Duncan, C. H., and Weissman, S. M. (1981). *Nucleic Acids Res.* **9,** 1171–1189.

Erikson, E., Erikson, R. L., Henry, B., and Pace, N. R. (1973). *Virology* **53,** 40–46.

Evans, G. A., Margulies, D. H., Camerini-Otero, R. D., Ozato, K., and Seidman, J. G. (1982). *Proc. Natl. Acad. Sci. U.S.A.* **79,** 1994–1998.

Fanning, T. G. (1982). *Nucleic Acids Res.* **10,** 5003–5013.

Fanning, T. G. (1983). *Nucleic Acids Res.* **11,** 5073–5091.

Feagin, J. E., Setzer, D. R., and Schimke, R. T. (1983). *J. Biol. Chem* **258,** 2480–2487.

Flavell, R. A., van den Berg, F. M., and Grosveld, G. C. (1977). *J. Mol. Biol.* **115,** 715–741.

Fornwald, J. A., Kuncio, G., Peng, I., and Ordahl, C. P. (1982). *Nucleic Acids Res.* **10,** 3861–3876.

Fowlkes, D. M., and Shenk, T. (1980). *Cell* **22,** 405–413.

Fox. G. M., Hess, J. F., Shen, C.-K. J., and Schmid, C. W. (1982). *Cold Spring Harbor Symp. Quant. Biol.* **47,** 1131–1134.

Frederiksen, S., Pedersen, I. R., Hellung-Larsen, P., and Engberg, J. (1974). *Biochim. Biophys. Acta* **340,** 64–76.

Fritsch, E. F., Shen, C. K. J., Lawn, R. M., and Maniatis, T. (1980). *Cold Spring Harbor Symp. Quant. Biol.* **45,** 761–775.
Fuhrmann, S. A., Deininger, P. L., LaPorte, P., Friedmann, T., and Geiduschek, E. P. (1981). *Nucleic Acids Res.* **9,** 6439–6456.
Fukumaki, Y., Collins, F., Kole, R., Stoeckert, C. J., Jagadeeswaran, P., Duncan, C. H., and Weissman, S. M. (1982). *Cold Spring Harbor Symp. Quant. Biol.* **47,** 1079–1086.
Galli, G., Hofstetter, H., and Birnstiel, M. L. (1981). *Nature (London)* **294,** 626–631.
Gauss, D. H., and Sprinzl, M. (1983). *Nucleic Acids Res.* **11,** r1–r103.
Gebhard, W., and Zachau, H. G. (1983a). *J. Mol. Biol.* **170,** 255–270.
Gebhard, W., and Zachau, H. G. (1983b). *J. Mol. Biol.* **170,** 567–573.
Gebhard, W., Meitinger, T., Höchtl, J., and Zachau, H. G. (1982). *J. Mol. Biol.* **157,** 453–471.
Georgiev, G. P., Kramerov, D. A., Ryskov, A. P., Skryabin, K. G., and Lukanidin, E. M. (1982). *Cold Spring Harbor Symp. Quant. Biol.* **47,** 1109–1121.
Gilboa, E., Sudha, W., Mitra, S. G., and Baltimore, D. (1979). *Cell* **18,** 93–100.
Glover, D. M. (1981). *Cell* **26,** 297–298.
Goddard, J. P., Squire, M., Bienz, M., and Smith, J. D. (1983). *Nucleic Acids Res.* **11,** 2552–2562.
Grimaldi, G., and Singer, M. F. (1982). *Proc. Natl. Acad. Sci. U.S.A.* **79,** 1497–1500.
Grimaldi, G., Queen, C., and Singer, M. F. (1981). *Nucleic Acids Res.* **9,** 5553–5568.
Gundelfinger, E. D., Krause, E., Melli, M., and Dobberstein, B. (1983). *Nucleic Acids Res.* **11,** 7363–7374.
Hall, B. D., Clarkson, S. G., and Tocchini-Valentini, G. (1982). *Cell* **29,** 3–5.
Hamada, H., and Kakunaga, T. (1982). *Nature (London)* **298,** 396–398.
Hamada, H., Petrino, M. G., and Kakunaga, T. (1982), *Proc. Natl. Acad. Sci. U.S.A.* **79,** 6465–6469.
Hammarström, K., Westin, G., and Pettersson, U. (1982). *EMBO J.* **1,** 737–739.
Harada, F., and Kato, N. (1980). *Nucleic Acids Res.* **8,** 1273–1285.
Harada, F., Kato, N., and Hoshino, H. (1979). *Nucleic Acids Res.* **7,** 909–917.
Harrington, C. A., and Chikaraishi, D. M. (1983). *Nucleic Acids Res.* **11,** 3317–3332.
Harris, S. E., Mansson, P.-E., Tully, D. B., and Burkhart, B. (1983). *Proc. Natl. Acad. Sci. U.S.A.* **80,** 6460–6464.
Harshey, R. M., and Bukhari, A. I. (1981). *Proc. Natl. Acad. Sci. U.S.A.* **78,** 1090–1094.
Hasan, G., Turner, M. J., and Cordingley, J. S. (1984). *Cell* **37,** 333–341.
Hayashi, K. (1981). *Nucleic Acids Res.* **9,** 3379–3388.
Haynes, S. R., and Jelinek, W. R. (1981). *Proc. Natl. Acad. Sci. U.S.A.* **78,** 6130–6134.
Haynes, S. R., Toomey, T. P., Leinwand, L., and Jelinek, W. R. (1981). *Mol. Cell. Biol.* **1,** 573–583.
Hendrick, J. P., Wolin, S. L., Rinke, J., Lerner, M., and Steitz, J. A. (1981). *Mol. Cell. Biol.* **1,** 1138–1149.
Hess, J. F., Fox, M., Schmid, C., and Shen, C-K. J. (1983). *Proc. Natl. Acad. Sci. U.S.A.* **80,** 5970–5974.
Höchtl, J., and Zachau, H. G. (1983). *Nature (London)* **302,** 260–263.
Hollis, G. F., Hieter, P. A., McBride, O. W., Swan, D., and Leder, P. (1982). *Nature (London)* **296,** 321–325.
Houck, C. M., Rinehart, F. P., and Schmid, C. W. (1979). *J. Mol. Biol.* **132,** 289–306.
Hurst, H. C., and Parker, M. G. (1983). *EMBO J.* **2,** 769–774.
Jagadeeswaran, P., Forget, B. G., and Weissman, S. M. (1981). *Cell* **26,** 141–142.
Jagadeeswaran, P., Tuan, D., Forget, B. G., and Weissman, S. M. (1982). *Nature (London)* **296,** 469–470.
Jelinek, W. R., and Leinwand, L. (1978). *Cell* **15,** 205–214.
Jelinek, W. R., and Schmid, C. W. (1982). *Annu. Rev. Biochem.* **51,** 813–844.

Jelinek, W., Evans, R., Wilson, M., Salditt-Georgieff, M., and Darnell, J. E. (1978). *Biochemistry* **17,** 2776–2782.

Jelinek, W. R., Toomey, T. P., Leinwand, L., Duncan, C. H., Biro, P. A., Choudary, P. V., Weissman, S. M., Rubin, C. M., Houck, C. M., Deininger, P. L., and Schmid, C. W. (1980). *Proc. Natl. Acad. Sci. U.S.A.* **77,** 1398–1402.

Kalb, V. F., Glasser, S., King, D., and Lingrel, J. B. (1983). *Nucleic Acids Res.* **11,** 2177–2184.

Karin, M., and Richards, R. I. (1982). *Nature (London)* **299,** 797–802.

Kato, K., Hayashizaki, Y., Takahashi, Y., Himeno, S., and Matsubara, K. (1983). *Nucleic Acids Res.* **11,** 8197–8203.

Kelly, F., and Condamine, H. (1982). *Biochim. Biophys. Acta* **651,** 105–141.

King, B. O., and Yao, M. C. (1982). *Cell* **31,** 177–182.

Kinniburgh, A. J., and Martin, T. E. (1976). *Biochem. Biophys. Res. Commun.* **73,** 718–726.

Kole, L. B., Haynes, S. R., and Jelinek, W. R. (1983). *J. Mol. Biol.* **165,** 257–286.

Kominami, R., Muramatsu, M., and Moriwaki, K. (1983). *Nature (London)* **301,** 87–89.

Konkel, D. A., Maizel, J. V., and Leder, P. (1979). *Cell* **18,** 865–873.

Kramerov, D. A., Grigoryan, A. A., Ryskov, P., and Georgiev, G. P. (1979). *Nucleic Acids Res.* **6,** 697–713.

Kramerov, D. A., Lekakh, I. V., Samarina, O. P., and Ryskov, A. P. (1982). *Nucleic Acids Res* **10,** 7477–7491.

Krayev, A. S., Kramerov, D. A., Skryabin, K. G., Ryskov, A. P., Bayev, A. A., and Georgiev, G. P. (1980). *Nucleic Acids Res.* **8,** 1201–1215.

Krayev, A. S., Markusheva, T. V., Kramerov, D. A., Ryskov, A. P., Skryabin, K. G., Bayev, A. A., and Georgiev, G. P. (1982). *Nucleic Acids Res.* **1O,** 7461–7475.

Krolewski, J. J., and Rush, M. G. (1984). *J. Mol. Biol.* **174,** 31–40.

Krolewski, J. J., Bertelsen, A. H., Humayun, M. Z., and Rush, M. G. (1982). *J. Mol. Biol.* **154,** 399–415.

Krolewski, J. J., Schindler, C. W., and Rush, M. G. (1984). *J. Mol. Biol.* **174,** 41–54.

Kvist, S., Roberts, L., and Dobberstein, B. (1983). *EMBO J.* **2,** 245–254.

Lalanne, J. L., Bregegere, F., Delarbe, C., Abastado, J. P., Gachelin, G., and Kourilsky, P. (1982). *Nucleic Acids Res.* **10,** 1039–1049.

Land, H., Grez, M., Hauser, H., Lindenmaier, W., and Schütz, G. (1981). *Nucleic Acids Res.* **9,** 2251–2266.

Larhammar, D., Hammerling, U., Denaro, M., Lund, T., Flavell, R. A., Rask, L., and Peterson, P. A. (1983). *Cell* **34,** 179–188.

Lee, M.G.-S., Lewis, S. A., Wilde, C. D., and Cowan, N. J. (1983). *Cell* **33,** 477–487.

Lehman, I. R. (1974). *Science* **186,** 790–797.

Leicht, M., Long, G. L., Chandra, T., Kurachi, K., Kidd, V. J., Mace, M., Davie, E. W., and Woo S. L. C. (1982). *Nature (London)* **297,** 655–659.

Lemischka, I., and Sharp, P. A. (1982). *Nature (London)* **300,** 330–335.

Lerman, M. I., Thayer, R. E., and Singer, M. F. (1983). *Proc. Natl. Acad. Sci. U.S.A.* **80,** 3966–3970.

Li, W.-Y., Reddy, R., Henning, D., Epstein, P., and Busch, H. (1982). *J. Biol. Chem.* **257,** 5136–5142.

Liebermann, D., Hoffman-Liebermann, B., Weinthal, J., Childs, G., Maxson, R., Mauron, A., Cohen, S. N., and Kedes, L. (1983). *Nature (London)* **306,** 342–347.

Limbach, K. J., and Wu, R. (1983). *Nucleic Acids Res.* **11,** 8931–8950.

Lomedico, P., Rosenthal, N., Efstradiadis, A., Gilbert, W., Kolodner, R., and Tizard, R. (1979). *Cell* **18,** 545–558.

Long, E. O., and Dawid, I. B. (1979). *Cell* **18,** 1185–1196.

Lopata, M. A., Havercroft, J. C., Chow, L. T., and Cleveland, D. W. (1983). *Cell* **32,** 713–724.

Lueders, K. K., and Paterson, B. M. (1982). *Nucleic Acids Res.* **10,** 7715–7729.
Lueders, K., Leder, A., Leder, P., and Kuff, E. (1982). *Nature (London)* **295,** 426–428.
McGrath, J. P., Capon, D. J., Smith, D. H., Chen, E. Y., Seeburg, P. H., Goeddel, D. V., and Levinson, A. D. (1983). *Nature (London)* **304,** 501–506.
MacLeod, A. R., and Talbot, K. (1983). *J. Mol. Biol.* **167,** 523–537.
Maeda, N., Bliska, J. B., and Smithies, O. (1983). *Proc. Natl. Acad. Sci. U.S.A.* **80,** 5012–5016.
Malissen, B., Malissen, B., and Jordan, B. R. (1982). *Proc. Natl. Acad. Sci. U.S.A.* **79,** 893–897.
Malissen, M., Hunkapiller, T., and Hood, L. (1983). *Science* **221,** 750–754.
Manley, J. L., and Colozzo, M. T. (1982). *Nature (London)* **300,** 376–379.
Manser, T., and Gesteland, R. F. (1982). *Cell* **29,** 257–264.
Martin, S. L., Voliva, C. F., Burton, F. H., Edgell, M. H., and Hutchison, C. A. (1984). *Proc. Natl. Acad. Sci. U.S.A.* **81,** 2308–2312.
Mason, A. J., Evans, B. A., Cox, D. R., Shine, J., and Richards, R. I. (1983). *Nature (London)* **303,** 300–307.
Masters, J. N., Yang, J. K., Cellini, A., and Attardi, G. (1983). *J. Mol. Biol.* **167,** 23–36.
Mattaj, I. W., and Zeller, R. (1983). *EMBO J.* **2,** 1883–1891.
Mayfield, J. E., McKenna, J. F., and Lessa, B. S. (1980). *Chromosoma* **76,** 277–294.
Miesfeld, R., Krystal, M., and Arnheim, N. (1981). *Nucleic Acids Res.* **9,** 5931–5947.
Milner, R. J., Bloom, F. E., Lai, C., Lerner, R. A., and Sutcliffe, J. G. (1984). *Proc. Natl. Acad. Sci. U.S.A.* **81,** 713–717.
Minty, A. J., Alonso, S., Guenet, J.-L., and Buckingham, M. E. (1983). *J. Mol. Biol.* **167,** 77–101.
Miyake, T., Migita, K., and Sakaki, Y. (1983). *Nucleic Acids Res.* **11,** 6837–6846.
Molloy, G. R., Thomas, W. L., and Darnell, J. E. (1972). *Proc. Natl. Acad. Sci. U.S.A.* **69,** 3684–3688.
Monroe, S. S., and Schlesinger, S. (1983). *Proc. Natl. Acad. Sci. U.S.A.* **80,** 3279–3283.
Monson, J. M., Friedman, J., and McCarthy, B. J. (1982). *Mol. Cell. Biol.* **2,** 1362–1371.
Monstein, H.-J., Hammarstrom, K., Westin, G., Zabielski, J., Philipson, L., and Pettersson, U. (1983). *J. Mol. Biol.* **167,** 245–257.
Moore, K. W., Sher, B. T., Sun, Y. H., Eakle, K. A., and Hood, L. (1982). *Science* **215,** 679–682.
Moos, M., and Gallwitz, D. (1983). *EMBO J.* **2,** 757–761.
Moyzis, R. K., Bonnet, J., Li, D. W., and Ts'o, P. O. P. (1981). *J. Mol. Biol.* **153,** 871–896.
Mroczka, D. L., Cassidy, B., Busch, H., and Rothblum, L. I. (1984). *J. Mol. Biol.* **174,** 141–162.
Munz, P., Amstutz, H., Kohli, J., and Leupold, U. (1982). *Nature (London)* **300,** 225–231.
Murphy, D., Brickell, P. M., Latchman, D. S., Willison, K., and Rigby, P. W. J. (1983). *Cell* **35,** 865–871.
Murphy, S., Altruda, F., Ullu, E., Tripodi, M., Silengo, L., and Melli, M. (1984). *J. Mol. Biol.* **177,** 575–590.
Myers, J. C., Dickson, L. A., de Wet, W. J., Bernard, M. B., Chu, M-L., Di Liberto, M., Pepe, G., Sangiorgi, F. O., and Ramirez, F. (1983). *J. Biol. Chem.* **258,** 10128–10135.
Nishioka, Y., Leder, A., and Leder, P. (1980). *Proc. Natl. Acad. Sci. U.S.A.* **77,** 2806–2809.
Nojima, H., and Kornberg, R. D. (1983). *J. Biol. Chem.* **258,** 8151–8155.
Ohshima, Y., Okada, N., Tani, T., Itoh, Y., and Itoh, M. (1981). *Nucleic Acids Res.* **9,** 5145–5158.
Ohtsuka, E., Nishikawa, S., Sugiura, M., and Ikehara, M. (1976). *Nucleic Acids Res.* **3,** 1613–1623.
Ono, M., Cole, M. D., White, A. T., and Huang, R. C. C. (1980). *Cell* **21,** 465–473.
Orgel, S. E., and Crick, F. H. C. (1980). *Nature (London)* **284,** 604–606.
Orkin, S. H., and Michelson, A. (1980). *Nature (London)* **286,** 538–540.
Ottolenghi, S., and Giglioni, B. (1982). *Nature (London)* **300,** 770–771.
Ou, C.-Y., Boone, L. R., and Yang, W. K. (1983). *Nucleic Acids Res.* **11,** 5603–5620.

Page, G. S., Smith, S., and Goodman, H. M. (1981). *Nucleic Acids Res.* **9,** 2087–2104 (1981).
Pan, J., Elder, J. T., Duncan, C. H., and Weissman, S. M. (1981). *Nucleic Acids. Res.* **9,** 1151–1170.
Paolella, G., Lucero, M. A., Murphy, M. H., and Baralle, F. E. (1983). *EMBO J.* **2,** 691–696.
Pederson, T., and Bhorjee, J. S. (1979). *J. Mol. Biol.* **128,** 451–480.
Piechaczyk, M., Lelay-Taha, M. N., Sri-Widada, J., Brunel, C., Liautard, J. P., and Jeanteur, P. (1982). *Nucleic Acids Res.* **10,** 4627–4640.
Posakony, J. W., Flytzanis, C. N., Britten, R. J., and Davidson, E. H. (1983). *J. Mol. Biol.* **167,** 361–389.
Potter, S. S. (1984). *Proc. Natl. Acad. Sci. U.S.A.* **81,** 1012–1016.
Potter, S. S., and Jones, R. S. (1983). *Nucleic Acids Res.* **11,** 3137–3153.
Proundfoot, N. J., Gil, A., and Maniatis, T. (1982). *Cell* **31,** 553–563.
Qasba, P. K., and Safaya, S. K. (1984). *Nature (London)* **308,** 377–380.
Rae, P. M. M. (1981). *Nucleic Acids Res.* **9,** 4997–5010.
Rae, P. M. M., Kohorn, B. D., and Wade, R. P. (1980). *Nucleic Acids Res.* **8,** 3491–3504.
Reddy, R., Henning, D., Tan, E., and Busch, H. (1983). *J. Biol. Chem.* **258,** 8352–8356.
Reyes, A. A., Schöld, M., and Wallace, R. B. (1982). *Immunogenetics* **16,** 1–9.
Rinehart, F. P., Ritch, T. G., Deininger, P. L., and Schmid, C. W. (1981). *Biochemistry* **20,** 3003–3010.
Ro-Choi, T. S., Reddy, R., Henning, D., Takano, T., Taylor, C. W., and Busch, H. (1972). *J. Biol. Chem.* **247,** 3205–3222.
Roe, B. A., Anandaraj, M. P. J. S., Chia, L. S. Y., Randerath, E., Gupta, R. C., and Randerath, K. (1975). *Biochem. Biophys. Res. Commun.* **66,** 1097–1105.
Rogers, J. (1983a). *Nature (London)* **301,** 460.
Rogers, J. (1983b). *Nature (London)* **305,** 101–102.
Rogers, J. (1983c). *Nature (London)* **306,** 113–114.
Rogers, J. H. (1985). *Biochim. Biophys. Acta* (in press).
Roiha, H., and Glover, D. M. (1981). *Nucleic Acids Res.* **9,** 5521–5532.
Roiha, H., Miller, J. R., Woods, L. C., and Glover, D. M. (1981). *Nature (London)* **290,** 749–753.
Roop, D. R., Kristo, P., Stumph, W. E., Tsai, M. J., and O'Malley, B. W (1981). *Cell* **23,** 671–680.
Ryskov, A. P., Ivanov, P. L., Kramerov, D. A., and Georgiev, G. P. (1983) *Nucleic Acids Res.* **11,** 6541–6558.
Saffer, J. D., and Lerman, M. I. (1983). *Mol. Cell. Biol.* **3,** 960–964.
Scarpulla, R. C., and Wu, R. (1983). *Cell* **32,** 473–482.
Schmid, C. W., and Jelinek, W. R. (1982). *Science* **216,** 1065–1070.
Schimenti, J. C., and Duncan, C. H. (1984). *Nucleic Acids Res.* **12,** 1641–1655.
Schon, E. A., Cleary, M. L., Haynes, J. R., and Lingrel, J. B. (1981). *Cell* **27,** 359–369.
Schuler, L. A., Weber, J. L., and Gorski, J. (1983). *Nature (London)* **305,** 159–160.
Seeburg. P. H. (1982). *DNA* **1,** 239–249.
Shafit-Zagardo, B., Brown, F. L., Zavodny, P. J., and Maio, J. J. (1983) *Nature (London)* **304,** 277–280.
Shapiro, S. G., Schon, E. A., Townes, T. M., and Lingrel, J. B. (1983). *J. Mol. Biol.* **169,** 31–52.
Sharp, P. A. (1983). *Nature (London)* **301,** 471–472.
Shen, C.-K. J., and Maniatis, T. (1982). *J. Mol. Appl. Genet.* **1,** 343–360.
Shen, S. H., and Smithies, O. (1982). *Nucleic Acids Res.* **10,** 7809–7818.
Shimizu, Y., Yoshida, K., Ren, C.-S., Fujinaga, K., Rajagopalan, S., and Chinnadurai, G. (1983). *Nature (London)* **302,** 587–590.
Shimotohno, K., and Temin, H. M. (1982). *Nature (London)* **299,** 265–268.

Singer, M. F. (1982a). *Cell* **28,** 433–434.
Singer, M. F. (1982b). *Int. Rev. Cytol.* **76,** 67–112.
Singer, M. F., Thayer, R. E., Grimaldi, G., Lerman, M. I., and Fanning, T. G. (1983a). *Nucleic Acids. Res.* **11,** 5739–5745.
Singer, D. S., Lifshitz, R., Abelson, L., Nyirjesy, P., and Rudikoff, S. (1983b). *Mol. Cell. Biol.* **3,** 903–913.
Smith, G. P. (1976). *Science* **191,** 528–535.
Soriano, S., Meunier-Rotival, M., and Bernardi, G. (1983). *Proc. Natl. Acad. Sci. U.S.A..* **80,** 1816–1820.
Spohr, G., Reith, W., and Sures, I. (1981). *J. Mol. Biol.* **151,** 573–592.
Spohr, G., Reith, W., and Crippa, M. (1982). *Dev. Biol.* **94,** 71–78.
Staden, R. (1982). *Nucleic Acids Res.* **10,** 2951–2961.
Stein, J. P., Munjaal, R. P., Lagace, L., Lai, E. C., O'Malley, B. W., and Means, A. R. (1983). *Proc. Natl. Acad. Sci. U.S.A.* **80,** 6485–6489.
Steinmetz, M., Moore, K. W., Frelinger, J. G., Sher, B. T., Shen, F.-W., Boyse, E. A., and Hood, L. (1981). *Cell* **25,** 683–692.
Straus, N. A., and Birnboim, H. C. (1974). *Proc. Natl. Acad. Sci. U.S.A.* **71,** 2992–2995.
Streisinger, G., Okada, Y., Emrich, J., Newton, J., Tsugita, J., Terzaghi, A., Terzaghi, E., and Inouye, M. (1966). *Cold Spring Harbor Symp. Quant. Biol.* **31,** 77–84.
Stumph, W. E., Kristo, P., Tsai, M.-J. and O'Malley, B. W. (1981). *Nucleic Acids Res.* **9,** 5383–5397.
Stumph, W. E., Baez, M., Beattie, W., Tsai, M.-J., and O'Malley, B. W. (1983). *Biochemistry* **22,** 306–315.
Sun, L., Paulson, K. E., Schmid, C. W., Kadyk, L., and Leinwand, L. (1984). *Nucleic Acids Res.* **12,** 2669–2690.
Sutcliffe, J. G., Milner, R. J., Bloom, F. E., and Lerner, R. A. (1982). *Proc. Natl. Acad. Sci. U.S.A.* **79,** 4942–4946.
Sutcliffe, J. G., Milner, R. J., Gottesfeld, J. M., and Lerner, R. A. (1984). *Nature (London)* **308,** 237–241.
Szostak, J. W., and Blackburn, E. H. (1982). *Cell* **29,** 245–255.
Takahashi, H., Hakamata, Y., Watanabe, Y., Kikuno, R., Miyata, T., and Numa, S. (1983). *Nucleic Acids Res.* **11,** 6847–6857.
Tani, T., Watanabe-Nagasu, N., Okada, N., and Ohshima, Y. (1983). *J. Mol. Biol.* **168,** 579–594.
Tautz, D., and Renz, M. (1984). *Nucleic Acids Res.* 12, 4127–4138.
Tchurikov, N. A., Naumova, A. K., Zelentsova, E. S., and Georgiev, G. P. (1982). *Cell* **28,** 365–373.
Temin, H. M. (1980). *Cell* **21,** 599–600.
Thayer, R. E., and Singer, M. F. (1983). *Mol. Cell. Biol.* **3,** 967–973.
Tokarskaya, O. N., Tchurikov, N. A., Ivanov, P. L., Kramerov, D. A., and Ryskov, A. P. (1980). *Nucleic Acids Res.* **8,** 425–440.
Tsukada, T., Watanabe, Y., Nakai, Y., Imura, H., Nakanishi, S., and Numa, S. (1982). *Nucleic Acids Res.* **10,** 1471–1479.
Ueda, S., Nakai, S., Nishida, Y., Hisajima, H., and Honjo, T. (1982). *EMBO J.* **1,** 1539–1544.
Uhler, M., Herbert, E., D'Eustachio, P., and Ruddle, F. D. (1983). *J. Biol. Chem.* **258,** 9444–9453.
Ullu, E., Murphy, S., and Melli, M. (1982a). *Cell* **29,** 195–202.
Ullu, E., Esposito, V., and Melli, M. (1982b). *J. Mol. Biol.* **161,** 195–201.
Van Arsdell, S. W., and Weiner, A. M. (1984). *Nucleic Acids Res.* **12,** 1463–1471.
Van Arsdell, S. W., Denison, R. A., Bernstein, L. B., Weiner, A. M., Manser, T., and Gesteland, R. F. (1981). *Cell* **26,** 11–17.

Vanin, E. F., Goldberg, G. I., Tucker, P. W., and Smithies, O. (1980). *Nature (London)* **286,** 222–226.

Varmus, H. E. (1982). *Science* **216,** 812–820.

Walter, P., and Blobel, G. (1982). *Nature (London)* **299,** 691–698.

Watanabe, Y., Tsukada, T., Notake, M., Nakanishi, S., and Numa, S. (1982). *Nucleic Acids Res.* **10,** 1459–1469.

Watanabe-Nagasu, N., Itoh, Y., Tani, T., Okano, K., Koga, N., Okada, N., and Ohshima, Y. (1983). *Nucleic Acids Res.* **11,** 1791.

Weiner, A. M. (1980). *Cell* **22,** 209–218.

Wilde, C. D., Crowther, C. E., Cripe, T. P., Lee, M. G.-S., and Cowan, N. J. (1982a). *Nature (London)* **297,** 83–84.

Wilde, C. D., Crowther, C. E., and Cowan, N. J. (1982b). *Science* **217,** 549–552.

Wilson, R., and Storb, U. (1983). *Nucleic Acids Res.* **11,** 1803–1817.

Wilson, S. H., and Kuff, E. D. (1972). *Proc. Natl. Acad. Sci. U.S.A.* **69,** 1531–1536.

Winoto, A., Steinmetz, M., and Hood, L. (1983). *Proc. Natl. Acad. Sci. U S.A.* **80,** 3425–3429.

Young, P. R., Scott, R. W., Hamer, D. H., and Tilghman, S. M. (1982). *Nucleic Acids Res.* **10,** 3099–3116.

Zieve, G., Benecke, B.-J., and Penman, S. (1977). *Biochemistry* **16,** 4520–4525.

INTERNATIONAL REVIEW OF CYTOLOGY, VOL. 93

Transposable Elements in Eukaryotes

D. J. FINNEGAN

Department of Molecular Biology, University of Edinburgh, Edinburgh, Scotland

I. Introduction

The discovery that both prokaryotic and eukaryotic genomes are less stable than classical genetic analyses had indicated is one of the most interesting insights to have come from the application of recombinant DNA techniques to genetic problems. Several geneticists, particularly McClintock working with maize and Green working with *Drosophila,* detected transposable genetic elements in eukaryotes many years ago. At the time these elements were generally regarded as being rare components of the genome. We now realize that they were just the tip of a genetic iceberg.

In this review I shall classify transposable elements according to their structure. Some elements with similar structures are related but this is not always the case. In Sections II–V I shall review some of the information available concern-

ISBN 0-12-364493-3

ing transposable elements of each type. In the last section I shall discuss some questions raised by transposable elements in general, particularly with regard to their biological significance and their possible use as experimental tools. More detailed discussion of many of the topics mentioned here can be found in the book "Mobile Genetic Elements" edited by Shapiro (1983).

II. Transposable Elements with Long Direct Terminal Repeats

A. *Copia*-LIKE ELEMENTS IN *Drosophila melanogaster*

1. *General Properties*

Ten to fifteen percent of the *D. melanogaster* genome consists of moderately repetitive DNA and about one-half of this is made up of 30 to 50 families of transposable elements with very similar properties. Some of the best studied of these are the *copia, 412, 297, mdg1, mdg3,* and *B104* families, These are named after the first member of each family to be cloned and studied in detail (Finnegan *et al.,* 1978; Strobel *et al.,* 1979; Ilyin *et al.,* 1980a,b; Scherer *et al.,* 1982; Ikenaga and Saigo, 1982). They are usually referred to collectively as "*copia*-like elements." The members of each family are well conserved and are located at between 20 and 100 sites distributed throughout the genome. In general there is only one element at each site but the number and location of these sites vary between strains of *D. melanogaster* and between embryonic and tissue culture cell DNA (Ilyin *et al.,* 1978; Potter *et al.,* 1979).

The first indication that *copia*-like elements might be transposable came from comparisons of the location of these elements in chromosomes of different strains of *D. melanogaster.* These were determined by *in situ* hybridization experiments in which probes containing particular *copia*-like elements were hybridized to polytene chromosomes from larval salivary glands. The number of copies of each element was about the same in each strain but their chromosome locations were very different. This could reasonably be explained only if these elements could transpose from one location in the genome to another. Analogous results were obtained when genomic DNAs from the same strains were digested with restriction enzymes and hybridized, in Southern transfer experiments (Southern, 1975), with a fragment from one end of a *copia*-like element (Potter *et al.,* 1979). Many restriction fragments from each strain hybridized to the probe but the patterns of hybridization were different in each case, again indicating that the distribution of elements differed from strain to strain. The results of an experiment of this type using a probe from a member of the *412* family of *copia*-like elements are shown in Fig. 1.

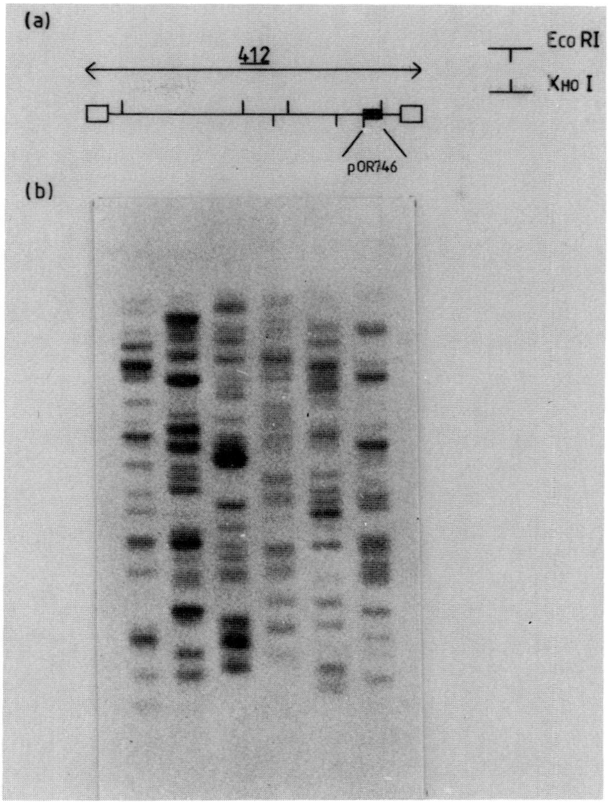

FIG. 1. Hybridisation of a *412* probe to DNA from different strains of *D. melanogaster*. DNAs from unrelated strains of *D. melanogaster* were digested with the restriction enzyme *Eco*RI and then electrophoresed through a 0.7% agarose gel. The DNA was transferred to nitrocellulose and then hybridized with the plasmid pOR746 which had been labeled with ^{32}P. This palsmid contains a fragment from near the right-hand end of a *412* element. Fragments hybridizing to this probe were detected by autoradiography. In this experiment each *412* element gives rise to one fragment hybridizing to the probe. The size of the fragment is dependent on the location of the element within the genome. DNAs from different strains give different patterns of hybridization indicating that the chromosomal distribution of *412* elements is different in each case. The strains used were, from left to right, Cha, se f 8, HJ30, Lu, Canton S, and Be. This is an unpublished experiment of Dr. H. Sang.

Direct evidence for transposition has come from studies of molecular lesions associated with mutations at a variety of loci. Many spontaneous and induced mutations in *D. melanogaster* are due to insertion of *copia*-like elements and it seems likely that this will prove to be generally the case (Zachar and Bingham, 1982; Rubin *et al.,* 1982; Bender *et al.,* 1983; Modellel *et al.,* 1983). The first example of such a mutation was the *white* apricot mutation. This is due to insertion of a *copia* element within the *white* gene (Gehring and Paro, 1980; Bingham and Judd, 1981; Bingham *et al.,* 1981).

The most characteristic feature of *copia*-like elements is that they have direct repeat sequences a few hundred nucleotides long at each end. These are usually called long terminal repeats or LTRs. The lengths and sequences of these repeats are specific for each family of elements although the LTRs of *copia* (Levis *et al.,* 1980), *mdg1* (Kulguskin *et al.,* 1981), *412* (Will *et al.,* 1981), *B104* (Scherer *et al.,* 1982), and possibly *mdg3* (Bayev *et al.,* 1980, see footnotes to Table I) elements all begin with the sequence T-G-T and end with A-C-A. These tri-nucleotide sequences are part of short inverted repeats at the end of each LTR (Fig. 2a). The LTRs of *17.6* and *mdg4/gypsy* elements are unusual in that they begin AGT and end ATT (Kugimiya *et al.,* 1983; Inouye *et al.,* 1984; Bayev *et al.,* 1984).

In general the LTRs at the ends of any one copy of an element are identical whereas LTRs of different copies of the same element may vary by a few base pairs. The LTRs at the ends of the *297* and *17.6* elements appear to be exceptions since they show several differences (Ikenaga and Saigo, 1982; Kugimiya *et al.,* 1983). These could arise by mutation or by genetic exchange between LTRs of different members of the same family. Some of the differences reported may be sequencing errors since much of the relevant sequence was determined on one strand only. The LTRs of *copia* (Levis *et al.,* 1980), *mdg1* (Kulgushin *et al.,* 1981), and *412* (Will, 1983) elements are rarely, if ever, present in the genome on their own as free repeats. In this respect they differ from the LTRs of Ty elements in yeast (see Section II,D). Shepherd and Finnegan (1984) have cloned a free *412* LTR. This was probably formed by recombination between the LTRs at the ends of a complete *412* element.

At the extreme ends of each element are short inverted repeats, about 10 bp long, which, with the possible exception of *mdg3* elements (see footnotes to Table I), are found at both ends of the LTR. Immediately before and after each element are direct repeats a few base pairs long. The length of these direct repeats is the same for all members of the same family but their sequences are unrelated (Dunsmuir *et al.,* 1980; Bayev *et al.,* 1980; Will *et al.,* 1981; Scherer *et al.,* 1982; Ikenaga and Saigo, 1982). In the case of *copia, 297,* and *412* elements comparisons have been made between pairs of DNA fragments from equivalent positions in the genome, one containing an element, a ''full'' site, and the other lacking the element, the corresponding ''empty'' site. The bases repeated on either side of an element in a full site are present only once at the equivalent position in an empty site, suggesting that they are duplicated during

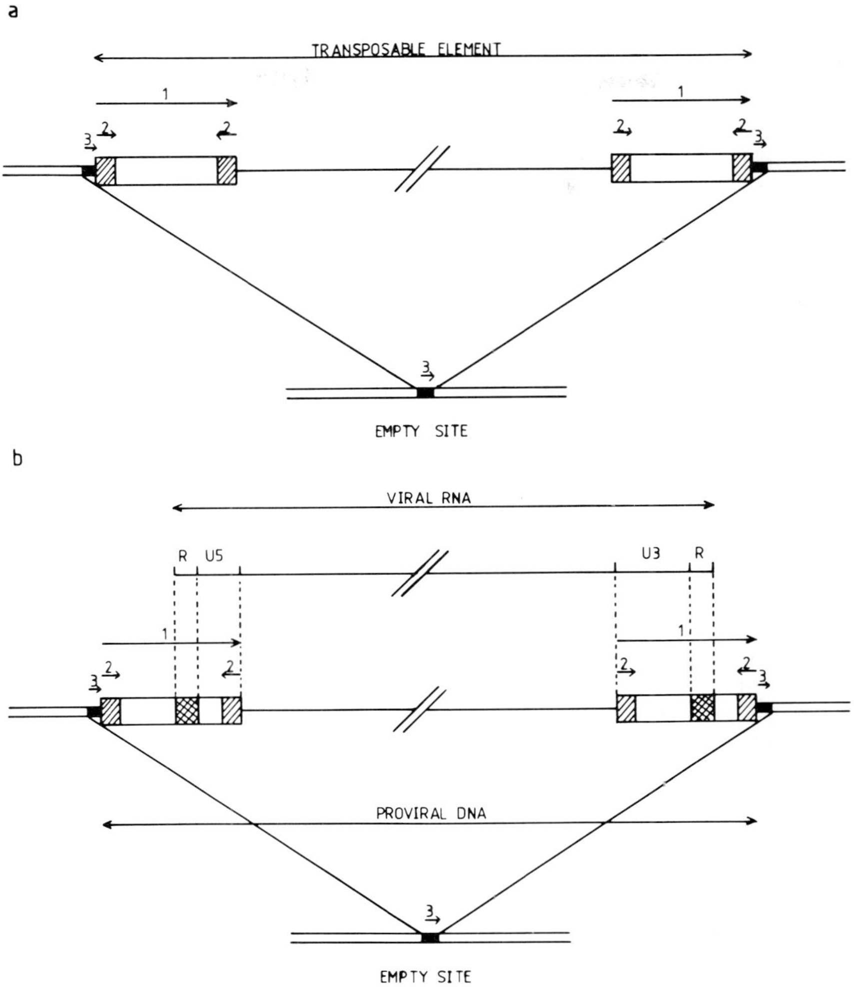

FIG. 2. (a) Diagram of a *copia*-like element, indicating the positions and orientations of the various repeat sequences. The relationship between a full and an empty site is shown. The repeats are, 1, long direct repeat or LTR; 2, short inverted repeat; 3, direct repeat of a host sequence present only once at the empty site. (b) Diagram showing the relationships between sequences in retroviral RNA and proviral DNA and between an integrated provirus and a site into which it has inserted. U5, R, and U3 are described in the text. Repeat sequences are described in (a) above.

transposition. This is a common feature of almost all transposable elements so far described. Figure 2a shows a typical *copia*-like element and Table I lists some of the properties of particular elements.

Copia-like elements are probably not distributed at random on *D. melanogaster* chromosomes. Some chromosome regions, particularly those containing heterochromatin, have a higher concentration of these sequences than others (Ananiev *et al.*, 1978; Tchurikov *et al.*, 1980; Young and Schwarz, 1981; Montgomery and Langley, 1983). These may be hot-spots for insertion either because of their base sequence or because chromatin structure in these regions favors insertion. On the other hand, they may simply be regions which can accumulate transposable elements because they do not contain essential genes. Dunsmuir *et al.* (1980) compared the insertion targets of three different members of the *copia* family and could find no obvious relationship between them. Similarly there was no homology between these sequences and the termini of *copia* elements themselves (Levis *et al.*, 1980). Although sites for insertion of *copia* elements are not specific sequences they may be defined by the DNA sequence in a more general way. *297* and *17.6* elements insert preferentially at the sequence ATAT (Ikenaga and Saigo, 1982; E. Young and G. M. Rubin, cited in Rubin, 1983; Inouye *et al.*, 1984), and *412* elements at the sequence NTNG, where N is any base (Shepherd and Finnegan, 1984).

There is no direct evidence that *copia*-like elements can excise, either precisely or imprecisely. The *white-apricot* mutation is stable as are two other *white* gene mutations associated with *copia* insertions (Rubin *et al.*, 1982). This suggests that *copia* elements do not normally excise precisely at high frequency. The heterogeneous distribution of *copia*-like elements in different strains of D. *melanogaster* suggests that infrequent excision does occur. This need not be precise.

2. *Transcription of Copia-like Elements*

Many *copia*-like elements were first detected because they are complementary to abundant polyadenylated RNAs in *D. melanogaster* tissue culture cells (Rubin *et al.*, 1976; Finnegan *et al.*, 1978; Ilyin *et al.*, 1978; Scherer *et al.*, 1982). Transcripts from *copia* elements may make up as much as 3% of the polyadenylated RNA in the cytoplasm of these cells (Finnegan *et al.*, 1978; Falkenthal and Lengyel, 1980) even though most *copia* RNA is confined to the nucleus (Schwartz *et al.*, 1982).

Transcripts complementary to the full length of *copia* (Carlson and Brutlag, 1978; Young and Schwartz, 1981), *mdg1*, *mdg3* (Ilyin *et al.*, 1980d), *412* (Finnegan *et al.*, 1978; Young and Schwartz, 1981), and *B104* (Scherer *et al.*, 1982) elements have been found. Most *copia* transcripts initiate within one LTR (Flavell *et al.*, 1981) and *B104* RNAs can both initiate and terminate within LTRs (Scherer *et al.*, 1982). Several smaller *copia* RNAs have been detected in

TABLE I

PROPERTIES OF SOME FAMILIES OF *copia*-LIKE ELEMENTS

Family of elements	Approximate number per haploid genome	Length of complete element	Length of LTR	Length of terminal inverted repeat	Length of target site duplication	References[a]
copia	60	5.0 kb	276 bp	13/17 bp	5 bp	1, 2, 3, 4
412	40	7.5 kb	481 bp[b]	8/10 bp	4 bp	1, 2, 5
297	30	7.0 kb	412/415 bp[c]	3 bp[c]	4 bp[c]	2, 6, 7
*mdg*1	25	7.2 kb	442/444 bp	13/16 bp	4 bp	8, 9
*mdg*3	15	5.5 kb	283 or 268 bp[d]	15/18 or 2 bp[d]	5 or 3 bp[d]	10, 11, 12
B104	80	8.7 kb	429 bp	3 bp	5 bp	13
17.6	40	6.9 bp	512 bp	0 bp	4 bp	14, 17
mdg 4/gypsy	10	7.3 kb	479 bp	4/5 bp	4 bp	15, 16

[a]References: 1. Finnegan *et al.* (1978); 2. Potter *et al.* (1979); 3. Levis *et al.* (1980); 4. Dunsmuir *et al.* (1980); 5. Will *et al.* (1981); 6. Ikenaga and Saigo (1982); 7. Young and Rubin cited in Spradling and Rubin (1981) and Rubin (1983); 8. Ilyin *et al.* (1980c); 9. Kulguskin *et al.* (1981); 10. Ilyin *et al.* (1980a); 11. Ilyin *et al.* (1980b); 12. Bayev *et al.* (1980); 13. Scherer *et al.* (1982); 14. Kugimiya *et al.* (1983); 15. Modelell *et al.* (1983); 16. Bayev *et al.* (1984); 17. Inouye *et al.* (1984).

[b]A *412* element with LTRs 571 bp long has been found. LTRs of this length are probably rare in most strains of *D. melanogaster* (Will *et al.*, 1981).

[c]Ikenaga and Saigo (1982) have described a *297* element associated with duplication of the 6 bp sequence TATATA. Their data are also consistent with there being duplication of the 4 bp sequence ATAT. The latter interpretation agrees with the data of Young and Rubin.

[d]Bayev *et al.* (1980) have reported the sequence of one *mdg3* element. They interpreted their results as indicating that 5 bp of target site DNA had been duplicated and that the LTRs were 269 bp long with 13/15 bp imperfect inverted repeats at the ends of the complete *mdg3* element but not at both ends of the LTRs. Rubin (1983) has pointed out that their data can be reinterpreted as indicating that *mdg3* has 268 bp LTRs with 2 bp terminal inverted repeats and that 3 bp of target site DNA have been duplicated.

addition to the full length 5 kb RNA. The most abundant of these is 2 kb long (Carlson and Brutlag, 1978) but smaller transcripts have also been found (Young and Schwartz, 1981; Flavell *et al.*, 1980). The 5 and 2 kb RNAs are both capped at their 5′ ends and polyadenylated at their 3′ ends (Falkenthal and Lengyel, 1980; Flavell *et al.*, 1980; Young and Schwartz, 1981). Both strands of *mdg1* and *mdg3* are transcribed but in each case transcripts from one strand are at least 10 times more abundant than those from the other (Ilyin *et al.*, 1980b).

Only a small proportion of the transcripts from *copia*-like elements is bound to polysomes and most are associated with smaller ribonucleoprotein particles (Ilyin *et al.*, 1980d; Falkenthal and Lengyel, 1980; Young and Schwartz, 1981). Purified *copia* RNA can be translated inefficiently *in vitro* to yield a major 51,000-Da polypeptide and several smaller molecules (Falkenthal and Lengyel, 1980; Flavell *et al.*, 1980; Shiba and Saigo, 1983). Synthesis of these polypeptides seems to be directed preferentially by the 2 kb RNA (Flavell *et al.*, 1980).

Copia, 412, B104, and *297* RNAs have been found at various stages of *D. melanogaster* development. These RNAs are not expressed coordinately and different types of element are expressed at different times (Rubin *et al.*, 1981; Scherer *et al.*, 1981; Schwartz *et al.*, 1982). There is no evidence at present to indicate how many members of each family are expressed at any one time, nor is it known whether their expression is controlled autonommously or is merely a reflection of the state of adjacent genes. Meyerowitz and Hogness (1982) have found a strain of *D. melanogaster* with a *copia*-like element close to the *sgs3* glue protein gene. This element, known as "*roo,*" is a member of the *B104* family (Scherer *et al.*, 1982). These sequences are present at position 68C on the third chromosome. This region is puffed during the last half of the third instar when *sgs3* is expressed at a high level in salivary glands. Transcripts complementary to *roo* elements can be found in embryos but not larval salivary glands even in the strain with a *roo* element at 68C and at the stage at which this region is puffed and *sgs3* is expressed. Either this *roo* element is incapable of being expressed or it is regulated independently of *sgs3*.

3. *Genetic Interactions between Copia-like Elements and Other Genes*

Bender *et al.* (1983) have found that several mutations of the bithorax complex in *D. melanogaster* are associated with insertions of *copia*-like elements. Six of them are due to insertions of members of a family of elements which they called "*gypsy*". Members of this family had been discovered previously by Tchurikov *et al.* (1981) who called them *mdg4* elements (Bayev *et al.*, 1984). These are 7.3 kb long and have 0.5 kb terminal direct repeats. Some of the mutations associated with *gypsy* elements are unstable. One of these mutations, *bx34e,* has given a partial revertant with an altered restriction map indicating a rearrangement of the *gypsy* element. Two other mutations, *bx3* and *bxd1*, have given revertants with the wild-type phenotype. In this case the *gypsy* elements have apparently undergone recombination between their terminal repeats leaving one copy of the repeat within the gene (Bender *et al.*, 1983). The phenotypic effect of *gypsy* insertions in the bithorax complex can be altered by mutations elsewhere in the genome. Several extragenic suppressors have been identified in *D. melanogaster,* the best studied of which is the *suppressor of Hairy Wing, su(Hw)* (Lewis, 1949). This can suppress the phenotypic effects of particular mutations at at least 11 loci scattered throughout the genome, including all mutations of the bithorax complex associated with *gypsy* insertions (Bender *et al.*, 1983). Suppression occurs without any alteration to *gypsy* elements themselves.

Modelell *et al.* (1983) have suggested that all mutations suppressible by *su(Hw)* might be due to *gypsy* insertions. They have tested this by hybridizing a *gypsy* probe to chromosomes carrying one or more mutations suppressible by *su(Hw).* Fourteen alleles at 8 loci were tested, and in all but two cases there was hybridization to the position of the gene carrying the suppressible mutation. The exceptions were two alleles of the *rudimentary* gene which are fully suppressed by *su(Hw)* at 18°C but only partially suppressed at 25°C. These may be associated with a part of the *gypsy* element too small to be detected in this experiment or their suppression may be indirectly associated with, or independent of, *gypsy* elements (Modelell *et al.*, 1983). These results support the suggestion that most, if not all, mutations suppressible by *su(Hw)* are due to *gypsy* insertions. They do not indicate what proportion of mutations caused by *gypsy* insertions is suppressable by *su(Hw).*

The effect of the *su(Hw)* mutation on *gypsy* insertions is probably just one of many interactions between transposable elements and other components of the genome. The phenotypic effect of the *white-apricot* mutation is reduced by the *suppressor of white-apricot* mutation, su(w^a), and is increased by *suppressor of forked su(f)* (Green, 1959). This presumably reflects interactions between the products of these genes and the *copia* element present in *white apricot.* Similar observations have been made in yeast where the phenotypic effect of Ty element insertions may be modified by mutations elsewhere in the genome (see Section II,D,3,a).

4. *Copia-like Elements in Other Drosophila Species*

Little is known about the genome organization of other *Drosophila* species but *copia*-like elements are probably not confined to *D. melanogaster.* Moderately repetitive DNA in general, and *copia*-like elements in particular, appears to have changed rapidly during evolution of *Drosophila* species and is a very flexible component of the genome. Disperse repetitive sequences are generally less frequently repeated in *D. simulans* and *D. mauritiana* than in *D. melanogaster* (Meselson *et al.*, 1980; Young and Schwartz, 1981; Dowsett and Young, 1982) and *D. simulans* has only 2–4 copies of sequences homologous to *copia* elements (Martin and Schedl, quoted in Spradling and Rubin, 1981). This is surprising since the chromosomes of *D. melanogaster* differ from those of *D. simulans* and *D. mauritiana* by only one large and three small inversions and these species can be crossed to give viable progeny (Tartof, 1979; Dowsett and Young, 1982).

Some species more distantly related to *D. melanogaster* contain *copia* sequences but *copia* elements are well conserved only in its sibling species. Particular families of *copia*-like elements may be absent altogether in some species. There are no sequences homologous to *mdg1* or *mdg3* in *D. virilis* (Tchurikov *et al.*, 1981) and no sequences homologous to *copia* and *412* in *D. hydei* (Martin *et al.*, 1983). Some moderately repetitive sequences in *D. simulans* have some of the properties of transposable elements (Young and Schwartz, 1981) but their structure has not been determined.

5. *Similarities between Copia-like Elements and Proviruses*

Several authors have noted the similarities between the sequence organization of *copia*-like elements and of proviruses of vertebrate retroviruses (Levis *et al.*, 1980; Bayev *et al.*, 1980; Finnegan, 1981; Young and Schwarz, 1981; Scherer *et al.*, 1982). Proviruses (1) are of similar length to *copia*-like elements, (2) are bounded by direct repeat sequences a few hundred base pairs long, (3) are flanked on each side by a small number of bases which occur once at the site of insertion in the absense of the element, and (4) are transcribed into full length RNAs starting in one LTR and ending in the other (Fig. 2b). The similarity even extends to the first and last two bases of each element. All proviruses that have been analyzed so far start with the sequence T-G and end with C-A (Weiss *et al.*, 1982) as do most *copia*-like elements.

The terminal direct repeats of proviruses, also known as LTRs, are formed during the events which take place after a retrovirus has infected a host cell. They contain sequences from both ends of the viral RNA. At the left-hand end of an LTR is a unique sequence from the 3′ end of the viral RNA. Adjacent to this is a sequence which is repeated at both ends of the viral RNA and a unique sequence from its 5′ end. These are designated U3, R, and U5, respectively, in Fig. 2b. Viral RNA molecules are first reverse-transcribed into linear double-stranded DNAs with an LTR at each end. These are converted to circular molecules with either one LTR, or two LTRs in tandem and separated by four base pairs. It is not known which of these DNAs integrates into the host chromosomes but some evidence favors a circular form (Ramareddy *et al.*, 1975; Shoemaker *et al.*, 1980). This process has been discussed in detail by Varmus (1983).

Synthesis of the progeny viral RNA is presumed to start at the beginning of R in the left-hand LTR of the provirus and to terminate near the end of the right-hand LTR. Putative promoter and polyadenylation signals have been found at appropriate positions in all LTRs which have been sequenced. Synthesis of the first DNA strand is primed by a transfer RNA hydrogen bonded to the viral RNA by its 3′ terminal 18 nucleotides. The sequence to which this tRNA hybridizes, the primer binding site, can be found two base pairs to the right of the left-hand LTR. A conserved purine rich sequence adjacent to the right-hand LTR is believed to be involved in priming synthesis of the second DNA strand.

The genomes of most, if not all, vertebrates contain endogenous proviruses which can make up a substantial proportion of their DNA. These sequences comprise about 0.1% of the genomes of primates and rodents, for example (Todaro *et al.*, 1980). Proviruses not only resemble *copia*-like elements in structure, they also appear to be transposable. The chromosomal locations of endogenous proviruses can differ between individuals in populations of mice and chickens (Cohen and Varmus, 1979; Hughes *et al.*, 1979) and integration of both exogenous and endogenous proviruses can cause mutations. Proviral sequences have been found associated with mutations which inhibit the expression of genes affecting coat color (Jenkins *et al.*, 1981) or embryonic development (Jaenisch *et al.*, 1983) of mice and which activate a cellular oncogene, *c-myc,* in chickens (Hayward *et al.*, 1981; Payne *et al.*, 1981). The last is particularly interesting since activation of *c-myc* by avian leukosis virus can occur when the provirus is inserted upstream of the gene and in the opposite orientation with respect to transcription (Payne *et al.*, 1981). This may indicate that the provirus contains an enhancer sequence (Benoist and Chambon, 1981) able to stimulate transcription by a mechanism independent of its orientation.

If *copia*-like elements and proviruses really are related one would expect to find within *copia*-like elements some of the conserved functional sequences characteristic of proviruses. This is the case, although not all elements have all of these sequences. All *copia*-like elements for which sequence information is available have purine-rich sequences, very similar to those of proviruses, adjacent to their right-hand LTRs and sequences similar to primer binding sites following their left hand LTRs. The putative primer binding sites of *297, 17.6, mdg1,* and *412* elements are similar to that of avian leukosis virus while that of *B104* is similar to that of mouse sarcoma virus (Kugimiya et al., 1983; Scherer *et al.*, 1982). Polyadenylation signals and promoter-like sequences are present in the LTRs of *297, 17.6, B104,* and *412* elements. Kugimiya *et al.* (1983) have made detailed comparisons of all these regions and conclude that *B104, 297, 17.6, mdg1,* and 412 elements resemble proviruses more closely than do *copia* and *mdg3* elements (Shiba and Saigo, 1983).

Since circular molecules containing proviral sequences are normal intermediates in the retroviral life cycle Flavell and Ish Horowicz (1981) wondered whether the circular DNAs present in *D. melanogaster* cells (Stanfield and Helinski, 1976) include molecules derived from *copia* elements. They purified circular DNAs from *D. melanogaster* tissue culture cells and found that they included molecules with all the sequences from the body of a *copia* element and either one LTR or two LTRs in tandem. Shepherd and Finnegan (1984) have found circular DNAs containing *412* sequences in a similar way. These included molecules with all the internal sequences of a *412* element plus one LTR, but no molecules with two LTRs in tandem. Junakovic and Ballario (1984) have screened DNA from *D. melanogaster* tissue culture cells for circular molecules containing sequences from several *copia*-like elements. They found evidence for circular *copia, 412, 297,* and *mdg1* elements, but could detect no *B104* circles.

Reverse transcription is not the only possible source of these circular molecules. They could be derived from chromosomal elements by recombination. A molecule with one LTR could be produced by recombination between LTRs at the ends of an element, and a molecule with two LTRs by recombination between sequences on either side of an element. The latter mechanism is supported by base sequence analysis of *copia* circles with two LTRs in tandem. Flavell and Ish Horowicz (1983) have found that the LTRs are separated by 0–15 bp of variable sequence instead of the 4 bp sequence T-T-A-A which is characteristic of circular proviruses. Nevertheless, a proportion of *copia* circles probably is produced by the action of reverse transcriptase. Flavell (1984) has found that extrachromosomal linear and circular *copia* molecules can incorporate a density label more rapidly than chromosomal DNA. This can continue in the presence of an inhibitor of the main cellular DNA polymerase, suggesting that reverse transcription may be involved.

The most convincing evidence for a link between *copia*-like elements and retroviruses would be the discovery of virus particles containing RNA from these elements. No infectious retroviruses have been found in *D. melanogaster* although virus-like particles have been seen in tissues and cells in culture (Schneider and Blumenthal, 1978). Heine *et al.* (1980) were able to isolate small amounts of subcellular particles associated with high-molecular-weight RNA and reverse transcriptase activity from *D. melanogaster* tissue culture cells and Shiba and Saigo (1983) have purified noninfectious virus-like particles, VLPs, from sonicated tissue culture cell nuclei. These particles were associated with two classes of RNA, low-molecular-weight 4–6 S RNAs and heterogeneous high-molecular-weight RNAs about 5 kb long. Molecules containing *copia* sequences were included in the latter class. Strangely they could not detect *412, 297,* or *17.6* sequences in this RNA.

Retroviruses code for at least some of their coat proteins and the same appears to be true of *D. melanogaster* VLPs. Shiba and Saigo found that VLP RNA stimulated *in vitro* synthesis of polypeptides, some of which reacted with anti-VLP antisera. Most of the coding capacity of the VLP RNA was due to *copia* sequences since prehybridization of the RNA with *copia* DNA almost completely abolished its template activity. Shiba and Saigo also obtained preliminary evidence that reverse transcriptase activity copurified with VLPs. These results are consistent with the notion that a substantial proportion of VLPs are *copia* particles and that *copia*-like elements are related to retroviruses.

Temin (1980) has cogently argued that the similarity between proviruses and transposable elements with long terminal direct repeats indicates that retroviruses evolved from transposable elements. Transposable elements might equally well be degenerate proviruses. These possibilities are not necessarily mutually exclusive and at present there is no way to distinguish between them.

B. Intracisternal A Type Particles in Mice

Intracisternal A type particles, or IAPs, are noninfectious virus-like particles found in mouse oocytes and preimplantation embryos (Callarco and Szollosi, 1973; Yotsuganagi and Szollosi, 1981) and certain mouse tumor cells (Kuff *et al.*, 1972; Lueders *et al.*, 1977; Lueders and Kuff, 1977). They contain several discrete polyadenylated RNAs which range in size from 29 S to 35 S and comprise nearly 8% of polyadenylated RNA in some cells. These RNAs are related in sequence and code for the major IAP structural protein of 73,000 Da (Lueders *et al.*, 1977; Paterson *et al.*, 1978).

DNA sequences complementary to IAP RNA are repeated about 1000-fold in the genome of *Mus musculus* (Lueders and Kuff, 1977) and can also be detected in DNA from rat, gerbil, and Syrian hamster cells (Lueders and Kuff, 1981). Fragments of *Mus musculus* DNA containing these sequences have been cloned and studied (Ono *et al.*, 1980; Lueders and Kuff, 1980; Kuff *et al.*, 1981; Shen-Ong and Cole, 1982). The majority of IAP complementary sequences are present in two types of element, Type I and Type II elements, which comprise 75 and 10% of IAP complementary sequences, respectively. Type I and Type II elements are well conserved although several subtypes can be distinguished. Type I elements are about 7.3 kb long. Type II elements are about 4.8 kb long but are not simply deletion derivatives of Type I elements (Shen-Ong and Cole, 1982).

These elements have several properties in common with *copia*-like elements. They have long direct repeats at each end and are complementary to full-length RNAs. There is no evidence as yet to indicate how many IAP elements are transcriptionally active in any particular tissue or tumor cell line. Different RNAs predominate in different tumors and this presumably reflects differential expression of particular IAP variants (Paterson *et al.*, 1978). The long direct repeats at the ends of IAP elements are variable in length but related in sequence (Kuff *et al.*, 1981; Cole *et al.*, 1981). The repeats at both ends of one cloned IAP element have been sequenced (Kuff *et al.*, 1983a). They have all the characteristics of the LTRs of proviruses and *copia*-like elements. These IAP LTRs are both 338 bp long and differ at only 7 of the 330 known positions; eight bases have yet to be determined. Both sequences start T-G and end C-A and have 4 bp inverted repeats at each end. Within the LTRs there are putative promoter and polyadenylation signals. There is a putative tRNA primer binding site just after the LTR at the end of the element complementary to the 5′ end of IAP RNA, and a purine-rich sequence just before the LTR at the other end. Immediately before and after the IAP element is a direct repeat of 6 bp. This suggests that the IAP element entered the DNA by a mechanism akin to that of proviruses and *cop-*

ia-like elements, although in this case it is not known whether the 6 bp were present in one or two copies prior to insertion.

IAP elements can transpose and have been found associated with three mutations in mouse cells. Kohler and Shulman (1980) have isolated mutant derivatives of a mouse hybridoma, two of which do not produce immunoglobulin κ light chains specific for the hapten with which the parental mouse was immunized. These mutations are associated with insertion of IAP sequences within the κ chain gene. The elements involved have inserted in opposite orientations. One is in the intron separating the leader and Vk coding regions while the other is in the intron separating the Vk and Ck regions (Hawley *et al.*, 1982; Kuff *et al.*, 1983a). Presumably these IAP elements prevent expression of the κ chain gene. The third example of a mutation associated with an IAP element is more difficult to explain. In this case a cellular oncogene, *c-mos,* has been activated in a mouse myeloma, apparently by insertion of an IAP element within the gene (Rechavi *et al.*, 1982; Kuff *et al.*, 1983b). This cannot be due to activation of *c-mos* by the putative promoter in the LTR of the IAP element since transcription from this promoter would be in the opposite direction from that of *c-mos*.

There is no information available concerning the frequency of transposition of IAP elements. Kuff *et al.* (1981) have hybridized labeled IAP RNA to digests of genomic DNA from several laboratory and ferral strains of *Mus musculus* and have found homology to fragments of very similar size in each case. This contrasts with the pattern of hybridization seen when *copia*-like elements are hybridized to digests of DNA from different *D. melanogaster* strains (Potter *et al.*, 1979). This does not indicate that IAP elements have the same chromosomal distribution in these strains. The bands of hybridization seen in this experiment probably all correspond to fragments from internal regions of IAP elements which would be repeated in the genome. Unique fragments from the ends of each element would probably not have been detected because of the complexity of the mouse genome, and it is only by comparing these that one can compare the distribution of elements. The frequency of IAP transposition may differ from tissue to tissue and there may be a link between transposition and IAP RNA levels. The mutations associated with IAP elements arose in myeloma or hybridoma cells which have high levels of IAP RNA and Shen-Ong and Cole (1982) have found that two myelomas which express an RNA colinear with a variant Type II IAP element have more copies of the same variant in their DNA. This may indicate that transposition involves reverse transcription of a full length IAP RNA. This will be discussed more extensively in Section VI.

C. A *Copia*-like Transposable Element from a Lepidopteran

The baculovirus *Autographica californica* nuclear polyhedrosois virus, or ACNPV, can infect over 25 species of Lepidoptera and is used in the United States as a biological pesticide. It has a circular double-stranded DNA genome

128 kb long which replicates in the nuclei of infected cells. This is contained in a rod-shaped capsid which can package various lengths of DNA (Carstens, 1980; Miller, 1981).

Miller and Miller (1982) have isolated mutants of ACNPV after 25 passages of growth on tissue culture cells of the Lepidopteran *Trichopulsia ni*. One of these mutants is associated with insertion of a 7.3 kb element within the ACNPV genome. This element, which they have called TED, does not contain viral sequences but apparently came from the genome of the *T. ni* cells. It has several properties in common with *copia*-like elements. It is repeated and conserved in the *T. ni* genome and is present at higher copy number in DNA from tissue culture cells as compared with larval DNA. The 7.3 kb element is bounded by direct repeat sequences 0.27 kb long. It is presumably a transposable element since it was able to move from the cellular to the viral genome even though the viral DNA does not normally integrate into the chromosomes of host cells.

D. Ty Elements in *Saccharomyces cerevisiae*

1. *Structure of Ty Elements*

The transposable element Ty1 of *S. cerevisiae* was discovered shortly after the *copia*-like elements of *D. melanogaster* (Cameron *et al.*, 1979). It was first detected as a sequence which was present near the *SUP4* locus in some strains of *S. cerevisiae* but not others. Cameron *et al.* (1979) digested DNA from several strains of *S. cerevisiae* with restriction enzymes and hybridized them, in a Southern transfer experiment, with a probe containing Ty1 sequences. About the same number of fragments hybridized to the probe in each case but the patterns of hybridization differed from strain to strain suggesting that the Ty1 element is transposable. This has been confirmed by the discovery that many mutations are due to insertion of Ty elements within or adjacent to genes (Farabaugh and Fink, 1980; Errede *et al.*, 1980; Williamson *et al.*, 1981).

The Ty1 element is a member of a family of related transposable elements. It is about 6.2 kb long including terminal direct repeats, known as δ sequences, about 330 bp long. Most laboratory strains of *S. cerevisiae* contain 30–35 complete Ty elements and about 100 additional copies of the δ sequence per haploid genome. Delta sequences separate from complete Ty elements are called "solo δs" (Cameron *et al.*, 1979). Natural isolates of *S. cerevisiae* have variable numbers of Ty elements and a strain with as few as 5 copies has been found (Eibel *et al.*, 1981). Several Ty elements and solo δs have been analyzed at the base sequence level. Ty elements are flanked by 5 bp direct repeats one copy of which is generated during Ty transposition (Farabaugh and Fink, 1980; Roeder *et al.*, 1980). The 5 bp duplications associated with 9 different Ty elements have been compared by Roeder and Fink (1983). These sequences are not obviously related apart from being A/T rich, suggesting that there is little sequence specificity associated with transposition. Nearly all Ty associated mutations analyzed so far are due to insertions into regulatory regions. Only two instances of insertions into structural regions have been reported. One is in the LYS2 gene (Simchen *et al.*, 1984) and the other into the URA3 gene (Rose and Winston, 1984). This bias toward insertion into regulatory regions is probably partly because regulatory mutations have been deliberately selected in some experiments. It may also reflect some sequence or structural preference on the part of Ty elements themselves.

The δ sequences associated with 9 Ty elements have been sequenced. They show considerable sequence heterogeneity although, in general, δs at the ends of any one Ty element are identical or very similar (Roeder and Fink, 1983). They all start with the sequence T-G and end with C-A, these bases constituting 2 bp terminal inverted repeats. Two classes of δ sequence have been recognized and these may differ by as many as 52 out of 333 nucleotides (Roeder and Fink, 1983). The sequences of solo δs are even more variable (Tschumper and Carbon, 1982). This may indicate that the δs associated with Ty elements have some functional significance which solo δs do not. They may be required for transposition, for example. The δ sequences at the ends of a Ty element may be closely related because they are generated from a single δ during transposition or because they exchange information by gene conversion events.

Ty elements can also be divided into two classes, the members of which differ by two large substitutions (Roeder *et al.*, 1980; Kingsman *et al.*, 1981). The major class usually exists in about 25–30 copies per haploid genome. The Ty1 element is a member of this class as is the Ty element, Ty912, inserted at the *HIS4* locus in the *his4-912* mutation. The second class has about 6 copies per haploid genome and includes the Ty element, Ty917, inserted at HIS4 in the *his4-917* mutation (Roeder and Fink, 1983). The members of each class are heterogeneous and can be distinguished by restriction site polymorphisms.

2. *Transcription of Ty Elements*

Two abundant polyadenylated RNAs complementary to Ty elements can be found in *S. cerevisiae* cells. The major species is 5.7 kb long and the minor species 5.0 kb. The 5′ end of the 5.7 kb RNA lies 236–240 bases from one end of the element. Transcription proceeds through the entire internal region of the element and terminates 33–46 bases from the other end. This transcript thus contains δ sequences at both ends, approximately 45 bases of which are repeated. The 5.0 kb transcript has the same 5′ end but terminates at an internal site(s) (Elder *et al.*, 1981; Elder *et al.*, unpublished results, cited in Roeder and Fink, 1983). It is not known how many Ty elements are expressed in any one cell nor whether both 5.7 and 5.0 kb RNAs can be transcribed from any one element.

Transcription of Ty elements is affected by mating type. Ty RNA comprises 5–10% of polyandenylated RNA in cells of *a*, α, *a*/*a* and α/α mating type. This is reduced by about one fifth in log phase, and about one-twentieth in stationary phase cells of *a*/α mating type. The level of 5.0 kb transcript is reduced in stationary phase cells whatever their mating type (Elder *et al.*, 1981).

3. *Mutations Caused by Ty Elements*

a. *Mutations Affecting the Level of Gene Expression.* Insertion of Ty elements upstream of structural genes may increase or decrease their expression. The *his4–912* and *his4–917* mutations are due to insertions of different Ty elements in opposite orientations and at different positions within the regulatory region of the *HIS4* gene. Both insertions prevent transcription of *HIS4* and yet transcription of Ty912 would be in the same direction as *HIS4* while transcription of Ty917 would be in the opposite direction (Roeder and Fink, 1983).

Roeder and Fink (1982) have generated a series of derivatives of *his4–912* and *his4–917* in which different Ty elements have substituted for Ty912 and Ty917. The new Ty elements are present at the same positions and in the same orientations as Ty912 and Ty917 but can have very different effects on *HIS4* expression. The *his4–917* mutation has yielded His$^-$, weakly His$^+$, and cold-sensitive His$^+$ derivatives the expression of which is affected by mating type. The *his4–912* mutation has yielded His$^-$ and His$^+$ derivatives but none that is under the control of mating type.

The effects of Ty912 and Ty917 on *HIS4* expression are to some extent determined by the products of at least seven genes elsewhere in the *S. cerevisiae* genome (Winston *et al.*, 1984). These were originally referred to as *SPM* genes (Roeder *et al.*, 1980; Roeder and Fink, 1983), but are now called *SPT* genes (Winston *et al.*, 1984). Mutations in these genes were isolated by selecting for revertants of the His$^-$ phenotype of *his4–912, his4–917,* or *his4–912*δ. The last mutation contains a single δ element at *HIS4,* and was derived from *his4–912* by recombination between the δ sequences at the ends of Ty912. Mutations in each *SPT* gene suppress *his4–912*δ. The *his4–917* mutation is suppressed by *spt1–1, spt2–1, spt3–1,* and *spt6–140,* but not by *spt4–3, spt5–194,* and *spt7–159.* The His$^-$ phenotype of *his4–912* is suppressed only by *spt1–1.* Simchen *et al.* (1984) have found that *spt2* and *spt3* mutations can also suppress the effects of some Ty and δ insertions at *LYS2.* The mechanism of *spt* suppression is not known, although suppression of *his4–912* by *spt1, spt2,* and *spt3* mutations has been found to affect its transcription (Silverman and Fink, reported in Winston *et al.*, 1984). The apparent effects of *spt* mutations on δ–δ recombination, reported by Roeder *et al.* (1980) are now known to have been an experimental artifact (Winston *et al.*, 1984).

Insertion of a Ty element adjacent to a gene may increase expression of that gene. Events of this type were first detected in screens for mutations increasing expression of the *CYC7* gene, which codes for iso-2-cytochrome *c* (Errede *et al.*, 1980, 1981) and the *ADR2* gene which codes for alcohol dehydrogenase II (Williamson *et al.*, 1981; Young *et al.*, 1982). Constitutive mutants were isolated and found to have Ty elements inserted in the 5′ regulatory region of these genes. The level of constitutive expression of these mutants depended on mating type, being high in *a*, α, *a*/*a*, and α/α cells and low in *a*/α cells. These mutations have been called ROAM mutations (regulated overproducing alleles responding to mating type). The simplest explanation for this phenotype is that expression is initiated within the Ty element since Ty expression is also affected by mating type. This cannot always be true since Ty transcription is in the opposite direction to that of the affected gene in some ROAM mutations (Young *et al.*, 1982; Roeder and Fink, 1983) and in one case transcription is known to start at the normal position within the affected gene (Young *et al.*, 1982). This may indicate that some Ty elements contain enhancer sequences.

b. *Mutations Affecting Genome Organization.* Recombination events involving the large number of Ty elements and δ sequences distributed throughout the *S. cerevisiae* genome can affect the expression of adjacent genes and lead to gross chromosomal rearrangements. The least disruptive event is recombination between δ sequences at the ends of a Ty element. This generates a circular molecule containing one δ sequence and leaves a second δ sequence within the chromosome. This is the most likely source of solo δs since δ sequences themselves cannot transpose (Roeder and Fink, 1983). Excision of an entire Ty element or sequence, which could occur by recombination with the flanking 5bp direct repeats, has never been observed (Roeder and Fink, 1983).

Recombination between Ty elements on the same chromosome will invert or delete DNA between them depending on their relative orientation. Recombination between elements in the same orientation will delete this DNA while recombination between elements in the opposite orientation will invert it. One revertant of *his4–912* is associated with large inversion of the left arm of chromosome III, the chromosome carrying the *HIS4* gene. Ty elements lie at both breakpoints of the inversion as expected if it arose by Ty–Ty recombination (Chaleff and Fink, 1980; Roeder and Fink, 1980). Liebman *et al.* (1979, 1981) have found a strain of *S. cerevisiae,* known as *DEL1,* in which the activities of three genes, *CYC1, OSM1,* and *RAD7,* are lost spontaneously at a frequency of 10^{-5}–10^{-6} per cell generation. In this strain these genes are contained on a 13 kb piece of DNA flanked by Ty elements. The mutant derivatives of *DEL1* have lost this 13 kb region as a result of Ty–Ty recombination. In *del1* strains no deletions occur and there are no Ty elements in the *CYC1* region (Liebeman *et al.*, 1981). Recombination between solo δs can also result in deletion. The *SUP4* gene is surrounded by solo δs in the same orientation (Cameron *et al.*, 1979). Deletion of *SUP4* by recombination between these δs occurs with a frequency of about 3×10^{-7} (Rothstein, 1979).

Recombination between Ty elements in the same orientation but at different sites on homologous chromosomes will produce a duplication in one chromosome and a deletion in the other (Fig. 3). Events of this kind occur between Ty elements at the HIS4 and LEU2 loci, which are 40 kb apart on chromosome III, in about 1% of meioses (Roeder and Fink, 1983). Similar duplication/deficien-

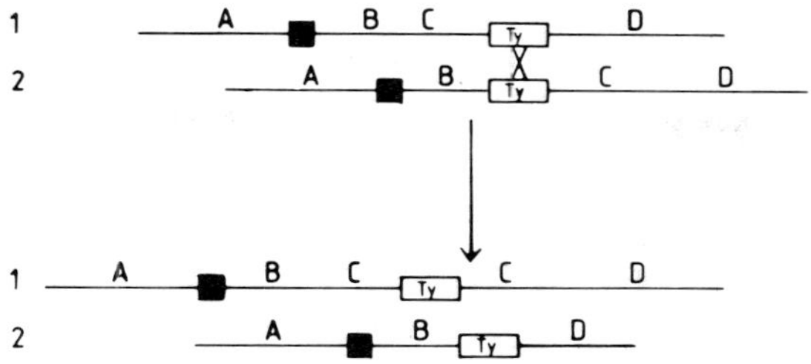

FIG. 3. Recombination between Ty elements in homologous chromosomes can cause deletions and duplications. Chromosomes 1 and 2 are homologues. Each contains a Ty element in the same orientation with respect to the centromere (■) but integrated at different sites. DNA from region C, which lies between these sites, has been deleted from chromosome 2 and duplicated in chromosome 1 as a result of recombination between the Ty elements.

cies have been found at the *white* locus of *D. melanogaster* due to recombination between *copia* elements associated with different *white* mutations (Goldberg *et al.*, 1983).

Recombination between Ty elements on different chromosomes should produce reciprocal translocations if the elements involved are in the same orientation with respect to the centromere of the chromosome carrying them. Recombination between Ty elements in the same orientation would produce a dicentric chromosome. Two revertants of the *his4–912* mutation are associated with translocations (Chaleff and Fink, 1980) and Ty elements lie at the translocation breakpoints in both (Roeder and Fink, 1980; Farabaugh and Fink, cited in Roeder and Fink, 1983).

Ty elements can apparently exchange information by nonreciprocal gene conversion events as well as by recombination. Roeder and Fink (1982) substituted Ty917 in *his4–917* with a Ty element containing the *URA3* gene in a *ura3* strain. They were able to isolate several Ura$^-$ derivatives of this strain in which the Ty-*URA3* element had been replaced by other Ty elements without altering flanking DNA. They believe that this is the result of gene conversion although double recombination cannot strictly be excluded.

III. Transposable Elements with Long Inverted Repeats

A. FOLD-BACK ELEMENTS IN *D. melanogaster*

The genomes of all eukaryotes contain inverted repeat sequences. These can be detected if their DNA is denatured and allowed to reanneal at low concentration. The first sequences to become double stranded are those that occur as inverted repeats. These repeats may be immediately adjacent to each other or up to several thousand base pairs apart (Wilson and Thomas, 1974; Chec and Hearst 1975; Schmid *et al.*, 1975). Since this is an intramolecular reaction structures formed in this way are called "snap-back" or "fold-back" structures (Fig. 4). Schmid *et al.* (1975) have estimated that there are 2000–4000 pairs of inverted repeats in the genome of *D. melanogaster* and that these make up about 3% of the total genome (Schmid *et al.*, 1975; Baker and Thomas, 1977). These inverted repeats and the DNA separating them include highly repetitive, moderately repetitive, and single copy DNA, and the inverted repeats themselves are 2- to 3-fold enriched for moderately repetitive sequences (Schmid *et al.*, 1975). Baker and Thomas (1977) have calculated that the total complexity of the inverted repeats is 5×10^5 and that, on average, each sequence is represent 10 to 20 times in this fraction of the genome.

Potter and his colleagues have made a detailed study of one family of inverted repeat structures in *D. melanogaster* and have found that they are transposable. Potter *et al.* (1980) screened a library of cloned fragments of *D. melanogaster* DNA to find those which would hybridize preferentially to a probe enriched for the stems of fold-back structures. They obtained one clone containing an inverted repeat and used it to detect others with related fold-back structures. They have called these "FB" elements. FB elements are very heterogeneous, in contrast to *copia*-like elements. The lengths of the inverted repeats vary from a few hundred to a few thousand base pairs and repeats at the ends of any one element need not be identical (Fig. 5). The DNA separating the inverted repeats also varys in

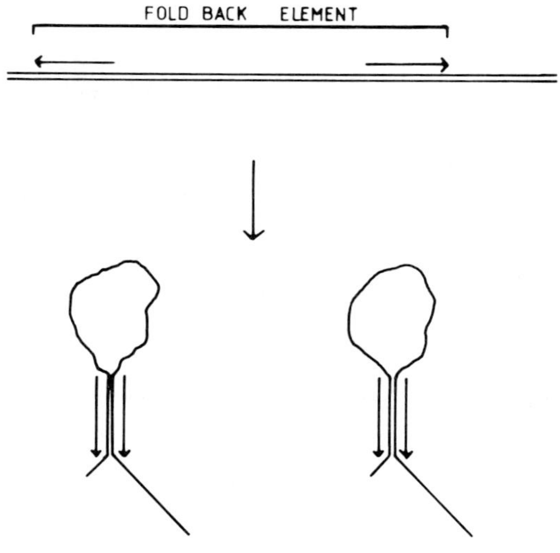

FIG. 4. Formation of a "fold-back" structure as a result of denaturation and renaturation of a DNA fragment containing an inverted repeat sequence.

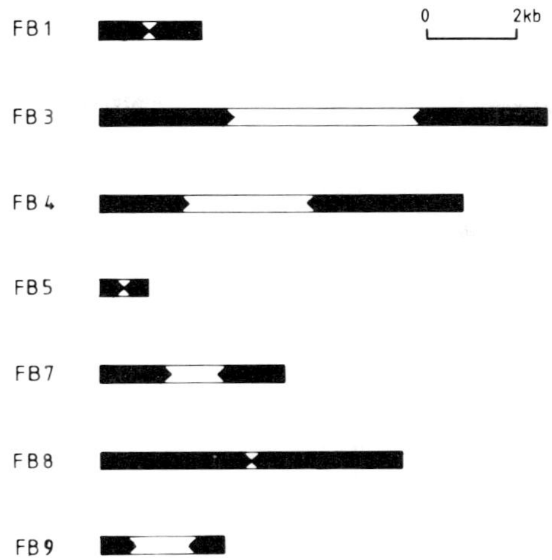

FIG. 5. Structures of different members of the FB family of transposable elements. The structures of 7 FB elements described by Truett *et al.*, (1981) are shown. The filled line indicates the extent of the inverted repeat sequences. FB1, FB5, and FB8 consist entirely of inverted repeat sequence. The FB4 element has been sequenced by Potter (1982).

length and sequence from one FB element to another. In most cases the central region is related to the inverted repeats themselves (Truett *et al.*, 1981) but FB elements containing unrelated sequences have also been found (Potter, 1982; Goldberg *et al.*, 1982; Levis *et al.*, 1982).

The inverted repeats of FB elements consist of moderately repetitive sequences. These are present at the chromocentre and at about 20–30 sites on the chromosme arms but not necessarily as inverted repeats in each case. The specific locations of these sequences varies from one *D. melanogaster* strain to another, suggesting that FB elements are transposable (Truett *et al.*, 1981; Levis *et al.*, 1982). Analysis of molecular lesions associated with mutations of the *white* gene of *D. melanogaster* has provided direct evidence that FB elements can transpose. The *white-crimson* mutation was obtained as a partial revertant of the *white-ivory* mutation by Green (1967). The *white-ivory* mutation is due to duplication of part of the *white* locus (Karess and Rubin, 1982) and *white-crimson* has an FB element inserted within this duplication (Collins and Rubin, 1982; Levis *et al.*, 1982). The *white-crimson* allele is unstable and reverts to *white-ivory* at a frequency of about 10^{-3} (Green, 1967). This is due to precise excision of this FB element (Collins and Rubin, 1983).

The inverted repeats of FB elements are not only repeated within the *D. melanogaster* genome but are also internally repetitious (Truett *et al.*, 1981; Levis *et al.*, 1982). Potter (1982) has sequenced the whole of the FB4 element and has found that its inverted repeats have a sequence organization similar to that of satellite DNA. Near the outside termini of the inverted repeats are multiple imperfect copies of a 10 bp sequence which are separated by regions of A/T-rich DNA. At nucleotide 222 the 10 bp repeat is expanded to 20 bp repeats which are also imperfect and separated by A/T-rich DNA. At nucleotide 499 the 20 bp repeat is expanded to 31 bp repeats which continue as tandem repeats for the rest of the inverted repeats. There are five variants of the 31 bp sequence and these alternate to make a 155 bp unit which can be recognized easily as it contains one site for the restriction nuclease *Taq*I.

The two inverted repeats of FB4 differ by single base changes, by the presence/absence of particular copies of the 31 bp repeat and in the number of 155 bp repeats. These differences in repeat number could arise from interrepeat or intrarepeat recombination events. This is probably the reason why the inverted repeats of most FB elements are of different lengths. The extreme outside termini of FB elements appear to be highly conserved (Truett *et al.*, 1981; Levis *et al.*, 1982) and end with exact inverted repeats 31 bp long (Potter, 1982; Collins and Rubin, 1983).

FB elements are flanked by short direct repeats like other transposable elements. In this case transposition is associated with duplication of 9 bp of target site DNA (Truett *et al.*, 1981; Collins and Rubin, 1983).

B. TE Elements: The Largest Eukaryotic Transposable Elements

Ising and Ramel (1976) first identified the TE transposing element in a strain of *D. melanogaster* in which the *white* region of the X chromosome had moved from its normal location to a site on the second chromosome. The element involved in this event, TE1, carried the *white-apricot* allele and the adjacent gene, *roughest*. These genes were not stable at this position but could transpose to new sites or be lost altogether, both events occurring with a frequency of about 10^{-3} (Ising and Block, 1981). A series of related elements have been derived from TE1 by subsequent transposition events. Each TE must contain several hundred kilobases of DNA and in favorable cases can be detected as visible insertions in polytene chromosomes (Ising and Block, 1981). About 150 different transpositions have been identified and about 90 of these have been mapped cytologically.

TEs are unstable and may change their genetic content. This is often associated with transposition. For example, TEs have been found to have a near wild-type *white* allele, to have duplicated or lost the entire *white* gene, or to have incorporated new genes (Ising and Block, 1981; Paro *et al.*, 1983). Paro *et al.* (1983) have used recombinant DNA techniques to examine both ends of TE77 and TE98 and one end of TE28. They found that both TE77 and TE98 are composite transposable elements containing a central region, including the *white* and *roughest* genes, flanked by FB elements. The FB elements at the ends of either TE77 or TE98 are not identical. The FB element at one end of TE28 appears to be the same as that responsible for the *white-crimson* mutation (Levis *et al.*, 1982; Paro *et al.*, 1983). This allele is itself transposable (Green, 1969a,b), possibly because it contains this FB element.

Another transposable element flanked by FB elements has been identified near the *white* locus. Bingham (1980, 1981) has isolated an unstable *white* mutation, w^{DZL}, due to insertion of a 13 kb element adjacent to the *white* gene. This element is made up of two FB elements flanking a 6.5 kb central region. The DNA in the central region is nonrepetitive and originated from near the tip of the second chromosome in the strain in which the w^{DZL} mutation arose (Levis *et al.*, 1982).

These results suggest that any segment of DNA flanked by FB elements can transpose. This might explain the changes in genetic content of TEs associated with their transposition. The original TE, TE1, may have contained an internal FB element as illustrated in Fig. 3. If TE1 were to transpose to a new site flanked by FB elements then two new TEs could be formed, one having the FB element within TE1 as its left-hand boundary and the other having this element as its right-hand boundary (Fig. 6).

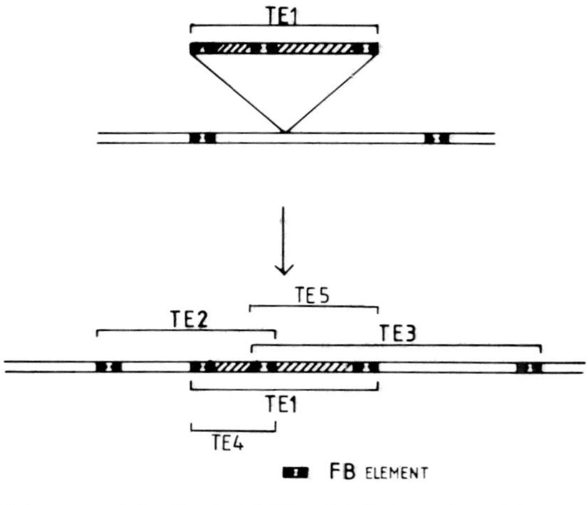

FIG. 6. A model to account for the instability of TE elements and the generation of new elements. TE1, containing an internal FB element, has inserted at a site flanked by chromosomal FB elements. Transposition events utilizing the internal FB element could generate new TE elements TE2, TE3, TE4, and TE5. TE2 has the internal FB element as its right-hand boundary and a flanking FB element as its left-hand boundary. It has lost DNA from the right-hand end of TE1 and has aquired chromosomal DNA lying to the left of it. TE3 has similarly lost DNA from the left-hand end of TE1 and has aquired chromosomal DNA lying to the right of it. TE4 and TE5 each retain one end of TE1 but use the internal FB element as the other. They just consist of a portion of TE1.

Composite transposable elements with FB elements at each end probably do not excise precisely from the genome. Ising and Ramel (1976) have found several examples of transposition of a TE element being associated with induction of a recessive lethal mutation at, or near, the site of insertion. The *white* gene carried by these TEs may subsequently be lost but the lethal mutation remains. This suggests that part of a TE element can be lost without the element excising precisely. An alternative, but less likely, explanation is that these lethal mutations arose independently of TE insertion. The composite element responsible for the w^{DZL} mutation is also unable to excise precisely. Levis and Rubin (1982) have analyzed 12 phenotypic revertants of the w^{DZL} mutation. Each of them has lost part of the w^{DZL} element but in no case has it excised precisely. The FB element responsible for the *white-crimson* mutation can excise precisely (Collins and Rubin, 1982) and this may also be true of FB elements at the ends of composite transposable elements. Excision of TE elements may be rare because of their composite structure since interactions between the ends of different FB elements are probably much less frequent than interactions between the ends of a single element.

B. FB-like Elements in Other Species

Liebermann *et al.* (1983) have found a transposable element with long inverted repeats, in the genome of the sea urchin *Strongylocentrotus purpuratus*. It was first identified as an insertion within a histone H2B gene where it was flanked by an 8 bp target site duplication. They have called this element TU1. There are 200–400 copies of TU1 in the genome of *S. purpuratus,* and their chromosomal distribution differs from one individual to another. The inverted repeats of the copy of TU1 which has been studied in most detail are about 840 bp long, and are made up of two domains. The outer domain comprises tandem direct repeats of a basic 15 bp sequence while the inner domains are of nonrepeating sequences, Despite the similarity between TU1 and FB elements there is no detectable homology between them.

Inverted repeats can be detected in rapidly reannealing DNA from many species. The bulk of inverted repeat structures in human DNA are comprised of two copies of the 300 bp ''Alu'' sequence in inverted orientation and separated by variable lengths of unrelated DNA (Deininger and Schmid, 1976). Individual Alu sequences may be able to transpose (Jagadeeswaran *et al.,* 1981; Rogers, this volume) but there is no indication that they behave like FB elements. This does not mean that humans and other vertebrates do not contain transposable elements with long inverted repeats. Even in *D. melanogaster* FB elements make up only a few percent of inverted repeat DNA and there is no evidence that the rest of this fraction of the genome is transposable.

IV. Transposable Elements with Short Inverted Terminal Repeats

A. ELEMENTS CONTROLLING HYBRID DYSGENESIS IN *D. melanogaster*

Hybrid dysgenesis is the name given to the appearance of a set of unusual characteristics in the progeny of crosses between certain strains of *D. melanogaster*. These characteristics include lowered fertility, recombination in males, and increased frequencies of mutations and chromosome aberrations. Two excellent reviews of hybrid dysgenesis have been published recently by Bregliano and Kidwell (1983) and Kidwell (1983a) and I shall only summarize them briefly here.

Two independent systems of hybrid dysgenesis, P-M and I-R, are known and strains of *D. melanogaster* may be classified as being of one or other of two types in each system. In the P-M system hybrid dysgenesis is seen in the progeny of crosses between *M (Maternal)* strain females and *P (Paternal)* strain males. In the I-R system the effect if seen in the progeny of *I (Inducer)* strain males and *R (Reactive)* strain females. In both cases the progeny of reciprocal crosses are normal.

The dysgenic traits produced by the I-R and P-M interactions are similar and in each case seem to be confined to cells of the germline. There are some differences, the most obvious of which is that the I-R interaction affects only females whereas the P-M interaction affects both sexes. Both interactions result in a degree of infertility but with different physiological characteristics in each case. The gonads of P-M dysgenic individuals fail to develop whereas I-R dysgenic females lay eggs which stop developing prior to blastoderm formation. Kidwell (1979) has classified several strains with respect to both P-M and I-R dysgenesis and has found that the two systems are genetically independent.

The characteristics of *P* and *I* strains are determined by transposable genetic determinants known as *P* factors and *I* factors, respectively. Transposition of these elements is controlled in a complex way. *P* factors and *I* factors are stable in *P* and *I* strains but transpose when introduced into the cytoplasmic background of *M* or *R* strains, repsectively, in a dysgenic cross. Many P-M induced mutations are unstable in individuals subject to P-M dysgenesis (Engels, 1979; Rubin *et al.*, 1982) leading to the suggestion that mutations induced by P-M dysgenesis are due to insertions of *P* factor DNA (Green, 1977; Golubovsky *et al.*, 1977; Simmons and Lim, 1980). This has been confirmed by Rubin and his colleagues (Rubin *et al.*, 1982). They compared the structure of the *white* gene in DNA from flies carrying P-M induced white mutations with that of the parental wild-type allele. The mutations which they studied apparently contained *P* factor sequences since they were unstable in P-M dysgenic individuals. Each mutation had foreign DNA inserted into the *white* gene as expected. These insertions ranged in size from 0.5 to 1.4 kb and were of related sequences. Rubin *et al.* (1982; O'Hare and Rubin, 1983) argued that these insertions were too short to be fully functional *P* factors, which probably code at least for a transposase and a regulatory molecule, but that they might be inactive *P* elements derived from *P* factors by internal deletion. Presumably these *P* elements can transpose if a functional *P* factor is available to provide the necessary transposase.

O'Hare and Rubin (1983) used one of these putative *P* elements as a probe with which to screen a library of recombinant phages containing DNA from a *P* strain. They recovered several clones containing a consereved 2.9 kb sequence which they presumed to be the complete *P* factor. The best evidence that these cloned sequences are indeed functional *P* factors would be to demonstrate that they can stimulate P-M dysgenesis when introduced into the cytoplasmic environment of an *M* strain. Spradling and Rubin (1982) have tested this directly by injecting clones of a putative *P* factor into *M* strain embryos to see whether or not the resulting flies exhibit dysgenesis. The *M* strain embryos which they injected carried a mutation of the *singed* locus, sn^w (Engels, 1979). This is a P-M induced mutation which is extremeley unstable in dysgenic flies. If the injected DNA were to induce hybrid dysgenesis this would be revealed by mutations of sn^w to sn^+ or more extreme *singed* alleles. This was scored in the progeny of the flies resulting from the injected embryos. Injection of a putative *P* factor did destabilize the sn^w mutation confirming its identity and showing that hybrid dysgenesis can be produced in this way. The *P* factor was itself affected by dysgenesis and transposed from plasmid DNA to chromosomes of the injected *M* strain embryos. It integrated by transposition rather than recombination since the complete *P* factor had inserted into the chromosomes without any flanking DNA.

Restriction mapping and base sequence data indicate that *P* factors are highly conserved and that *P* elements can be derived from *P* factors by simple deletions. Each *P* factor and *P* element is flanked by 8 bp duplications of target site DNA. *P* elements do not insert at random since 3 of the 5 P-M induced *white* gene mutations studied by Rubin *et al.* (1982) were due to insertions at exactly the same position. O'Hare and Rubin (1983) have been able to derive a consensus sequence from the 8 bp duplications generated by *P* elements inserted at 8 different sites but they could find no obvious homology between the sequences flanking these sites.

Chromosome rearrangements occur at high frequency in the germline of P-M dysgenic individuals (Berg *et al.*, 1980; Simmons and Lim 1980; Engels and Preston, 1981). The breakpoints of these rearrangements are not randomly distributed in any one strain but occur at particular hotspots. Berg *et al.* (1980) and Engels and Preston (1981) mapped the hotspots on one X chromosome and suggested that they were the chromosomal locations of *P* factors. Bingham *et al.*

(1982) have confirmed that there are *P* factor sequences at these hotspots although it is not known whether these are of *P* factors or *P* elements.

O'Hare and Rubin (1983) have determined the base sequence of one complete *P* factor, and of several *P* elements. The *P* factor is 2907 bp long and is bounded by perfect inverted repeats of 31 bp. One strand of the *P* factor contains four open reading frames of 297, 714, 792, and 654 bp, respectively. Karess and Rubin (1984) have made mutations in each reading frame, and have tested the ability of the resulting mutant *P* factors to destabilize sn^{w} after injection into *M* strain embryos. Mutations in any one reading frame blocked this trans-acting function of *P*, but did not affect its ability to transpose when coinjected with a wild-type *P* factor. These open reading frames must determine a single function since *P* factors carrying mutations in different reading frames were unable to complement each other in pair-wise combinations. The most likely function affected by these mutations is that of a transposase.

The *I* factor which controls I-R hybrid dysgenesis has also been identified by examination of the molecular lesions associated with I-R induced *white* mutations. Five white mutations induced by I-R dysgenesis are due to insertions of apparently identical 5.4 kb transposable elements (Finnegan *et al.*, 1983; Bucheton *et al.*, 1984; Sang *et al.*, 1984). One of these mutations is very tightly linked to a genetically active *I* factor (Pelisson, 1981) indicating that the 5.4 kb element is the *I* factor. The DNA of two *I* factors has been cloned. No homology can be detected by hybridization between the ends of the *I* factor indicating that it does not contain long direct or inverted repeats.

P elements can excise either precisely or imprecisely in P-M dysgenic individuals (Rubin *et al.*, 1982; O'Hare and Rubin, 1983; Voelker *et al.*, 1984) and the same is probably true of *P* factors. There is no evidence that *I* factors can excise but they are unstable in I-R dysgenic individuals. Pelisson (1981) has found a mutation of the *white* gene which is due to insertion of an *I* factor and which determines a brown eye color. This allele does not revert in I-R dysgenic individuals but does mutate to alleles determining a more extreme *white* phenotype.

The simplest explanation for the difference between *P* and *M* strains or between *I* and *R* strains is that *P* elements are present in *P* strains but not *M* strains and that *I* factors are present in *I* strains but not *R* strains. All *P* strains contain *P* factor sequences and all *I* strains contain *I* factor sequences, as expected (Rubin *et al.*, 1982; Finnegan *et al.*, 1983; Bucheton *et al.*, 1984). The DNAs of some *M* strains lack detectable homology to *P* factor DNA but this is not always the case (Bingham *et al.*, 1982; Kidwell, 1983b). All *R* strains so far tested contain *I* factor sequences and in Southern transfer experiments DNAs from unrelated *R* strains show remarkably similar patterns of hybridization to *I* factor probes (Finnegan *et al.*, 1983; Bucheton *et al.*, 1984). Presumably these sequences in *M* and *R* strains represent genetically inactive *P* and *I* elements, respectively.

B. Controlling Elements in Maize

The first transposable genetic elements to be discovered were the controlling elements of *Zea mays*. They were detected because they can cause unstable mutations affecting pigmentation of the aleurone layer of the maize kernel. The first detailed studies of these unstable color mutations were carried out by Emerson (1914) and Rhoades (1938) but it was McClintock (1946, 1950) who first demonstrated that they are due to insertion of transposable elements. This work has been thoroughly reviewed by Fincham and Sastry (1974) and Fedoroff (1983).

McClintock coined the term ''controlling element'' for elements of this type as they appeared to control gene expression. In the simplest case this occurs when a controlling element inserts within, or adjacent to, a gene and inhibits its expression. This inhibition is relieved when the controlling element is excised. This may occur in either somatic or germline cells and may be correlated with insertion of the controlling element elsewhere in the genome. If the controlling element has inserted near or within a gene required for pigmentation of the aleurone layer this will result in colorless kernels. Excision of the element in aluerone cells will produce colored spots on a colorless background. Excision in germ cells can be detected in the phenotype of plants of the next generation. Some controlling elements can be detected even when they do not affect gene expression since they are sites of frequent chromosome breaks (McClintock, 1951).

The simplest controlling elements are said to be autonomous because they not only affect the expression of adjacent genes but also mediate their own excision and/or transposition. Many controlling elements are nonautonomous and can only transpose if a related autonomous element is present elsewhere in the genome. Fincham and Sastry (1974) have termed nonautonomous elements ''receptor elements'' and autonomous elements ''regulator elements.'' Regulator elements can promote their own excision and transposition as well as that of appropriate receptor elements. It is believed that receptors are derived from regulators by mutation (McClintock, 1962; Fincham and Sastry, 1974), perhaps by deletion in the same way that *P* elements are derived from *P* factors (O'Hare and Rubin, 1983). The receptor–regulator relationship is highly specific and four such systems are known at present (Peterson, 1977).

One of the best studied receptor–regulator pairs is the *Ac-Ds* system. The *Ds,* or *Dissociation,* element was first identified as a site of chromosome breakage. It cannot transpose autonomously and requires the presence of a second, autonomous, element called *Ac,* or *Activator.* An *Ac* element is also required for chromosome breakage at a *Ds* element. In some cases *Ac* elements have been shown to give rise to *Ds*-like derivatives suggesting that *Ac* and *Ds* elements are structurally related (McClintock, 1955, 1956, 1962, 1963).

Several groups have started to investigate controlling elements at the molecular level (Burr and Burr, 1981, 1982; Döring *et al.*, 1981; Courage-Tebbe *et al.*, 1982; Federoff *et al.*, 1983a; Peacock *et al.*, 1983). Osterman and Schwartz (1981) have isolated an unstable mutant of the alcohol dehydrogenase gene, *Adh1*, resulting from insertion of a *Ds* element. Peacock and his colleagues (Peacock *et al.*, 1983; Sutton *et al.*, 1984) have compared DNA in the region of the *Adh1* gene in the parental, mutant and revertant strains in order to identify the *Ds* element. The mutant contains a 405 bp insertion in the 5′ untranslated region of the gene. This insertion is not present in the parental or revertant strains and so is presumably the *Ds* element. It has several characteristics of a transposable element. It is flanked by 8 bp direct repeats of target site DNA and terminates in perfect inverted repeats 11 bp long.

Two revertants have been examined. In each case the 405 bp *Ds* element has been excised but the 8 bp duplication remains in a slightly altered form. These imprecise excisions could result from base pairing of the terminal inverted repeats to form a cruciform structure during DNA replication. Replication across the base of the cruciform would delete the *Ds* element but leave the 8 bp duplication in place. Short transposable elements, like this *Ds* element, are probably more likely to excise by this mechanism than longer ones.

Fedoroff *et al.* (1983b) have cloned an *Ac* element, *Ac9*, and two *Ds* elements, *Ds9* and *Ds6*, from mutant alleles of the *waxy* gene. The *Ac9* element is 4.3 kb long. The *Ds9* element was derived from *Ac9* (McClintock, 1963) and differs from it by a 0.2 kb deletion near the middle. This presumably affects the production of a transposase since *Ac9* can transpose autonomously but *Ds9* cannot. The *Ds6* element is only 2.0 kb long but is closely related to *Ac9*. Heteroduplex experiments indicate that it is homologous to 1.0 kb at either end of *Ac9* and lacks the central region. This is supported by similarities between their restriction maps. These *Ac* and *Ds* elements do not have extensive internal repetition and do not form fold-back structures after denaturation (Fedoroff *et al.*, 1983b).

Pohlman *et al.* (1984) have sequenced the whole of *Ac9*. It is 4563 bp long and contains two open reading frames encoding polypeptides of 839 and 210 amino acids. These are read outward from a short intergenic region containing a promoter-like sequence. The larger open reading frame probably codes for a transposase since the 0.2 kb deletion which distinguishes *Ds9* from *Ac9* removes part of this region. *Ac9* ends with an imperfect 11 bp inverted repeat. The left hand copy of the repeat is identical to that of the 405 bp *Ds* element studied by Peacock *et al.* (1983, Sutton *et al.*, 1984). The right-hand copy differs in one position. Like this *Ds*, *Ac9* is flanked by an 8 bp target site duplication. Pohlman *et al.* have studied one revertant of the *waxy* mutation associated with *Ac9*. It has retained six bases of the 8 bp duplication. This is similar to the situation found for revertants of the *Ds* associated mutation of *Adh1* (Sutton *et al.*, 1984).

Three groups have studied unstable mutants of the *Shrunken* gene, *Sh1* (Burr and Burr, 1981, 1982; Doring *et al.*, 1981; Courage-Tebbe *et al.*, 1983). These mutations are associated with *Ds* activity and were derived from a strain carrying a defective *Ds* element in an adjacent chromosomal region. This *Ds* element could not transpose but acted as a local mutagen (McClintock 1956, Courage-Tebbe *et al.*, 1983). The molecular changes associated with these mutations are more complicated than those described above. One of these mutations, *sh-m5933,* is associated with insertion of about 30 kb of DNA within the *Shrunken* gene and duplication of one end of the gene and part of the insertion. The other three mutations are associated with large insertions or inversions (Courage-Tebbe *et al.*, 1982, 1983; Fedoroff *et al.*, 1983a). The right-hand end of the large insertion in *sh-m5933* has been cloned (Geiser *et al.*, 1982). It contains two interspersed pairs inverted repeats a few hundred base pairs long, one of which appears to lie next to the point of insertion. The left-hand end of the insertion has not been cloned. Restriction analysis indicates that it is similar to the right-hand end but is in the inverted orientation and has lost the outermost inverted repeat (Courage-Tebbe *et al.*, 1982, 1983).

It is not clear exactly how this complex structure relates to the *Ac* and *Ds* elements cloned by Peacock *et al.* (1983) and Fedoroff *et al.* (1983a), but it is unlikely that the whole insertion in *sh-m5933* is composed of *Ds* sequences. The insertion may be flanked by *Ds* sequences so that the whole structure is a composite element similar to a TE element in *D. melanogaster*. Döring *et al.* (1984a) have determined the base sequence of 4 kb at the right-hand end of the *sh-m5933* insertion. This consists of two identical copies of a 2040 bp element in opposite orientation. One element is inserted within the center of the other giving rise to the two pairs of interspersed inverted repeats detected by restriction analysis. This 2 kb element is probably responsible for the *Ds* activity of *sh-m5933*. It has 11 bp terminal inverted repeats identical to those of the 0.4 kb *Ds* element identified by Peacock *et al.* (1983) and, like it, the central 2 kb element is flanked by 8bp direct repeats.

Weck *et al.* (1984) have studied another unstable mutation at the *shrunken* locus, *sh-m6233*. This is also associated with *Ds* activity and is due to insertion of apparently the same 4 kb double *Ds* element as is present at the right-hand end of the 30 kb *sh-m5933* insertion. This 4 kb element is flanked by an 8 bp direct repeat of target site DNA as expected from the structure of the double *Ds* itself (Döring *et al.*, 1983a). Weck *et al.* (1984) have analyzed one shrunken$^+$ revertant of *sh-m6233*. It has lost the 4 kb insertion, but retains six bases of the 8 bp duplication. This is similar to the revertants of the *Adh1* mutation studied by Sutton *et al.* (1984) and of the *waxy* mutation studied by Pohlman *et al.* (1984).

The heterogeneity of *Ds* elements is further demonstrated by the discovery of a 1.3 kb element present at another *Ds* induced mutation of the *Adh1* gene (Döring *et al.*, 1984b). This element has not yet been cloned but appears to have some sequence homology to the 2 kb element at the end of the *sh-m5933* insertion. *Ds* elements may not all be simple deletion derivatives of *Ac* elements, as appears to be the case for *Ac9* and *Ds9*, since, apart from the 11 bp terminal inverted repeats, there is little homology between the 0.4 kb *Ds* element of Peacock *et al.* (1983) and the 2 kb putative 2 kb *Ds* sequence at the end of the *sh-m5933* insertion (Döring *et al.*, 1984a).

All strains of maize so far tested contain sequences hybridizing to *Ac* and *Ds* elements even if they have no detectable *Ac* or *Ds* activity (Geiser *et al.*, 1982; Peacock *et al.*, 1983; Fedoroff *et al.*, 1983b). Geiser *et al.* (1982) found that DNAs from several strains of maize each contain approximately 40 *Bam*HI restriction fragments which can hybridize to the putative *Ds* sequence at the right-hand end of the *sh-m5933* insertion. Peacock *et al.* (1983) obtained similar results using DNAs from different strains of maize. They found that the 0.4 kb *Ds* element from *AdhI* hybridized to 30–40 restriction fragments.

Fedoroff *et al.* (1983b) have tried to estimate the number of *Ac* and *Ds* elements in various strains of maize. They digested maize DNA with *Bst*EII, an enzyme which does not cut within the *Ac* and *Ds* elements which they found at the *waxy* locus, and hybridized them separately with the whole *Ds6* element and a fragment from the central region of *Ac9*. Two of the strains they used were known to contain *Ds* elements, one had *Ac* activity and one had no detectable *Ac* or *Ds* elements. About 20 fragments hybridized to *Ds6* in each case and at least four fragments hybridized to the central fragment of *Ac9*. The sequences which hybridized to these probes in DNA from the strain lacking *Ac* and *Ds* activity must represent defective elements.

Saedler and his colleagues have started a molecular analysis of a second controlling element system in maize. The autonomous elements in this system were called *Spm* (Suppressor-mutator) by McClintock (1955) and *En* (Enhancer) by Peterson (1953). The corresponding nonautonomous elements were not named by McClintock but were called *I* (Inhibitor) by Peterson. Schwarz-Sommer *et al.* (1984) have studied a mutation of the *waxy* gene which is unstable in the presence of *Spm*/*En* and is therefore associated with a nonautonomous element. This mutation, wx^{m-8}, has three insertions compared with the wild-type *waxy* locus and one of these is excised in the presence of *Spm*/*En*. They have named this element *Spm-I8*. It is about 2 kb long and, like *Ac*/*Ds*, occurs as a repeated sequence in the maize genome. It has a terminal inverted repeat of 13 bp and appears to cause a 3 bp duplication of target site DNA.

C. Tam1: A Transposable Element of *Antirrhinum majus*

Transposable elements are also associated with unstable mutations in *Antirrhinum majus* and are responsible for the patterns of color seen on the petals of some variagated strains (Harrison and Carpenter, 1973). Saedler *et al.* (1983)

have found that one unstable mutant, *Nivea recurrens,* is due to the activity of a 17 kb transposable element which they have called Tam1. In plants carrying the niv^{rec} mutation this element is present in the promoter region of the chalcone synthase gene and it is absent in niv^{+} revertants (Wienand *et al.*, 1982). Tam1 appears to have caused a 3 bp target site duplication in the chalcone synthase gene. Two bases of this duplication have been retained in a niv^{+} revertant of niv^{rec}, in a manner reminiscent of some revertants of controlling element induced mutations in maize (Bonas *et al.*, 1984b).

Tam1 has perfect 13 bp inverted repeats at its termini. Remarkably, the sequence of this repeat is identical in 12/13 positions to that of *Spm-I8,* and the terminal 5 bp are identical to the corresponding bases of a putative transposable element in soybean (Vodkin *et al.*, 1983). This may indicate that there has been horizontal transfer of transposable elements between these species, or reflect similar transposition mechanisms in each case.

D. Tc1: A Transposable Element of *Caenorhabditis elegans*

A putative transposable element has recently been identified in the genome of the nematode *Caenorhabditis elegans*. It was first detected when the DNAs of two different laboratory strains were compared. One of these strains, called Bristol, was originally isolated from an English soil sample while the other, Bergerac, was isolated in France. Emmons *et al.* (1983) and Liao *et al.* (1983) found that several restriction fragments from the Bristol strain were smaller than the corresponding fragment from Bergerac. In each case the difference was due to a single element 1.7 kb long. This element, which has been called Tc1, is repeated and highly conserved in the genomes of both strains. There are about 200 copies per haploid genome in Bergerac but only about 20 copies per genome in Bristol.

Rosenzweig *et al.* (1983a) have determined the complete base sequence of one copy of Tc1, together with that of its target site. The internal region has a single promoter-like sequence, and transcripts initiated from this point could code for polypeptides of 273 and 112 amino acids. The exact length of Tc1 is not certain. The sequence of the first copy to be studied suggested that it is 1610 bp long, with exact terminal inverted repeats of 54 bp, and flanked by a duplication of the target site, TA. These results are consistent with Tc1 being a transposable element, but no transposition events have been observed directly as yet.

Rosenzweig *et al.* (1983b) have sequenced the boundaries and target site of a second copy of Tc1. It had also inserted at, and had duplicated, the sequence TA. This raises two possibilities. Firstly, that Tc1 is 1610 bp long and inserts preferentially at the sequence TA. Second, that Tc1 is 1612 bp long, has 55 bp terminal inverted repeats, and inserts without causing any target site duplication. If the latter is true, then Tc1 may have an unusual mechanism of transposition.

Insertions of Tc1 appear to be unstable in the Bergerac strain. Emmons *et al.* (1983) have found that excision of one copy of Tc1 takes place continuously in each generation and yet is not lost from the population during subculture. This suggests that excision occurs predominantly in somatic cells.

Emmons *et al.* (1983) and Liao *et al.* (1983) have used Southern transfer experiments to look for Tc1 sequences in the DNA of several strains of *C. elegans* recently isolated from the wild. The Tc1 probe hybridized to DNA from each strain. Most strains gave patterns of hybridization which were like that of Bristol suggesting that they contain about similar number of Tc1 elements at the same chromosomal locations. The DNA from one isolate gave an extremely complex pattern of hybridization like that of Bergerac.

These results are reminiscent of those found when an *I* factor probe is hybridized to restriction digests of DNAs from different strains of *D. melanogaster* (Section IV,A), the Bergerac and Bristol strains of *C. elegans* being analogous to *I* and *R* strains of *D. melanogaster,* respectively. Despite this the Tc1 element is probably not part of a hybrid dysgenesis system since Emmons *et al.* (1983) and Liao *et al.* (1983) could find no dysgenic effects when they crossed the Bergerac and Bristol strains. The Bergerac strain may have accumulated a large number of Tc1 elements because one of them over produces a transposase or other product necessary for Tc1 transposition. The Bergerac strain apparently carries several mutations and Emmons *et al.* (1983) have suggested that these are due to Tc1 insertions.

V. Transposable Elements without Terminal Repeats

The genomes of all higher eukaryotes contain highly repetitive, moderately repetitive, and nonrepetitive sequences. In most species, but not *D. melanogaster,* much of the moderately repetitive DNA is arranged in short stretches about 300 bp long interspersed with longer stretches of nonrepetitive DNA (Davidson *et al.*, 1975). A large proportion of these short moderately repetitive sequences in the human genome are members of a single family of sequences. This is known as the ''Alu'' family since many members of the family contain a single site for the restriction enzyme *Alu*I (Houck *et al.*, 1979; Jelinek *et al.*, 1980)

Alu sequences are one example of a group of elements which resemble the transposable elements discussed above in that they are dispersed throughout the genome and are flanked by short direct repeats. They are distinguishable from these elements in that they do not have conserved terminal inverted repeats and the short direct repeats flanking them differ in length from one copy of the element to another. Other examples of such elements are pseudogenes complementary to rat and human small nuclear RNAs (van Arsdell *et al.*, 1981), a human tubulin pseudogene (Wilde *et al.*, 1982), a human immunoglobulin pseudogene (Hollis *et al.*, 1982), and a putative transposable sequence in *D. melanogaster* (Dawid *et al.*, 1981). Various properties of these elements suggest that they may transpose via a double-stranded intermediate generated by reverse transcriptase (Jagadeeswaran *et al.*, 1981; van Arsdell *et al.*, 1981). I shall not discuss them further as they are the subject of the article by Rogers in this volume.

VI. General Comments

A. Experimental Uses of Transposable Elements

1. *Transposable Elements as Genetic Markers*

The chromosomal distribution of elements which transpose infrequently may be an easily scored phenotype with which to identify different populations of a particular species. The distribution of elements in different populations could be scored by means of Southern transfer experiments as outlined in Section II,A,1. The pattern of bands hybridizing to a probe containing the transposable element would be diagnostic for each strain. Goman *et al.* (1982) have cloned a putative transposable sequence from *Plasmodium falciparum,* the causative agent in human malaria. This sequence hybridizes to different sets of restriction fragments in DNAs from different isolates and strains of *P. falciparum* (Langsley, personal communication). Goman *et al.* (1982) have pointed out that this could be a way of distinguishing between isolates. This might be particularly useful to clinicians and epidemiologists since strains of *P. falciparum* can vary in virulence or resistance to therapeutic agents, and it may be important to be able to identify the source of particular infections.

Restriction site polymorphisms are quite frequent in human populations (Jeffreys, 1979; Tuan *et al.*, 1979; Murray *et al.*, 1982). Botstein *et al.* (1980) and Southern (1981) have proposed that these polymorphisms could be used as markers with which to construct a genetic linkage map of the human genome, and that restriction site variants linked to mutations responsible for particular inherited diseases might be used to distinguish carrier from noncarrier adults or affected from unaffected foetuses *in utero*. Kan and Dozy (1978) have already used this method to diagnose sickle cell trait *in utero*. Transposable elements change the restriction pattern of regions of the genome into which they insert, either because they carry sites for the enzymes in question, or because they increase the distance between existing sites. In human populations there may be considerable heterogeneity between individuals with regard to the distribution of transposable elements. A set of cloned DNA fragments containing copies of a transposable element should thus contain unique sequences from regions of the genome likely to contain polymorphisms which could be used in these studies. A difference in the distribution of Tc1 elements between the Bristol and Bergerac strains of *C. elegans* has already allowed a cluster of actin genes to be mapped on linkage group V (Files *et al.*, 1983).

2. *Transposable Elements as Mutagens*

Insertion of transposable elements is one source of spontaneous mutations in eukaryotes and in certain circumstances it may be possible to use them in this way experimentally. Transposition of *P* elements in P-M hybrid dysgenesis and of *I* factors in I-R hybrid dysgenesis has already been shown to increase mutation frequencies at several loci in *D. melanogaster* (Green, 1977; Picard *et al.*, 1978). Transposition of the Tc1 element in the Bergerac strain of *C. elegans* might also be used in this way (S. W. Emmons, personal communication).

Mutations due to insertion of transposable elements have some advantages over those resulting from base substitutions or the addition or deletion of a small number of bases. If segments of DNA containing the gene in question have already been cloned then the position of an insertion can be mapped by restriction or heteroduplex analysis. This may help to locate genes on restriction maps or distinguish structural and regulatory regions within genes. Zachar and Bingham (1982) have used this method to distinguish at the molecular level the structural and regulatory regions of the *white* locus. These had been identified previously by genetic methods.

Insertion mutations can also aid in gene cloning. If mutations due to insertion of particular transposable elements can be recognized then the element in question can be used as a probe with which to screen for DNA fragments containing the mutant gene. Bingham *et al.* (1981) have cloned the *white* gene of *D. melanogaster* by making use of the fact that the *white-apricot* mutation is due to insertion of a *copia* element (Gehring and Paro, 1980; Bingham and Judd, 1981) and Modelell *et al.* (1983) have cloned the *scute* and *cut* genes using alleles associated with *gypsy* elements.

3. *Transposable Elements as Transformation Vectors*

The ability to reintroduce cloned DNA into chromosomes of eukaryotic cells is an important aspect of recombinant DNA technology. It is the best way of testing conclusions regarding the function of cloned segments of DNA and will be essential if this technology is to be used to manipulate plant or animal genomes directly. Endogenous transposable elements can insert themselves at new sites within chromosomes. If the same is true of transposable elements added to cells exogenously then they might be used as transformation vectors. This has already been demonstrated by Rubin and Spradling (1982) for the P element of *D. melanogaster*. They inserted a fragment of DNA carrying the *rosy* gene, within a *P* element. They then injected a plasmid carrying the *P-rosy* element into embryos of an *M* strain homozygous for a deletion of the *rosy* gene. A second plasmid containing a complete *P* factor was injected simultaneously to provide any functions necessary for transposition. Up to 50% of the flies resulting from the injected embryos produced Rosy$^+$ progeny indicating that the *rosy* gene had been incorporated into chromosomes in at least some germ-cells. The *P-rosy* element could insert at many sites in the genome without altering the structure of the *rosy* gene. This was the result of transposition and not recombination since the complete *P-rosy* element had inserted into the chromosomes without any of the flanking plasmid DNA. *P-rosy* elements can integrate at a large number of sites without seriously affecting expression of the *rosy* gene (Spradling and Rubin, 1983). *D. melanogaster* embryos have been successfully transformed with *P* elements carrying chorion (Rubin and Spradling, 1982), dopa decarboxylase (Scolnick *et al.*, 1983), and alcohol dehydrogenase genes (Goldberg *et al.*, 1983), and there is probably no restriction on sequences which can be used in these experiments.

Ac and *Ds* elements in maize resemble *P* factors and *P* elements, respectively, of *D. melanogaster* (Section IV,B) and it may be possible to use them as transformation vectors in a similar way (Peacock *et al.*, 1983). This would be particularly useful as the only transformation system currently available for plants makes use of the tumour inducing plasmid of *Agrobacterium tumefaciens* (Barton *et al.*, 1983). This will only transform dicotyledenous plants and cannot be used for maize and other cerials which are monocots.

B. Aspects of Transposition

1. *Rates of Transposition*

Transposition frequencies are not easy to estimate in eukaryotes. Most eukaryotic transposable elements do not carry detectable genetic markers and as a result their transposition can only be detected by the relatively laborious techniques of Southern transfer experiments and *in situ* hybridization. The issue is further complicated by the fact that these elements occur in families, different members of which may transfer at different rates. The small amount of data available concerning *copia*-like elements suggests that they normally transpose infrequently in laboratory populations. Pierce and Luchessi (1981) could find no changes in the locations of copies of the putative *copia*-like element *Dm25* on the X chromosome of one strain of *D. melanogaster* over a 4-month period. Young and Schwartz (1981) detected one possible *copia* transposition event on an X chromosome but this could have occurred anytime during the past 50 years. Transposition rates may well vary under different environmental or genetic conditions. The very heterogeneous distribution of *copia*-like elements in different strains of *D. melanogaster* could indicate that transposition frequencies are higher in natural populations or that transposition occurs in bursts. Gerosimova *et al.*

(1984) have recently obtained evidence suggesting that transposition of several transposable elements is stimulated by outcrossing. The frequencies with which *P* factors and *I* factors transpose is certainly under genetic control. These elements are stable in *P* strains and *I* strains, respectively, but transpose at high frequency if introduced into the cytoplasmic environment of *M* or *R* strains. There is even an indication that *copia* elements may transpose more frequently in P-M dysgenic flies (Rubin *et al.*, 1982).

2. *Mechanism of Transposition*

Transposition of bacterial transposons is believed to be a replicative event which does not require excision (Kleckner, 1981). There is no direct evidence to indicate whether or not this is true of eukaryotic elements. The fact that all eukaryotic transposable elements, with the exception of the composite elements of *D. melanogaster,* are comprised of repeated sequences suggests that transposition is replicative. *P* factors can apparently increase in number during hybrid dysgenesis (Bingham *et al.*, 1982; Bregliano and Kidwell, 1983) and most *copia*-like elements are more abundant in *D. melanogaster* tissue culture cells than in flies, presumably as a result of transposition (Potter *et al.*, 1979; Ilyin *et al.*, 1980b; Kulguskin *et al.*, 1981). Several eukaryotic elements can excise but this is not necessarily associated with transposition even in the case of *P* elements which both transpose and excise at high frequency in P-M dysgenic flies.

Different transposition mechanisms may be used by different elements and any element may have more than one possibility open to it. The structural similarities between proviruses and *copia*-like, Ty, and IAP elements suggest one possible mechanism for their transposition. Some of the full length RNAs transcribed from these elements may be copied into double-stranded linear and circular DNAs by reverse transcriptase and these may integrate at new sites in the genome. Flavell and Ish Horowicz (1981, 1983) have suggested that the circular *copia* DNAs which they have found in *D. melanogaster* tissue culture cells are transposition intermediates, but think that only a small proportion of them are produced by reverse transcription. Transposable elements which do not have LTRs must transpose by a different mechanism(s), possibly related to those proposed for bacterial transposons (Kleckner, 1981).

C. Origin of Transposable Elements

Nothing is known about the origin of transposable elements. They must either evolve from sequences already present in the genome or be aquired from outside, possibly by a viral infection. Transposable elements could arise from DNA sequences which can, by chance, make use of endogenous cellular enzymes to replicate and insert at new sites in the genome (Fig. 7a and b). The transposable elements without terminal inverted repeats described in Section V may be se-

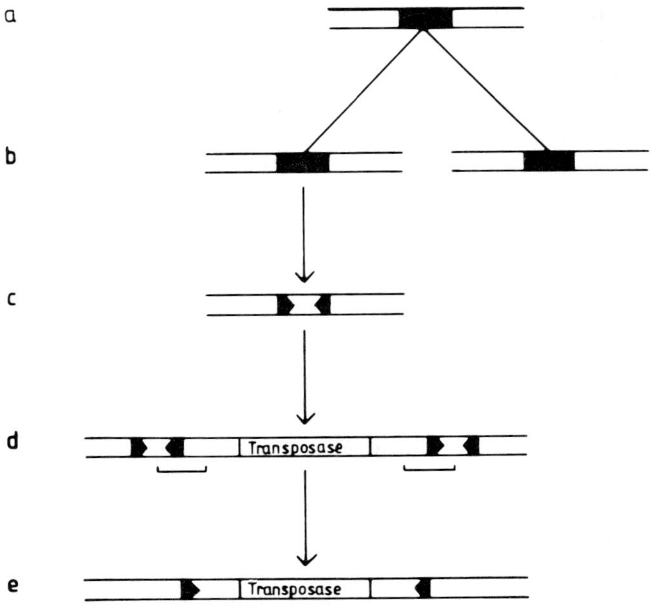

FIG. 7. Evolution of a transposable element. A chromosomal DNA sequence (▭) may be able to transpose fortuitously, making use of chromosomally coded enzymes (a,b). This primitive transposable element may accumulate mutations allowing it to transpose more efficiently (c). These might, for example, create short inverted terminal repeats. An unstable, composite transposable element could be formed if two copies of the primitive transposable element were to insert on either side of a gene coding for an enzyme required for transposition (d). This, in turn, could give rise to a stable, mature transposable element if deletions removed the internal termini of the primitive transposable element (e).

quences of this type. A primitive transposable element of this sort might accumulate mutations which allow it to transpose more frequently. These mutations could, for example, create terminal inverted repeats demarcating the end of the element (Fig. 7e). A more complex transposable element coding for an enzyme required for transposition could then evolve if two copies of the primitive element were to insert on either side of a gene coding such an enzyme (Fig. 7d). Occasionally a transposition event might include both primitive elements plus the intervening DNA, producing a composite element coding for a transposase. This would be similar to the composite transposable elements in *D. melanogaster* which are flanked by FB elements as described in section III.B. Deletions which subsequently remove the internal ends of the primitive transposable elements would generate a mature transposable element (Fig. 7e).

Mature transposable elements will be inactivated by mutation and eukaryotic genomes should contain degenerate, as well as primitive and mature, elements.

Several lines of evidence support this. The genomes of many strains of *D. melanogaster* contain defective *P* elements which can be derived from functional *P* factors by internal deletions (Section IV,A) and sequences complementary to *Ac* and *Ds* elements have been detected in one strain of maize with no *Ac* or *Ds* activity (Section IV,B). Wensink *et al.* (1979) have found regions of the *D. melanogaster* genome containing clusters of moderately repetitive sequences in scrambled arrays. Some of them contain sequences complementary to *412* elements and could have been produced by several cycles of insertion of one transposable element into another. Degenerate transposable elements which have been immobilized by mutation should eventually be lost from the genome since they can no longer replicate by transposition.

D. Biological Role of Transposable Elements

In several instances DNA rearrangements are instrumental in producing changes in gene expression which are important in development. These include rearrangements associated with activation of immunoglobulin genes (Tonegawa, 1983), mating-type switching in *S. cerevisiae* (Hicks *et al.*, 1979; Nasmyth and Tatchell, 1980), and production of antigenic variants in *Trypanosoma brucei* (Borst and Cross, 1982). There is no evidence that transposable elements of the sort discussed in this review play any role in development. These events differ from transposition in that they are nonrandom, and the sequences involved only undergo one cycle of rearrangement. Transposable elements transpose repeatedly and can distribute themselves around the genome at least semirandomly.

The most obvious effect that eukaryotic transposable elements have on an organism is to cause mutations. All the changes in gene structure and function required for evolution of a species can be promoted, either directly or indirectly, by transposable elements with the possible exception of base substitutions. Insertion of transposable elements can increase or decrease the level of gene expression, while recombination between different copies of an element can delete, duplicate, invert, or translocate whole segments of DNA. Some transposable elements, particularly those controlling hybrid dysgenesis in *D. melanogaster*, may also contribute to the genetic isolation of populations prior to speciation (Bingham *et al.*, 1982; Dover *et al.*, 1982; Bregliano and Kidwell, 1983).

Transposable elements are not necessarily maintained in a species because of these potentially advantageous consequences of transposition. As a rule transposition is disadvantageous and transposable elements are probably maintained simply because the rate at which their numbers are reduced by selection is balanced by the rate at which they are increased by transposition (Finnegan *et al.*, 1982; Maynard Smith, 1982). The term "selfish DNA" has been coined to describe sequences which have no obvious function but which can be accumulate within a genome by one means or another (Doolittle and Sapienza, 1980; Orgel and Crick, 1980). Selfish is perhaps too anthropomorphic a word to use in this context and "parasitic DNA" might be a more appropriate expression. Transposable elements are clearly sequences of this type.

Questions relating to the long-term effects of transposable elements on genes, genomes, individuals, populations, and species can be answered only by the combined efforts of evolutionary and molecular biologists. Molecular biologists have led the way so far. Some attempts are being made to adapt theories of population genetics to these issues (Ohta, 1981, 1983; Ohta and Kimura, 1981; Charlesworth and Charlesworth, 1983; Langley *et al.*, 1983) but much remains to be done.

Acknowledgments

I am grateful to Adrian Bird, Renato Paro, and Helen Sang for having read and criticized drafts of this review. I am also indebted to all those who sent me published and unpublished material for this review. Some of the work described in this review was supported by a project grant from the Medical Research Council.

References

Ananiev, E. V., Gvozdev, V. A., Ilyin, Y. V., Tchurikov, N. A., and Georgiev, G. P. (1978). *Chromosoma* **70,** 1–17.

Baker, R. F., and Thomas, C. A. (1977). *In* "DNA Insertion Elements, Plasmids and Episomes" (A. I. Bukhari, J. A. Shapiro, and S. L. Adhya, eds.), pp. 463–470. Cold Spring Harbor Laboratory, Cold Spring Harbor, New York.

Barton, K. A., Binns, A. N., Matzke, A. J. M., and Chilton, M. (1983). *Cell* **32,** 1033–1043.

Bayev, A. A., Krayev, A. S., Lyubomirskaya, N. V., Ilyin, Y. V., Skryabin, K. G., and Georgiev, G. P. (1980). *Nucleic Acids Res.* **8,** 3263–3273.

Bayev, A. A., Lyubomirskaya, N. V., Dzhumagaliev, E. B., Ananiev, E. V., Afmiantova, I. G., and Ilyin, Y. V. (1984). *Nucleic Acids Res.* **12,** 3707–3723.

Bender, W., Akam, M., Karch, F., Beachy, P. A., Spierer, P., Lewis, E. B., and Hogness, D. S. (1983). *Science* **221,** 23–29.

Benoist, C., and Chambon, P. (1981). *Nature (London)* **290,** 304–31O.

Berg, R. L., Engels, W. R., and Krebs, R. A. (1980). *Science* **210,** 427–429.

Bingham, P. M. (1980). *Genetics* **95,** 341–353.

Bingham, P. M. (1981). *Cold Spring Harbor Symp. Quant. Biol.* **45,** 519–525.

Bingham, P. M., and Judd, B. H. (1981). *Cell* **25,** 705–711.

Bingham, P. M., Levis, R., and Rubin, G. M. (1981). *Cell* **25,** 693–704.

Bingham, P. M., Kidwell, M. G., and Rubin, G. M. (1982). *Cell* **29,** 995–1004.

Bonas, V., Sommer, H., Harrison, B. J., and Saedler, H. (1984a). *Mol. Gen. Genet.* **194,** 138–143.

Bonas, V., Sommer, H., and Saedler, H. (1984b). *EMBO J.* **3,** 1015–1019.

Borst, P., and Cross, G. M. (1982). *Cell* **29,** 291–303.

Botstein, D., White, R. L., Skolnick, M., and Davis, R. W. (1980). *Am. J. Hum. Genet.* **32,** 314–331.

Bregliano, J. C., and Kidwell, M. G. (1983). *In* "Mobile Genetic Elements" (J. Shapiro, ed.), pp. 363–410. Academic Press, New York.
Bucheton, A., Paro, R., Sang, H. M., Pelisson, A., and Finnegan, D. J. (1984). *Cell* **38,** 153–163.
Burr, B., and Burr, F. A. (1981). *Cold Spring Harbor Symp. Quant. Biol.* **45,** 463–465.
Burr, B., and Burr, F. A. (1982). *Cell* **29,** 977–986.
Callarco, P. G., and Szollosi, D. (1973). *Nature (London) New Biol.* **243,** 91–93.
Cameron, J. R., Loh, E. Y., and Davis, R. W. (1979). *Cell* **16,** 739–751.
Carlson, M., and Brutlag, D. L. (1978). *Cell* **15,** 733–742.
Carstens, E. (1980). *Trends Biochem. Sci.* **52,** 107–110.
Cech, T. R., and Hearst, J. E. (1975). *Cell* **5,** 429–446.
Chaleff, D. T., and Fink, G. R. (1980). *Cell* **21,** 227–237.
Charlesworth, B., and Charlesworth, D. (1983). *Genet. Res.* (*Cambridge*) **42,** 1–27.
Cohen, J. C., and Varmus, H. E. (1979). *Nature (London)* **278,** 418–423.
Cole, M. D., Ono, M., and Huang, R. C. (1981). *J. Virol.* **38,** 680–687.
Collins, M., and Rubin, G. M (1982). *Cell* **30,** 71–79.
Collins, M., and Rubin, G. M. (1983). *Nature (London)* **303,** 259–260.
Courage-Tebbe, U., Döring, H. P., Geiser, M., Starlinger, P., Tillman, E., Weck, E., and Werr, W. (1982). *In* "Manipulation and Expression of Genes in Eukaryotes" (A. Nagely, A. W. Linnane, W. J. Peacock, and J. A. Pateman, eds.), pp. 221–228. Academic Press, New York.
Courage-Tebbe, U., Döring, H. P., Fedoroff, N., and Starlinger, P. (1983). *Cell* **34,** 383–393.
Davidson, E. H., Galeau, G. A., Angerer, R. C., and Britten, R. J. (1975). *Chromosoma* **51,** 253–259.
Dawid, I. B., Olong, E. O., DiNocera, P. P., and Pardue, M. L. (1981). *Cell* **25,** 399–408.
Deininger, P. L., and Schmid, C. W. (1976). *J. Mol. Biol.* **106,** 773–790.
Doolittle, W. F., and Sapienza, C. (1980). *Nature (London)* **284,** 601–603.
Döring, H. P., Geiser, M., and Starlinger, P. (1981). *Mol. Gen. Genet.* **184,** 377–380.
Döring, H. P., Tillmann, E., and Starlinger, P. (1984a). *Nature (London)* **307,** 127–130.
Döring, H. P., Freeling, M., Halne, S., Johns, M. A., Kurze, R., Merekelbuch, A., Salamini, F., and Starlinger, P. (1984b). *Mol. Gen. Genet.* **193,** 199–204.
Dover, G., Brown, S., Coen, E., Dallas, J., Strachan, T., and Trick, M. (1982). *In* "Genome Evolution" (G. A. Dover and R. B. Flavell, eds.), pp. 343–372. Academic Press, New York.
Dowsett, A. P., and Young, M. W. (1982). *Proc. Natl. Acad. Sci. U.S.A.* **79,** 4570–4574.
Dunsmuir, P., Brorien, W. J., Simon, M. A., and Rubin, G. M. (1980). *Cell* **21,** 576–579.
Eibel, H., Gafner, J., Stotz, A., and Philippsen, P. (1981). *Cold Spring Harbor Symp. Quant. Biol.* **45,** 609–617.
Elder, R. T., St. John, T. P., Stinchcomb, D. T., and Davis, R. W. (1981). *Cold Spring Harbor Symp. Quant. Biol.* **45,** 581–584.
Emerson, R. A. (1914). *Am. Nat.* **48,** 87–115.
Emmons, S. W., Yesner, L., Ruan, K., and Katzberg, D. (1983). *Cell* **32,** 55–65.
Engels, W. R. (1979). *Proc. Natl. Acad. Sci. U.S.A.* **79,** 4011–4015.
Engels, W. R., and Preston, G. R. (1981). *Cell* **26,** 421–428.
Errede, B., Cardillo, T. S., Sherman, F., and Dubois, E. (1980). *Cell* **25,** 427–236.
Errede, B., Cardillo, T. S., Wever, G., and Sherman, F. (1981). *Cold Spring Harbor Symp. Quant. Biol.* **45,** 593–602.
Falkenthal, S., and Lengyel, J. (1980). *Biochemistry* **19,** 5842–5850.
Farabaugh, P. J., and Fink, G. R. (1980). *Nature (London)* **286,** 352–356.
Fedoroff, N. V. (1983). *In* "Mobile Genetic Elements" (J. A. Shapiro, ed.), pp. 1–65. Academic Press, New York.
Fedoroff, N. V., Mauvais, J., and Chaleff, D. (1983a). *J. Mol. Appl. Genet.* **2,** 11–29.
Fedoroff, N. V., Wessler, S., and Shure, M. (1983b). *Cell* **35,** 235–242.

Files, J. G., Carr, S., and Hirsh, D. (1983). *J. Mol. Biol.* **164,** 355–375.
Fincham, J. R. S., and Sastry, G. R. K. (1974). *Annu. Rev. Genet.* **8,** 15–50.
Finnegan, D. J. (1981). *In* "Chromosomes Today" (M. D. Bennett, M. Bobrow, and G. Hewitt, eds.), pp. 84–91. Allen & Unwin, London.
Finnegan, D. J., Rubin, G. M., Young, M. W., and Hogness, D. S. (1978). *Cold Spring Harbor Symp. Quant. Biol.* **42,** 1053–1063.
Finnegan, D. J., Will, B. M., Bayev, A. A., Bowcock, A. M., and Brown, L. (1982). *In* "Genome Evolution" (G. A. Dover and R. B. Flavell, eds.), pp. 29–39. Academic Press, New York.
Finnegan, D. J., Bucheton, A., and Sang, H. M. (1983). *In* "Genetic Rearrangements" (K. F. Chater, C. A. Cullis, D. A. Hopwood, A. W. B. Johnston, and H. W. Woolhouse, eds.), pp. 75–92. Croom Helm, London.
Flavell, A. J. (1984). *Nature (London)* **310,** 514–516.
Flavell, A. J., and Ish Horowicz, D. (1981). *Nature (London)* **292,** 591–574.
Flavell, A. J., and Ish Horowicz, D. (1983). *Cell* **34,** 415–419.
Flavell, A. J., Ruby, S. W., Toole, J. J., Roberts, B. E., and Rubin, G. M. (1980). *Proc. Natl. Acad. Sci. U.S.A.* **77,** 7107–7111.
Flavell, A. J., Levis, R., Simon, M. A., and Rubin, G. M. (1981). *Nucleic Acids Res.* **9,** 6279–6291.
Geisen, M., Weck, E., Döring, H. P., Werr, W., Courage-Tebbe, V., Tillmann, E., and Starlinger, P. (1982). *EMBO J.* **1,** 1455–1460.
Gehring, W. J., and Paro, R. (1980). *Cell* **19,** 897–904.
Gerasimova, T. I., Mizrokhi, L. J., and Georgiev, G. P. (1984). *Nature (London)* **309,** 714–716.
Goldberg, D., Posakony, J. W., and Maniatis, T. (1983). *Cell* **34,** 59–73.
Goldberg, M. L., Paro, R., and Gehring, W. J. (1982). *EMBO J.* **1,** 93–98.
Goldberg, M. L., Sheen, J., Gehring, W. J., and Green, M. M. (1983). *Proc. Natl. Acad. Sci. U.S.A.* **80,** 5017–5021.
Golubovsky, M. D., Ivanov, Y. N., and Green, M. M. (1977). *Proc. Natl. Acad. Sci. U.S.A.* **74,** 2973–2975.
Goman, M., Langsley, G., Hyde, J. E., Yankovsky, N. K., Zolg, J. W., and Scaife, J. G. (1982). *Mol. Biochem. Parasitol.* **5,** 391–400.
Green, M. M. (1959). *Heredity* **13,** 302–315.
Green, M. M. (1967). *Genetics* **56,** 467–482.
Green, M. M. (1969a). *Genetics* **61,** 423–428.
Green, M. M. (1969b). *Genetics* **61,** 429–441.
Green, M. M. (1977). *Proc. Natl. Acad. Sci. U.S.A.* **74,** 3490–3493.
Harrison, B., and Carpenter, R. (1973). *Heredity* **31,** 309.
Hawley, R. G., Shulman, M. J., Murialdo, H., Gibson, D. M., and Hozumi, N. (1982). *Proc. Natl. Acad. Sci. U.S.A.* **79,** 7425–7429.
Hayward, W. S., Neel, B. G., and Astrin, S. M. (1981). *Nature (London)* **290,** 475–480.
Heine, C. W., Kelly, D. C., and Avery, R. J. (1980). *J. Gen. Virol.* **49,** 385–395.
Hicks, J., Strathearn, J., and Klar, A. (1979). *Nature (London)* **282,** 478–483.
Hollis, G. F., Heiter, P. A., McBride, O. W., Swan, D., and Leder, P. (1982). *Nature (London)* **296,** 321–325.
Houck, C. M., Rinehart, F. P., and Schmid, C. W. (1979). *J. Mol. Biol.* **132,** 289–306.
Hughes, S. H., Payvor, F., Spector, D., Schimke, R. T., Robinson, H. L., Payne, G. S., Bishop, J. M., and Varmus, H. E. (1979). *Cell* **18,** 347–359.
Ikenaga, H., and Saigo, K. (1982). *Proc. Natl. Acad. Sci. U.S.A.* **79,** 4143–4147.
Ilyin, Y. V., Tchurikov, N. A., Ananiev, E. V., Ryskov, A. P., Yenikolopov, G. N., Limborska, S. A., Maleeva, N. E., Gvozdev, V. A., and Georgiev, G. P. (1978). *Cold Spring Harbor Symp. Quant. Biol.* **42,** 959–969.

Ilyin, Y. V., Chmeliauskaite, V. G., Ananiev, E. V., and Georgiev, G. P. (1980a). *Chromosoma* **81,** 27–53.

Ilyin, Y. V., Chmeliauskaite, V. G., and Georgiev, G. P. (1980b). *Nucleic Acids Res.* **8,** 3439–3457.

Ilyin, Y. V., Chmeliauskaite, V. G., Ananiev, E. V., Lyubomirskya, N. V., Bayev, A. A., and Georgiev, G. P. (1980c). *Nucleic Acids Res.* **8,** 5333–5346.

Ilyin, Y. V., Chmeliauskaite, V. G., Kulguskin, V. V., and Georgiev, G. P. (1980d). *Nucleic Acids Res.* **8,** 5347–5361.

Inouye, S., Yuki, S., and Saigo, M. (1984). *Nature (London)* **310,** 332–333.

Ising, G., and Block, K. (1981). *Cold Spring Harbor Symp. Quant. Biol* **45,** 527–549.

Ising, G., and Ramel, C. (1976). *In* "The Genetics and Biology of Drosophila" (M. Ashburner and E. Novitski, eds.), Vol. 1c, pp. 947–954. Academic Press, New York.

Jaenisch, R., Harbers, K., Schnieke, A., Lohler, J., Chumakov, I., Jahner, D., Grotkopp, D., and Hoffman, E. (1983). *Cell* **32,** 209–216.

Jagadeeswaran, P., Forget, B. G., and Weissman, S. M. (1981). *Cell* **26,** 141–142.

Jeffreys, A. J. (1979). *Cell* **18,** 1–10.

Jelinek, W. R., Toomey, T. P., Leinwand, L., Duncan, C. H., Biro, P. A., Choudary, P. V., Weissman, S. M., Rubin, C. M., Houk, C. M., Deininger, P. L., and Schmid, C. W. (1980). *Proc. Natl. Acad. Sci. U.S.A.* **77,** 1398–1402.

Jenkins, N. A., Copeland, N. G., Taylor, B. A., and Lee, B. K. (1981). *Nature (London)* **293,** 370–374.

Junakovic, N., and Ballario, P. (1984). *Plasmid* **11,** 109–115.

Kan, Y. W., and Dozy, A. M. (1978). *Lancet* **2,** 910–912.

Karess, R. E., and Rubin, G. M. (1982). *Cell* **30,** 63–69.

Karess, R. E., and Rubin, G. M. (1984). *Cell* **38,** in press.

Kidwell, M. G. (1979). *Genet. Res.* **33,** 205–217.

Kidwell, M. G. (1983a). *In* "Genetics and Biology of Drosophila" (M. Ashburner, H. L. Carson, and J. N. Thompson, eds.), Vol. 3c, pp. 125–153. Academic Press, New York.

Kidwell, M. G. (1983b). *Proc. Natl. Acad. Sci. U.S.A.* **80,** 1655–1659.

Kingsman, A. J., Gimlich, R., Clarke, L., Chinault, C., and Carbon, J. (1981). *J. Mol. Biol.* **145,** 619–632.

Kleckner, N. (1981). *Annu. Rev. Genet.* **15,** 341–404.

Kohler, G., and Shulman, M. J. (1980). *Eur. J. Immunol.* **10,** 467–476.

Kuff, E. L., Wivel, N. A., and Lueders, K. K. (1972). *Proc. Natl. Acad. Sci. U.S.A.* **69,** 218–222.

Kuff, E. L., Smith, L. A., and Lueders, K. K. (1981). *Mol. Cell Biol.* **1,** 216–227.

Kuff, E. L., Feenstra, A., Lueders, K., Smith, L., Hawley, R., Hozumi, N., and Shulman, M. (1983a). *Proc. Natl. Acad. Sci. U.S.A.* **80,** 1992–1996.

Kuff, E. L., Feenstra, A., Lueders, K., Rechavi, G., Givol. D., and Canaani, E. (1983b). *Nature (London)* **302,** 547–548.

Kugimiya, W., Ikenaga, H., and Saigo, K. (1983). *Proc. Natl. Acad. Sci. U.S.A.* **80,** 3193–3197.

Kulguskin, V. V., Ilyin, Y. V., and Georgiev, G. P. (1981). *Nucleic Acids Res.* **9,** 3451–3463.

Langley, C. H., Brookfield, J. F. Y., and Kaplan, N. (1983). *Genetics* **104,** 457–471.

Levis, R., and Rubin, G. M. (1982). *Cell* **30,** 543–550.

Levis, R., Dunsmuir, P., and Rubin, G. M. (1980). *Cell* **21,** 581–588.

Levis, R., Collins, M., and Rubin, G. M. (1982). *Cell* **30,** 551–565.

Lewis, E. B. (1949). *Drosophila Inf. Ser.* **23,** 59–60.

Liao, L. W., Rosenzweig, B., and Hirsh, D. (1983). *Proc. Natl. Acad. Sci. U.S.A.* **80,** 3585–3589.

Liebermann, D., Hoffman-Liebermann, B., Weinthal, J., Childs, G., Moxson, A., Cohen, S. N., and Kedes, L. (1983). *Nature (London)* **306,** 342–347.

Liebman, S. W., Singh, A., and Sherman, F. (1979). *Genetics* **92,** 783–802.

Liebman, S. W., Shalit, P., and Picologous, S. (1981). *Cell* **26,** 401–409.
Lueders, K. K., and Kuff, E. L. (1977). *Cell* **12,** 963–972.
Lueders, K. K., and Kuff, E. L. (1980). *Proc. Natl. Acad. Sci. U.S.A.* **77,** 3571–3575.
Lueders, K. K., and Kuff, E. L. (1981). *Nucleic Acids Res.* **9,** 5917–5929.
Lueders, K. K., Segal, S., and Kuff, E. L. (1977). *Cell* **11,** 83–94.
McClintock, B. (1946). *Carnegie Inst.* of Washington *Year Book* **45,** 176–186.
McClintock, B. (1950). *Proc. Natl. Acad. Sci. U.S.A.* **36,** 344–355.
McClintock, B. (1951). *Cold Spring Harbor Symp. Quant. Biol.* **16,** 13–47.
McClintock, B. (1955). *Carnegie Inst. Washington Year Book* **54,** 245–255.
McClintock, B. (1956). *Carnegie Inst. Washington Year Book* **55,** 323–332.
McClintock, B. (1962). *Carnegie Inst. Washington Year Book* **61,** 448–461.
McClintock, B. (1963). *Carnegie Inst. Washington Year Book* **62,** 486–493.
Martin, G., Wiernasz, D., and Schedl, P. (1983). *J. Mol. Evol.* **19,** 203–213.
Maynard Smith, J. (1982). *In* "Genome Evolution" (G. A. Dover and R. B. Flavell, eds.), pp. 375–382. Academic Press, New York.
Meselson, M., Dunsmuir, P., Schweber, M., and Bingham, P. (1980). *In* "Genes, Cells and Behavior: A View of Biology Fifty Years Later" (H. Horowitz and E. Hutchings, eds.), pp. 88–92. Calif. Inst. of Technology, Pasadena.
Meyerowitz, E. M., and Hogness, D. S. (1982). *Cell* **28,** 165–176.
Miller, D. W., and Miller, L. K. (1982). *Nature (London)* **299,** 562–564.
Miller, L. K. (1981). *In* 'Genetic Engineering in the Plant Sciences" (N. Panopoulos, ed.), Praeger, New York.
Modelell, J., Bender, W., and Meselson, M. (1983). *Proc. Natl. Acad. Sci. U.S.A.* **80,** 1678–1682.
Montgomery, E. A., and Langley, C. H. (1983). *Genetics* **104,** 473–483.
Murray, J. M., Davies, K. E., Harper, P. S., and Meredith, L. (1982). *Nature (London)* **300,** 69–71.
Nasmyth, K., and Tatchell, K. (1980). *Cell* **19,** 753–764.
O'Hare, K., and Rubin, G. M. (1983). *Cell* **34,** 25–35.
Ohta, T. (1981). *Nature (London)* **292,** 648–649.
Ohta, T. (1983). *Genet. Res.* **41,** 1–16.
Ohta, T., and Kimura, M. (1981). *Proc. Natl. Acad. Sci. U.S.A.* **78,** 1129–1132.
Ono, M. M., Cole, D., White, A. T., and Huang, R. C. (1980). *Cell* **21,** 465–473.
Orgel, L. E., and Crick, F. H. C. (1980). *Nature (London)* **284,** 604–607.
Osterman, J. C., and Schwartz, D. (1981). *Genetics* **99,** 267–273.
Paro, R., Goldberg, M. L., and Gehring, W. J. (1983). *EMBO J.* **2,** 853–860.
Paterson, B. M., Segal, S., Lueders, K. K., and Kuff, E. L. (1978). *J. Virol.* **27,** 118–126.
Payne, G. S., Courtneidge, S. A., Crittenden, L. B., Fadly, H. M., Bishop, J. M., and Varmus, H. E. (1981). *Cell* **23,** 311–322.
Peacock, W. J., Dennis, E. S., Gerlach, W. L., Llewellyn, D., Lorz, H., Pryor, A. J., Sachs, M. M., Schwartz, D., and Sutton, W. D. (1983). *Proc. Miami Winter Symp., 15th.* Academic Press, New York.
Pelisson, A. (1981). *Mol. Gen. Genet.* **183,** 123–129.
Peterson, P. A. (1953). *Genetics* **38,** 682–683.
Peterson, P. A. (1977). *In* "DNA Insertion Elements, Plasmids and Episomes" (A. I. Bukhari, J. A. Shapiro, and S. L. Adhya, eds.), pp. 427–435. Cold Spring Harbor Laboratory, Cold Spring Harbor, New York.
Picard, G., Bregliano, J. C., Bucheton, A., Levige, J. M., Pelisson, A., and Kidwell, M. G. (1978). *Genet. Res.* **32,** 275–287.
Pierce, D. A., and Luchessi, J. C. (1981). *Chromosoma* **82,** 471–492.
Pohlman, R. F., Fedoroff, N. V., and Messing, J. (1984). *Cell* **37,** 635–643.
Potter, S. S. (1982). *Nature (London)* **297,** 201–204.

Potter, S. S., Brorien, W. J., Dunsmuir, P., and Rubin, G. M. (1979). *Cell* **17,** 415–427.
Potter, S. S., Truett, M., Phillips, M., and Maher, A. (1980). *Cell* **20,** 639–647.
Ramareddy, V. G., Mahy, B. W. J., Bishop, J. M., and Varmus, H. E. (1975). *Nature (London)* **253,** 507–511.
Rechavi, G., Givol, D., and Canaani, E. (1982). *Nature (London)* **300,** 607–611.
Rhoades, M. M. (1938). *Genetics* **23,** 377–397.
Roeder, G. S., and Fink, G. R. (1980). *Cell* **21,** 239–249.
Roeder, G. S., and Fink, G. R. (1982). *Proc. Natl. Acad. Sci. U.S.A.* **79,** 5621–5625.
Roeder, G. S., and Fink, G. R. (1983). *In* "Mobile Genetic Elements" (J. A. Shapiro, ed.), pp 299–328. Academic Press, New York.
Roeder, G. S., Farabaugh, P. J., Chaleff, D. T., and Fink, G. R. (1980). *Science* **209,** 1375–1380.
Rose, M., and Winston, F. (1984). *Mol. Gen. Genet.* **193,** 557–560.
Rosenzweig, B., Liao, L. W., and Hirsh, D. (1983a). *Nucleic Acid Res.* **11,** 4201–4209.
Rosenzweig, B., Liao, L. W., and Hirsh, D. (1983b). *Nucleic Acids Res.* **11,** 7137–7140.
Rothstein, R. (1979). *Cell* **17,** 185–190.
Rubin, G. M. (1983). *In* "Mobile Genetic Elements" (J. A. Shapiro, ed.). pp. 329–361. Academic Press, New York.
Rubin, G. M., and Spradling, A. C. (1982). *Science* **218,** 348–353.
Rubin, G. M., Finnegan, D. J., and Hogness, D. S. (1976). *Prog. Nucleic Acid Res.* **19,** 221–226.
Rubin, G. M., Brorien, W. J., Dunsmuir, P., Flavell, A. J., Strobel, J. J., Toole, J. J., and Young, E. (1981). *Cold Spring Harbor Symp. Quant. Biol.* **45,** 619–628.
Rubin, G. M., Kidwell, M. G., and Bingham, P. M. (1982). *Cell* **29,** 987—994.
Saedler, H., Bonas, V., Deumling, B., Gupta, M., Hahlbrook, K., Harrison, B., Kreuzaler, F., Peterson, P. A., Reif, J., Schwarz-Sommer, Z., Shepherd, N., Sommer, H., Ubben, D., and Wienand, U. (1983). *In* "Genetic Rearrangements" (K. F. Chater, C. A. Cullis, D. A. Hopwood, A. W. B. Johnston, and H. W. Woolhouse, eds.), pp. 107–115. Croom Helm, London.
Sang, H. M., Pelisson, A., Bucheton, A., and Finnegan, D. J. (1984). *EMBO J.,* **3,** 3079–3085.
Scherer, G., Telford, J., Baldari, C., and Pirrotta, V. (1981). *Dev. Biol.* **86,** 438–447.
Scherer, G., Tschudi, C., Perera, J., Delius, H., and Pirrotta, V. (1982). *J. Mol. Biol.* **157,** 435–452.
Schmid, C. W., Manning, J. E., and Davidson, N. (1975). *Cell* **5,** 159–172.
Schneider, I., and Blumenthal, A. B. (1978). *In* "Genetics and Biology of Drosophila" (M. Ashburner and T. R. F. Wright, eds.), Vol. 2a, pp. 266–316. Academic Press, New York.
Schwartz, H. E., Lockett, T. J., and Young, M. W. (1982). *J. Mol. Biol.* **157,** 49–58.
Schwarz-Sommer, Z., Grierl, A., Klorgen, R. B., Wienand, V., Peterson, P. A., and Saedler, A. (1984). *EMBO J.* **3,** 1021–1028.
Scolnick, S. B., Morgan, B. A., and Hirsh, J. (1983). *Cell* **34,** 37–45.
Shapiro, J. A., ed. (1983). "Mobile Genetic Elements." Academic Press, New York.
Shen-Ong, G. L., and Cole, M. D. (1982). *J. Virol.* **42,** 411–421.
Shepherd, B. M., and Finnegan, D. J. (1984). *J. Mol. Biol.,* **180,** 21–40.
Shiba, T., and Saigo, K. (1983). *Nature (London)* **302,** 119–124.
Shoemaker, C., Goff, S., Gilboa, E., Paskind, M., Mitra, S. W., and Baltimore, D. (1980). *Proc Natl. Acad. Sci. U.S.A.* **77,** 3932–3936.
Simchen, G., Winston, F., Styles, C. A., and Fink, G. R. (1984). *Proc. Natl. Acad. Sci. U.S.A.* **81,** 2431–2434.
Simmons, M. J., and Limm, J. K. (1980). *Proc. Natl. Acad. Sci. U.S A.* **77,** 6042–6046.
Southern, E. M. (1975). *J. Mol. Biol.* **98,** 503–517.
Southern, E. M. (1981). *In* "Chromosomes Today" (M. D. Bennett, M. Bobrow, and G. M. Hewitt, eds.), Vol. 7, pp. 55–61. Allen & Unwin, London.
Spradling, A. C., and Rubin, G. M. (1981). *Annu. Rev. Genet.* **15,** 219–264.

Spradling, A. C., and Rubin, G. M. (1982). *Science* **218,** 341–347.

Spradling, A. C., and Rubin, G. M. (1983). *Cell* **34,** 47–57.

Stanfield, S., and Helinski, D. R. (1976). *Cell* **9,** 333–345.

Strobel, E., Dunsmuir, P., and Rubin, G. M. (1979). *Cell* **17,** 429–439.

Sutton, W. D., Gerlach, W. L., Schwarz, D., and Peacock, W. J. (1984). *Science* **223,** 1265–1268.

Tartof, K. (1979). Cell **17,** 607–614.

Tchurikov, N. A., Ilyin, Y. V., Ananiev, E. V., and Georgiev, G. P. (1978). *Nucleic Acids Res.* **6,** 2169–2187.

Tchurikov, N. A., Zelentosova, E. S., and Georgiev, G. P. (1980). *Nucleic Acids Res.* **8,** 1243–1258.

Tchurikov, N. A., Ilyin, Y. V., Skryabin, K. G., Ananiev, E. V., Bayev, A. A., Krayev, A. J., Zelentrova, E. S., Kulgushin, V. V., Lyubomivskaya, N. V., and Georgiev, G. P. (1981). *Cold Spring Harbor Symp. Quant. Biol.* **45,** 655–665.

Temin, H. M. (1980). *Cell* **21,** 599–600.

Todaro, G. J., Callahan, R., Rapp, V. R., and de Larco, J. E. (1980). *Proc. R. Soc. London Ser. B* **210,** 367–385.

Tonegawa, S. (1983). *Nature (London)* **302,** 575–581.

Truett, M. A., Jones, R. S., and Potter, S. S. (1981). *Cell* **24,** 753–763.

Tschumper, G., and Carbon, J. (1982). *J. Mol. Biol.* **156,** 293–307.

Tuan, D., Biro, P. B., de Riel, J. K., Lazarus, H., and Forget, B. G. (1979). *Nucleic Acids Res.* **6,** 2519–2544.

Van Arsdell, S. W., Denison, R. A., Bernstein, L. B., Weiner, A. M., Manser, T., and Gesteland, R. F. (1981). *Cell* **26,** 11–17.

Varmus, H. E. (1983). *In* "Mobile Genetic Elements" (J. A. Shapiro, ed.), pp. 411–505. Academic Press, New York.

Vodkin, L. O., Rhodes, P. R., and Goldberg, R. B. (1983). *Cell* **34,** 1023–1031.

Voelker, R. A., Greenleaf, A. L., Gyurkovics, H., Wisely, G. B., Huong, S., and Searles, L. L. (1984). *Genetics* **107,** 279–294.

Weck, E., Courage, V., Döring, H. P., Fedoroff, N., and Starlinger, P. (1984). *EMBO J.* **3,** 1713–1716.

Weiss, R., Teich, N., Varmus, H., and Coffin, J. (1982). "Molecular Biology of RNA Tumor Viruses." Cold Spring Harbor Laboratory, Cold Spring Harbor, New York.

Wensink, P. W., Tabata, S., and Pachi, C. (1979). *Cell* **18,** 1231–1246.

Wienand, U., Sommer, H., Schwarz, Z., Shepherd, N., Saedler, H., Kreuzaler, H., Ragg, H., Fautz, E., Hahlbrock, K., Harrison, B., and Peterson, P. A. (1982). *Mol. Gen. Genet.* **187,** 195–201.

Wilde, C. D., Crowther, C. E., Cripe, T. P., Lee, M. G., and Cowans, N. J. (1982). *Nature (London)* **297,** 83–84.

Will, B. M., Bayev, A. A., and Finnegan, D. J. (1981). *J. Mol. Biol.* **153,** 897–915.

Williamson, V. M., Young, E. T., and Ciracy, M. (1981). *Cell* **23,** 605–614.

Wilson, D. A., and Thomas, C. W. (1974). *J. Mol. Biol.* **84,** 115–144.

Winston, F., Chaleff, D. T., Valent, B., and Fink, G. R. (1984). *Genetics* **107,** 179–197.

Yotsuganagi, Y., and Szollosi, D. (1981). *J. Natl. Cancer Inst.* **67,** 677–685.

Young, M. W. (1979). *Proc. Natl. Acad. Sci. U.S.A.* **76,** 6274–6278.

Young, M. W., and Schwartz, H. E. (1981). *Cold Spring Harbor Symp. Quant. Biol.* **45,** 629–640.

Young, T., Williamson, V., Taguchi, A., Smith, M., Sledziewski, A., Russell, D., Osterman, J., Denis, C., Cox, D., and Beier, D. (1982). *In* "Genetic Engineering of Microorganisms for Chemicals" (A. Hollaender, R. D. De Moss, S. Kaplan, J. Konisky, D. Savage, and R. S. Wolfe, eds.), pp. 335–361. Plenum, New York.

Zachar, Z., and Bingham, P. M. (1982). *Cell* **30,** 529–541.

INTERNATIONAL REVIEW OF CYTOLOGY, VOL. 93

Splicing in Adenovirus and Other Animal Viruses

EDWARD B. ZIFF

Kaplan Cancer Center and Department of Biochemistry, New York University Medical Center, New York, New York

I. Introduction

Viruses provide especially good models for the study of the splicing reaction, and its potential for influencing gene expression in animal cells. The splicing mechanism removes intronic sequences from nuclear transcripts, and thus brings together RNA sequences encoded by distant positions on the genome. Therefore, the effect of the splicing mechanism is intimately related to genome topology. This fact is especially important for splicing in viruses. Viral genomes have fixed sizes, and viruses have evolved to maximize the genetic utility of each segment

ISBN 0-12-364493-3

of their DNA or RNA. Thus, viruses often employ complex transcription units, with multiple promoters, poly(A) sites, and intricate splicing arrangements. In some circumstances, multiple mRNA species are produced from a single primary transcript. The versatility provided by splicing allows the encoding of proteins with novel structural relationships, such as common amino and carboxy terminal peptide sequences, but different internal peptides. Novel petide structural changes may take place at specific stages of the viral life cycle through changes in mRNA splice patterns in a manner which suggests splicing control. Because splicing of a single viral transcript can often yield multiple mRNAs with different coding capacities, the splicing mechanism must be coordinated with the mechanism of mRNA translation by eukaryotic ribosomes, in particular the rules for selection of AUGs for translation initiation.

Splicing was first discovered to occur with RNAs from the human adenovirus-2 (Berget *et al.*, 1977; Chow *et al.*, 1977a; Klessig, 1977; Kitchingman *et al.*, 1977). Since this discovery, splicing has also been found in the processing of RNAs from the papovaviruses, papilloma viruses, and the herpes viruses (although many herpes mRNAs are synthesized without the benefit of splicing) (for a review see Broker, 1983). Thus, RNA splicing is a common feature of DNA tumor virus gene expression. Retroviruses also employ splicing, which operates on RNAs encoded by the retroviral integrated genome (see Weiss *et al.*, 1982). Notably, the influenzae viruses also employ splicing, although their genomes, which serve as template for mRNA precursors, are single stranded RNA (see Lamb and Choppin, 1983; Broker, 1983). Thus, the splicing mechanism is not a specialized feature of DNA encoded transcripts.

Although the features of splicing in viruses vary somewhat from virus type to type, a great number of the most interesting aspects are to be found in adenovirus. This review will focus upon splicing of adenovirus transcripts rather than attempt to review the details of all viral splicing. However, other viruses will be briefly reviewed, and particularly relevant features will be noted.

As with any new biological phenomenon, the terminology is not complete for describing the different features of the splicing mechanism, or its RNA substrate and the underlying DNA gene structure. For the purposes of this review, sequences conserved in mRNA will be called "exons," and those which lie between cap and poly(A) and which are removed by the splicing mechanism "introns." Unfortunatley, with adenovirus these terms must be used with great care since certain sequences may be either intronic or exonic depending upon the fate of the particular transcript molecule. Also, in this review the splice boundary on the 5′ side of an intron will be called a "splice donor," and on the 3′ side a "splice acceptor." These terms will be used for convenience, and are not intended to prejudice the interpretation of the splicing mechanism. Other recent reviews of adenovirus gene expression are Ziff (1980), Nevins and Chen-Kiang (1981), and Logan and Shenk (1982).

Adenovirus splicing will be considered as it serves the virus during the two main stages, the early and the late, of infection. During the early stage of infection, splicing processes RNA from early transcription units, often to allow the encoding of several related mRNAs and gene products from a single "gene." In contrast, at the late stage of infection, splicing facilitates the expression of the major late transcription unit, which utilizes an unusual processing pathway, with splicing a crucial feature, to produce upwards of 20 different mRNA species from a single major late primary RNA transcript. The review will be organized according to these aspects of adenovirus splicing.

II. The Early Stage of the Adenovirus Life Cycle

Adenovirus-2 has a linear DNA genome about 35,000 nucleotides long, and the virus productively infects human cells. The adenovirus-2 life cycle may be divided into two main stages, the early and the late, each of which is complex. These will be briefly described to place adenovirus splicing in the context of the viral life cycle. The adenovirus genome is depicted in Fig. 1, with the mRNA map given in panel b. During the early stage, adenovirus genes are regulated primarily by transcriptional control (Berk *et al.*, 1979; Jones and Shenk, 1979; Nevins, 1981). During the first 4 hours of infection, which constitutes a period of transcription activation, five early promoters for early transcription units EIa, EIb, and EII–EIV are activated (Nevins *et al.*, 1979). The promoters (Baker and Ziff, 1981) and poly(A) sites (Fraser *et al.*, 1982) of early transcription units have been mapped and sequenced. The major late promoter, at 16.6 map units, is also active (see below). One of the promoters, the EIa promoter, appears to function constituitively, through the action of cis-acting "enhancer like" elements (Hearing and Shenk, 1983). The EIa gene in turn encodes products which activate the other early transcription units through a *trans* acting mechanism. At 4 hours postinfection, the peak of early transcription, the promoters for the early transcription units EIa, EIb, EII, EIII, EIV and the major late promoter are all active (Nevins *et al.*, 1979; Shaw and Ziff, 1980). Between 4 and 6 hours postinfection, the transcription rate from these early regions decreases. The *ts* mutant, *ts125* in the 72-kDa DNA binding protein, DBP, which is the EIIa gene product, fails to shut off transcription from region EIV (Nevins and Winkler, 1980). This suggests that the DBP represses EIV.

The major late promoter accounts for the majority of RNA Polymerase II transcripts synthesized in the infected cell at late times, and was initially considered to be a "late specific" promoter. However, the major late promoter is active at early times, albeit at a low level (Shaw and Ziff, 1980; Nevins and Wilson, 1981). The primary transcript encoded by the major late transcription unit, in its early mode, appears to terminate in the coordinate 40–60 region of the

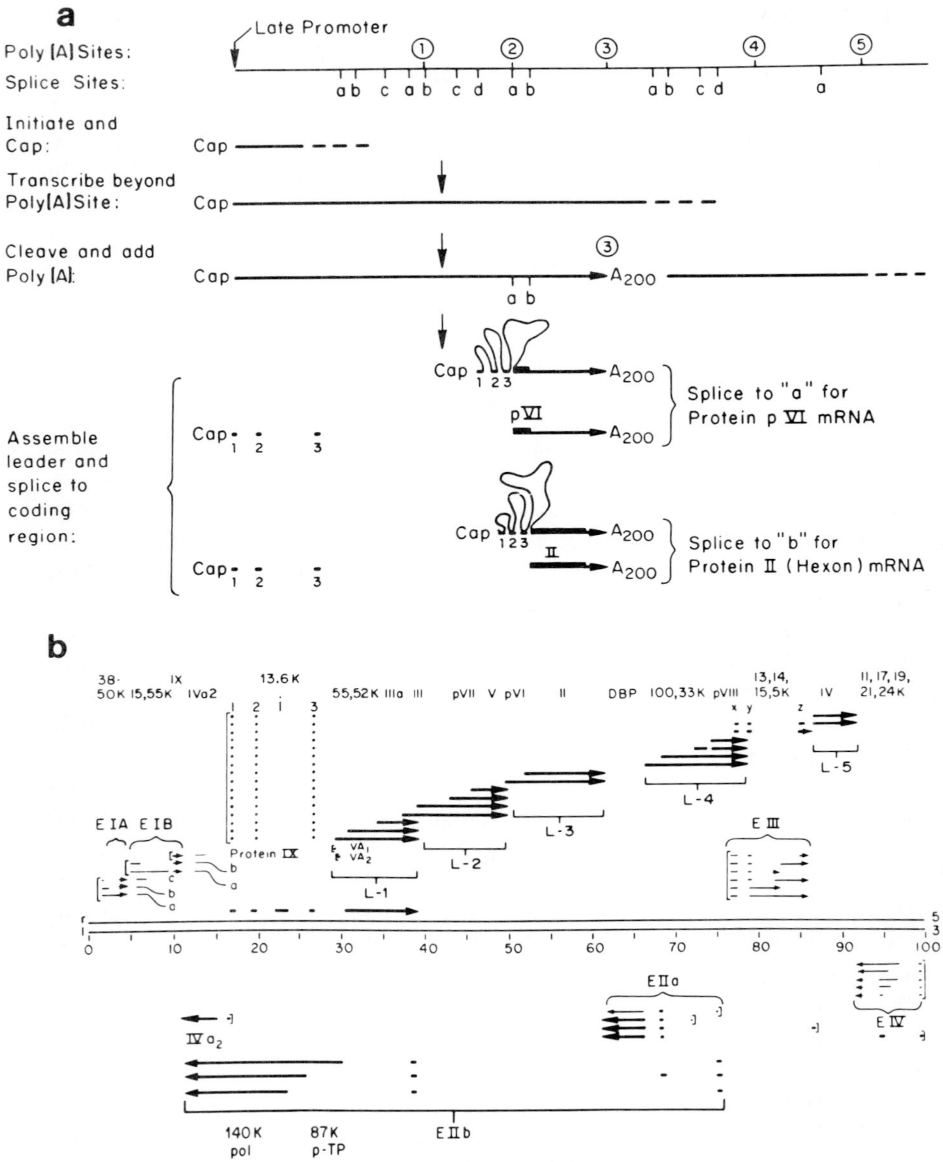

FIG. 1. Mechanism of synthesis and physical map of messenger RNA of adenovirus-2. (a) Transcription and processing of messenger RNA from the late transcription unit of adenovirus-2. (b) Adenovirus-2 messenger RNA map. The physical locations of early (light line) and late (heavy line) mRNAs and molecular weights of encoded proteins are given. This figure is adapted from Ziff (1980) and is based on data provided by L. Chow, T. Broker, and J. Lewis (personal communication).

genome (Shaw and Ziff, 1980), and mRNA synthesis at early times is primarily restricted to production of the first family of major late transcription unit mRNAs, the L-1 family, which uses the poly(A) site at 38 map units (Chow *et al.*, 1979).

Although transcription regulation is a major feature of adenovirus early gene expression, splicing also plays a crucial role. If we consider the structures of the mRNAs encoded by the various early transcription units, we may deduce the role splicing plays in adenovirus gene expression during the early stage.

A. Early Region EIa

EIa is one of two early transcription units which lie at the left end of the adenovirus genome. Together with EIb this region encodes the products necessary and sufficient for transformation. The genomic locations of these transcription units is given in Fig. 1b, and the arrangement of coding sequences and splice points is shown in Fig. 2. At early times, EIa encodes two principal messenger species, which are 12 S and 13 S in sedimentation (Chow *et al.*, 1977b; Berk and Sharp, 1978; Kitchingman *et al.*, 1977). Deletion mutants, such as d1312, which lack the EIa promoter and thus do not yield EIa proteins, also fail to express other early transcripts at wild-type levels indicating that EIa plays a function in transcription activation (Jones and Shenk, 1979; Berk *et al.*, 1979; Nevins, 1981). The two EIa mRNAs are 5′ and 3′ coterminal, and have a common splice acceptor (designated A1 in Fig. 2) but differ in the position of the donor splice site (designated D2 and D3), which defines a variable boundary for the single EIa intron (Perricaudet *et al.*, 1979).

This arrangement of donor splice sites illustrates an important feature of adenovirus splicing that is not a common aspect of cellular splicing. Two alternative splice sites may function for the same transcript. The versatility provided by multiple splicing requires that some step in the splicing pathway can proceed to form two alternative products. From inspection of the open reading frames found in cDNA clones of the 12 S and 13 S mRNAs (Perricaudet *et al.*, 1979; van der Eb *et al.*, 1980) (shown in Fig. 2), one may deduce that the proteins encoded by these two messengers will be amino and carboxy coterminal, but will differ in an internal peptide. The splicing flexibility for EIa mRNA allows the virus to encode two closely related polypeptides, which differ by the presence or absence of a 46 amino acid internal peptide. Mutagenesis studies suggest that this internal peptide sequence is crucial to the transcription activation function of the 13 S translation product (Riccardi *et al.*, 1981; Montell *et al.*, 1982).

At late times in infection, EIa encodes a third shorter mRNA shown in Fig. 2, a 9 S species (mRNA c of Fig. 1b) (Spector *et al.*, 1978; Chow *et al.*, 1979; Spector *et al.*, 1980). The cap and poly(A) sites of this mRNA are identical to those of the 12 S and 13 S mRNAs which are predominant at early times, but the

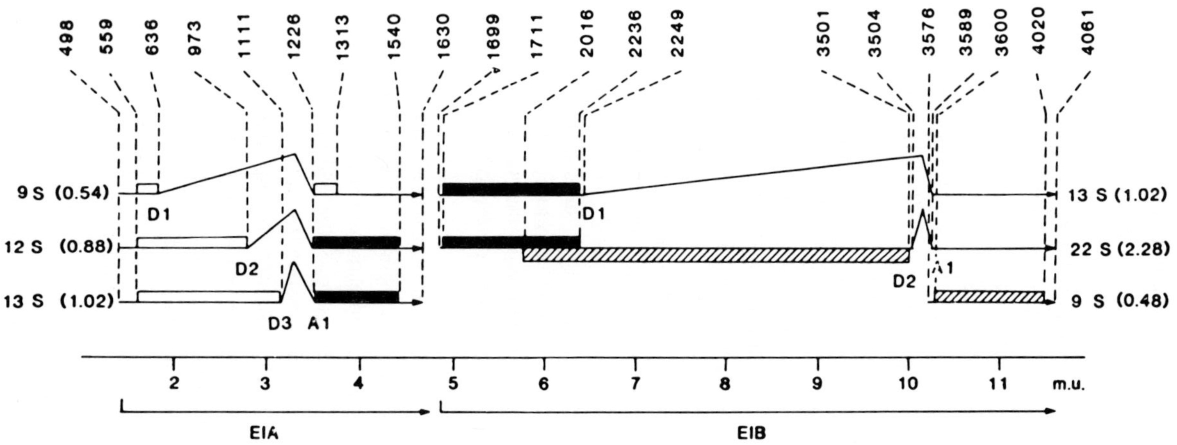

FIG. 2. Messenger RNAs from the EIa and EIb transcription units of adenovirus-2. The map positions of the 9 S, 12 S, and 13 S mRNAs of the EIa transcription unit, and the 13 S and 22 mRNAs of the EIb transcription unit, as well as the 9 S mRNA of the pIX transcription unit are given. D and A are respectively donor and acceptor splice sites. Numbers above the figure are nucleotide residue positions within Ad-2 DNA, and open, solid, and hatched boxes are open reading frames of the coding regions. This figure is from Pettersson *et al.* (1983).

9 S species has a new donor splice site, designated D1, which is shifted 5′ to the 12 S mRNA donor site, D2. Thus this new late specific EIa mRNA has a shorter first exon. Also, the splice to the acceptor site A1 dictates a new reading frame for the second exon. The translation product of the 9 S mRNA is predicted to be amino coterminal with the aforementioned EIa early proteins, but to have a different carboxy-terminal sequence.

As we shall discuss in greater detail below, changes in the steady-state splicing patterns of mRNA from EIa are apparent if species at early times are compared with those at late times. These changes are similar to the splicing changes seen with mRNAs from several other adenovirus mRNA families. Thus the EIa transcription unit, as well as others, is capable of a stage specific alteration in the protein products which it encodes. This change clearly depends on alterations in the spliced structures of the steady state mRNA products. The question of whether or not this constitutes splicing control will also be discussed below.

If one considers possible mechanisms for recognition of splice site boundaries, a problem arises with multiple donor sites such as found with EIa. Consider the possibility that splicing has a "polarity" and operates through a processive mechanism which scans the RNA for introns or splice junctions. Let us examine the consequence for producing the early EIa mRNAs, the 12 S and 13 S species. If scanning were in the 5′ to 3′ direction, and started by locating a donor splice site (5′ boundary of intron), we might expect that the 12 S mRNA would be the exclusive product from EIa. This would be expected at early times because its donor splice site D2 would be encountered before D3, the 13 S donor site, if scanning were 5′ to 3′. Likewise, if scanning were 3′ to 5′, and the first event was the locating of the splice acceptor site of EIa transcripts, A1 (3′ boundary of exon), and the next event were a search for a donor site, the production of the 13 S mRNA alone would be expected. Neither predicted outcome is actually observed since both mRNAs are produced at approximately equal levels. The problem is complicated by the stage-specific production of the 9 S species, with its additional donor splice site. Although this complexity arises with multiple alternative *donors* in the case of EIa, multiple *acceptors* are found with major late transcription unit mRNAs (see below). Scanning models might account for the efficiency and accuracy of intron removal, but simple processive hypotheses utilizing scanning are not adequate to explain the pattern of mRNA species actually observed. Perhaps another feature, such as secondary structure, may be involved.

The EIa region has been the target of direct mutagenesis studies, and two of these studies of altered splice site sequences are relevant to this discussion. Solnick (1981) used nitrous acid mutagenesis to introduce base changes into the DNA residues encoding the donor splice site of the 12 S mRNA (as well as other neighboring positions). This splice donor sequence lies within the first exon of the 13 S mRNA, and its mutation introduced a chain terminating codon which truncates the peptide from the 13 S mRNA. In the absence of a functional donor splice site for the 12 S mRNA, only the 13 S species is produced. This mutant virus produces mRNA from the EIa, EIb, and EIV transcription units, but not from EII or EIII (Solnick and Anderson, 1982) reflecting the requirement for the EIa 13 S mRNA product for full early transcription activation. Montell *et al.* (1982) have performed an analogous construction in which the 12 S mRNA donor splice site was made nonfunctional through site-directed mutation. This base change was introduced by means of a mutagenic oligonucleotide primer which altered a residue critical to the 12 S mRNA donor splice sequence, but also introduced a synonymous codon in the 13 S mRNA first exon. Thus the change did not alter the 13 S protein translation product but destroyed the function of the 12 S mRNA splice donor sequence. Once again, the production of 12 S mRNA was eliminated without apparent effect upon 13 S mRNA synthesis.

It is not known whether the mRNA nuclear precursors encoded by these mutants normally destined to be processed to the 12 S mRNA species were degraded in the nucleus after a failure to complete formation of the 12 S splice, or whether the potential precursor to 12 S RNA was processed to yield 13 S mRNA instead.

B. Early Region EIb

EIb constitutes the second block of genes at the left end of adenovirus-2, and encodes functions essential to the transforming capability of the virus. One gene product from EIb, a 58-kDa polypeptide, binds to a cellular p53 protein (Esche *et al.*, 1980; Sarnow *et al.*, 1982). p53 levels increase within the cell, possibly by the 58-kDa peptide stabilizing p53 from degradation. Thus, this EIb product may function through interaction with cellular factors. EIb also encodes a 19-kDa protein with unknown function (Ross *et al.*, 1980; Esche *et al.*, 1980).

At early times, the major EIb mRNA is a 22 S species, designated species a in Fig. 1b (Chow *et al.*, 1979). The structure is shown in greater detail in Fig. 2. The 22 S mRNA has a long main body sequence followed by a short splice, a second exon and a poly(A) tail. Sequence determination of cDNA has shown that the 22 S mRNA contains two large overlapping reading frames, both within the first exon (Bos *et al.*, 1981). These are also shown in Fig. 2.

Production of multiple proteins from a single transcription unit occurs with EIb but does not rely on splicing as was the case with EIa. The 22 S EIb mRNA is an example of a single mRNA which encodes two protein products. In fact, the arrangement of coding sequences in the 22 S mRNA presents an apparent problem to the ribosome. The problem derives from the mechanism used by the ribosome to choose the initiator triplet and the consequent reading frame for mRNA translation. Translation of the 22 S mRNA starting at the first, cap-proximal, AUG dictates a reading frame (shown in Fig. 2) which yields the 19-

kDa polypeptide. However, to translate the major tumor antigen of adenovirus, the 58-kDa protein, the ribosome must initiate at an internal AUG which lies downstream from the AUG of the 19-kDa peptide initiator (Bos *et al.*, 1981). As discussed by Kozak (1978) the cap-proximal AUG is the functional initiator in most eukaryotic messengers. The internal initiator employed for translation of the 58-kDa protein is followed by a large open reading frame, also shown in Fig. 2, which overlaps the 3′ segment of the 19-kDA protein reading frame. Splicing plays no apparent role in the choice between 19- and 58-kDa protein translation. In fact, the single intron of the 22 S mRNA lies downstream from the translation terminator of the 58-kDa protein reading frame, and the one mRNA encodes both proteins. A potentially similar situation is found with the VP2 and VP3 mRNAs of the papovaviruses.

The second exon of 22 S mRNA contains the coding sequences for a stuctural protein, pIX. However, pIX is expressed from a different mRNA species altogether, as discussed below.

The EIb transcription unit mRNAs undergo a splicing pattern shift which is analagous to the EIa species shift and which occurs after the start of the late phase of infection (Spector *et al.*, 1978; Chow *et al.*, 1979). As with EIa, a shorter mRNA predominates at late times, in this case a 13 S EIb mRNA species. In the 13 S EIb messenger, the donor EIb splice D1, shown in Fig. 2, is utilized. This splice falls just beyond the 3′ end of the first open reading frame, and presumably this mRNA (as well as the 22 S species) can be translated to yield the 19-kDa protein.

C. The Protein IX mRNA

The 9 S mRNA which encodes pIX is from a transcription unit wholly contained within EIb (Pettersson and Mathews, 1977; Aelstrom *et al.*, 1980; Petersson and Mathews, 1977). The pIX promoter lies within the EIb 22 S mRNA intron, and the mRNA is 3′ coterminal with EIb mRNAs (Fraser and Ziff, 1978). The mRNA is abundant at late times, but may be detected toward the end of the early period (Persson *et al.*, 1978). Most strikingly, the mRNA contains no introns and is transported to the cytoplasm without removal of intervening sequences. It also lacks the EIb common splice acceptor sequence, A1 of Fig. 2.

The lack of a splice makes the pIX mRNA unique in adenovirus-2. pIX is the only adenovirus transcription unit which lacks an intron and produces mRNA without splicing. Because pIX lies wholly within EIb, splice signals in the pIX transcription unit could potentially interfere with splicing of EIb transcripts. One other adenovirus transcription unit, which lies wholly within a second transcription unit, is EIII. Both the 5′ and 3′ limits of EIII fall within the major late transcription unit. In this case signals from the EIII transcription unit apparently *do* affect transcripts of the other overlapping unit (see below).

D. The IVa2 mRNA

This mRNA, shown in Fig. 1b, is from a promoter which is adjacent to the major late promoter, but yields leftward transcripts (Baker and Ziff, 1981). The mRNA 3′ end is positioned just overlapping the EIb poly(A) site (Fraser *et al.*, 1982). A single intron is removed from IVa2 during processing (Chow *et al.*, 1979), however this splice squence does not appear to function in EII transcripts initiated upstream. The IVa2 protein is a minor structural component of the virion, but its function is unknown.

E. Early Region EII

The EII transcription unit is complex. That is, EII produces mRNAs with two alternative poly(A) sites, one located near map unit 60 and the other near map unit 11. This permits EII to yield two separate families of mRNAs, respectively EIIa and EIIb shown in Fig. 1b. Electron microscopy and *in vitro* translation support the conclusion that there are no introns within EII coding sequences (Stillman *et al.*, 1981).

The first family, EIIa, encodes the DNA binding protein (DBP), a 72-kDa species which coats single strands of DNA that are displaced as intermediates during adenovirus DNA replication (Van der Vliet and Levine, 1973). The DBP also represses transcription from region IV (Nevins and Winkler, 1980), and, pertainent to this review, modulates mRNA stability and possibly also splicing patterns (Babich and Nevins, 1981).

At early times, EIIa transcripts are initiated at a promoter at 75 map units (Baker *et al.*, 1979), and RNA processing forms a two part leader with second exon at 68 map units, which is spliced to the main body (Chow *et al.*, 1979; Goldenberg and Raskas, 1979). At late times, the promoter at 75 map units is repressed, but a second late specific promoter at 72 map units is activated (Chow *et al.*, 1979; Baker and Ziff, 1981). A late leader for EIIa mRNA is encoded at 72 map units and is spliced to the same (that is, identical at the resolution of electron microscope) second exon acceptor as is used for early mRNA. This in turn is joined to the same mRNA body produced at early times. Thus, by virtue of splicing, the early and late EIIa mRNAs are nearly identical, although their promoters are separated by approximately 1000 nucleotides. The mRNAs remain nearly constant in structure and protein encoding capacity, despite a stage specific change in promoter. Note also that the donor splice site of the 72 map unit leader does not appear to function early, although its sequence is present in the primary transcript.

Although only a single principal mRNA species is produced from EIIa, at least three (and possibly more) EIIb mRNAs are synthesized, generating a 3′ coterminal family (Stillman *et al.*, 1981). Messengers belong to a 3′ coterminal family when they share a common 3′ poly(A) site, but the splice acceptors which mark the start of their coding sequences are staggered on the 5′ side. This relationship is satisfied by EIIb mRNAs. The arrangement of coding sequences into 3′ coterminal families transcribed from a single promoter is the distinctive feature of the adenovirus major late transcription unit (see below).

Two proteins encoded by EIIb are the 87kDa preterminal protein, p-TP, which serves as primer for adenovirus DNA initiation, and the adenovirus 140-kDa DNA polymerase (Stillman *et al.*, 1981; Lichy *et al.*, 1982). Functions encoded by EIIb therefore pertain to DNA replication. Because the mRNAs from EIIb have not yet been cloned and sequenced, the precise arrangement of coding sequences is still speculative. Differential use of the 60 map unit and the 11 map unit poly(A) sites, or termination of transcription downstream from 60 map units, may regulate the relative levels of EIIa and EIIb mRNA. Indeed, EIIa mRNAs are at least 10 times more abundant than EIIb (Stillman *et al.*, 1981).

F. Early Regions III and IV

Early regions III and IV yield complex groups of mRNAs which differ in their splicing patterns. Like EII, EIII has two poly(A) sites and therefore is a complex transcription unit (Chow *et al.*, 1979; Fraser *et al.*, 1982). Although the DNA sequence of this region of adenovirus-2 has been determined, the precise mapping of the various spliced species is not complete. However, the mRNA structures have been investigated in some detail by the heteroduplex method (Chow *et al.*, 1979). When the coding properties of these mRNAs is established, additional contributions by splicing to adenovirus gene control will undoubtably be revealed.

Unlike the cases of EIa and EIb, the question of stage-specific splicing patterns does not arise for EIII and EIV because these transcription units are repressed at late times (Nevins and Winkler, 1980). However, the EIII transcription unit lies completely within the major late transcription unit, and its splice signals may also be recognized in transcripts initiated at the major late promoter (see below).

III. The Major Late Transcription Unit

The products from the major late transcription unit are initiated at the major late promoter at map unit 16.6 (Ziff and Evans, 1978). The mRNAs from the major late transcription unit are characterized by (1) a common (with some variation) tripartite leader; (2) the use of multiple poly(A) sites; and (3) the formation of 3′ coterminal families through the use of multiple alternate splice acceptor sequences. The most common form of the late leader consists of three

RNA segments encoded at 16, 19, and 26 map units (Berget *et al.*, 1977; Chow *et al.*, 1977a, Klessig, 1977; Zain *et al.*, 1979; Akusjarvi and Pettersson, 1979a,b). In genomic DNA, these exons are separated by ~1700 and ~2500 nucleotides respectively. The fully assembled tripartite leader lacks any AUG translation initiation sequence. Therefore it is noncoding.

Although the major late transcription unit products are most predominant at late times, the major promoter also functions prior to DNA replication. The primary transcript from the major late promoter at early times is, however, shorter than the one produced during the late stage (Shaw and Ziff, 1980; Nevins and Wilson, 1981). Because the 16 map unit - 40 map unit region of the late transcription unit which encodes L-1 mRNA is active during both the early and the late stages, stage specific splicing changes in the L-1 species is possible.

A. The Major Late Transcription Unit as Expressed in the Early Stage

At early times, the primary transcript from the major late transcription unit terminates somewhere in the coordinate 40–60 region of the genome (Shaw and Ziff, 1980). Should termination in fact occur in the vicinity of 40 m.u., the transcript would extend beyond the first poly(A) site at 38 map units (which serves the L-1 family), but it would not reach the second poly(A) site (for the L-2 family at 50 m.u.). This would permit the production of mRNAs from the L-1 family, but not for other families such as L-2 or L-3, whose poly(A) sites would not be traversed by the transcript. The transcript may however extend beyond 40, to map unit 60 (Nevins and Wilson, 1981), and in this case the L-2 and L-3 poly(A) sites would be transcribed. The production of mRNA actually observed at early times greatly favors L-1 over L-2 (Shaw and Ziff, 1980; Nevins and Wilson, 1981), and L-3 mRNA is hardly detected at all. Were the transcript to extend as far as 60 map units, the low expression of L-2 and L-3 mRNAs relative to L-1 mRNAs would require some form of posttranscriptional control, e.g., a preference for L-1 during the polyadenylation or splicing steps.

Note that an L-1 preference is not seen at late times (Nevins and Darnell, 1978). With the major L-1 mRNA synthesized early, the tripartite leader is ligated to a main body coding sequence at 29 map units that encodes a doublet of proteins of 52,000 and 55,000 molecular weight (Chow *et al.*, 1979; Lewis and Mathews, 1980; Miller *et al.*, 1980). The function of these proteins is unknown. No other late transcription unit mRNA (with the exception of i leader species discussed below) is abundant early.

B. The Early to Late Shift

At approximately 6 to 8 hours postinfection, adenovirus DNA replication commences. Replication starts after a period of transcription repression. Indeed,

EIII and EIV are turned off at the end of the early period, never to be reactivated (Nevins and Winkler, 1980). EII transcription from the 75 map unit promoter is also repressed at late times, but the downstream promoter at 72 map units is activated and permits continued EII expression at late times (Baker and Ziff, 1981).

One major transcription event following the start of DNA replication is the augmentation of the rate of transcription from the major late promoter (Shaw and Ziff, 1980). This increase is due in part to an increase in the number of templates. However, the intrinsic activity of the promoter may also increase. The second event is the antitermination of the major late transcript shown in Fig. 3 (Shaw and Ziff, 1980; Nevins and Wilson, 1981).

The fact that RNA from a single promoter dominates at late times, yet upwards of 20 late species are produced, is made possible by complex splicing.

Messenger RNAs from EIa and EIb continue to be made at late times (see below).

C. The Late Mode of the Major Late Transcription Unit

At late times, the major late transcription unit produces a remarkable collection of messengers all sharing the spliced tripartite leader (or variants thereof) (see Ziff, 1980). Each messenger falls into one of five 3′ coterminal families. These families are designated L-1 through L-5 in Fig. 1b. Within each family, different members have the late leader appended to a different splice acceptor in the main mRNA body.

The production of the major late mRNAs represents perhaps the most elegant coordination of RNA processing steps thus far discovered amongst eukaryotic transcripts. The steps in late processing are outlined in Fig. 1a.

The primary transcript serves as a substrate for the processing machinery is a long transcript, encoding up to 30,000 bases (Bachenheimer and Darnell, 1975; Fraser *et al.*, 1979). The transcript may never exist as an intact molecule extending from cap to extending from cap to termination sequence, because it is soon acted upon by the poly(A) synthesis machinery. Poly(A) addition requires endonucleolytic cleavage of the RNA at one of five possible poly(A) sites (Nevins and Darnell, 1978a). The cleavage occurs soon after the polymerase passes the particular poly(A) site which will be the target for the endonucleolytic clip. Poly(A) is then rapidly added. The product of this processing step is an RNA molecule with a cap at one end and poly(A) at the other (Nevins and Darnell, 1978b). As we shall see, this RNA still contains a large proportion of the intronic sequences which will be eventually removed by splicing.

The net effect of splicing on the capped and polyadenylated large nuclear late transcript, is the juxtaposition of the late tripartite leader to one of several splice acceptors, each proximal to the poly(A) (Ziff and Fraser, 1978; McGrogan and Raskas, 1978; Nevins and Darnell, 1978a). This pathway is shown in Fig. 1a.

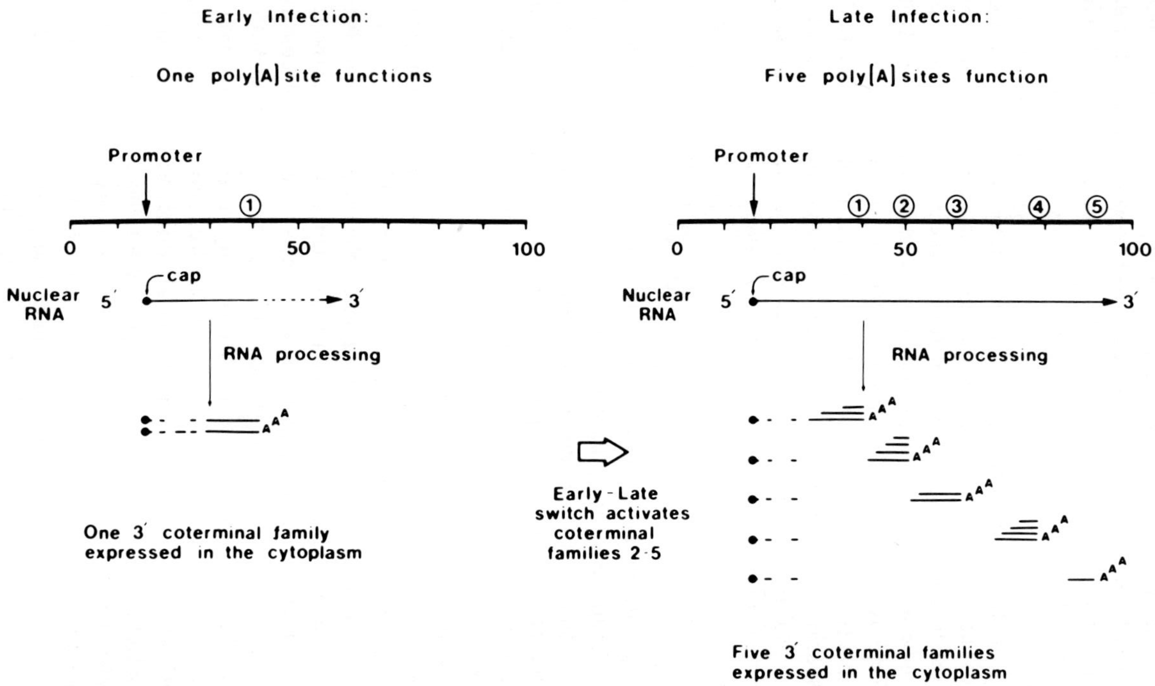

FIG. 3. Activation of expression of adenovirus-2 coterminal families L-2 through L-5. Messenger RNA species from the major late transcription unit produced during the early and late stages of infection are shown. This figure is from Shaw and Ziff (1980).

This figure shows an example in which poly(A) is added to the third poly(A) site. In this case the nuclear transcript is committed to be processed to an mRNA from the L-3 family. Splicing of this precursor can follow more than one course. The L-3 family is composed of mRNAs with at least two acceptor sites, a and b, seen in Fig. 1a. The splicing machinery can use either of these as a splice acceptor for the leader.

An intriguing feature of splicing of late adenovirus transcripts is the ability of the splicing machinery to deliver the mature tripartite leader to a splice acceptor which is closely associated with the position of the poly(A). This property is required for the function of complex transcription units. Thus, for the L-1 family, the leader may be spliced at late times to one of 3 possible acceptors, enabling formation of three L-1 coterminal mRNAs. However, if poly(A) is added at the L-2 poly(A) site, the leader does not get ligated to these L-1 acceptor sites in the final mRNA product. Instead it is transferred to an L-2 acceptor, of which there are three which function at late times. Likewise, the leader is not joined to an L-1, L-2, or L-3 family splice acceptor when poly(A) has been added to the L-4 poly(A) acceptor. This specificity creates an "intron–exon paradox" discussed below and is but one aspect of the rather complex requirements for faithful production of adenovirus late messenger RNA. The 3′ coterminal mRNAs produced from the major late transcription unit express different proteins by virtue of the translation properties of animal cell ribosomes. The properties are summarized in "Kozak's Rules" (Kozak, 1978), which, as applied to adenovirus, are as follows: The late tripartite leader (which lacks an AUG, and thus cannot initiate translation) is spliced to the 5′ side of a coding sequence present in the mRNA body. Ribosomes, which scan the RNA 5′ to 3′ starting from the cap, will pass the leader and enter the main body region adjacent to the splice acceptor, and initiate translation, in general from the first AUG. Translation of this main body open reading frame will continue until the ribosome encounters an in-phase terminator. By forming coterminal mRNAs through splicing, the leader is juxtaposed to different AUGs, and the ribosome gains access to different reading frames (Akusjarvi and Pettersson, 1978, 1979; Akusjarvi *et al.*, 1981). All of these reading frames lie in overlapping mRNAs which share the same polyadenylated 3′ end. The larger mRNAs contain coding sequences for more than one protein (pVI and hexon in the example in Fig. 1a) but only the first cistron (pVI in our example) is translated because initiation is at the cap-proximal AUG. The second reading frame (coding for hexon) is silent in the longer mRNA, but is expressed in the shorter mRNA in which the AUG of the second reading frame is proximal to the cap. The formation of 3′ coterminal mRNAs based on this sort of organization of coding sequences is found for mRNAs from four of the five late families (L-5 is excepted from multiple protein encoding) permitting the synthesis of over 15 different late proteins from the major late transcription unit.

IV. The Splicing Mechanism

Because adenovirus-2 transcripts are produced in large abundance, and because adenovirus applies the splicing mechanism in an unusual fashion, the virus offers novel opportunities for studying how splicing relates to the overall mechanism of RNA processing. Nevins and Darnell (1978b) have used the uridine pulse-labeling technique to analyze the mRNA precursors in the nucleus which are the substrates for the splicing machinery. In their experiments, adenovirus infected cells were pulse labeled with tritiated uridine during the late stage of infection. Nuclear RNA was isolated after various labeling periods, fractionated by size or poly(A) content, and then hybridized to viral DNA fragments to measure the progress of the primary transcript from nascent molecules to mature cytoplasmic mRNA. These workers showed that polyadenylated nuclear RNA sequences adjacent to each of the major late transcription unit poly(A) sites are labeled with equal kinetics in time periods less than required to complete synthesis of the transcript (i.e., for the polymerase to travel from the late promoter to coordinate 100). This finding implied that nascent RNA was the poly(A) acceptor. Furthermore, transcription of the major late transcription unit is equimolar. That is, the rate of RNA synthesis at all positions is equivalent (except for the DNA immediately adjacent to the promoter—see below). Thus, these workers concluded that all polymerases must continue transcription to the end of the genome before transcription is terminated. This conclusion implied that primary transcripts were endonucleolytically cleaved to expose an RNA 3′ terminus, and that this terminus was the site of addition of poly(A). It ruled out termination as the mechanism for forming the poly(A) acceptor. Thus, in this system, poly(A) sites are generated by cleavage. These workers also demonstrated that only one poly(A) tract was added per primary transcript. That is, there are essentially no nonproductive polyadenylation events, and receipt of poly(A) by a nuclear molecule is a guarantee of transport to the cytoplasm.

The nascent polyadenylated molecules detected in these studies were large, and their size implied that they retained essentially all of their intervening sequences. Therefore, it was likely that the nuclear molecules acquired their poly(A) soon after transcription of the poly(A), but prior to the removal of introns from the transcript by splicing. These data provide evidence that with adenovirus late RNA, polyadenylation precedes splicing. However, with these experiments it was difficult to exclude the possibility that short stretches of intronic RNA are in fact removed prior to polyadenylation. Furthermore, the fact that polyadenylation preceded splicing in these experiments may not have been the result of a direct requirement of the mechanism of RNA processing. Instead, it could simply reflect the kinetics of the two steps, with polyadenylation rapid, and splicing relatively slow.

More recent experiments support the latter interpretation. Zeevi *et al.* (1981) have further examined the requirement for polyadenylation prior to splicing using the drug cordycepin, which blocks poly(A) addition. They have shown that region EIIa mRNA can be correctly spliced, even in the presence of cordycepin when the nuclear precursors are devoid of poly(A). Although splicing generally occurs on polyadenylated molecules, it evidently can proceed even when poly(A) addition is blocked.

When the major features of adenovirus late mRNA production are considered, it is evident that two posttranscriptional "choices," outlined in Fig. 1a, are required for the production of a late transcription unit mRNA. The first "choice" is the selection of the poly(A) site from one of five, and the second is the "choice" of the splice acceptor (at the mRNA main body) for the tripartite late leader. This second choice is confined to splice acceptors proximal to the poly(A). When the existence of extra exonic sequences (see below) including the "i leader" and the X, Y, and Z leaders is taken into account, the "choices" become somewhat more complex, although the two above remain the fundamental ones.

A. The Intron–Exon Paradox in Adenovirus

Introns are generally considered to be sequences which, in RNA, are efficiently removed from transcripts by splicing. Likewise, exons are sequences which are conserved in the mature cytoplasmic mRNA product. However, when these terms are considered in the light of late adenovirus mRNA production, a paradox is evident. Within the adenovirus late transcription unit, the sequences which are conserved as mRNA main body are those which are proximal to the poly(A) site. Hexon coding sequences are conserved when poly(A) is added at polyadenylation site 3. However, the hexon coding sequences will be retained in the mRNA product if and only if this particular proximal polyadenylation site (the L-3 site) is employed. Thus, when the L-3 site is used, the hexon mRNA body sequences will constitute exons. However, when a different polyadenylation site is employed, such as the L-4 site, these same hexon coding sequences will be dutifully excised and consigned to the category of introns (or "downstream RNA" if the newly considered poly(A) site is upstream, as with L-2). Thus, with the adenovirus major late transcription unit, a sequence will on one occasion be exonic, and on another, intronic.

Consider what this intron–exon duality implies for the splicing mechanism. A leader is a noncoding RNA segment at the 5′ end of an mRNA. With adenovirus, the major form of the late leader consists of three separate RNA segments joined by splicing. Suppose that this leader is assembled by linking leader 1 to leader 2, and that this product is then joined to leader 3. The assembled leader now has a very large number of potential splice acceptors to which it may be ligated. There are three such positions which belong to late family L-1, and four for L-2 etc (see Fig. 1b). When we consider all such splice acceptor positions upstream from the final poly(A) site, L-5, the number of acceptors approaches 15 or more.

To enable the splicing flexibility seen in the major late transcription unit, the splicing mechanism must deliver the leader to an appropriate splice acceptor. This, however, presents a problem. If the poly(A) is added to poly(A) acceptor site 2, as is required for making L-2 mRNA, it is also necessary that the leader reach a L-2 proximal splice acceptor. However, if the poly(A) is added to poly(A) acceptor site 3, as is required for making L-3 mRNA, the leader must skip over the acceptors appropriate in the previous example (good only for making L-2 mRNA) and come to rest at an L-3 splice acceptor site. When the multitude of poly(A) sites and splice acceptors at mRNA main boundaries is considered, the problem of coordinating leader with acceptor becomes overwhelming. (Note also that formally, the donor boundary for leader segment 2 might be expected to be capable of bypassing the acceptor for leader segment 3 and reaching the main body acceptors itself. This, and many other possible aberrations, in fact do not occur.) Thus, the control of intron versus exon in late mRNA production is, so far, difficult to explain mechanistically, although it is a crucial aspect of late mRNA production.

B. Variable Exonic RNA Segments

We have previously discussed the fact that transcripts from the major late transcription unit may be processed to give different mRNA species belonging to the various 3′ coterminal families. The production of multiple mRNA products depends on the possibility for multiple splicing pathways for RNA molecules with the same 5′ and 3′ termini. A different type of splicing variability allows the inclusion or exclusion of specific variable RNA leader exonic segments. These variable leader segments may lie either *within* the common tripartite leader, or *between* this leader and the main body.

The first extra exon type, which maps within the standard leader, is the "i leader," a 439 base long RNA segment which resides, in some mRNAs, between leader segments 2 and 3 (Chow *et al.*, 1979; Virtanen *et al.*, 1982; Falvey and Ziff, 1983). The second extra exon type is represented by the X, Y, and Z leaders, which fall between the third leader segment and the coding sequences of the L-5 mRNA, the fiber mRNA (Chow *et al.*, 1979; Zain *et al.*, 1979). The existence of these variable exons adds another level of complexity to the structure of adenovirus late mRNAs, and with the i leader, provides a novel change in late mRNA coding capacity. These exons are detected during the course of normal infections of HeLa cells, but a potential parallel is also seen in infections of monkey cells as discussed below.

C. The "i Leader"

Chow *et al.* (1979) have discovered an extra RNA segment which lies between the second and third leaders of the major late mRNAs. This exon, called the "i leader" was first seen in mRNA from the L-1 message family isolated at early times of infection. The relationship of the i leader to other leaders is given in Fig. 4. The i leader maps between coordinate 22 and 22.5 and was estimated to be approximately 400 nucleotides long on the basis of electron microscopy. Virtanen *et al.* (1982) and Falvey and Ziff (1983) have mapped the positions of the splice points of the i leader in genomic DNA and determined its sequence. The i leader contains an AUG initiator triplet 26 residues from its 5′ boundary. Because the i leader potentially initiates translation at this AUG it differs from the standard tripartite leader, which lacks any AUG and thus cannot code for any polypeptide. The AUG of the i leader is followed by an in-phase, uninterrupted open reading frame which, in the i leader itself, is 414 nucleotides long, and continues for the full extent of the i leader. The open frame runs across the 3′ boundary of the i leader where in genomic DNA it soon encounters a terminator. These data suggest that the i leader is capable of encoding protein and in fact is the 5′ segment of a gene.

Virtanen *et al.* (1982) have cloned a cDNA copy of an L-1 family mRNA which contains the i leader. They have determined its sequence, and compared it with the genomic DNA. In mRNA, the open reading frame of the i leader continues across the splice point into the third leader, where it soon reaches a terminator. Thus a peptide is predicted from the open frame in the i leader. To determine whether such a peptide actually can be translated, Virtanen *et al.*

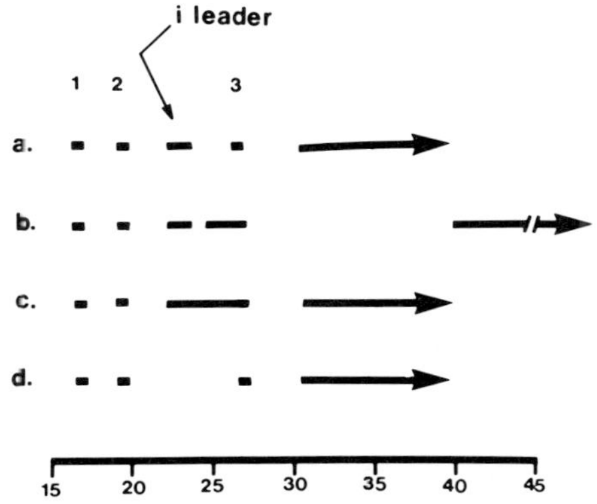

Fig. 4. Messenger RNA species containing the adenovirus-2 i leader. Species a, b, and c are mRNAs containing the i leader or variants of this leader. Species d has the standard tripartite leader. This figure is redrawn from Chow *et al.* (1979).

(1982) enriched mRNA from adenovirus infected cells for species containing the i leader, and tested the coding capacity by *in vitro* translation. A product was obtained corresponding to the size predicted by the open frame for an i leader peptide.

Falvey and Ziff (1983) and Chow *et al.* (1979) also report that the i leader can be detected on mRNAs encoded by the major late transcription unit during the late stage of infection, and possibly on mRNAs from several of the late families. Thus it is necessary to consider the implications of the presence of the i leader for the coding capacity of the entire collection of late mRNA species. Because the AUG of the i leader lies upstream from the AUGs of all other late transcription unit genes, the i leader AUG would be expected to preempt the translational capacity of any main body sequences contained in the messenger. The i leader AUG would be the first AUG encountered by the ribosome, and it would be the favored initiator, whatever coding sequences lay downstream. Thus, the translation product predicted by Kozak's Rules (1978) would be the i leader peptide.

In addition to the most common form of the i leader discussed above, Chow *et al.* (1979) noted several other less common forms shown in Fig. 4. These include a species in which all RNA between the i leader and leader three is retained. In this case, a ribosome initiating at the i leader AUG would terminate at a UGA codon just beyond the i leader 3′ side. The predicted protein translation product would have an altered carboxy terminus (relative to the standard i leader peptide). An additional form of i leader seen by electron microscopy and given in Fig. 4, but which has not yet been cloned or sequenced, utilizes a splice between the i leader donor site and a position on the 5′ side of the leader 3 acceptor. This form suggests encoding of another related peptide, with an alternative change in the peptide's carboxy-terminal sequence.

It is tempting to speculate about the possible function of a product encoded by the i leader. Why might a gene or gene family be positioned such that its coding sequences appear in late transcription unit mRNA, upstream from main-body coding sequences? The level of expression of such a gene might be in direct proportion to the level of production of late mRNA. This would be expected if i leader retention in mRNA were a random process, resulting in the presence of the i leader in a fixed fraction of late transcription unit messengers. Such expression would be advantageous if the i leader gene product participated in the synthesis or translation of late transcription unit mRNAs, and were required in stoichiometric quantity. Recently, Thimmappaya *et al.* (1982) have provided evidence that viral mRNA translation in the late infected cell is dependent upon a small adenovirus RNA transcribed by RNA polymerase III called VA RNA. Thus translation at late times has properties distinct from the translation mechanism as it takes place in the uninfected cell. It is not known, however, whether viral protein products are required to institute either this translation difference or other late gene expression steps which are special to the virus, such as late transcription unit splicing. Virus-specific late expression steps however could proceed by specialized mechanisms made possible through the action of a viral peptide such as encoded by the i leader.

D. The X, Y, and Z Leaders

Like the i leaders, these leaders are additional short RNA segments which may appear in late transcription unit messengers. Unlike the i leader, they do not lie between the standard leader segments, but instead they lie between leader three and main body coding sequences. The X, Y, and Z leaders all map within a short segment of the genome, and in fact, their templates, shown in Fig. 1b, lie exclusively within DNA corresponding at early times to the EIII transcription unit (Chow *et al.*, 1979). Thus, they lie downstream from the right-most splice acceptor of the L-4 family of mRNAs. This position prevents the inclusion of the X, Y, or Z leaders in any late mRNAs except those with the L-5 main body. Indeed when L-5 mRNA (which encodes the fiber protein) is analyzed by Northern blot analysis, multiple bands are seen (Dunn *et al.*, 1978). These correspond to fiber mRNAs with the standard tripartite leader and combinations of the X, Y, and Z leaders (Chow and Broker, 1978). When the map positions of the X, Y, and Z leaders, as determined by electron microscopy, are compared with EIII mRNA structures, certain X, Y, and Z leader and EIII mRNA donor and acceptor splice boundaries appear to coincide. Unlike the case of the i leader, the messengers with or without these extra RNA segments have equivalent coding capacities. This was determined by comparing the translation products of the different forms of the fiber mRNAs. Apparently the X, Y, and Z leaders lack any AUG triplet. Zain *et al.* (1979) have cloned a L-5 (fiber protein) late mRNA leader which contains a fourth 151 nucleotide leader segment. The nucleotide sequence of this fourth extra exon leader lacks any AUG initiator.

Because of the apparent lack of translational capacity, it is more difficult to propose a role for the X, Y, and Z leaders than for the i leader. It is possible that these RNA segments are simply included in processed mRNAs as a consequence of a mechanistic requirement of splicing. The EIII transcription unit, to which they are apparently related, is transcribed in the same direction as the major late transcription unit. Thus splicing signals which function in processing EIII mRNA will also be present in major late transcription unit mRNA precursors. The appearance of the X, Y, and Z leaders might therefore reflect a requirement by the splicing mechanism to utilize these splice boundaries when L-5 mRNA is produced. Indeed, only one other transcription unit in adenovirus contains within it the entirety of a second transcription unit. This is the EIb transcription unit, which contains the whole protein IX transcription unit (see above). However, unlike EIII mRNA, pIX mRNA is unspliced. Thus the pIX transcription unit does not contain splice signals, and cannot present the possibility of modifying the splicing pattern of EIb mRNAs. If the X, Y, and Z leaders do result from an interaction of the splicing of L-5 mRNA with the splicing of EIII, the requirement for such interaction cannot be absolute. Fiber mRNA lacking these leader structures is a predominant product (Dunn *et al.*, 1978; Chow and Broker, 1978). Unlike families L-1 through L-4, the L-5 "family" contains only a single coding sequence. Perhaps the X, Y, and Z leaders give L-5 the same splicing complexity as found in the true coterminal families, but without enabling it to encode other proteins.

E. Early versus Late Splice Patterns

From the foregoing discussion it is evident that the splice patterns of steady-state mRNA from several transcription units change as the cell enters the late stage of infection. This is true for EIa, EIb, and L-1 mRNAs. In fact it is true for each mRNA family which is expressed at both early and late times, except EII. The EII transcription unit, however, switches its promoter from 75 to 72 map units at early versus late times. A splicing pattern change may yet be found for EIIb mRNAs, which have not yet been completely characterized.

The most striking stage-dependent splicing pattern changes observed are the increase in the steady state levels of two left-end species—the 9 S EIa mRNA and the 13 S EIb mRNA—and the increased levels of two shorter members of the L-1 3′ coterminal family. These changes show a consistent pattern. In each case the splice pattern favors larger introns (and hence shorter mRNA species) at late times. For EIa and EIb, mRNAs with shifted splice *donor* boundaries increase in proportion at late times, while with L-1 the change is in the abundance of mRNAs with shifted splice *acceptors*.

In principle, these changes could result from either of two mechanisms. First, the splicing reaction could change its specificity, and the newly abundant shorter mRNAs could result from the activation of new splicing sequences. Splice donors or acceptors, which were not functional early, could be recruited at late times by a change in the specificity of the splicing machinery. Alternatively, splicing could proceed in exactly the same manner at all stages of infection, but the late-specific species could be stabilized at late times, increasing their proportion within the steady-state mRNA population. Wilson *et al.* (1980) have presented evidence supporting the latter hypothesis. However the matter still awaits a final resolution.

F. Cellular Transcripts

If indeed there is a stage specific alteration in either the splicing mechanism itself, or in the stability of specific spliced mRNAs, it may be related to a dramatic change in the fate of cell transcripts at late times. Cell mRNA transport is essentially completely blocked after DNA replication (see Beltz and Flint,

1979). Cellular transcription and polyadenylation continue (Beltz and Flint, 1979), but cellular RNAs are apparently degraded in the nucleus subsequent to their polyadenylation, but prior to their transport to the cytoplasm. This destruction of host mRNA precursors seems to be activated in parallel to the viral mRNA splicing pattern changes discussed above.

G. The Monkey Cell Block

When adenovirus-2 infects certain strains of monkey cells, the efficiency of viral plaque formation is greatly reduced relative to HeLa cells, implying an impediment to successful completion of the infectious cycle. This impediment is known as the monkey cell block, and represents a host-range limitation upon the infectivity of adenovirus-2. The effect is readily observed with the monkey kidney cell line, CV-1.

It has also been known for some time that the monkey cell block to adenovirus may be overcome by coinfection of the monkey cells with SV40 (Rabson *et al.*, 1964). The result of SV40 coinfection is the production of a normal yield of adenovirus infectious particles, and a return to high plaquing efficiency. The contribution made by SV40 to completion of the adenovirus life cycle is known as the "SV40 helper function," with an action called the "SV40 helper effect."

Analysis of viral proteins made during the late stage of monkey cell infections reveals that many species are produced in subnormal quantities (Klessig and Anderson, 1975). This decrease evidently results in part from subnormal production of adenovirus late messenger RNAs. Certain species are affected to a greater extent than others. Hexon and fiber mRNA are substantially decreased, while 100K protein mRNA is produced in near normal levels.

Klessig and Chow (1980) have used electron microscopy to map the structures of mRNAs made in monkey cells from the L-4 and L-5 families. They have identified molecules for which the L-5 poly(A) site has been employed by the processing machinery, but RNA sequences corresponding to main body sequences of L-4 m RNAs are found in the polyadenylated cytoplasmic transcripts. These extra segments, shown in Fig. 5, are sequences which are, in HeLa cell infections, intronic for L-5 mRNAs. That is, they are normally removed by the splicing machinery. These results suggest that one aspect of the defect suffered by adenovirus-2 in monkey cells pertains to mRNA processing, and possibly to the splicing mechanism.

Klessig and Chow (1980) made a second observation about the aberrant RNAs produced in monkey cells. They compared Northern blot analyses of polyadenylated cytoplasmic RNAs from HeLa and monkey cell infections using a fiber mRNA main body sequence probe. Not only were the mRNAs which contained the fiber body altered by the inclusion of upstream sequences, but their levels were also greatly (100-fold or more) decreased.

It appears that in monkey cells, the mechanism for establishing the normal

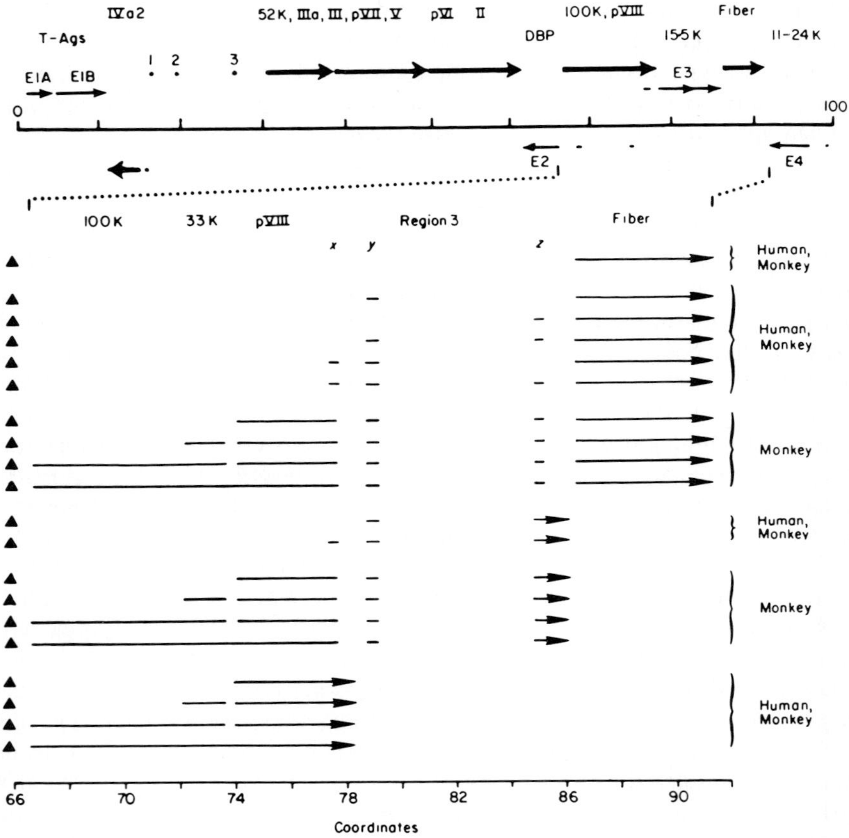

FIG. 5. Comparison of Ad2 RNA in HeLa and CV-1 cells. Species are shown which contain extra RNA segments as a consequence of the monkey cell block. This figure is from Klessig and Chow (1980).

cytoplasmic level of fiber mRNA does not function. In one model for the processing defect, the aberrant species found at low levels could result from incorrect processing of precursors to fiber mRNA and thereby the retention of intronic sequences in the cytoplasmic mRNA product. Thus fiber mRNA precursor might be shunted, in monkey cells, to the production of defective mRNA. Alternatively, the aberrant mRNAs seen by Klessig and Chow (1980) might also be produced in HeLa cells, but they might go unnoticed because they represent a minority fraction of the population. In this latter case, the monkey cell block would be unrelated to the aberrant mRNA, and would be the outcome of a virtually complete failure of the late-infected monkey cell to produce fiber mRNA (rather than the processing of the fiber nuclear precursor mRNA to give an incorrect product); the aberrant mRNA would be a minor side product present in either cell type. To take into account the observed effects in monkey cells, this model would require that other mRNAs are also subject to the processing failure which blocks fiber mRNA, but generally to a lesser extent.

Although the basis for the monkey cell block is unknown, the block may be overcome through an action of the SV40 large T antigen protein. When CV-1 monkey cells are coinfected with both adenovirus-2 and SV40, adenovirus late mRNA synthesis apparently proceeds normally. Certain adeno-SV40, hybrid viruses are also nondefective in late mRNA production (Kelly and Lewis, 1973; Lewis *et al.*, 1973; Morrow *et al.*, 1973). These hybrids have acquired the helper function from SV40 by recombination, and often are adenoviruses in which EIII sequences are substituted by early SV40 DNA. The relevant SV40 early gene region is the DNA that encodes the carboxy-terminal segment of T antigen. Mutants of SV40 with deletions at the C-terminus lack the SV40 helper function, confirming the genomic location of the helper function (Cole *et al.*, 1979). Microinjection of a fusion protein which contains the amino-terminus of the adeno fiber linked to the carboxy-terminus of SV40 T antigen also overcomes the block (Tjian *et al.*, 1978). Although the helper effect has been conclusively mapped to the T antigen corboxy-terminus, the mechanism of its action is unknown.

Additional clues to the nature of the monkey cell block come from host range mutants of adenovirus which are fully permissive in monkey cells. Marker rescue experiments have mapped the locus of the compensating adenovirus mutation to a site within the EIIa transcription unit, which encodes the 72-kDa DNA binding protein, DBP (Klessig and Grodzicker, 1979). A single base change in the DBP gene appears capable of extending the host range.

One interpretation of these results is that in the course of RNA processing, the DNA binding protein must interact with a human cellular factor, and that the equivalent monkey cell factor interacts incorrectly with DBP. In monkey cells, this would lead to a disruption of the processing of late mRNA. The compensating host range mutation would alter the DBP to permit correct interaction with the putative monkey cell factor.

H. The DBP and RNA Metabolism

The DBP is a multifunctional protein, and there is evidence (independent of that cited above) that one of the functions of the DBP pertains to RNA metabolism. The DBP is known to participate in viral DNA replication, during which it binds to displaced single strands of viral DNA which are replication intermediates (van der Vliet *et al.*, 1975). It also acts as a repressor of EIV transcription during the close of the early transcription phase (Nevins and Winkler, 1980). The evidence for an RNA metabolism function comes from studies of mRNA stability in adenovirus-infected cells, and in cells infected with temperature-sensitive mutants of the DBP.

During the early stage of infection, EIa and EIb viral messenger RNAs have half lives of about 30 minutes, and therefore are relatively unstable (Babich and Nevins, 1981). The stability is however greatly increased with infections at the nonpermissive temperature by a temperature sensitive mutant of the DBP, ts125. The half-lives of adenovirus EIa mRNAs increase 3-fold at the nonpermissive temperature to about 95 minutes, and EIb 5-fold to about 150 minutes (Babich and Nevins, 1981). The nonpermissive temperature also leads to DBP degradation. These data suggest that the DBP can decrease the half life of the mRNA, perhaps through a direct protein–RNA interaction occurring in the cytoplasm. Because the DBP is known to interact with single-stranded DNA, an interaction with RNA seems plausible.

I. *In Vitro* Splicing

Reconstruction of the splicing reaction *in vitro* from purified components is still in its infancy. However, several important advances have been achieved with adenovirus. The first *in vitro* demonstration of splicing utilized polyadenylated nuclear precursors to the EIIa mRNA that were snythesized and labeled *in vivo* (Blanchard *et al.,* 1978). Nuclei were isolated, and incubated with "cytoplasmic extracts," and a time-dependent change in the polyadenylated mRNA from the unspliced form to the mature spliced structure was detected. In a similar system, splicing of EII transcripts in nuclei was shown (Yang and Flint, 1979). Using antisera to the U-1 containing small nuclear RNP, this splicing was blocked, implicating the RNP in the splicing mechanism (Yang *et al.,* 1981). More recently, *in vitro* transcripts from cloned late promoter DNA have been successfully spliced *in vitro* to yield the tripartite late leader (Padgett *et al.,* 1983). Undoubtedly adenovirus will make additional contributions to *in vitro* splicing.

J. Splicing in Other Viruses

A detailed discussion of splicing in other viruses is beyond the scope of this review. However, the major features of splicing of SV40 and polyoma viruses and the retrovirus Rous Sarcoma virus will be given and briefly related to adenovirus splicing.

Splicing in the papovaviruses SV40 and polyma has been reviewed (Ziff, 1980). The mRNAs of SV40 and polyoma are shown respectively in Figs. 6a and 7a. With both viruses, shifts in splice donors or splice acceptors enable the production of multiple early mRNAs from a single early transcription unit. The products, with both viruses, are tumor antigen peptides which are amino coter-

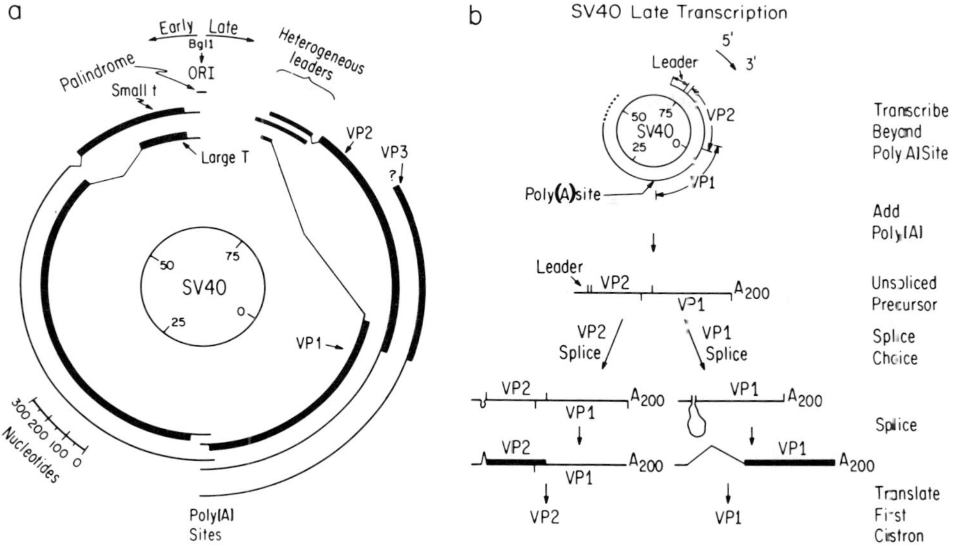

FIG. 6. SV40 messenger RNAs. (a) Physical map of early and late mRNA species. (b) Pathway of processing of late mRNA. This figure is modified from Ziff (1980).

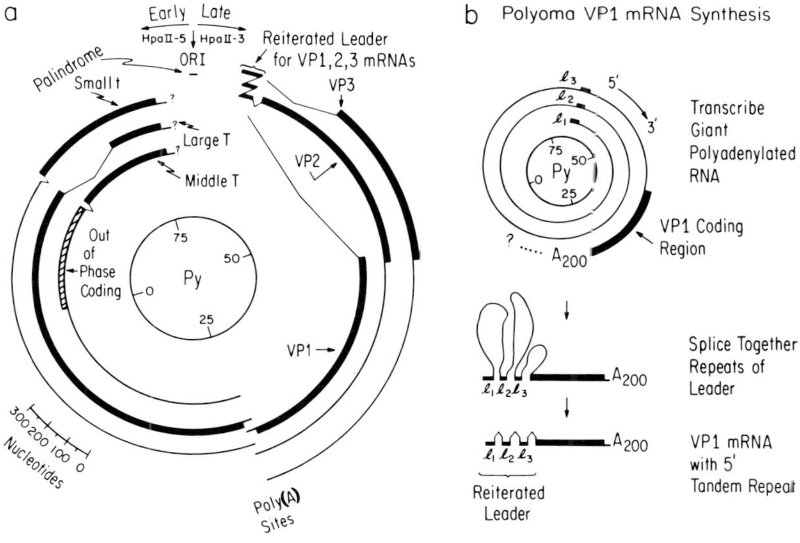

FIG. 7. Polyoma virus messenger RNAs. (a) Physical map of early and late mRNA species. (b) Pathway of processing of late messenger RNA. This figure is modified from Ziff (1980).

minal but which differ in their carboxy termini. Versatility in splicing of early mRNAs is also a feature of the adenovirus EIa transcription unit.

At late times, SV40 and polyoma both encode the major virion capsid proteins, VP1, and two peptides, related to each other, VP2 and VP3 shown in Figs. 6 and 7. In both viruses, 3′ coterminal mRNAs are used to express these proteins (Reddy *et al.,* 1979; Hsu and Ford, 1977; Aloni *et al.,* 1977). But very significant differences are found in the mRNA processing pathways for late mRNAs when polyoma and SV40 are compared.

The principal late SV40 mRNAs shown in Fig 6a are 16 S and 19 S species which share a common polyadenylated 3′ end. Both have complex 5′ leader splicing patterns (Aloni *et al.,* 1977; Ghosh *et al.,* 1978; Reddy *et al.,* 1978; Hsu and Ford, 1977). The 19 S mRNA encodes VP2. Translation is initiated at an AUG that commences an open reading frame which is uninterrupted by introns (Ghosh *et al.,* 1978). VP3 is encoded either by the same mRNA, or by a species closely related in size. The AUG for VP2 gene and lies in the same frame as VP2 (Contreras *et al.,* 1977). Potentially an internal initiation is required for VP3 synthesis, as is the case for the adenovirus EIb polypeptide. VP1 mRNA has a larger intron, with splice acceptor shifting to remove the VP2 and VP3 initiators, allowing VP1 initiation (Aloni *et al.,* 1977). These features of late SV40 mRNAs resemble other 3′ coterminal mRNA families. Intron removal for late SV40 mRNA is shown in Fig. 7b.

In many respects, polyoma late mRNAs shown in Fig. 7a are equivalent to their SV40 counterparts. VP2, VP3, and VP1 coding sequences occupy corresponding positions on the genome, and rely on similar splicing steps for their expression. However, the polyoma mRNAs possess a reiterated leader structure which is made possible by a very novel application of splicing.

Although SV40 nuclear polyadenylated transcripts are almost always shorter than genome length (Ford and Hsu, 1978), polyoma synthesizes giant polyadenylated nuclear RNA (Legon *et al.,* 1979). These giant RNAs are encoded by polymerases which circle the genome up to eight times before poly(A) addition takes place. The formation of giant RNA probably depends on inefficient polyadenylation. The polymerase must pass the very same poly(A) site on each circuit, but the polyadenylation mechanism functions only once per up to 8 possible opportunities, thus yielding giant RNA (Legon *et al.,* 1979). In this respect, polyoma late nuclear RNA polyadenylation resembles the use of polyadenylation by the major late adenovirus transcription unit, where only one of five poly(A) sites functions per primary transcript molecule.

To splice a late messenger from polyoma giant nuclear RNA, the splicing machinery transfers a leader to a downstream site adjacent to the second occurrence of the leader, approximately 1 genome distance away. This is shown in Fig. 7b. The second leader repeats this splice, and then the third, until each copy of the later leader, in the tandemly repeated giant RNA, is joined together. The tandem leader produced by splicing then splices to the main body coding sequence. Thus, the use of the polyoma circular genome to encode a tandemly repeated precursor, and by splicing together each occurrence of the leader in the precursor, a reiterated RNA structure is formed from a sequence which occurs only once in the genome.

The retrovirus Rous Sarcoma virus yields mRNAs which also constitute a classical 3′ coterminal family. These are reviewed in Weiss *et al.* (1982). A 5′ leader is spliced to multiple splice acceptors to yield mRNAs sharing a common poly(A) acceptor. The transcription unit which yields these mRNAs also yields the viral genomic RNA which is packaged into virions. This dual role may explain why the retroviral transcription unit has not evolved *multiple* 3′ coterminal families, as found with the adenovirus major late transcription unit. If additional poly(A) sites were present *within* the retroviral genome, they might impair production of full length genomic RNA by diverting the transcripts to the formation of mRNAs in the second family. Use of a single poly(A) site to form one coterminal family would ensure production of full length RNAs.

K. Summary

Splicing provides viruses with great genetic versatility. It is still too early to say whether this versatility is derived from ingeneous mechanisms evolved by necessity by the viruses, or whether the viruses indeed mimic cellular mechanisms. In any event, it is unlikely that cells will provide a single genomic cluster of genes that utilize splicing in such diverse ways as adenovirus, or the other viruses discussed here. And we may speculate that when the full role of splicing in adenovirus gene expression program is known, its import will continue to be a source of amazement!

Acknowledgments

I thank M. Greenberg, D. Ross, R. Stein, D. Smith, A. Velcich, and R. Vosatka for comments on the manuscript and the American Cancer Society for a Faculty Research Award.

References

Akusjarvi, G., and Pettersson, U. (1978). *Proc. Natl. Acad. Sci. U.S.A.* **75,** 5822–5826.
Akusjarvi, G., and Pettersson, U. (1979a). *Cell* **16,** 841–850.
Akusjarvi, G., and Pettersson, U. (1979b). *J. Mol. Biol.* **134,** 143–158.
Akusjarvi, G., Zabielski, J., Perricaudet, M., and Pettersson, U. (1981). *Nucleic Acids Res.* **9,** 1–17.

Aelstrom, P., Akusjarvi, G., Perricaudet, M., Mathews, M. B., Klessig, D. F., and Pettersson, U. (1980). *Cell* **19,** 671–681.
Aloni, Y., Dhar, R., Laug, O., Horowtiz, M., and Khoury, G. (1977). *Proc. Natl. Acad. Sci. U.S.A.* **74,** 3686–3690.
Babich, A., and Nevins, J. R. (1981). *Cell* **26,** 371–379.
Bachenheimer, S., and Darnell, J. E. (1975). *Proc. Natl. Acad. Sci. U.S.A.* **72,** 4445–4449.
Baker, C. C., Herisse, J., Courtois, G., Galibert, F., and Ziff, E. (1979). *Cell* **18,** 569–580.
Baker, C. C., and Ziff, E. B. (1981). *J. Mol. Biol.* **149,** 189–221.
Beltz, G. A., and Flint, S. J. (1979). *J. Mol. Biol.* **131,** 353–373.
Berget, S. M., Moore, C., and Sharp, P. A. (1977). *Proc. Natl. Acad. Sci. U.S.A.* **74,** 3171–3175.
Berk, A. J., and Sharp, P. A. (1978). *Cell* **14,** 695–711.
Berk, A. J., Lee, F., Harrison, T., Williams, J., and Sharp, P. A. (1979). *Cell* **17,** 935–944.
Blanchard, J. -M., Weber, J., Jelinek, W., and Darnell, J. E. (1978). *Proc. Natl. Acad. Sci. U.S.A.* **75,** 5344–5348.
Box, J. L., Polder, L. J., Bernards, R., Schrier, P. I., van den Elsen, P. J. van der Eb, A. J., and Ormondt, H. (1981). *Cell* **27,** 121–131.
Broker, T. R. (1984). *In* "Processing of RNA" (D. Apirion, ed.), pp. 181–212. Chemical Rubber Co. Press, Boca Raton, Florida.
Chow, L. T., and Broker, T. R. (1978). *Cell* **15,** 497–510.
Chow, L. T., Gelinas, R. E., Broker, T. R., and Roberts, R. J. (1977a). *Cell* **12,** 1–8.
Chow, L. T., Roberts, J. M., Lewis, J. B., and Broker, T. R. (1977b). *Cell* **11,** 819–836.
Chow, L. T., Broker, T. R., and Lewis, J. B. (1979). *J. Mol. Biol.* **134,** 265–303.
Cole, C. N., Crawford, L. V., and Berg, P. (1979). *J. Virol.* **30,** 683–691.
Contreras, R., Rogiers, R., Van de Voorde, A., and Fiers, W. (1977). *Cell* **12,** 529–538.
Dunn, A. R., Mathews, M. B., Chow, L. T., and Sambrook, J. (1978). *Cell* **15,** 511–526.
Esche, H., Mathews, M. B.,and Lewis, J. B. (1980). *J. Mol. Biol.* **142,** 399–417.
Falvey, E., and Ziff, E. B. (1983). *J. Virol.* **45,** 185–191.
Ford, J. P., and Hsu, M. T. (1978). *J. Virol.* **28,** 795–801.
Fraser, N., and Ziff, E. (1978). *J. Mol. Biol.* **124,** 27–51.
Fraser, N. W., Nevins, J. R., Ziff, E. B., and Darnell, J. E., Jr. (1979). *J. Mol. Biol.* **129,** 643–656.
Fraser, N. W., Baker, C. C., Moore, M. A., and Ziff, E. B. (1982). *J. Mol. Biol.* **155,** 207–233.
Ghosh, P. K., Reddy, V. B., Swinscoe, J., Choudary, P. V., Lebowitz, P., and Weissman, S. M. (1978). *J. Biol. Chem.* **253,** 3643–3647.
Goldenberg, C. J., and Raskas, H. J. (1979). *Cell* **16,** 131–138.
Hearing, P., and Shenk, T. (1983). *Cell* **33,** 695–703.
Hsu, M. -T., and Ford, J. (1977). *Proc. Natl. Acad. Sci. U.S.A.* **74,** 4982–4985.
Jones, N., and Shenk, T. (1979). *Proc. Natl. Acad. Sci. U.S.A.* **76,** 3665–3669.
Kelly, T. J. Jr., and Lewis, A. M., Jr. (1973). *J. Virol.* **12,** 643–652.
Kitchingman, G. R., Lai, S. -P., and Westphal, H. (1977). *Proc. Natl. Acad. Sci. U.S.A.* **74,** 4392–4395.
Klessig, D. F. (1977). *Cell* **12,** 9–21.
Klessig, D. F., and Anderson, C. W. (1975). *J. Virol.* **16,** 1650–1668.
Klessig, D. F., and Chow, L. T. (1980). *J. Mol. Biol.* **139,** 221–242.
Klessig, D. F., and Grodzicker, T. (1979). *Cell* **17,** 957–966.
Kozak, M. (1978). *Cell* **15,** 1109–1123.
Lamb, R. A., and Choppin, P. W. (1983). *Annu. Rev. Biochem.* **52,** 467–506.
Legon, S., Flavell, A. J., Cowie, A., and Kamen, R. (1979). *Cell* **16,** 373–388.
Lewis, A. M., Jr., Levine, A. S., Crumpacker, C. S., Levin, M. J., Samaha, R. J., and Henry, P. H. (1973). *J. Virol.* **11,** 655–664.
Lewis, J. B., and Mathews, M. B. (1980). *Cell* **21,** 303–313.

Lichy, J. H., Field, J., Horowitz, M. S., and Hurwitz, J. (1982). *Proc. Natl. Acad. Sci. U.S.A.* **79,** 5225–5229.

Logan, J. S., and Shenk, T. (1982). *Microbiol. Rev.* **46,** 377–383.

McGrogan, M., and Raskas, H. J. (1978). *Proc. Natl. Acad. Sci. U.S.A.* **75,** 625–629.

Miller, J. S., Ricciardi, R. P., Roberts, B. E., Paterson, B. M., and Mathews, M. B. (1980). *J. Mol. Biol.* **142,** 455–488.

Montell, C., Fisher, E. F., Caruthers, M. H., and Berk, A. J. (1982). *Nature (London)* **295,** 380–384.

Morrow, J. F., Berg, P., Kelly, T. J., Jr., and Lewis, A. M., Jr. (1973) *J. Virol.* **12,** 653–658

Nevins, J. R. (1981). *Cell* **26,** 213–220.

Nevins, J. R., and Chen-Kiang, S. (1981). *Adv. Virus Res.* **26,** 1–31.

Nevins, J. R., and Darnell, J. E., Jr. (1978a). *J. Virol.* **25,** 811–823.

Nevins, J. R., and Darnell, J. E., Jr. (1978b). *Cell* **15,** 1477–1493.

Nevins, J. R., and Wilson, M. C. (1981). *Nature (London)* **290,** 113–118.

Nevins, J. R., and Winkler, J. J. (1980). *Proc. Natl. Acad. Sci. U.S.A.* **77,** 1893–1897.

Nevins, J. R., Ginsberg, H. S., Blanchard, J.-M., Wilson, M. C., and Darnell, J. E., Jr. (1979). *J. Virol.* **32,** 727–733.

Padgett, R. A., Hardy, S. F., and Sharp, P. A. (1983). *Proc. Natl. Acad. Sci. U.S.A.* **80,** 5230–5234.

Perricaudet, M., Akusjarvi, G., Virtanen, A., and Pettersson, U. (1979). *Nature (London)* **281,** 694–696.

Persson, H., Pettersson, U., and Mathews, M. B. (1978). *Virology* **90,** 67–79.

Pettersson, U., and Mathews, M. B. (1977). *Cell* **12,** 741–750.

Pettersson, U., Virtanen, A., Perricaudet, M., and Akusjarvi, G. (1983). *In* "Current Topics in Microbiology and Immunology: Volume 109: The Molecular Biology of Adenoviruses 1" (W. Doerfler, ed.), pp. 107–123. Springer-Verlag, Berlin and New York.

Rabson, A. S., O'Conor, G. T., Berezesky, I. K., and Paul, F. J. (1964). *Proc. Soc. Exp. Biol. Med.* **116,** 187–190.

Reddy, V. B., Ghosh, P. K., Lebowitz, P., and Weissman, S. M. (1978). *Nucleic Acids Res.* **5,** 4195–4213.

Reddy, V. B., Ghosh, P. K., Lebowitz, P., Piatak, M., and Weissman, S. M. (1979). *J. Virol.* **30,** 279–296.

Ricciardi, R. P., Jones, R. L., Cepko, C. L., Sharp, P. A., and Roberts, B. E. (1981). *Proc. Natl. Acad. Sci. U.S.A.* **78,** 6121–6125.

Ross, S. R., Flint, S. J., and Levine, A. J. (1980). *Virology* **100,** 419–432.

Sarnow, P., Ho, Y. S., Williams, J., and Levine, A. J. (1982). *Cell* **28,** 387–394.

Shaw, A. R., and Ziff, E. B. (1980). *Cell* **22,** 905–916.

Solnick, D. (1981). *Nature (London)* **291,** 508–510.

Solnick, D., and Anderson, M. A. (1982). *J. Virol.* **42,** 106–113.

Spector, D. J., McGrogan, M., and Raskas, H. J. (1978). *J. Mol. Biol.* **126,** 395–414.

Spector, D. J., Crossland, L. D., Halbert, D. N., and Raskas, H. J. (1980). *Virology* **102,** 218–221.

Stillman, B. W., Lewis, J. B., Chow, L. T., Mathews, M. B., and Smart, J. E. (1981). *Cell* **23,** 497–508.

Thimmappaya, B., Weinberger, C., Schneider, R. J., and Shenk, T. (1982). *Cell* **31,** 543–551.

Tjian, R., Fey, G., and Graessmann, A. (1978). *Proc. Natl. Acad. Sci. U.S.A.* **75,** 1279–1283.

van der Eb, A. J., van Ormondt, H., Schrier, P. I., Lupker, J. H., Jochemsen, H., van den Elsen, P. J., DeLeys, R. J., Maat, J., van Beveren, C. P., Dijkema, R., and de Waard, A. (1980). *Cold Spring Harbor Symp. Quant. Biol.* **44,** 383–399.

Van der Vliet, P. C., and Levine, A. J. (1973). *Nature (London) New Biol.* **246,** 170–174.

Van der Vliet, P. C., Levine, A. J., Ensinger, M. J., and Ginsberg, H. S. (1975). *J. Virol.* **15,** 348–354.

Virtanen, A., Alestrom, P., Persson, H., Katze, M. G., and Pettersson, U. (1982). *Nucleic Acids Res.* **10,** 2539–2548.

Weiss, R., Teich, N., Varmus, H., and Coffin, J., eds. (1982). "RNA Tumor Viruses, Molecular Biology of Tumor Viruses," 2nd Ed. Cold Spring Harbor Laboratory, Cold Spring Harbor, New York.

Wilson, M. C., Nevins, J. R., Blancharcd, J. M., Ginsberg, H. S., and Darnell, J. E., Jr. (1980). *Cold Spring Harbor Symp. Quant. Biol.* **44,** 447–455.

Yang, V. W., and Flint, S. J. (1979). *J. Virol.* **32,** 394–403.

Yang, V. W., Lerner, M. R., Steitz, J. A., and Flint, S. J. (1981). *Proc. Natl. Acad. Sci. U.S.A.* **78,** 1371–1375.

Zain, S., Sambrook, J., Roberts, R. J., Keller, W., Fried, M., and Dunn, A. R. (1979). *Cell* **16,** 851–861.

Zeevi, M., Nevins, J. R., and Darnell, J. R., Jr. (1981). *Cell* **26,** 39–46.

Ziff, E. B. (1980). *Nature (London)* **287,** 491–499.

Ziff, E. B., and Evans, R. M. (1978). *Cell* **15,** 1463–1475.

Ziff, E. B., and Fraser, N. (1978). *J. Virol.* **25,** 897–906.

Index

Contents of Recent Volumes

Volume 74

Volume 75

Volume 76

Volume 77

Volume 78

Volume 79

Volume 80

Volume 81

Volume 82

Volume 83

Volume 84

Volume 85

Volume 86

Volume 87

Volume 88

Volume 89

Volume 90

Volume 91

Volume 92